Joachim Weiss

Ion Chromatography

Second Edition

Distribution:

VCH, P.O. Box 10 11 61, D-69469 Weinheim, Federal Republic of Germany

Switzerland: VCH, P.O. Box, CH-4020 Basel, Switzerland

United Kingdom and Ireland: VCH, 8 Wellington Court, Cambridge CB1 1HZ, United Kingdom

USA and Canada: VCH, 220 East 23rd Street, New York, NY 10010-4606, USA

Japan: VCH, Eikow Building, 10-9 Hongo 1-chome, Bunkyo-ku, Tokyo 113, Japan

ISBN 3-527-28698-5

Joachim Weiss

Ion Chromatography

Second Edition

Weinheim · New York · Basel · Cambridge · Tokyo

Dr. Joachim Weiss
Dionex GmbH
Am Wörtzgarten 10
D-65510 Idstein
Germany

Originally published under the German title Ionenchromatographie
by VCH Verlagsgesellschaft mbH, Weinheim 1991

2nd edition 1995

Published jointly by
VCH Verlagsgesellschaft mbH, Weinheim (Federal Republic of Germany)
VCH Publishers, Inc., New York, NY (USA)

Editorial Directors: Dr. Christina Dyllick, Dr. Barbara Böck
Translator: Dr. Jörg Mönig
Production Manager: Peter Biel

Library of Congress Card No. applied for
A CIP catalogue record for this book is available from the British Library

Die Deutsche Bibliothek – CIP-Einheitsaufnahme
Weiss, Joachim
Ion chromatography / Joachim Weiss. – 2. ed. – Weinheim; New York; Basel; Cambridge; Tokyo: VCH, 1995
Dt. Ausg. u.d.T.: Weiss, Joachim: Ionenchromatographie
ISBN 3-527-28698-5

Printed on acid-free and low chlorine paper

Composition: pagina media, D-69494 Hemsbach
Printing: Betzdruck GmbH, D-64291 Darmstadt
Bookbinding: Industrie- und Verlagsbuchbinderei Heppenheim GmbH, D-64646 Heppenheim

Printed in the Federal Republic of Germany

Foreword

Joachim Weiss's revised book of Ion Chromatography follows almost ten years after the first highly successful edition. In the last eight to ten years, we have witnessed not only the recognition of IC as a powerful, practical and indispensable tool in many diverse fields of analytical chemistry, but also a substantial evolution of separation media, detection, and most profoundly the improvement of suppressor performance. From this standpoint, Dr. Weiss's new book is long overdue.

The basic approach of this edition is little changed from the first: the emphasis is still on the review of methods and applications which are most useful for quantitative, analytical determination of ions in a wide variety of matrices. An ultimate practitioner of ion chromatography, the author has added a substantial amount of data from his own applications development work. The theoretical background description on various subjects of ion determination is short but informative, and is written so that a novice in the field will not only read and understand it, but also enjoy it. Experts in the field, on the other hand, will undoubtedly find Dr. Weiss's new text a useful reference for many applications and practical problems faced by an analytical chemist, ranging from the field of water purity analysis to the complex task of carbohydrate analysis of glycoproteins.

This new edition is proof that the analytical chemistry community has accepted ion chromatography as a routine, practical tool that brings quantitation, reproducibility, speed, and automation to the field of separation science.

Sunnyvale, CA, July 1994 — Nebojsa Avdalovic

Preface to the Second Edition

The first edition of this book, written during 1984 and 1985, was intended to widen the application of ion chromatography. This aim has been successfully met. Today, it is difficult to imagine the analysis of inorganic anions and cations without using ion chromatography. Moreover, ion chromatography is being increasingly applied to special substances – for example carbohydrates and glycoproteins – which do not immediately come into mind when talking about ions.

The rapid development of this method since the publication of the first edition, mainly in the analysis of organic ions, has resulted in the fact that this second edition can be classed a totally new text; as almost every Section has been revised or extended.

The table of contents has been compressed in order to organize the techniques, summarized within the term ion chromatography, as logically and as clearly as possible. It is based on the concept of my ion chromatography courses.

The well-established division of the separation methods into chapters on ion-exchange chromatography, ion-exclusion chromatography, and ion-pair chromatography is still used. The treatment of the detection methods and the discussion of quantitative analysis, the latter being basically unchanged, follow this Section.

The chapter on instrumentation has been omitted because a complete overview of the apparatus currently available for ion chromatography would have exceeded the limits of this book, and a selected – perhaps company-specific – description of the required components has also been avoided. Moreover, since the technical equipment is subject to such rapid change in design and specification, a specific chapter would have soon become outdated.

The chapter on the application possibilities has been greatly extended with new material and a large number of chromatograms. Within this Section, a special chapter has been incorporated covering the application of ion chromatography in environmental protection, the pharmaceutical industry, and the clinical industry.

With this second edition, the author addresses analysts wishing to work confidently with the various methods, as well as practitioners who employ these techniques on a day-to-day basis.

At this point, I would like to express sincere gratitude to all my colleagues for their help in preparing the second edition. Special thanks are dedicated to Hermann J. Möckel from the Hahn-Meitner-Institute in Berlin. I owe very much to his valuable stimulus and critical advice after having checked the manuscript, and also to Nebojsa Avdalovic from Dionex Corporation in Sunnyvale for the critical review of the English translation. I should be most grateful for any criticism or suggestions which could serve to improve future editions of this book.

Finally, I would like to thank my family who patiently endured the many evenings, weekends, and holidays that I spent writing and drawing.

Idstein, November 1994

Joachim Weiss

Contents

1 Introduction

1.1 Historical Perspective

"Chromatography" is the general term for a variety of physico-chemical separation techniques, all of which have in common the distribution of a component between a mobile phase and a stationary phase. The various chromatographic techniques are subdivided according to the physical state of these two phases.

The discovery of chromatography is attributed to Tswett [1,2], who in 1903 was the first to separate leaf pigments on a polar solid phase and to interpret this process. In the following years, chromatographic applications were limited to the distribution between a solid stationary and a liquid mobile phase (**l**iquid **s**olid **c**hromatography, LSC). In 1938, Izmailov and Schraiber [3] laid the foundation for **t**hin **l**ayer **c**hromatography (TLC). Stahl [4,5] refined this method in 1958 and developed it into the technique known today. In their noteworthy paper of 1941, Martin and Synge [6] proposed the concept of theoretical plates, which was adapted from the theory of distillation processes, as a formal measurement of the efficiency of the chromatographic process. This approach not only revolutionized the understanding of liquid chromatography, but set the stage for the development of both gas chromatography (GC) and paper chromatography.

In 1952, Martin and James [7] published their first paper on gas chromatography, initiating a rapid development of this analytical technique.

High **p**erformance **l**iquid **c**hromatography (HPLC) was derived from the classical column chromatography and, besides gas chromatography, is one of the most important tools of modern analytical chemistry today. The technique of HPLC flourished after it became possible to produce columns with packings made of very small beads (≈ 10 μm) and to operate them under high pressure. The development of HPLC and the theoretical understanding of the separation processes rest on the basic work of Horvath [8], Knox [9], Scott [10], Snyder [11], Guiochon [12], Möckel [13], and others.

Ion **c**hromatography (IC) was introduced in 1975 by Small, Stevens, and Baumann [14] as a new analytical method. Within a short period of time, ion chromatography developed from a new detection scheme for a few selected inorganic anions and cations to a versatile analytical technique for ionic species in general. For a sensitive detection of ions via their electrical conductance, the effluent from the separator column was passed through a "suppressor" column. This suppressor column chemically reduces the background conductance of the eluent, while at the same time increasing the electrical conductance of the analyte ions.

In 1979, Fritz et al. [15] described an alternative separation and detection scheme for inorganic anions, in which the separator column is directly coupled to the conductivity cell. As a prerequisite for this chromatographic setup, ion-exchange resins with low capacities have to be employed so that eluents with low ionic strengths can be used. In addition, the eluent ions should exhibit low equivalent conductances, thus enabling sensitive detection of the sample components.

At the end of the 1970s, ion chromatographic techniques were used to analyze organic ions for the first time. The requirement for a quantitative analysis of organic acids brought about an ion chromatographic method based on the ion-exclusion process, which was first described by Wheaton and Bauman [16] in 1953.

The last seven years witnessed the development of separator columns with high efficiencies, which resulted in a significant reduction of analysis time. In addition, separation methods based on the ion-pair process were introduced as an alternative to ion-exchange chromatography, since they allow the separation and determination of both anions and cations.

The scope of ion chromatography was considerably enlarged by newly designed electrochemical and spectrophotometric detectors. A milestone of this development was the introduction of a pulsed amperometric detector in 1983, allowing a very sensitive detection of carbohydrates [17].

A growing number of applications using post-column derivatization in combination with photometric detection opened the field of heavy and transition metal analysis for ion chromatography, thus providing a powerful extension to conventional atomic spectroscopy methods.

These developments make ion chromatography an integral part of both modern inorganic and organic analysis.

1.2 Types of Ion Chromatography

This book only discusses separation methods which can be summarized under the general term *Ion Chromatography*. Modern ion chromatography as part of liquid chromatography is based on three different separation mechanisms, which also provide the basis for the nomenclature in use.

Ion-Exchange Chromatography (HPIC)
(**H**igh **P**erformance **I**on **C**hromatography)

This separation method is based on an ion-exchange process occurring between the mobile phase and ion-exchange groups bonded to the support material. In ions with high polarizibility, additional non-ionic adsorption processes contribute to the separation mechanism. The stationary phase consists of a polystyrene resin copolymerized with divinylbenzene and modified with ion-exchange groups. Ion-exchange chromatography is used for the separation of both organic and inorganic anions and cations, respectively. Separation of anions is accomplished with quaternary ammonium groups attached to the polymer, whereas sulfonate groups are used as ion-exchange sites for the separation of cations. Chapter 3 deals with this type of separation method in greater detail.

Ion-Exclusion Chromatography (HPICE)
High **P**erformance **I**on **C**hromatography **E**xclusion

The separation mechanism in ion-exclusion chromatography is governed by Donnan exclusion, steric exclusion, and sorption processes. A totally sulfonated polystyrene/divinylbenzene-based cation exchange material with high capacity is employed as the stationary phase. Ion-exclusion chromatography is particularly useful for the separation of weak inorganic and organic acids from those acids which are completely dissociated at the eluent pH. All acids with high acid strengths are not retained and elute unresolved within the void volume. In combination with suitable detection systems, this separation method is also useful for determining amino acids, aldehydes, and alcohols. A detailed description of this separation method is given in Chapter 4.

Ion-Pair Chromatography (MPIC)
(**M**obile **P**hase **I**on **C**hromatography)

The dominating separation mechanism in ion-pair chromatography is adsorption. The stationary phase consists of a neutral porous divinylbenzene resin of low polarity and high specific surface area. Alternatively, chemically bonded silica phases of the octyl or octadecyl type with an even lower polarity can be used. The selectivity of the separator column is determined solely by the mobile phase. Besides an organic modifier, an ion-pair reagent is added to the eluent (water, aqueous buffer solution, etc.) depending on the chemical nature of the analytes. Ion-pair chromatography is particularly suited for the separation of surface-active anions and cations as well as transition metal complexes. A detailed description of this separation method is given in Chapter 5.

Alternative Methods

In addition to the three classical separation methods mentioned above, **r**eversed-**p**hase **l**iquid **c**hromatography (RPLC) is becoming increasingly popular for the separation of highly polar and ionic species, respectively. Long-chain fatty acids, for example, are separated on a chemically bonded octadecyl phase after protonation in the mobile phase with a suitable aqueous buffer solution. This separation mode is known as ion suppression [18].

Chemically bonded aminopropyl phases have also been successfully employed for the separation of inorganic ions. Leuenberger et al. [19] described the separation of nitrate and bromide in foodstuffs on such a phase using a phosphate buffer solution as the eluent. These alternative techniques are also described in Chapter 5. However, it is presently very difficult to comment on their potential for universal applicability, since they have mainly been used for the analysis of UV-absorbing species.

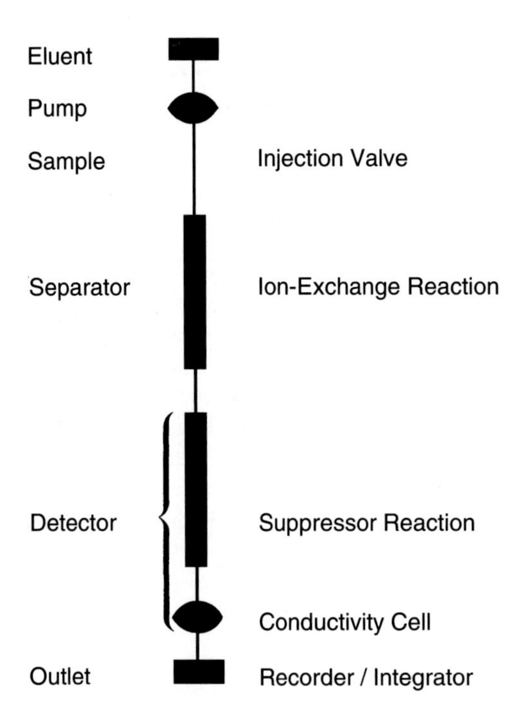

Fig. 1-1. Basic components of an ion chromatograph.

1.3 The Ion Chromatographic System

The basic components of an ion chromatograph are shown schematically in Fig. 1-1. It resembles the setup of conventional HPLC systems.

A pump delivers the mobile phase through the chromatographic system. In general, either single-piston or dual-piston pumps are employed. A pulse-free flow of the eluent is necessary for the sensitive UV/Vis and amperometric detectors. Therefore, pulse dampeners are used with single-piston pumps and electronic circuitry with dual-piston pumps.

The sample is injected into the system via a valve injector, as schematically shown in Fig. 1-2. A three-way valve is required, with two ports being connected to the sample loop. The sample loading is carried out at atmospheric pressure. After switching the injection valve, the sample is transported to the separator by the mobile phase. Typical injection volumes are between 10 μL and 100 μL.

The most important part of all chromatographic systems is the separator column. The choice of a suitable stationary phase (see Section 1.5) and the chromatographic conditions determine the quality of the analysis. The column tubes are manufactured from inert material such as Tefzec or epoxy resins. In general, separation is achieved at room temperature. Elevated temperatures are required only in very few cases, such as the analysis of carbohydrates or long-chain fatty acids.

The analytes are detected and quantified by a detection system. The performance of any detector is examined according to the following criteria:

- sensitivity,
- linearity,
- resolution (volume of the detector cell),
- noise (detection limit).

The most commonly employed detector in ion chromatography is the conductivity detector, which is used with or without a suppressor system. The main function of the suppressor system as part of the detection unit is to **chemically** reduce the high background conductivity of the electrolytes in the eluent, and to convert the sample ions into a more conductive form. In addition to conductivity detectors, UV/Vis, amperometric and fluorescence detectors are used, all of which are described in detail in Chapter 6.

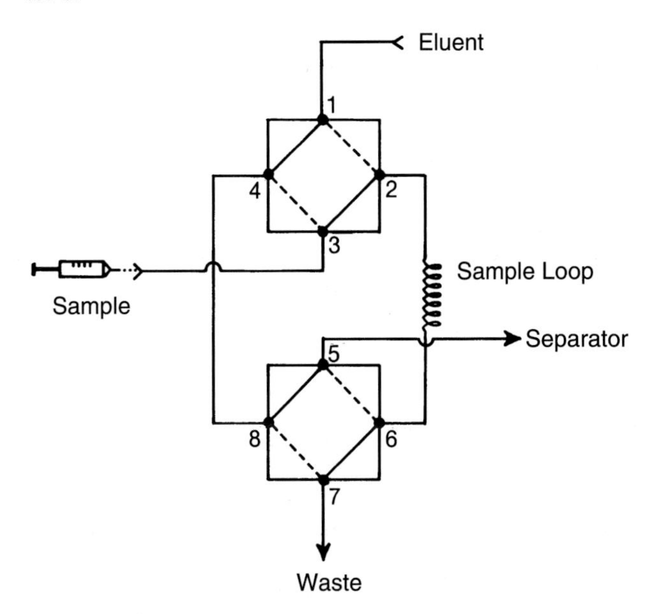

Fig. 1-2. Schematic representation of a loop injector.

The chromatographic signals can be displayed on a recorder. Quantitative results are obtained by evaluating the peak area or peak height, both of which are proportional to the concentration of the analyte over a wide range. This is typically performed using digital integrators connected directly to the analog output signal of the detector. Modern detectors feature an additional parallel interface (e.g., RS-232-C), which allows the connection to a personal computer or a host computer with a suitable chromatography software. The computer may also take on control functions, thus allowing a fully automated operation of the chromatography system.

Corrosive eluents such as diluted hydrochloric acid and sodium hydroxide solutions are often used in ion chromatography. Therefore, all parts of the chromatographic system being exposed to these liquids should be made of metal-free inert materials. Conventional HPLC systems with stainless steel tubings and pump heads are only partially suited for ion chromatography, since even stainless steel is eventually corroded by aggressive eluents. Considerable contamination problems would result, since metal ions exhibit a high affinity toward the stationary phase of ion exchangers, leading to a significant loss of separation efficiency. Metal parts in the chromatographic flow system would make the trace analysis of heavy and transition metals more difficult.

1.4 Advantages of Ion Chromatography

The determination of ionic species in solution is a classical analytical problem with a variety of solutions. Whereas in the field of cation analysis both fast and sensitive analytical methods are available (AAS, ICP, polarography, and others), the lack of corresponding, highly sensitive methods for anion analysis is noteworthy. The conventional wet-chemical methods such as titration, photometry, gravimetry, turbidimetry, and colorimetry are all labor-intensive, time-consuming, and occasionally troublesome. In contrast, ion chromatography offers the following advantages:

- speed,
- sensitivity,
- selectivity,
- simultaneous detection,
- stability of the separator columns.

Speed

The time necessary to perform an analysis has become an increasingly important aspect, since the number of samples has recently increased significantly due to both enhanced manufacturing costs for high quality products and additional environmental efforts.

The average analysis time could be reduced to about ten minutes. With the high efficiency separator columns introduced for ion-exchange, ion-exclusion, and ion-pair chromatography in recent years, today the analysis of the seven most important inorganic anions [20] with a baseline-resolved separation requires only three minutes. Thus, quantitative results are obtained in a fraction of the time previously necessary using traditional wet chemical methods, thus increasing the sample throughput.

Sensitivity

The introduction of microprocessor technology, in connection with modern stationary phases of high chromatographic efficiencies, makes it a routine task to detect ions in the medium and lower ppb concentration range without pre-concentration. The detection limit for simple inorganic anions and cations is about 10 ppb based on an injection volume of 50 μL. The total amount of injected sample lies in the lower ng range. Even ultrapure water, required for the operation of power plants or for the production of semiconductors, may be analyzed for its anion and cation content after preconcentration with respective concentrator columns. With these pre-concentration techniques, the detection limit could be lowered to the ppt range. However, it should be emphasized that the cost of instrumentation is extremely high for such analysis. In addition, high demands have to be met in the creation of suitable environmental conditions. The limiting factor for further lowering the detection limits is the contamination by ubiquitous chloride and sodium ions.

High sensitivities in the pmol range are also possible in amino acid analysis by using fluorescence detection after derivatization with *o*-phthalaldehyde.

Selectivity

The selectivity of ion chromatographic methods for analyzing inorganic and organic anions and cations is ensured by selecting suitable separation and detection systems. Regarding conductivity detection, the suppression technique is of paramount importance, since the respective counter ions of the analyte ions as a potential source of interference are exchanged against hydronium and hydroxide ions, respectively. A high degree of selectivity is achieved by using solute specific detectors such as a UV/Vis detector to analyze nitrite in the presence of high amounts of chloride. New developments in the field of post-column derivatization show that specific compound classes such as heavy metals, alkaline-earth metals, polyvalent anions, silicate, etc. can be detected with high selectivity. Such examples explain why sample preparation for ion chromatographic analyses usually involves only a simple dilution and filtration of the sample. This high degree of selectivity facilitates the identification of unknown sample components.

Simultaneous Detection

A major advantage of ion chromatography – especially in contrast to other instrumental techniques such as photometry and AAS – is its ability to simultaneously detect sample components. Anion and cation profiles may be obtained within a short time, which provide information about the sample composition, thus avoiding time-consuming tests. The ability of ion chromatographic techniques for simultaneous detection is limited, however, by extreme concentration differences between the various sample components. For example, the major and minor components in a waste water matrix may only be detected simultaneously in a few cases; i.e. they have to be analyzed in two different chromatographic runs using either different sensitivity settings on the detector or different dilutions of the sample.

Stability of the Separator Columns

The stability of the separator columns is highly dependent on the type of packing material being used. In contrast to silica-based separator columns commonly used in conventional HPLC, resin materials such as polystyrene/divinylbenzene copolymers prevail as support material in ion chromatography. The high pH stability of these resins allows the use of strong acids and bases as eluents, which is a prerequisite for the widespread application of this method. Strong acids and bases, on the other hand, can also be used for rinsing procedures. One disadvantage of organic polymers is their limited stability toward organic solvents, which, therefore, cannot be used for the removal of organic contaminants (see also Chapter 8). However, polymer-based stationary phases exhibit a low sensitivity toward complex matrices such as waste water, foods, or body fluids. Sample preparation in these cases is often solely a simple dilution of the sample with de-ionized water followed by filtration.

1.5 Selection of Separation and Detection Systems

As previously mentioned, a wealth of different separation techniques could be described under the term "ion chromatography." Therefore, what follows is a survey of the criteria for selecting stationary phases and detectors being suitable for solving a specific separation problem.

The analyst usually has some information regarding the nature of the ion to be analyzed (inorganic or organic), its surface activity, its valency, and its acidity or basicity, respectively. With this information and on the basis of the selection criteria outlined schematically in Table 1-1, it should not be difficult for the analytical chemist to select a suitable stationary phase and detection mode. In many cases, several procedures are feasible for solving a specific separation problem. In these cases, the choice of the analytical procedure is determined by the type of matrix, the simplicity of the procedure, and, increasingly, by financial aspects. Two examples illustrate this:

Various sulfur-containing species in the scrubber solution of a flue-gas desulfurization plant (see also Section 8.2) are to be analyzed. According to Table 1-1, non-polarizable ions such as sulfite and sulfate, with pK values below 7, are separated by HPIC using an anion exchanger and are detected via electrical conductivity. A suppressor system may be used to increase the sensitivity and the specificity of the procedure. Scrubber solutions also contain thiocyanate and thiosulfate in small concentrations. However, due to their polarizability, these ions exhibit a high affinity toward the stationary phase of conventional anion exchangers. Three different approaches are feasible for the analysis of such ions. A conventional anion exchanger may be used with a mobile phase of high ionic strength. Depending on the concentration of the ions to be analyzed, difficulties with the sensitivity of the subsequent conductivity detection may arise. Alternatively, special anion exchangers with hydrophilic functional groups may be employed. Polarizable anions are not adsorbed as strongly on these stationary phases and, therefore, elute in a shorter time. Taking into account that higher sulfur species such as dithionate may also have to be analyzed, ion-pair chromatography (MPIC) is much more suited, since this technique allows thiocyanate, thiosulfate, and dithionate to be separated in a single run. Today, all compounds of interest can be separated in a single run, using an anion exchanger of high chromatographic efficiency, applying a gradient elution technique. Detection is carried out via electrical conductivity. However, the required concentration gradient makes the use of a suppressor system inevitable.

Table 1-1: Scheme of the selection criteria for seperation and detection systems.

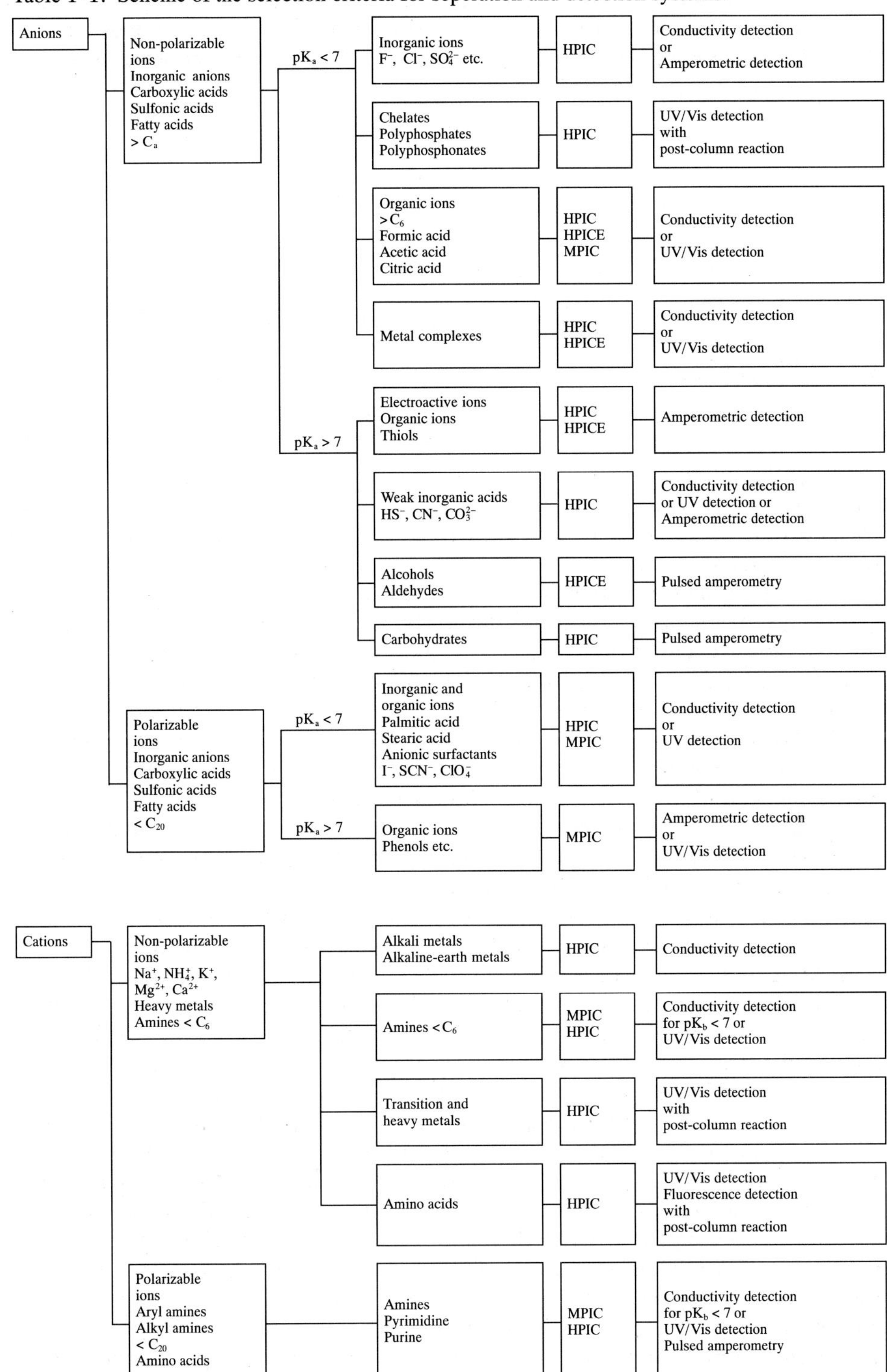

A second example is the determination of organic acids in coffee. According to Table 1-1, aliphatic carboxylic acids are separated by HPICE on a totally sulfonated cation exchange resin with subsequent conductivity detection. While this procedure is characterized by a high selectivity for aliphatic monocarboxylic acids with a small number of carbon atoms, sufficient separation cannot be obtained for the aliphatic open-chain and cyclic hydroxy acids, which are also present in coffee. Only after introducing a new stationary phase with specific selectivity for hydroxy acids did it become possible to separate the most important representatives of this class of compounds in such a matrix. Ion-exclusion chromatography is not suited for the separation of aromatic carboxylic acids, which are present in coffee in large numbers. Examples are ferulic acid, caffeic acid, and the class of chlorogenic acids. Due to π-π-interactions with the aromatic rings of the organic polymers used as support material for the stationary phase, aromatic acids are strongly retained and, thus, are not analyzable. Good separation is achieved by reversed-phase chromatography using chemically bonded octadecyl phases with high chromatographic efficiencies. These compounds are then detected by measuring their light absorption at 254 nm. Further details on the selection of separation and detection systems are given in Chapters 3 to 5.

2 Theory of Chromatography

2.1 Chromatographic Terms

The representation of chromatographic signals in the form of a chromatogram generally has the appearance similar to that in Fig. 2-1:

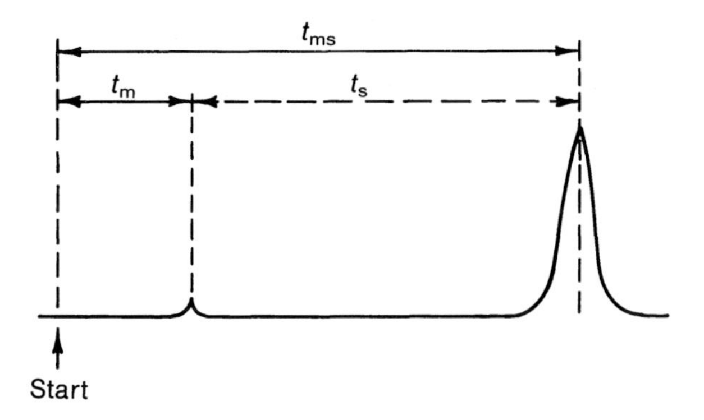

Fig. 2-1. General illustration of a chromatogram.

Two different components are separated in a chromatographic column only if they spend different times in or at the stationary phase, respectively. The time in which the components do not travel along the column, is called the solute retention time, t_s. The column dead time, t_m, is defined as the time necessary for a non-retained component to pass through the column. The gross retention time, t_{ms}, is calculated from the solute retention time and the column dead time:

$$t_{ms} = t_s + t_m \tag{1}$$

The chromatographic terms for the characterization of a separator column can be inferred from Fig. 2-1.

In a first approximation, the shape of a chromatographic peak is described by a Gaussian curve (Fig. 2-2).

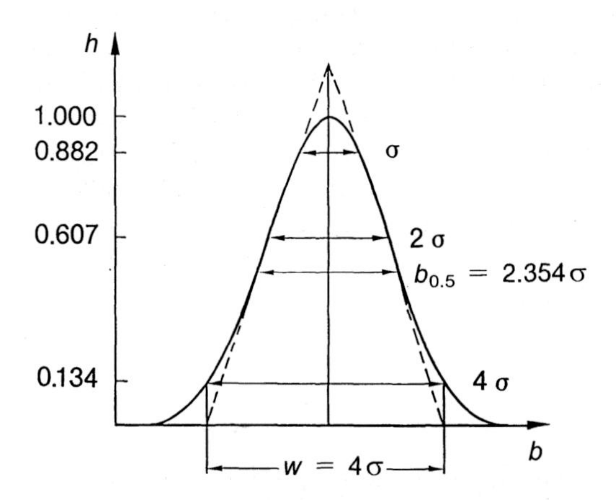

Fig. 2-2. The Gaussian curve.

σ Standard deviation
(half the peak width at the point of inflection)
$y = h$ Peak height at maximum
x Any particular point within the peak
μ Position of peak maximum

The peak height at any given position x can be derived from Eq. (2):

$$y = \frac{1}{\sigma \cdot \sqrt{2\pi}} \cdot e^{-\frac{(x-y)^2}{2\sigma^2}} \cdot Y_0 \tag{2}$$

Asymmetry Factor A_s

The signal (called the peaks) due to elution of a species from a chromatographic column is rarely perfectly Gaussian. Normally, the peaks are asymmetrical to some extent, which is expressed by Eq. (3) (see also Fig. 2-3):

$$A_s = \frac{b}{a} \tag{3}$$

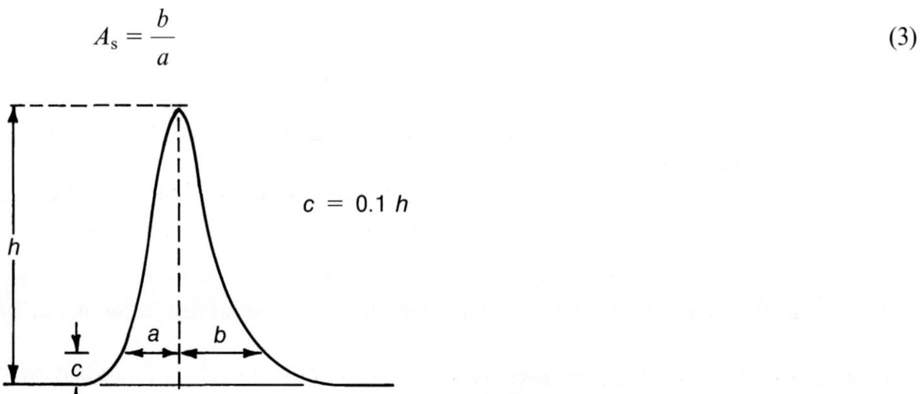

Fig. 2-3. Definition of the asymmetry factor.

At A_s-values higher than 1, this asymmetry is called "tailing." The peak shape is characterized by a fast increase of the chromatographic signal followed by a comparatively slow decrease. Adsorption processes are mainly responsible for such tailing effects. At A_s-values lower than 1, the asymmetry is called "leading" or "fronting", respectively. This effect is characterized by a slow increase of the signal followed by a fast decrease. The leading effect arises if the stationary phase does not possess a sufficient number of suitable adsorption sites and, thence, some of the sample molecules (or sample ions) pass the peak center. For practical applications, separator columns are considered to be good when the asymmetry factor lies between 0.9 and 1.2.

2.2 Parameters for Assessing the Quality of a Separation

Resolution

The objective of a chromatographic analysis is to resolve the components in a mixture into separate bands. The resolution R of two neighboring peaks is defined as the quotient of the difference in the absolute retention times and the arithmetic mean of their peak widths w at the respective peak base.

$$R = \frac{t_{ms_1} - t_{ms_2}}{\frac{w_1 + w_2}{2}} = \frac{2\Delta t_{ms}}{w_1 + w_2} \quad (4)$$

t_{ms_1}, t_{ms_2} Gross retention times for signal 1 and 2, respectively

w_1, w_2 Peak widths at the baseline as determined by the intersection points of the tangents drawn to the peak above its points of inflection

As shown in Fig. 2-4, these parameters can be obtained directly from the chromatogram. A resolution of $R = 2.0$ (corresponding to an 8σ separation) is sufficient for quantitative analysis, if the peaks exhibit a Gaussian peak shape. The two peaks are thus completely baseline resolved, since the peak width at the base is given by

$$w = 4\sigma \quad (5)$$

Higher values of R would result in excessively long analysis times. At a resolution of $R = 0.5$, it is still possible to recognize two sample components as separate peaks.

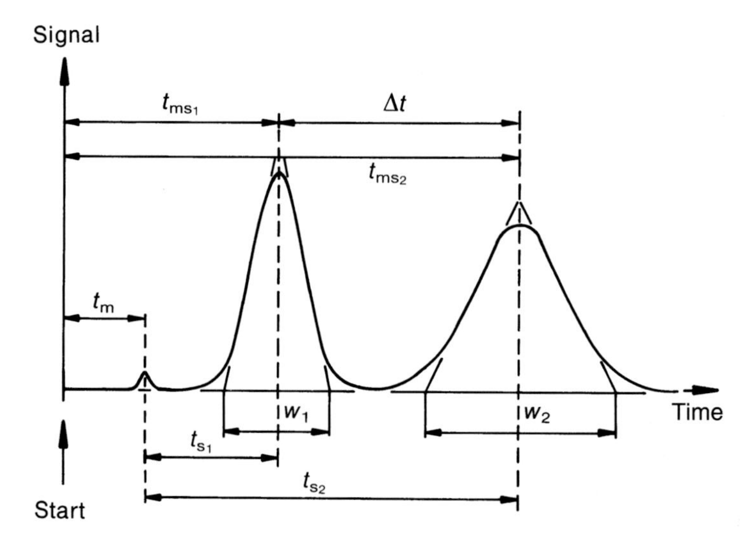

Fig. 2-4. Parameter for assessing resolution and selectivity.

Selectivity

The decisive parameter for the separation of two components is their relative retention, which is called selectivity α. The selectivity is defined as the ratio of the solute retention times of two different signals:

$$\alpha = \frac{t_{s_2}}{t_{s_1}} = \frac{t_{ms_2} - t_m}{t_{ms_1} - t_m} \tag{6}$$

According to Fig. 2-4, these parameters may also be obtained from the chromatogram. The selectivity is determined by the properties of the stationary phase. In case of HPLC, the selectivity is additionally affected by the mobile phase composition. If $\alpha = 1$, then there are no thermodynamic differences between two sample components at the given chromatographic conditions; therefore, no separation is possible. At equilibrium, which is established in reasonably close approximation in chromatography, the selectivity α is a thermodynamic quantity, which at constant temperature only depends on the specific properties of the sample components to be separated and on the properties of the mobile and stationary phase being used.

Capacity Factor

The capacity factor, k', is the product of the phase ratio ϕ between stationary and mobile phases in the separator column and the Nernst distribution coefficient, K:

$$k' = K \cdot \frac{V_s}{V_m} = \frac{C_s}{C_m} \cdot \frac{V_s}{V_m} = \frac{t_{ms} - t_m}{t_m} = \frac{t_s}{t_m} \tag{7}$$

K Nernst distribution coefficient
V_s Volume of the stationary phase
V_m Volume of the mobile phase
C_s Solute concentration in the stationary phase
C_m Solute concentration in the mobile phase

The capacity factor is independent of the equipment being used, and is a measure of the column's ability to retain a sample component. Small values of k' imply that the respective component elutes near the void volume; thus the separation will be poor. High values of k', on the other hand, are tantamount to long analysis times, associated peak broadening, and a decrease in sensitivity.

2.3 Column Efficiency

A fundamental disadvantage of chromatography is the broadening of the sample component zone during its passage through the separation system. The peak broadening is caused by diffusion processes and flow processes. A measure for this peak broadening is the plate number, N, or the plate height, H, which is independent of the column length, L.

The height of a theoretical plate, in which the distribution equilibrium of sample molecules between stationary and mobile phases is established, is related to the plate number via the length of the separator column:

$$H = \frac{L}{N} \tag{8}$$

N Plate number
L Length of the separator column

Based on chromatographic data, the theoretical plate height, H, which is defined as the ratio of the peak variance and column length L, can be calculated via Eq. (9):

$$H = \frac{\sigma^2}{L} = \frac{1}{16} \cdot \frac{w^2}{L} = \frac{1}{8 \ln 2} \left(\frac{b_{0,5}}{L} \right)^2 \tag{9}$$

$b_{0.5}$ Peak width at half the peak height
w Peak width at the baseline between the tangents drawn to the peak above the points of inflection

The term 8 ln 2 arises from the approximation of a peak as a Gaussian curve. With Eq. (8), it follows for the number of theoretical plates:

$$N = \frac{L^2}{\sigma^2} = \frac{t_{ms}^2}{\sigma_t^2} \tag{10}$$

Two sample components may only be separated from each other if their k' values differ. The effective plate number, N_{eff}, or the effective plate height, H_{eff}, is used to describe the separation efficiency of a column.

$$N_{eff} = 5.54 \cdot \left(\frac{t_s}{b_{0.5}} \right)^2 \tag{11}$$

$$H_{eff} = \frac{L}{N_{eff}} \tag{12}$$

The resolution of a column can be coupled to the parameter's efficiency of the separator column, selectivity, and capacity factor in a single equation according to:

$$R = \underbrace{\frac{1}{4} \sqrt{N}}_{a} \cdot \underbrace{\left(\frac{\alpha - 1}{\alpha} \right)}_{b} \cdot \underbrace{\left(\frac{k'}{k' + 1} \right)}_{c} \tag{13}$$

a Term for the column efficiency
b Term for the column selectivity
c Term for the capacity factor

Upon rearranging Eq. (13), the plate number required to afford the desired resolution may be calculated for any given values of k' and α:

$$N = 16R^2 \cdot \left(\frac{k'+1}{k'}\right) \cdot \left(\frac{\alpha}{\alpha - 1}\right)^2 \tag{14}$$

2.4 The Concept of Theoretical Plates (van Deemter Theory)

Based on the work of Martin and Synge [1], van Deemter, Zuiderweg, and Klinkenberg [2] introduced the concept of the theoretical plate height, H, as a measure for the relative peak broadening analogous to the terminology used in distillation technology. In a general form, the plate height H is given by

$$H = A + \frac{B}{v} + C \cdot v \tag{15}$$

(van Deemter equation)

v Linear flow velocity of the mobile phase in cm/s

The individual terms of the sum vary in their dependence in different ways on the velocity v of the mobile phase.

Accordingly, term A is independent of the flow velocity and characterizes the peak dispersion caused by the Eddy diffusion. This effect considers the different pathways for solute molecules in the column packing. The longitudinal diffusion is described by the term B/v. The term $C \cdot v$ comprises the lateral diffusion and the resistance to mass-transfer between mobile and stationary phases. These effects depend linearly on the flow velocity.

With the peak width expressed in terms of length units σ_1, it follows by taking Eq. (8):

$$N = \left(\frac{L}{\sigma_1}\right)^2 \tag{16}$$

or

$$H = \frac{\sigma_l}{L} \tag{17}$$

Various portions of σ_i contribute to the broadening of a peak. The sum of their variances σ_i^2 gives to the total band spreading:

$$\sigma_l^2 = \sum_{i=1}^{n} \sigma_i^2 \tag{18}$$

The portion of the peak broadening which arises due to the irregularity of the column packing is called Eddy diffusion. It is approximated by Eq. (19):

$$\sigma_E^2 = 2\lambda \cdot d_p \tag{19}$$

All molecules present in the mobile phase at time t_m may diffuse in and against the flow direction. The contribution of the longitudinal diffusion in the mobile phase is described by Eq. (20):

$$\sigma_L^2 = \frac{2\gamma_m \cdot D_m}{v} \tag{20}$$

γ_m Obstruction factor, which takes into account the obstruction of the free longitudinal diffusion due to collisions with particles of the column packing
D_m Diffusion coefficient in the mobile phase
v Linear velocity of the mobile phase in cm/s

Lateral diffusion and resistance-to-mass-transfer are the predominating effects for the total peak broadening and, thus, mainly determine the efficiency of the separator column. Any one sample molecule which interacts with the column packing, diffuses back and forth between the stationary and the mobile phase. It is retained at the stationary phase, and therefore, trails the center of the peak, which passes through the separator column. This effect is illustrated in Fig. 2-5 [3]. In the mobile phase, on the other hand, the sample molecule travels with the eluent. The mass transfer effect causes a peak broadening, since sample molecules pass through the column ahead of, as well as behind, the peak center. Due to the eluent flow through the column, the equilibrium between the solute concentration in the mobile phase and in the adjacent stationary phase is not attained. Both phases contribute to the resistance-to-mass-transfer. Therefore, two terms must be considered for the peak broadening. The peak broadening by the stationary phase is given by:

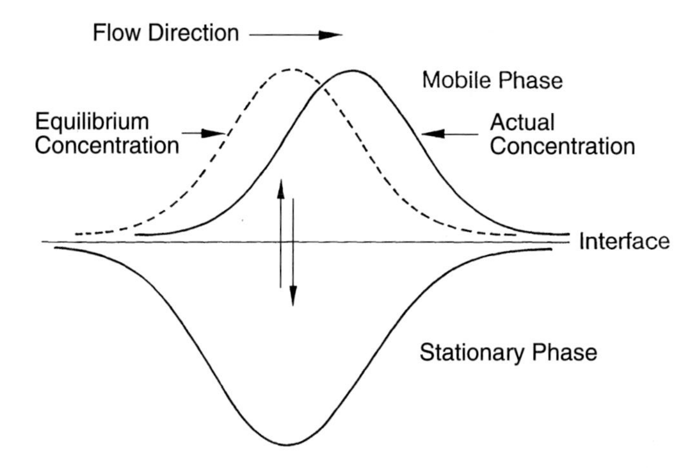

Fig. 2-5. Illustration of the mass transfer effect.

$$\sigma_s^2 = \frac{f \cdot k_e \cdot d_f^2}{D_s} \cdot v \quad (21)$$

f Form factor for the stationary phase
d_f Thickness of the stationary phase
D_s Diffusion coefficient of the solute in the stationary phase

Instead of the capacity factor, k', the capacity ratio, k_e, which is calculated by Eq. (22), is used in the above equation:

$$k_e = \frac{V_R - V_e}{V_e} \quad (22)$$

V_R Retention volume of the solute
V_e Exclusion volume of the column

The surface-functionalized column packings used in ion chromatography minimize the contribution to the total peak broadening caused by the stationary phase, since solute molecules are not able to penetrate into the packing material.

The dependence of the peak broadening on the mobile phase is given by Eq. (23):

$$\sigma_m^2 = \frac{\omega \cdot d_p^2}{D_m} \cdot v \quad (23)$$

The column coefficient, ω, is a measure for the regularity of the packing. The contribution σ_m^2 to the total peak broadening is significant, but may be substantially influenced by decreasing the particle diameter, d_p. The use of eluents with low viscosity also leads to a reduction in the peak broadening σ_m^2, since the diffusion coefficient of the solutes in the mobile phase increases accordingly.

The total plate height, H, may be expressed by combining Equations (19) to (23):

$$H = 2\lambda \cdot d_p + \frac{2\gamma_m \cdot D_m}{v} + \frac{f \cdot k_e \cdot d_f^2}{D_s} \cdot v + \frac{\omega \cdot d_p^2}{D_m} \cdot v \tag{24}$$

The first experimental results published by Keulemans and Kwantes [4] confirmed the applicability of the equation to gas chromatography, but it soon became apparent that the equation introduced by van Deemter et al. is only of limited validity for liquid chromatography.

In 1961, Giddings [5] proposed a HETP equation, which may be considered a special case of the van Deemter equation:

$$H = \frac{A}{1 + E/v} + \frac{B}{v} + C \cdot v \tag{25}$$

Giddings' main criticism of the van Deemter equation was that a finite contribution to the peak broadening by the Eddy diffusion term is predicted even for zero flow velocity. However, at the flow velocities encountered in practical applications, the equation proposed by Giddings reduces to the van Deemter equation, since all other terms remain the same.

The equation introduced in 1967 by Huber and Hulsman [6]

$$H = \frac{A}{1 + E/v^{1/2}} + \frac{B}{v} + C \cdot v + D \cdot v^{1/2} \tag{26}$$

accounts for this phenomenon. They introduced a coupling term, which causes the Eddy diffusion term to vanish if the flow velocity approaches zero. In contrast to van Deemter and Giddings, the resistance-to-mass-transfer in the mobile phase is described by an additional term $D \cdot v^{1/2}$. However, this factor resembles the coupling term proposed by Giddings in its physical interpretation and in its dependence on the flow velocity.

In the early 70s, Knox et al. [7,8] suggested another HETP equation based on their extensive data. Their equation differed significantly from the equation discussed thus far. This equation was derived by curve fitting on the author's extensive data:

$$H = A \cdot v^{1/3} + \frac{B}{v} + C \cdot v \tag{27}$$

Finally, Horvath and Lin [9,10] developed an equation very similar to the one introduced by Huber and Hulsman:

$$H = \frac{A}{1+E/v^{1/3}} + \frac{B}{v} + C \cdot v + D \cdot v^{2/3} \tag{28}$$

The only difference between the two equations is the description of the resistance-to-mass-transfer effect, which Horvath et al. interpret to depend on the square of the cubic root of the flow velocity instead of a quadratic root dependence.

Although the various HETP equations differ significantly from each other, Scott et al. [11] showed that, based on their extensive experimental data, the dependence of the theoretical plate height H on the linear flow velocity may be satisfactorily described by the van Deemter equation in the range between 0.02 cm/s and 1 cm/s. In Scott's tests, porous silicas with four different particle sizes were employed as stationary phases, on which nine solute compounds were separated using six different eluent mixtures.

Although the experimental data for H and v may be depicted by any hyperbolic function, not all of them provide a meaningful physical insight into the dispersion process. According to Scott et al. [11], the van Deemter equation in the form

$$H = 2\lambda \cdot d_p + \frac{2\gamma_m \cdot D_m}{v} + \frac{(a+b \cdot k_e + c \cdot k_e^2) \cdot d_p^2 \cdot v}{24(1+k_e)^2 \cdot D_m} \tag{29}$$

is applicable to liquid chromatography at normal operating conditions. The coefficients λ and γ are numbers to describe the quality of the packing; in the case of well-packed columns, they are between 0.5 and 0.8. The coefficients a, b, and c were calculated by the authors to be 0.37, 4.69, and 4.04, respectively.

The coupling term introduced by Giddings seems to be particularly significant only for surface-functionalized packings. This is due to the low porosity of these materials, which reduces the volume of mobile phase being retained in the packing. Hence, the term describing the resistance-to-mass-transfer is smaller, so that the mass transfer effect between the particles gains significance.

Instead of the total plate height H and the linear flow velocity v, often the reduced plate height, h,

$$h = \frac{H}{d_p} \tag{30}$$

and the reduced flow velocity, u,

$$u = \frac{v \cdot d_p}{D_m} \tag{31}$$

are used.

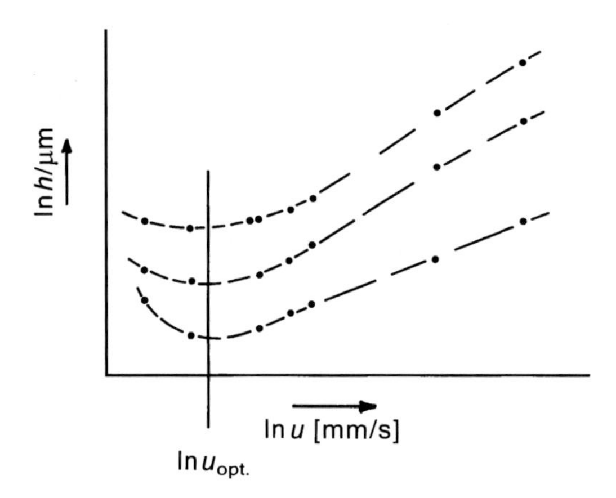

Fig. 2-6. General illustration of a Knox plot.

The graphical representation of ln h as function of ln u (Fig. 2-6) is known in the literature as the Knox plot. The dependence of the curve's position on the retention of the compound is disadvantageous. Minima in this kind of illustration are only obtained for compounds having no retention ($k' = 0$).

2.5 van Deemter Curves in Ion Chromatography

The terms for the various contributions to the peak broadening combined in Eq. (24) give the impression that they are independent of each other. In practice, however, an interdependence exists between these terms. This leads to a much smaller decrease in separation efficiency than predicted by the simplifying van Deemter theory.

By plotting the theoretical plate height calculated via Equations (8) and (10) over the flow rate, u, the dependence shown in Fig. 2-7 is obtained for an anion exchange column (IonPac AS4).

From Fig. 2-7 it is clear that the plate height is almost invariant at higher flow rates. Similar dependences have also been observed by Majors [12] for silica-based HPLC columns with smaller particle size. Such dependences show that higher flow rates may lead to drastically reduced analysis times, without any significant loss in chromatographic efficiency.

Cation separator columns exhibit a more pronounced dependence of the plate height on the flow rate (Fig. 2-8).

In this case, a compromise between separation efficiency and required analysis time has to be made. Flow rates between 2.0 mL/min and 2.3 mL/min have proved to be most suitable for practical applications.

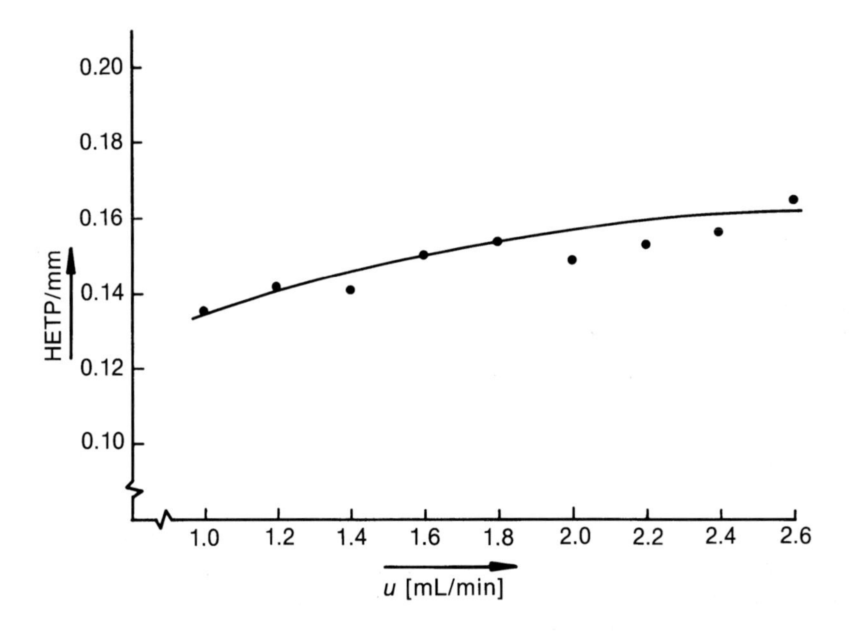

Fig. 2-7. HETP versus linear velocity u for an IonPac AS4 anion separator using t_{ms} (SO_4^{2-}).

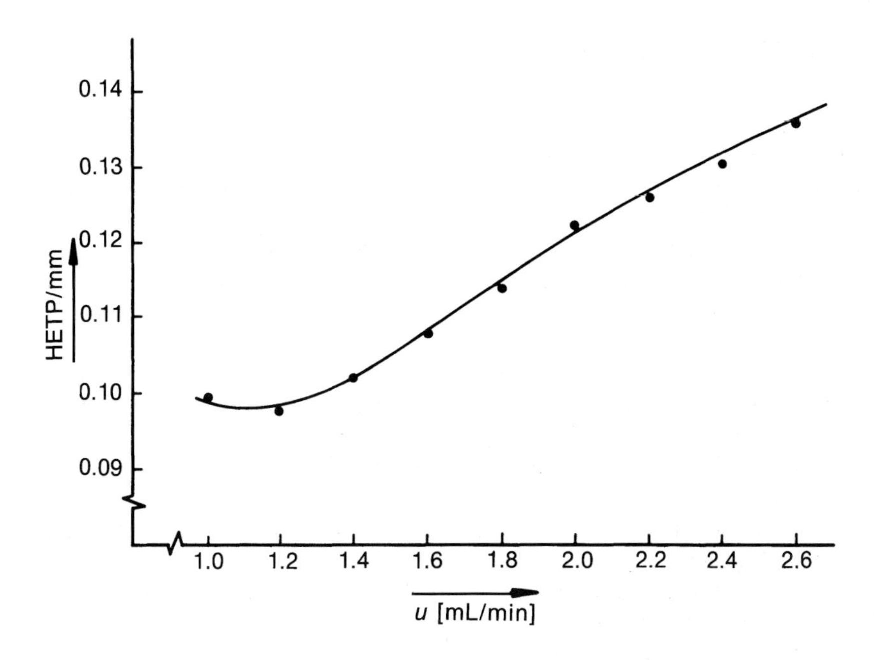

Fig. 2-8. HETP versus linear velocity u for an IonPac CS1 cation separator using t_{ms} (K^+).

3 Ion-Exchange Chromatography (HPIC)

3.0 General Remarks

Ion exchange is one of the oldest separation processes described in the literature [1,2]. The classical column chromatography on macroporous ion-exchange resins was a precursor of modern ion-exchange chromatography. The major differences between both processes are the method of sample introduction and the type of separation and detection systems being used. In classical ion-exchange chromatography, a column 10 cm to 50 cm long was typically filled with an anion or cation exchange resin, respectively, having a particle size between 60 and 200 mesh (0.075 to 0.25 mm). After the sample was applied to the top of the column, it migrated down the column driven by gravitational force, and became more or less separated. Individual fractions of the eluent were collected using a fraction collector, and subsequently analyzed in a separate workstep. Due to the high ion-exchange capacity of the columns high electrolyte concentrations were necessary to ensure the elution of the sample ions from the column. In many cases, several liters of eluent had to be worked up.

The enormous improvement in the capability of modern ion-exchange chromatography is attributed to the pioneering work of Small et al. [3]. Their major achievement was the development of ion-exchange resins of low capacities and *high chromatographic efficiencies*, which could be prepared reproducibly. The required injection volume was reduced to 10 μL to 100 μL, which resulted in an enhanced resolution due to very narrow peaks. Another important improvement was that of automated detection, which allowed continuous monitoring of the signal. The introduction of *conductivity detection* for ionic species added a new dimension to ion-exchange chromatography.

3.1 The Ion-Exchange Process

The resins employed in ion-exchange chromatography carry functional groups with a fixed charge. The respective counter ions are located near these functional groups, thus rendering the whole entity electrically neutral. In anion exchange chromatography, quaternary ammonium bases are generally used as ion-exchange groups; sulfonate groups are used in cation exchange chromatography.

When the counter ion of an exchange site is replaced by a solute ion, the latter is temporarily retained by the fixed charge. The various sample ions remain a different length of time within the column due to their different affinities toward the stationary phase, thus, separation is brought about.

For example, if a solution containing carbonate anions is passed through an anion exchange column, the quaternary ammonium groups attached to the resin are exclusively in their carbonate form. If a sample with the anions A^- and B^- is injected onto the column, these anions are exchanged for the carbonate ions according to the reversible equilibrium process given by Equations (32) and (33):

$$\text{Harz-}\overset{+}{N}R_3\,HCO_3^- + A^- \underset{}{\overset{K_1}{\rightleftharpoons}} \text{Harz-}\overset{+}{N}R_3\,A^- + HCO_3^- \tag{32}$$

$$\text{Harz-}\overset{+}{N}R_3\,HCO_3^- + B^- \underset{}{\overset{K_2}{\rightleftharpoons}} \text{Harz-}\overset{+}{N}R_3\,B^- + HCO_3^- \tag{33}$$

The separation of the anions is determined by their different affinities toward the stationary phase. The constant determining the equilibrium process is the selectivity coefficient, K, and is defined as follows:

$$K = \frac{[X^-]_s \cdot [HCO_3^-]_m}{[HCO_3^-]_s \cdot [X^-]_m} \tag{34}$$

$[X^-]_{m,s}$ Concentration of the sample ion in the mobile (m) or the stationary phase (s)

$[HCO_3^-]_{m,s}$ Carbonate concentration in the mobile (m) or the stationary phase (s)

The selectivity coefficient can be determined experimentally by adding a certain amount of resin material to a solution with known concentrations of X^- and HCO_3^-. The resulting concentration of the exchanged ions is determined in the mobile and stationary phase, respectively, after equilibrium is achieved. To precisely calculate the selectivity coefficient, the activities a_i have to be used instead of the concentrations c_i. As a prerequisite, the determination of the activity coefficient f_i according to Eq. (35) is required, which is difficult to perform in the matrix of an ion-exchange resin.

$$a_i = f_i \cdot c_i \tag{35}$$

In ion chromatography, the concentration of ions in the solutions to be analyzed is small, thus they may be equated with the activity coefficient in the first approximation. For the sake of simplicity, we only use the concentrations c_i within the scope of this discussion.

The efficiency of an ion as eluent for ion chromatography may be estimated on the basis of the selectivity coefficient. Ions with high selectivity coefficients are preferentially used as eluent ions, since they exhibit a high elution power even in diluted solutions. If the sample ions elute too quickly with a given eluent, the ionic strength must be lowered or another eluent ion with a smaller selectivity coefficient must be used. In the selection of eluents, in general, the selectivity coefficients of eluent ion and sample ion should be comparable.

Another measure for the affinity of a solute ion toward the stationary phase of an ion-exchange resin is the weight distribution coefficient D_g, which is defined for an exchanged ion X^- as follows:

$$D_g = \frac{[X^-]_s}{[X^-]_m} \tag{36}$$

Instead of the weight distribution coefficient, in most cases the capacity factor, k', is used, which is defined by Eq. (37):

$$k' = D \cdot \frac{m_{\mathrm{Resin}}}{V_{\mathrm{Solution}}} \tag{37}$$

It follows from Eq. (37) that the capacity factor can be derived from the weight distribution coefficient. However, the determination of k' from chromatographic data [see Eq. (7) in Section 2.2] is much more convenient.

3.2 Thermodynamic Aspects

Apart from pure ion-exchange processes, non-ionic interactions with the stationary phase are also observed with specific ionic species. The most important non-ionic interaction is adsorption.

If an organic polymer with aromatic backbone is used as substrate for the ion-exchange resin, in ions with aromatic and olefinic carbon skeleton the ion-exchange process is superposed by sorption interactions, which can be attributed to π-π-interactions with the aromatic resin backbone. Not only are adsorption-type interactions observed in aromatic or olefinic solutes, but generally with all polarizable inorganic and organic ions. In some cases, even the separation of simple inorganic anions such as bromide and nitrate is entirely attained by various non-ionic sorption properties. This effect may be demonstrated by a simple experiment. It is possible to block the adsorption sites on the surface of the stationary phase by adding *p*-cyanophenol, which is especially effective in this respect, to the carbonate/bicarbonate eluent. As shown in Fig. 3.1, bromide and nitrate co-elute under these conditions, although the exchange groups of the stationary phase remain unaffected.

Such sorption phenomena may also be characterized thermodynamically. It should be mentioned here that in ion chromatography the quantities k' and K can be defined as comprising the sum of ionic and non-ionic interactions. However, the following discussion refers only to non-ionic interactions. It is based on the general representation of the retention value of compound i in terms of the capacity factor k' according to Eq. (38).

$$k'_i = \frac{t^i_{\mathrm{ms}} - t^i_{\mathrm{m}}}{t^i_{\mathrm{m}}} \tag{38}$$

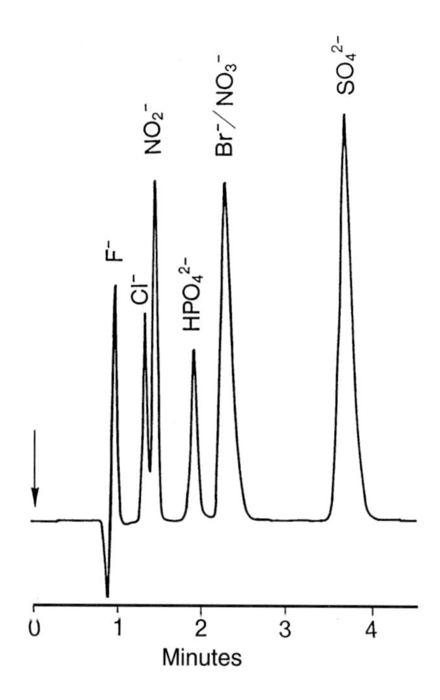

Fig. 3-1. Representation of the sorption effects in the separation of bromide and nitrate. – Separator: IonPac AS4; eluent: 0.0043 mol/L $NaHCO_3$ + 0.0034 mol/L Na_2CO_3 + 100 mg/L *p*-cyanophenol; flow rate: 2 mL/min; detection: suppressed conductivity; injection volume: 50 µL; solute concentrations: 3 ppm fluoride, 4 ppm chloride, 10 ppm nitrite, 10 ppm phosphate, 10 ppm bromide, 20 ppm nitrate, and 25 ppm sulfate.

The capacity factor is directly connected to the distribution coefficient, K, and thus with the thermodynamic quantities $\Delta H_{m \rightarrow s}$ and $\Delta S_{m \rightarrow s}$, which determine the chromatographic sorption process. (The indices m→s characterize the mass transfer from the mobile phase to the stationary phase.)

$$k'_i = K_i \cdot \Phi \tag{39}$$

$$= K_i \cdot \frac{v_{stat}}{v_{mob}}$$

Φ Phase volume ratio
v_{stat} Volume of the stationary phase
v_{mob} Volume of the mobile phase

Thermodynamically [4-6], the retention of a solute A can be expressed as a function of its free sorption enthalpy, $\Delta G_{m \rightarrow s}$ (A), and the phase volume ratio, Φ:

$$\ln k' = -\frac{\Delta G_{m \rightarrow s}(A)}{RT} + \ln \Phi \tag{40}$$

If A is a member of a homologous or quasi-homologous series of compounds, we get

$$\Delta G_{m \rightarrow s} = \Delta G^*_{m \rightarrow s} + \Delta\Delta G_{m \rightarrow s} \cdot n \tag{41}$$

with $\Delta\Delta G_{m \rightarrow s}$ corresponding to the change of free sorption enthalpy with each incremental step of the homologous series. The term $\Delta G^*_{m \rightarrow s}$ takes into account both the non-linearity of $\Delta G_{m\to s}$ with n for small members of the series and the contribution by the functional group. It then follows for ln k':

$$\ln k' = -\frac{\Delta G^{*}_{m\to s}}{RT} + \ln\Phi - \frac{\Delta\Delta G_{m\to s}}{RT}\cdot n \tag{42}$$

By plotting ln k' as function of n, a linear dependence is derived for each homologous series with the general form

$$\ln k' = a + b\cdot n \tag{43}$$

By substituting into Eq. (42) it follows for a und b:

$$a = -\frac{\Delta G^{*}_{m\to s}}{RT} + \ln\Phi \tag{44}$$

$$b = -\frac{\Delta\Delta G_{m\to s}}{RT} \tag{45}$$

Since

$$\Delta G = \Delta H - T\cdot\Delta S \tag{46}$$

it follows:

$$a = -\frac{\Delta H^{*}_{m\to s}}{RT} + \frac{\Delta S^{*}_{m\to s}}{R} + \ln\Phi \tag{47}$$

$$b = -\frac{\Delta\Delta H_{m\to s}}{RT} + \frac{\Delta\Delta S_{m\to s}}{R} \tag{48}$$

Therefore, the following expression is derived for the retention:

$$\ln k' = -\frac{\Delta H^{*}_{m\to s}}{RT} + \frac{\Delta S^{*}_{m\to s}}{R} - \left(\frac{\Delta\Delta H_{m\to s}}{RT} - \frac{\Delta\Delta S_{m\to s}}{R}\right)\cdot n + \ln\Phi \tag{49}$$

Both the ln k' values and the sorption enthalpies, $\Delta H_{m\to s}$, may be determined experimentally from the temperature dependence of retention. To calculate the sorption entropy, the phase volume ratio must be known. However, the thermodynamic data may be regarded simply as formal quantities, since the capacity factors correlate directly with $\Delta G_{m\to s}$ via the distribution coefficient according to Eq. (38), and since sorption exhibits both distributive and adsorptive character.

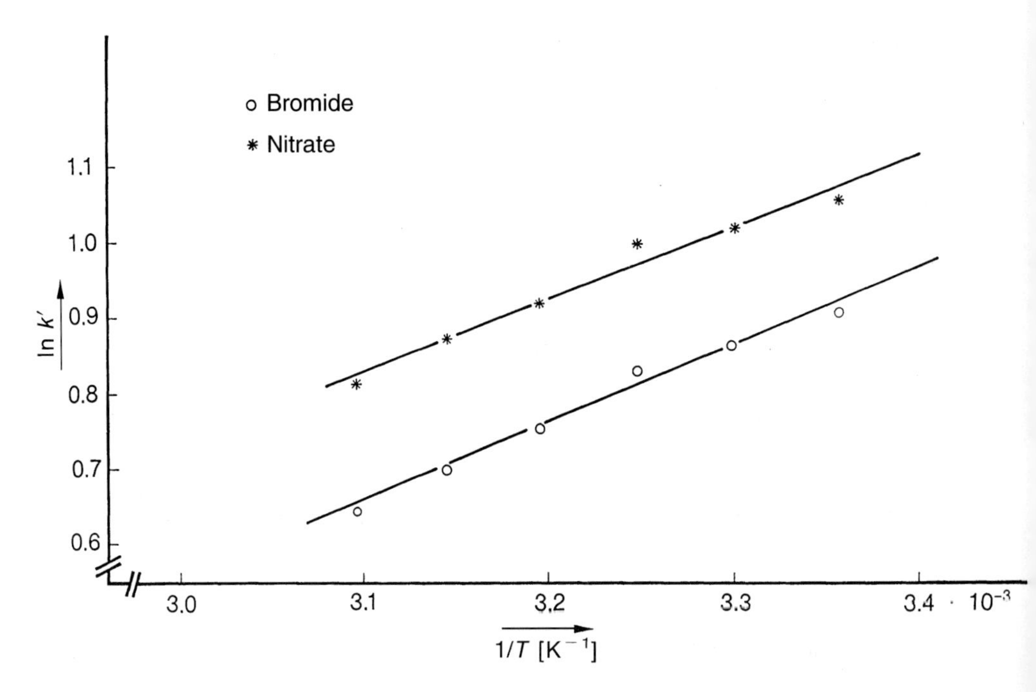

Fig. 3-2. Variation of the retention of bromide and nitrate with temperature. – Separator: IonPac AS3; eluent: 0.0028 mol/L $NaHCO_3$ + 0.0022 mol/L Na_2CO_3; flow rate: 2.3 mL/min; detection: suppressed conductivity; injection volume: 50 μL anion standard.

The retention model developed by Eon and Guiochon [7,8] to describe the adsorption effects at both gas-liquid and liquid-solid interfaces, which was later modified by Möckel et al. [6] to account for the retention at chemically bonded reversed-phase materials in HPLC, is not applicable to ion chromatography. But if the dependence of the capacity factors of various inorganic anions on the column temperature is studied, certain parallels with HPLC are observed. The linear dependences shown in Fig. 3-2 are obtained for the ions bromide and nitrate when the $\ln k'$ values are plotted versus the reciprocal temperature (van't Hoff plot). However, in the case of fluoride, chloride, nitrite, orthophosphate, and sulfate, the k' values were found to be constant within experimental error limits in the temperature range investigated. Upon linear regression of the values in Table 3-1, the following relations are derived for bromide and nitrate:

Table 3-1. $\ln k'$ values for bromide and nitrate at different temperatures.

T [K]	T^{-1} [K^{-1}]	$\ln k'$ Bromide	$\ln k'$ Nitrate
298	$3.356 \cdot 10^{-3}$	0.909	1.060
303	$3.300 \cdot 10^{-3}$	0.862	1.019
308	$3.247 \cdot 10^{-3}$	0.829	1.004
313	$3.195 \cdot 10^{-3}$	0.752	0.918
318	$3.145 \cdot 10^{-3}$	0.699	0.871
323	$3.096 \cdot 10^{-3}$	0.645	0.813

Bromide

$$\ln k' = 1036.03\ T^{-1} - 2.56 \qquad (50)$$
$$r = 0.994$$

Nitrate

$$\ln k' = 968.38\ T^{-1} - 2.17 \qquad (51)$$
$$r = 0.983$$

The heats of sorption $\Delta H_{m \to s}$ are derived from the slope b via

$$b = -\frac{\Delta H_{m \to s}}{R} \qquad (52)$$

With $R = 8.313$ J/(mol K) the following sorption enthalpies are calculated:

Bromide

$$\Delta H_{m \to s} = -8.6 \text{ kJ/(mol K)}$$

Nitrate

$$\Delta H_{m \to s} = -8.1 \text{ kJ/(mol K)}$$

The values thus derived are in the same order of magnitude as the sorption enthalpies determined by Möckel et al. [6] for various non-ionic inorganic and organic solutes. It is not possible to calculate the sorption entropy, $\Delta S_{m \to s}$, which is normally derived from the intercept a by using Eq. (53)

$$a = \frac{\Delta S_{m \to s}}{R} + \ln \Phi \qquad (53)$$

since the phase volume ratio is unknown. New investigations by Möckel et al. [6] revealed that the latter may be approximately estimated in reversed-phase chromatography. For that, the capacity factors of individual members of a homologous series are determined using two different eluent compositions, and are then plotted versus n. Since the free sorption enthalpy becomes zero for solutes with $K = 1$, the phase volume ratio may be directly taken via

$$\ln k' \approx \ln \Phi \qquad (54)$$

from the intersection of the two straight lines. However, suitable eluent systems are still required to calculate the phase volume ratio in ion chromatography analogously.

The sorption enthalpy $\Delta H_{m \to s}$ for fluoride, chloride, nitrite, orthophosphate, and sulfate is small or zero, respectively, since no relation between $\ln k'$ and the reciprocal temperature could be established. The change in the molar sorption enthalpy is therefore entirely due to an entropy increase. Ongoing investigations will reveal the contributions to the molar sorption entropy from changes in the mixing entropy, from the orientation of water caused by the shrinkage or swelling of hydration shells, or from the configuration entropy of the matrix.

3.3 Anion Exchange Chromatography

3.3.1 Stationary Phases

In contrast to the silica-based column packings used in classical HPLC, organic polymers are predominantly employed as support material in ion chromatography. These materials show a much higher stability toward extreme pH conditions. While silica-based HPLC columns can only be used within a pH range between 2 and 8, ion exchangers based on organic polymers are also stable in the alkaline region. Nonetheless, a couple of silica-based ion exchangers have recently been developed which exhibit a much higher chromatographic efficiency in comparison to organic polymers. However, the stationary phases used in anion exchange chromatography not only differ in the type of support material, they can also be classified according to their different pore sizes and ion-exchange capacities.

Ion-Exchange Capacity

The ion-exchange capacity of a resin is defined as the number of ion-exchange sites per weight equivalent of the column packing. It is typically expressed in terms of milli equivalent per gram resin (mequiv/g)*).

Generally, it holds:

The retention times for the analyte ions increase with increasing ion-exchange capacity of the resin.

This effect can be partly compensated by using eluents of higher ionic strength.

Independent of the type of the support materials, anion exchangers with comparatively high exchange capacities in the range of 1 mequiv/g only play a minor role. They are mentioned here only for the sake of completeness. Such materials are quite limited in applicability, since inorganic anions have to be eluted at high ionic strength. This renders conductivity detection very difficult, if not impossible. Therefore, anion exchangers with high capacities are only used in combination with amperometric [9] or photometric detection after derivatization of the column effluent [10]. For example, nitrite and nitrate can be separated on pellicular ion exchangers based on silica [11],

*) The correct unit is mmol/g referring to monovalent exchange groups. Here, we still employ the expression commonly used in the literature for the unit of the exchange capacity.

organic polymers [12], or cellulose [13], respectively. They are then detected by direct UV detection. Recent developments in the field of indirect detection methods [14,15] show that refractive index detection may provide an alternative to conductivity detection.

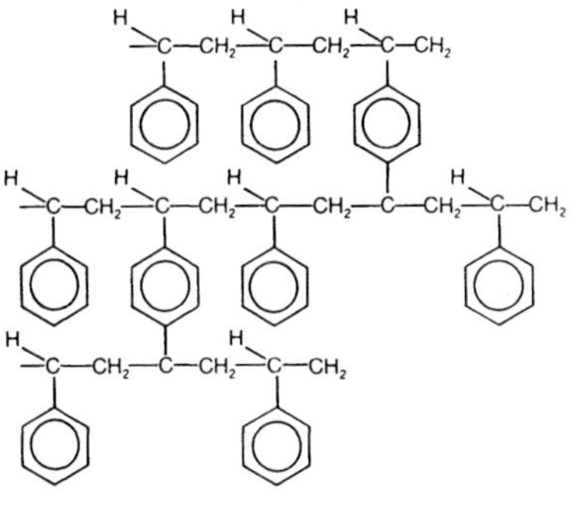

Fig. 3-3. Schematic illustration of the polystyrene/divinylbenzene skeleton.

Much more important are ion exchangers with low exchange capacities. The various types are described in detail as follows.

3.3.1.1 Polymer-Based Anion Exchangers

Styrene/divinylbenzene copolymers, polymethacrylate, and polyvinyl resins are the most important organic compounds that were tested for their suitability as substrate materials in the manufacturing process for polymer-based anion exchangers.

Styrene/Divinylbenzene Copolymers

Styrene/divinylbenzene copolymers are the most widely used substrate materials. Since they are stable in the pH range between 0 and 14, eluents with extreme pH values may be used. This allows the conversion of compounds such as carbohydrates, which are not ionic at neutral pH, into the anionic form, making them available for ion chromatographic analysis (see also Section 3.3.5.3).

The copolymerization of styrene with divinylbenzene (DVB) is necessary in order to obtain the required stability of the resin. Upon adding divinylbenzene to styrene, the two functional groups of DVB crosslink two polystyrene chains with each other. Part of the resulting skeleton is depicted schematically in Fig. 3-3. The percentage of DVB in the resin is indicated as "percent crosslinking". The degree of crosslinking determines the porosity of the resin, which is another characteristic for the classification of PS/DVB resins.

In general, one differentiates between microporous and macroreticular resins. The former is, by far, the more important one for anion exchange chromatography.

Microporous (gel-type) substrates are prepared by bead polymerization. In this process, the two monomeric compounds, styrene and divinylbenzene, are suspended as small droplets in water by rapid, even stirring. The polymerization is initiated by appropriate catalysts and leads to uniform particles, which have a microporous structure and a size distribution depending on the stirring speed. In general, a decrease of the average particle size is observed with increasing stirring speed.

The polarity of the particles is substantially augmented upon functionalization of the resin to the desired anion exchange material. Polar solvents, e.g., water or acetonitrile, hydrate the resin according to the relative number of functional groups. The associated swelling is less pronounced as the degree of crosslinking increases. Unpolar solvents lead to a dehydration of the resin and, thus, to its shrinkage. The column packing collapses as a result of the particle shrinkage, creating a void volume at the top end of the column, which is accompanied by a drastic loss of separation efficiency. An eluent change can also cause swelling and shrinkage, since the hydration of the ion-exchange resin depends on the counter ion in the mobile phase; i.e., the ionic form of the resin. Ion-exchange materials with a low degree of crosslinking are particularly prone to strong swelling and shrinkage if the ionic form of the resin changes significantly (e.g., transition from inorganic sodium hydroxide to organic potassium hydrogenphthalate). The optimal degree of crosslinking for a microporous PS/DVB resin is determined experimentally. Normally, this value lies between 2% and 8% which is a compromise between mechanical stability and chromatographic efficiency of the stationary phase. Higher values have been found to be disadvantageous, since ion-exclusion effects are increasingly observed with ions having large radii. High amounts of DVB also limit the mass-transfer, leading to a low column efficiency. On the other hand, resins which are crosslinked less than 2% are mechanically not stable enough.

Macroporous (macroreticular) [16,17] substrates are prepared by bead polymerization in polar solvents. To form such structures, a chemically inert solvent is added to the monomer suspension, in which the monomer is soluble but not the resulting polymer. When the polymerization process is complete, the solvent inclusions formed during the polymerization are washed out. Particles prepared by this method are spherical and have a high surface area between 25 m^2/g and 800 m^2/g depending on the type of polymer. In other methods, finely ground inorganic salts, such as calcium carbonate, are used instead of chemically inert solvents. These salts are extracted from the polymer after completion of the polymerization. Both methods result in the formation of copolymers having an average pore size in the range of 20 to 500 Å. These particles are called macroporous.

Macroporous resins exhibit a remarkable mechanical stability due to their high degree of crosslinking. They are preferably employed in anion exchange chromatography at conditions where the mobile phase contains significant amounts of organic solvents. Swelling and shrinkage, which could result from changing the polarity of the solvent, are also suppressed due to the high degree of crosslinking.

Functionalization of the Resin

Organic polymers are functionalized directly at their surface, with the exception of latex-based anion exchangers (see Section 3.3.1.2), where the totally porous latex particle acts as ion-exchange material. Surface-functionalized, "pellicular" substrates show a much higher chromatographic efficiency compared to the fully functionalized resins.

The organic polymers mentioned above are functionalized in a two-step process:

1. Addition of a chloromethyl group to the aromatic skeleton of the resin

$$\text{Resin}-C_6H_5 \xrightarrow[ZnCl_2]{ClCH_2OCH_3} \text{Resin}-C_6H_4-CH_2Cl \quad (55)$$

2. Amination of the choromethylated resin with a tertiary amine

$$\text{Resin}-C_6H_4-CH_2Cl \xrightarrow{NR_3} \text{Resin}-C_6H_4-\overset{+}{CH_2NR_3}\ Cl^- \tag{56}$$

In the past, the chloromethyl group was added to the aromatic skeleton of the resin via reaction with chloromethyl methyl ether. The polarity of the C-X bond in the primary halide was enhanced for the electrophilic attack by adding zinc chloride as Lewis acid to the reaction mixture. Since chloromethyl methyl ether is suspected of being carcinogenic and is no longer commercially available, Fritz et al. [18] devised a method in which chloromethyl methyl ether is prepared in situ and reacted with the resin.

However, the use of chloromethyl methyl ether can be avoided if the polymer is reacted with formaldehyde and hydrogen chloride in the presence of an acid:

$$\text{Resin}-C_6H_5 + HCHO + HCl \xrightarrow{ZnCl_2} \text{Resin}-C_6H_4-CH_2Cl + H_2O \tag{57}$$

The conjugated acid of formaldehyde, i.e., the hydroxymethylene cation, acts as an electrophile.

$$H-\underset{H}{C}=O + H^+ \longrightarrow \left\{ H-\underset{H}{C}=\overset{+}{O}H \longleftrightarrow H-\underset{H}{\overset{+}{C}}-OH \right\}$$

Hydroxymethylene Cation

In a second reaction step, the alcohol formed via electrophilic substitution is transformed by hydrochloric acid into the respective halogenide.

$$\text{Resin}-C_6H_5 + \overset{+}{C}H_2-OH \longrightarrow \text{Resin}-\overset{+}{C_6H_5}(H)-CH_2OH \xrightarrow{-H^+} \text{Resin}-C_6H_4-CH_2OH \xrightarrow{HCl} \text{Resin}-C_6H_4-CH_2Cl \tag{58}$$

The resin is subsequently aminated by reacting it with a tertiary amine. This method allows a good control of the ion-exchange capacity of the final anion exchange resin. Depending on the reaction pathway, ion-exchange capacities between 0.001 and 0.09 mequiv/g are possible.

Overview of Surface-Aminated Styrene/Divinylbenzene Copolymers

In connection with the introduction of a chromatographic system for the analysis of inorganic anions, in which the separator column was coupled directly to the conductivity cell, Fritz et al. [16,17] developed a series of macroporous, macroreticular anion exchangers of low capacity, featuring highly crosslinked PS/DVB particles as substrate (XAD-1, Rohm & Haas, Philadelphia, PA, USA). The substrate was chloromethylated and aminated with triethyl amine. Investigations into the dependence of the retention of inorganic anions on the ion-exchange capacity of the resin [19] revealed that the distribution coefficients decrease parallel to a decreasing ion-exchange capacity. On the other hand, the selectivity coefficients, which are decisive for the separation, remained constant within the investigated capacity range of 0.04 to 1.46 mequiv/g.

Polymers of this type are also commercially available. For example, Hamilton (Reno, NV, USA) introduced an anion exchange resin under the trade name PRP-X100. This resin features spherical PS/DVB particles, which are surface-aminated with trimethyl amine [20,21]. Similar stationary phases are also available from SYKAM (Gilching, Germany) under the trade name LCA A01.

The characteristic structural and technical properties of these columns are summarized in Table 3-2.

Fig. 3-4 shows a typical chromatogram of the separation of fluoride, chloride, and bromide on a surface-aminated XAD-1 resin. According to the literature, these stationary phases exhibit a low chromatographic efficiency (about 1000 to 2000 theoretical plates per meter). The signal in the void volume is characteristic for all ion chromatographic analyses without a suppressor system (see also Section 3.3.4.3). It results from the sample cations, which are not retained at the stationary phase. Therefore, the cations elute as a single band with the mobile phase and pass the conductivity detector together with the anions of the eluent. A positive peak results if the conducitvity of these anions and cations is higher than the conductivity of the eluent. In cases where the conductivity is lower than that of the eluent, a negative peak is observed, which renders the analysis of the signal more difficult.

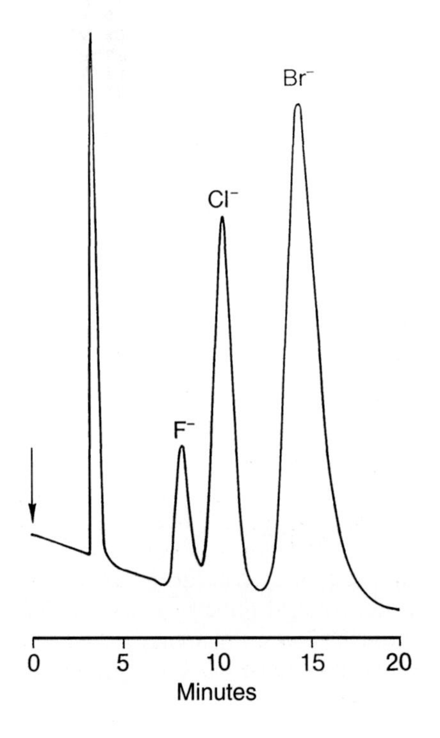

Fig. 3-4. Separation of fluoride, chloride, and bromide on a surface-aminated XAD-1-resin. – Separator: 1000 mm × 2 mm I.D. XAD-1 (0.04 mequiv/g); eluent: $6.5 \cdot 10^{-4}$ mol/L potassium benzoate (pH 4.6); flow rate: 2 mL/min; detection: direct conductivity; injection volume: 50 µL; solute concentrations: 4.8 ppm fluoride, 5.1 ppm chloride, and 26 ppm bromide (taken from [19]).

Table 3-2. Structural and technical properties of surface-aminated styrene/divinylbenzene copolymers.

Separator Column	PRP-X100	LCA A01	LCA A04
Dimensions (Length × ID) [mm]	100 × 4.1 150 × 4.1 250 × 4.1	200 × 4.0	200 × 4.0
Manufacturer	Hamilton	Sykam	Sykam
pH Range	1-13	1-14	1-14
Max. operating pressure [MPa]	35	25	25
Max. flow rate [mL/min]	8	3	3
Solvent stability [%]	100	10	10
Capacity [mequiv/g]	0.2	0.04	0.15
Particle size [μm]	10	12	7
Type of packing material	Sperical PS/DVB aminated with triethyl amine	PS/DVB with quaternary ammonium groups	PS/DVB with quaternary ammonium groups

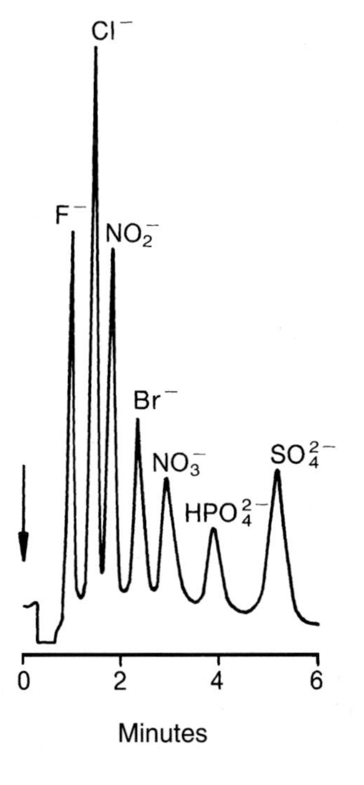

Fig. 3-5. Separation of various anions on PRP-X100. – Eluent: 0.004 mol/L sodium *p*-hydroxybenzoate (pH 8.6); flow rate: 3 mL/min; detection: direct conductivity; injection volume: 100 μL; solute concentrations: 20 ppm each; (taken from [21]).

For most applications, surface-aminated macroreticular anion exchange resins do not exhibit the required chromatographic efficiency. Thus, this type of stationary phase can only be employed when a high resolution is not required [22].

In comparison, the stationary phases manufactured by Hamilton and Sykam are much more efficient. As shown in Fig. 3-5, seven different inorganic anions, which are often referred to as standard anions, are separated on a PRP-X100 column within a short time. Sodium *p*-hydroxybonzoate was employed as the eluent. A slightly different elution order is obtained using an eluent mixture of sodium carbonate and sodium bicarbonate (Fig. 3-6). In addition, Sykam offers a stationary phase, under the trade name LCA A04, with which bromide and nitrate have a higher retention time than sulfate. The resulting separation is comparable to the chromatogram obtained with the Dionex AS2 column shown in Fig. 3-14 (see Section 3.3.1.2).

Polymethacrylate and Polyvinyl Resins

In 1983, Toyo Soda (Japan) introduced TSK Gel IC-PW, a high efficiency anion exchange resin, which is a hydrophilic, porous methacrylate polymer with an ion-exchange capacity of 0.03 mequiv/g being surface-aminated with methyldiethyl amine. The chromatographic efficiency of this 50 mm x 4.6 mm i.d. column is surprisingly high. For the nitrate peak, 20,000 theoretical plates per meter have been calculated. However, polymethacrylate resins are only stable at pH values between 1 and 12. Fig 3-7 shows the respective anion chromatogram obtained with potassium gluconate in a borate buffer solution as the eluent. Potassium hydroxide can also be used for the separation of monovalent anions [23]. The TSK Gel IC-PW is also traded by Waters Millipore under the name "IC Pak A". Much shorter analysis times are achieved using the acrylate-based stationary phase LCA A03, which is manufactured by Sykam, and IC-A1, which is manufactured by Shimadzu. The selectivities of these phases are clearly shown in Figs. 3-8 and 3-9, respectively.

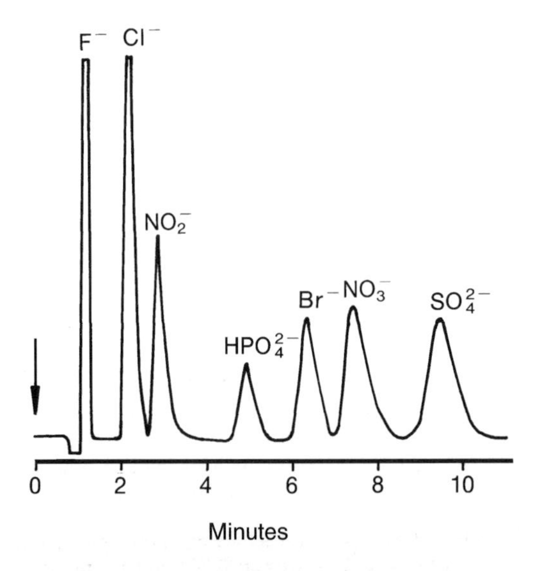

Fig. 3-6. Separation of various inorganic anions on LCA A01. – Eluent: 0.003 mol/L $NaHCO_3$ + 0.002 mol/L Na_2CO_3 in water/butanol (99.5:0.5 v/v); flow rate: 2.5 mL/min; detection: suppressed conductivity; injection volume: 50 µL; solute concentrations: 1 ppm each.

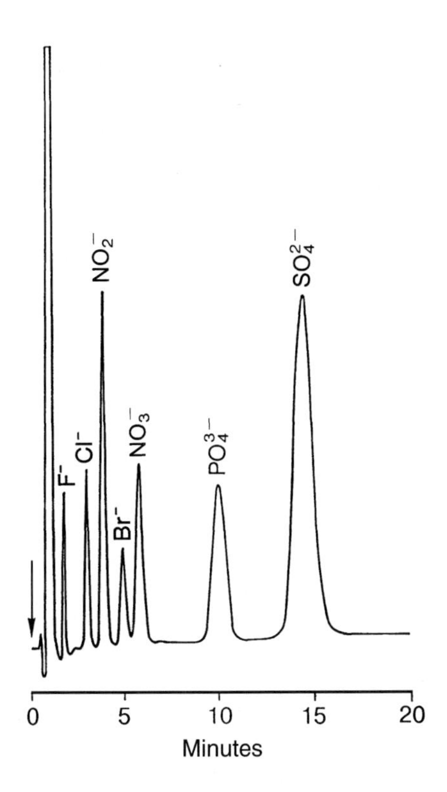

Fig. 3-7. Separation of various inorganic anions on TSK Gel IC-PW. – Eluent: 0.0013 mol/L $Na_2B_4O_7$ + 0.0058 mol/L H_3BO_4 + 0.0014 mol/L potassium gluconate (pH 8.5) + 120 mL/L acetonitrile; flow rate: 1.2 mL/min; detection: direct conductivity; injection volume: 100 μL; solute concentrations: 5 to 40 ppm; (taken from [66]).

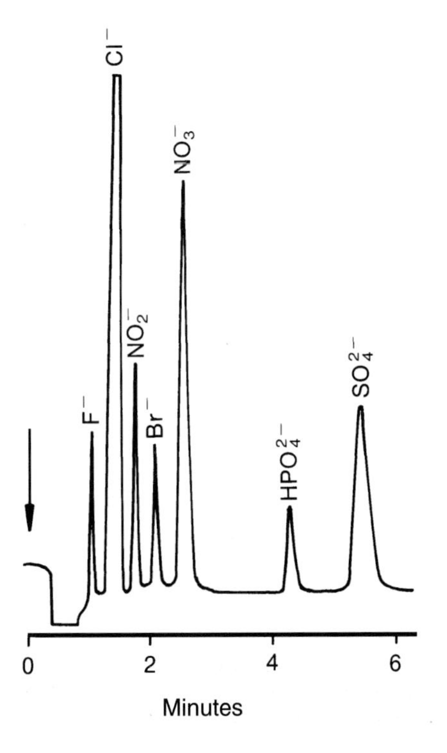

Fig. 3-8. Separation of various inorganic anions on LCA A03. – Eluent: 0.004 mol/L sodium *p*-hydroxybenzoate (pH 8.5); flow rate: 2 mL/min; detection: direct conductivity; injection volume: 50 μL.

Polyvinyl-based anion exchange resins have been available since 1984 from Interaction Chemicals (Mountain View, CA, USA) under the trade names "ION-100" and "ION-110", respectively [24]. According to the manufacturer, these macroreticular resins are stable at pH values between 0 and 14, thus allowing the use of a great number of eluents. In comparison to TSK Gel IC-PW, this stationary phase exhibits a much lower chromatographic efficiency [25]. As an example, the separation of simple inorganic anions on ION-110 with salicylate as the eluent is shown in Fig. 3-10. The characteristic structural and technical properties of theses columns are summarized in Table 3-3.

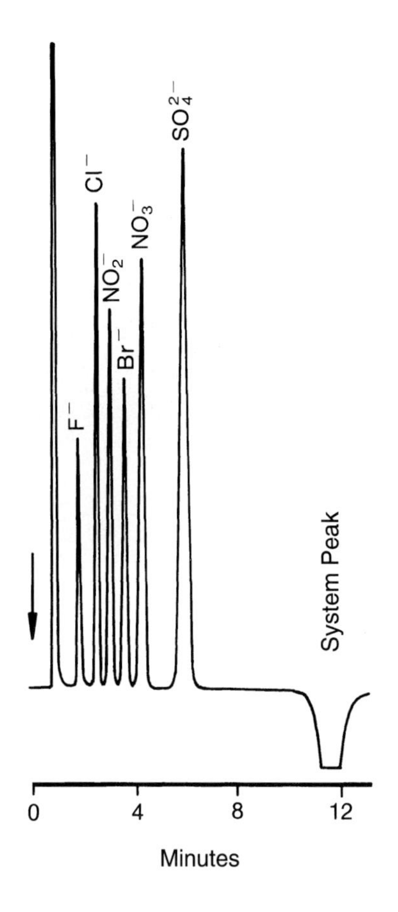

Fig. 3-9. Separation of various inorganic anions on Shimadzu IC-A1. – Column temperature: 40 °C; eluent: 0.0025 mol/L phthalic acid + 0.0024 mol/L Tris; flow rate: 1.5 mL/min; detection: direct conductivity; injection volume: 20 μL; solute concentrations: 5 ppm fluoride, 10 ppm chloride, 15 ppm nitrite, 20 ppm bromide, 30 ppm nitrate, and 40 ppm sulfate.

3.3.1.2 Latex-Agglomerated Anion Exchangers

A special type of pellicular anion exchangers was first introduced in 1975 by Small et al. [3] in their introductory paper on ion chromatography. These stationary phases, which are called latex-based anion exchangers, have been further developed by Dionex (Sunnyvale, CA, USA). The structure of these stationary phases is schematically depicted in Fig. 3-11.

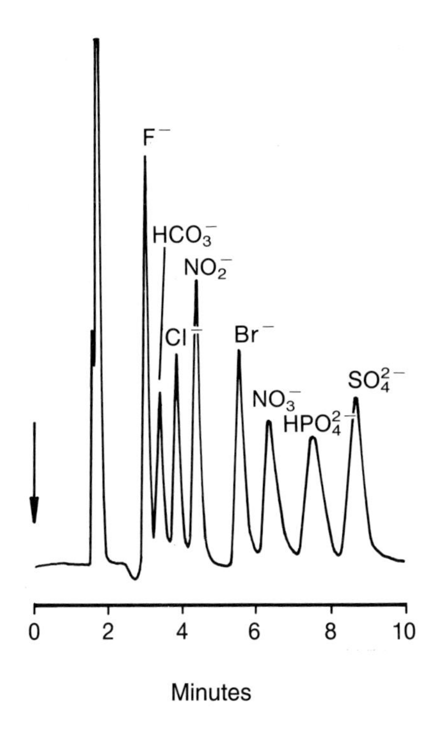

Fig. 3-10. Separation of various inorganic anions on ION-110. – Eluent: 0.003 mol/L salicylate (pH 8.2); flow rate: 0.75 mL/min; injection volume and solute concentrations: not specified.

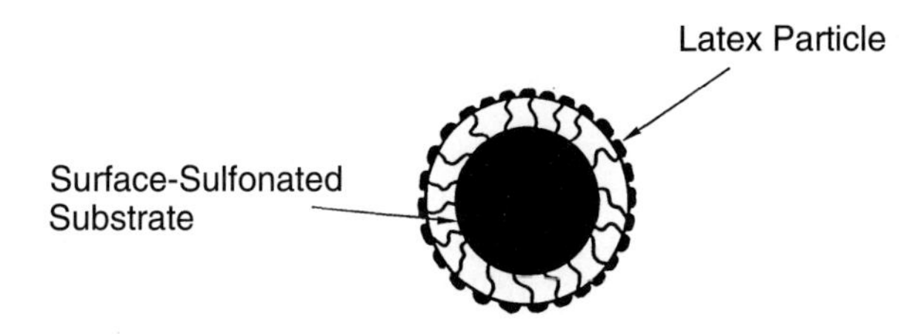

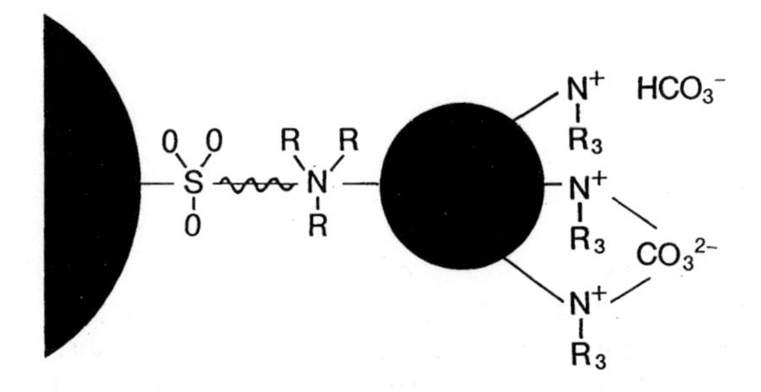

Fig. 3-11. Structure of a latex-based anion exchange resin.

Table 3-3. Structural and technical properties of surface-aminated polymethacrylate and polyvinyl resins.

Separator column	TSK Gel[1] IC-PW	ION-100 (ION-110)	LCA A03	Shimpack IC-A1
Dimensions (Length × ID) [mm]	50 × 4.6	100 × 3.0 (250 × 3.0)	100 × 4.0	100 × 4.6
Manufacturer	Toyo Soda	Interaction Chemicals	Sykam	Shimadzu
pH Range	1–12	0–14	2–12	2–11
Max. operating pressure [MPa]	7	10	20	5
Max. flow rate [mL/min]	2	1–2	5	2
Solvent stability [%]	20	10	10	10
Capacity [mequiv/g]	0.03	0.1	0.1	not specified
Particle diameter [μm]	10	10	10	10
Type of column packing	Polymethacrylate aminated with methyldiethyl amine	Polyvinyl resin with quaternary ammonium groups	Polymethacrylate with quaternary ammonium groups	Polymethacrylate with quaternary ammonium groups

[1] identical to Waters IC Pak A

Latex-based anion exchangers are comprised of a surface-sulfonated polystyrene/divinylbenzene substrate with particle diameters between 5 μm and 25 μm and fully aminated porous polymer beads of high capacity, which are called latex particles. The latter have a much smaller diameter (about 0.1 μm) and are agglomerated to the surface by both electrostatic and van-der-Waals interactions. A scanning electron micrograph of this material is shown in Fig. 3-12. Hence, the stationary phase features three chemically distinct regions:

- An inert and mechanically stable substrate,
- A thin coating of sulfonic acid groups which covers the substrate, and
- An outer layer of latex beads, which carry the actual anion exchange groups, $-\overset{+}{N}R_3$.

Although the latex polymer exhibits a very high exchange capacity due to its complete amination, the small size of the beads finally results in a low anion exchange capacity of about 0.03 mequiv/g. The pellicular structure of these anion exchangers is responsible for their high chromatographic efficiencies. The parameters

- Degree of surface sulfonation and
- Size of the latex beads

play a significant role in this respect. The surface sulfonation of the substrate prevents the diffusion of inorganic species into the inner part of the stationary phase via Donnan exclusion (see Section 4.1). Therefore, the diffusion process is dominated by the functional groups bonded to the latex beads. The bead size determines the length of the diffusion paths, and thus the rate of the diffusion process [26].

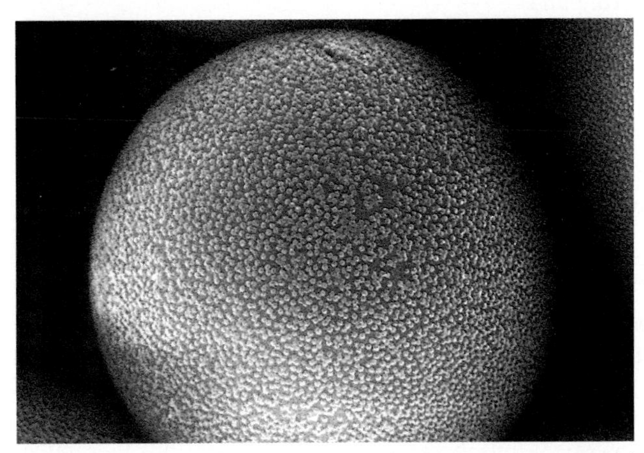

Fig. 3-12. Scanning electron micrograph of a latex-agglomerated anion exchange particle.

This complex system offers several advantages compared with other column packings such as silica-based anion exchangers (see Section 3.3.1.3) and directly aminated resins discussed above, which were developed by Fritz et al. [27]:

- The inner substrate provides mechanical stability and a moderate backpressure.
- Fast exchange processes and thus, a high chromatographic efficiency of the separator column are ensured by the small size of the latex beads.
- Swelling and shrinkage are considerably reduced due to the surface functionalization.

Latex-agglomerated anion exchangers are chemically very stable. Even sodium hydroxide at concentrations of c = 4 mol/L is unable to cleave the ionic bond between the substrate particle and the latex bead. The selectivity is altered by changing the chemical nature of the quaternary ammonium base. Since the latex material is synthesized in a separate step, it is possible to optimize the selectivity of a separator column for a specific analytical problem, by either varying the functional groups bonded to the latex beads or by changing the degree of crosslinking. For the users, economic eluents and especially the different selectivities are of special interest. Since the functionality and the substrate of latex-agglomerated exchangers are two separate entities, the Dionex family of anion exchangers is adequate for a discussion of the various parameters affecting the selectivity.

Review of different latex-agglomerated anion exchangers

At present, ten different anion exchangers (IonPac AS1 to AS7) with diverse selectivities are available from Dionex. The structural and technical characteristics of these separator columns are summarized in Table 3-4. Additionally, special columns for the fast separation of simple inorganic anions, for the separation of bromide, nitrate, and chlorate, as well as for the anion exchange chromatography of amino acids are available.

The different selectivities are mainly governed by the parameters

- Degree of the latex crosslinking and
- Type of functional groups attached to the latex bead.

The ion-exchange capacity, on the other hand, is determined by

- Substrate particle size,
- Size of the latex beads, and
- Degree of latex coverage on the substrate surface.

While the ion-exchange capacities of these separator columns are directly proportional to both the size of the latex beads and their degree of coverage on the substrate surface, they are inversely proportional to the particle size of the substrate.

Fig. 3-13 shows the influence of the above-mentioned parameters on the separation of simple inorganic anions. It is evident from looking at Fig. 3-13 that the elution order of anions, which are separated exclusively via anion exchange processes, cannot be altered. An exception are polyvalent ions, because their valency and thus their retention are a function of the pH value.

Table 3-4. Structural and technical properties of different latex-agglomerated anion exchangers.

Separator column	Particle diameter [μm]	Degree of cross-linking [%]	Latex particle size [nm]	Applicability
IonPac AS1	25	5	150	Universal anion exchanger
IonPac AS2	25	5	150	Anion exchanger with high affinity for bromide and nitrate
IonPac AS3	25	2	100	Anion exchanger for highly contaminated samples
IonPac AS4	15	3.5	75	High-performance separator for less contaminated samples
IonPac AS4A	15	0.5	180	Universal high-performance separator
IonPac AS5	15	0.5	180	Separator for polarizable anions
IonPac AS6 (CarboPac)	10	5	350	Separator for carbohydrates
IonPac AS6A	5	4	210	High-performance separator for carbohydrates
IonPac AS7	10	5	350	Separator for polyvalent anions

On the other hand, the elution order for anions such as bromide and nitrate, for which the separation is mainly based on a different adsorption behavior (see Section 3.2), is governed by the hydrophobicity of the exchange function. In this respect, a

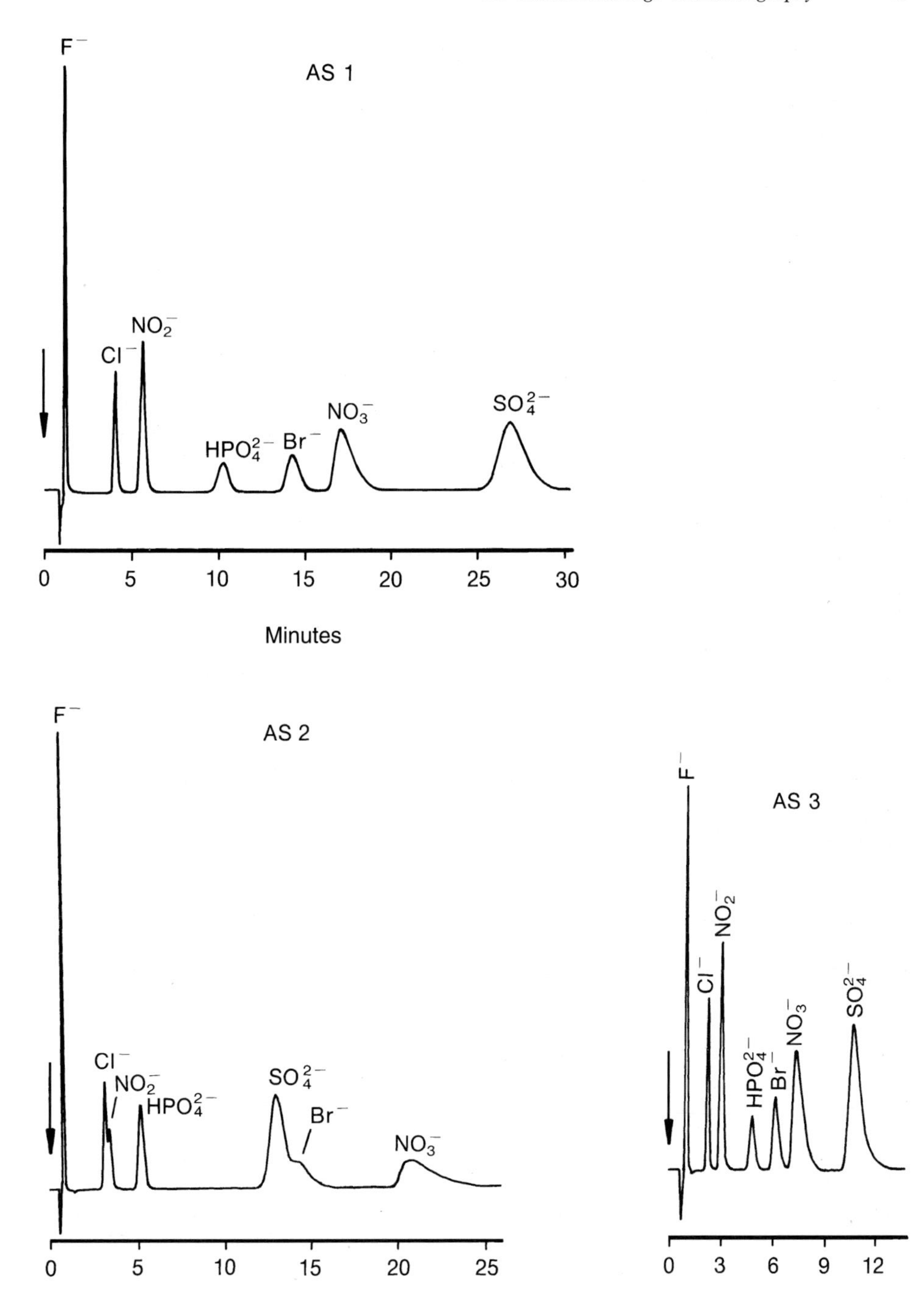

Fig. 3-13. Comparison of selectivity for the anion exchangers IonPac AS1 to AS3. – Eluent: 0.0028 mol/L $NaHCO_3$ + 0.0022 mol/L Na_2CO_3; flow rate: 2 mL/min; detection: suppressed conductivity; injection volume: 50 μL; solute concentrations: 3 ppm fluoride, 4 ppm chloride, 10 ppm nitrite, 10 ppm orthophosphate, 10 ppm bromide, 20 ppm nitrate, and 25 ppm sulfate.

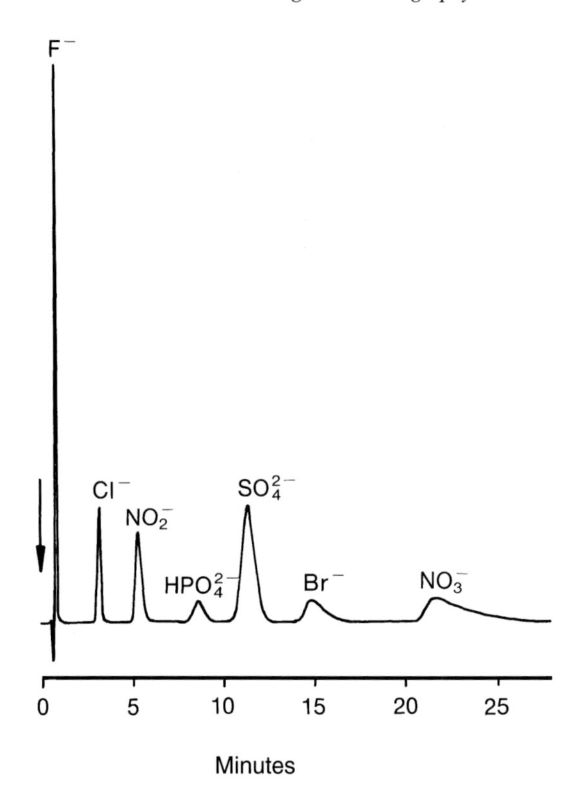

Fig. 3-14. Standard chromatogram of an AS2 separator column. – Eluent: 0.003 mol/L Na_2CO_3 + 0.002 mol/L NaOH; other chromatographic conditions: see Fig. 3-13.

comparison of the two separator columns, AS1 and AS2, is instructive. Both columns differ only in the type of their functional group. In comparison to the AS1, the functional group of the AS2 is considerably more hydrophobic which, in turn, yields a higher retention for these two anions. The resolution between chloride and orthophosphate, and between sulfate and bromide observed under these chromatographic conditions can be improved considerably by a slight modification of the eluent. The standard chromatogram of an AS2 column shown in Fig. 3-14 has been obtained using a carbonate/bicarbonate eluent.

The elution profile of the AS3 separator column with considerably shorter analysis times is obtained by reducing the degree of crosslinking and the size of the latex beads. The comparison between the AS1 and AS3 columns demonstrates impressively how the analysis time for a baseline-resolved separation of the seven standard anions has been reduced by about 60% just by improving the development of the latex technology. Yet the particle diameter of 25 µm was held constant. In terms of chromatographic efficiency, ion-exchange capacity, and resistance to column fouling, the IonPac AS3 is the final product in the development of latex-agglomerated anion exchangers of the *first* generation. It is particularly suited for the analysis of waste water and other heavily contaminated samples featuring large concentration differences between the individual sample components.

The objective for the *second* generation of latex-agglomerated anion exchangers was a further reduction of the retention times via a decrease of the substrate particle size. The concurrent increase of the chromatographic efficiency results in a significantly higher column backpressure. This development was initiated in 1982 with the introduction of the IonPac AS4. A respective elution profile is shown in Fig. 3-15. Although the chemical nature of the functional groups and the physical properties of the latex beads, such as their size and their degree of crosslinking, were altered only slightly, a reduction in the retention times by another 50% in comparison to the AS3 column was accomplished by decreasing the substrate particle size to 15 μm. Disadvantageous is the low exchange capacity and the less pronounced ruggedness of this high-performance separator column. Therefore, this column can only be used for the analysis of samples with low electrolyte contents.

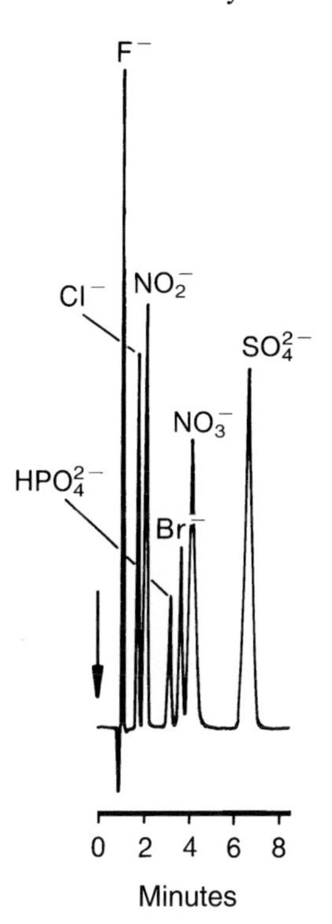

Fig. 3-15. Elution profile of the IonPac AS4 separator column for simple inorganic anions. – Chromatographic conditions: see Fig. 3-13.

This disadvantage was overcome four years later when a universal high-performance separator column – the IonPac AS4A – was introduced. This column combines the positive properties of both the IonPac AS3 and AS4 separators. While the particle size of the substrate is identical to the AS4 column, the loading capacity could be brought up to the capacity of the AS3 column by significantly reducing the degree of crosslinking. The tailing of the bromide and nitrate peaks, which is observed with the AS4 column (see Fig. 3-15), could be reduced by altering the chemical nature of the functional groups. A prerequisite for this is the stronger hydrophilicity of the functional groups of the AS4A separator. In comparison to the AS4 column, bromide and nitrate are less strongly retained by means of adsorption. In other ions, the modification in the functionality of the latex beads does not result in any notable change of the retention behavior. The only exception to this is orthophosphate, which elutes much later with the standard eluent mixture of sodium carbonate and sodium bicarbonate. As seen in Fig. 3-16 (left chromatogram), orthophosphate interferes with nitrate. However, the standard elution profile of the AS4A column (Fig. 3-16, chromatogram right-hand side) is obtained by slightly changing the concentration ratio of two eluent components.

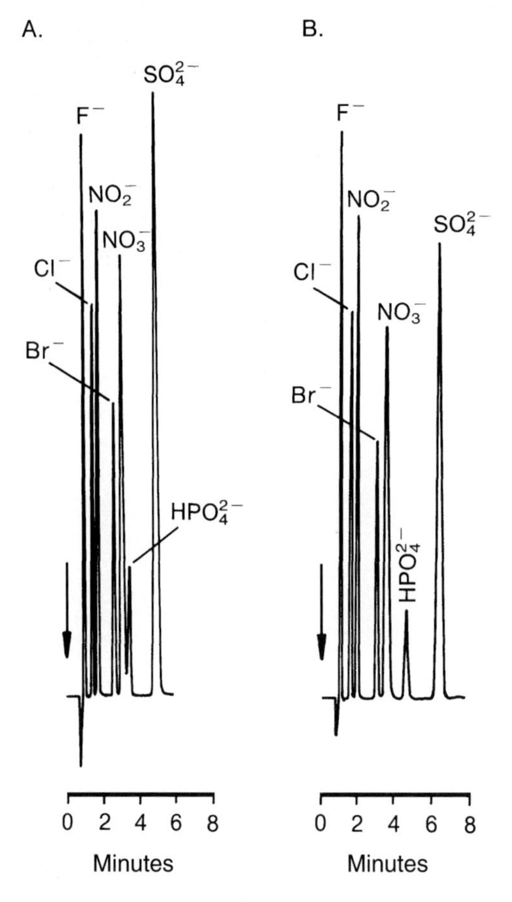

Fig. 3-16. Elution profile of the IonPac AS4A separator column for simple inorganic anions depending on the ratio of the two eluent components. – Eluent: (A) 0.0028 mol/L $NaHCO_3$ + 0.0022 mol/L Na_2CO_3, (B) 0.0017 mol/L $NaHCO_3$ + 0.0018 mol/L Na_2CO_3; flow rate: 2 mL/min; detection: suppressed conductivity; injection volume: 50 μL; solute concentrations: see Fig. 3-13.

The IonPac AS5, a special 15-μm anion exchanger, was developed for the analysis of strongly polarizable ions such as iodide, thiocyanate, and thiosulfate. A comparison of the selectivities of the AS4 and AS5 separators for the separation of simple inorganic anions is shown in Fig. 3-17. The resulting retention times are comparable, since the particle diameter is identical for both stationary phases. The different selectivity is based on the hydrophilic properties of the functional groups bonded to the latex beads of the AS5 column, which prohibits a separation of bromide and nitrate. However, this hydrophilicity is a prerequisite for the separation of polarizable anions, which, due to their large radii, are strongly retained at the stationary phase of anion exchangers by adsorptive forces and, thus, give rise to a strong tailing effect. Taking the separation of iodide and thiocyanate as an example, Fig. 3-18 shows that this tailing can be significantly reduced by adding *p*-cyanophenol to the eluent, which is especially suited in this respect.

Alternatively, these anions may be separated at a stationary phase which was originally developed for the chromatographic analysis of heavy and transition metals having a defined anion and cation exchange capacity (see Section 3.4.5.3). This separator column (IonPac CS5), is based on a surface-sulfonated PS/DVB substrate with a particle diameter of 13 μm and a degree of crosslinking of about 2%. In analogy to the AS5 column, the CS5 column also has latex beads of about 150 nm. However, the functionality of the AS5 column was replaced by another with a similar hydrophilicity. As shown in Fig. 3-19, iodide and thiocyanate are separated at this stationary phase using a mixture of sodium carbonate and sodium bicarbonate without any organic modifiers. There is little difference between the analysis time and peak shape of the chromatogram displayed in Fig. 3-18. Only with respect to its loading capacity is the CS5 column inferior to the AS5 column due to its higher degree of crosslinking.

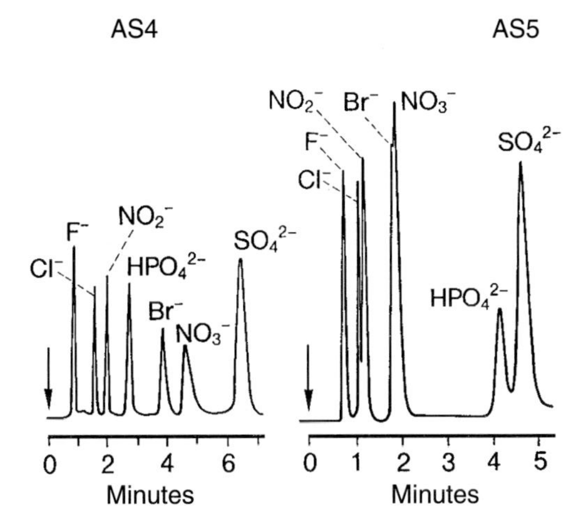

Fig. 3-17. Comparison of the selectivities between the IonPac AS4 and AS5. – See Fig. 3-13 for chromatographic conditions.

Two other anion exchangers, the IonPac AS6 (CarboPac) and AS7 separator columns, were developed especially for the analysis of carbohydrates and polyvalent anions, respectively. The stationary phase is based on a 10-μm support material to which relatively large latex beads are bonded. The comparatively high exchange capacity is necessary, since eluents with high ionic strength are required for the analysis of carbohydrates and polyvalent anions. In order to separate carbohydrates by anion exchange, they must first be converted into the ionic form. Sodium hydroxide in concentrations of c = 0.1 to 0.15 mol/L is employed as the eluent, since the pK values of carbohydrates and sugar alcohols lie in the range between 12 and 14 (see Section 3.3.5.3).

In the past, polyvalent anions such as polyphosphonic acids could not be analyzed by ion chromatography, because their retention times increase drastically with increasing charge number. Strong adsorptive forces, caused by the organic skeleton of these compounds, resulted in significant tailing effects. However, polyvalent anions elute as symmetrical peaks, if strongly acidic mobile phases of a high elution strength are utilized (see Section 3.3.5.2).

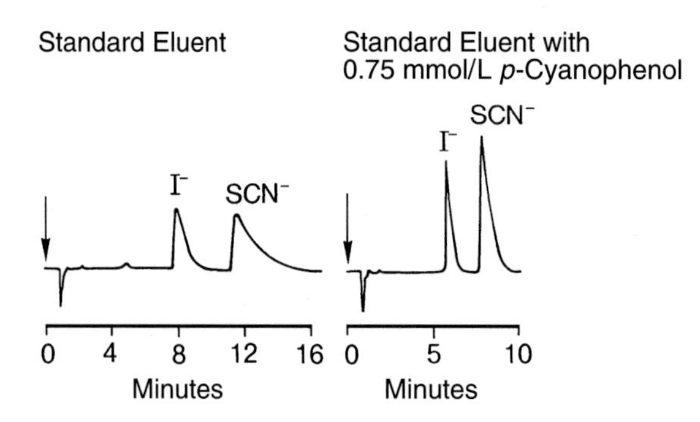

Fig. 3-18. Effect of *p*-cyanophenol on the tailing properties of iodide and thiocyanate. – Chromatographic conditions: see Fig. 3-1.

Both separator columns have in common a relatively low loading capacity resulting from the high degree of crosslinking of the substrate particles. The functionality of the latex beads corresponds to that of a conventional anion exchanger, for example, the AS3 column.

Following the development of the second generation separator columns with particle diameters between 10 μm and 15 μm, it was assumed that a further reduction in analysis time was not appropriate, since the loading capacity decreases with a further increase in the separation efficiency. Even samples with a low electrolyte content would have to be diluted in order to avoid an overloading of the separator column. The time required for the sample preparation would in no way justify a shorter analysis time. However, this train of thought is only valid, if the reduction of analysis time is achieved once again by a reduction of the particle diameter and by a miniaturization of the separator column. Experiments with 5-μm substrate particles and column dimensions of 35 mm × 3 mm I.D. were indeed carried out. The experimental results confirmed that such an approach is pointless, especially in view of optimizing the chromatographic system. Only

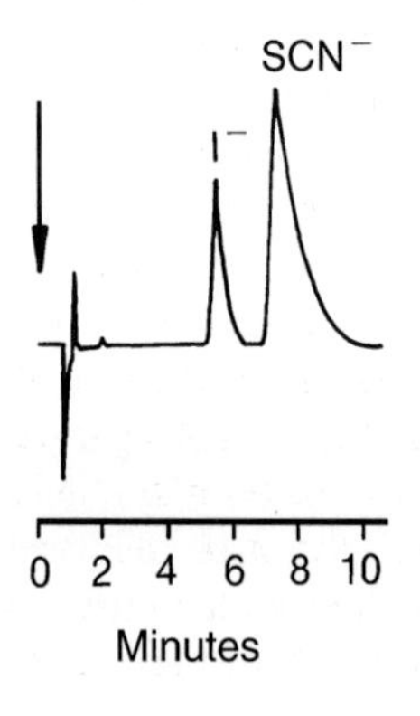

Fig. 3-19. Separation of iodide and thiocyanate on a separator column of the IonPac CS5 type. – Eluent: 0.001 mol/L $NaHCO_3$ + 0.006 mol/L Na_2CO_3; flow rate: 2 mL/min; detection: suppressed conductivity; injection volume: 50 μL; solute concentrations: 20 ppm iodide and 40 ppm thiocyanate.

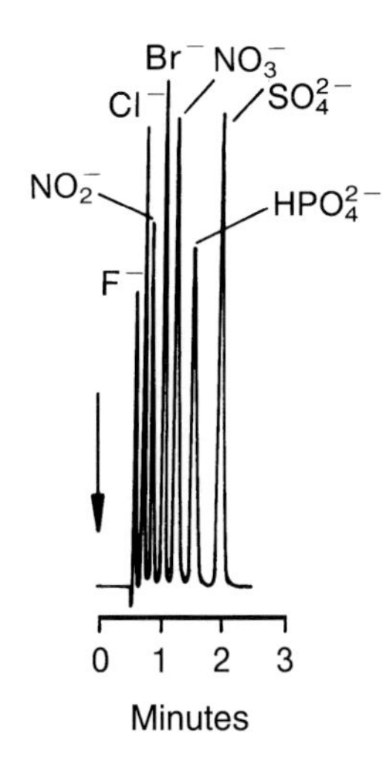

Fig. 3-20. Elution profile of a Fast-Sep anion exchange column. – Eluent: 0.00015 mol/L $NaHCO_3$ + 0.002 mol/L Na_2CO_3; flow rate: 2 mL/min; detection: suppressed conductivity; injection volume: 20 μL; solute concentrations: 1.5 ppm fluoride, 2.5 ppm chloride, 7.5 ppm nitrite and bromide, 10 ppm nitrate and sulfate, 15 ppm orthophosphate.

under laboratory conditions could the analysis time for the seven most important inorganic anions be reduced to less than two minutes. In addition, a chromatographic efficiency of about 50,000 theoretical plates per meter – related to sulfate – was achieved. However, because of handling and reliability problems, the utilization of this system in routine analysis is inconceivable.

Reducing the analysis times by increasing the flow rate and/or the eluent concentration, respectively, is also not feasible due to an insufficient sensitivity and ion-exchange capacity.

Only the use of various *acrylate-based* latex polymers aided the successful development of a separator column with analysis times shorter than three minutes. In 1987, an anion exchange column, the Fast-Sep Anion [28], was introduced. The substrate particle size and the elution order of the seven most important inorganic anions do not differ from the AS4A column, which represents the current state-of-the-art in chromatographic anion analysis. However, the amination was carried out using a different amine. Fig. 3-20 displays the elution profile of the Fast-Sep column, which is again obtained using a mixture of sodium carbonate and sodium bicarbonate as the mobile phase. To avoid overloading the column, the injection volume should be limited to 20 μL. Upon installation in an existing chromatographic system, a short tubing should be employed to obtain optimal resolution between the individual signals. However, such a column is only suitable for analyzing slightly contaminated samples (e.g., potable waters and surface waters) in which the main components (chloride, nitrate, bromide, and sulfate) can be separated to baseline. For samples with complex matrices, the separation efficiency of this column is usually insufficient to separate main and minor components. Compared to the conventional exchanger AS4A, the Fast-Sep column has the advantage of a higher sensitivity, especially for later eluting compounds such as orthophosphate and sulfate, that have a much shorter retention and, thus, elute as sharper signals. Although the ion-exchange capacity of the Fast-Sep column is only half that of the AS4A column, the values for the loading capacity – at least for monovalent anions – are comparable. The loading capacity is defined as the solute concentration at which the peak efficiency falls below 90% of its original value. As is seen in Table 3-5, both columns differ in their loading capacity for sulfate, which is twice as high for a conventional anion exchanger. This phenomenon clearly establishes the higher affinity of divalent anions toward styrene/divinylbenzene-based latex polymers.

Table 3-5. Comparison of the loading capacity between a Fast-Sep column and a conventional IonPac AS4A anion exchanger.

Solute ion	Fast-Sep Anion [ppm][1]	[ng]	IonPac AS4A [ppm][2]	[ng]
Chloride	30	600	12	600
Sulfate	35	700	30	1 500

[1] Injection volume: 20 μL
[2] Injection volume: 50 μL

Compared to conventional anion exchangers such as the AS4A column, the acrylate-based latex polymer of a Fast-Sep column exhibits a lower pH stability. However, long-term experiments with the Fast-Sep column showed no deterioration as long as the pH value of the mobile phase was kept between 2 and 11. The pH value of the sample to be analyzed should not exceed 13.

By agglomerating acrylate-based latex beads on polystyrene/divinylbenzene substrates, a separation of chlorate and nitrate can also be achieved, which was previously not possible using conventional ion exchangers. These two anions exhibit the same interactions with both latex-based anion exchangers and directly aminated substrates and, thus, co-elute at these stationary phases. Therefore, ion-pair chromatography was used for this separation, which allows baseline separation of chlorate and nitrate due to their

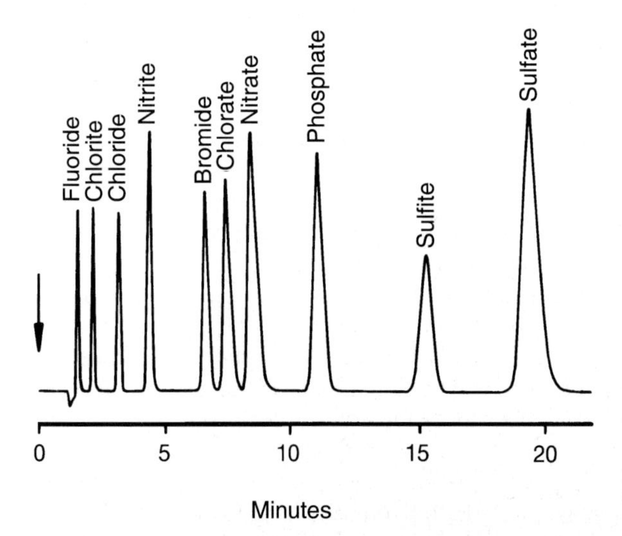

Fig. 3-21. Elution profile of the IonPac AS9 latex anion exchanger. – Eluent: 0.00075 mol/L $NaHCO_3$ + 0.002 mol/L Na_2CO_3; flow rate: 1 mL/min; detection: suppressed conductivity; injection volume: 50 μL; solute concentrations: 1 ppm fluoride, 5 ppm chlorite, 1.5 ppm chloride, 6 ppm nitrite, 10 ppm bromide, 15 ppm chlorate and nitrate, 20 ppm orthophosphate and sulfite, and 25 ppm sulfate.

different sizes. This selectivity problem was solved with the IonPac AS9 anion exchanger, developed especially for the separation of chlorate and nitrate. Fig. 3-21 shows a standard chromatogram obtained using a mixture of sodium bicarbonate and sodium carbonate as the eluent. It is advisable to use an eluent flow rate of 1 mL/min to obtain an optimal resolution between bromide, chlorate, and nitrate, although the analysis time for sulfate increases to about 20 minutes in this case. With this stationary phase, chlorite is also much better resolved from chloride, which allows the determination of comparatively low concentrations of chlorite in the presence of high amounts of chloride. This analytical problem becomes increasingly important in the analysis of potable water that is disinfected with chlorine dioxide (see Fig. 8-7 in Section 8.1). As can also be seen in Fig. 3-21, a much better separation between sulfite and sulfate is obtained with the IonPac AS9 compared to the IonPac AS4A. Even large concentration differences between these two anions do not aggravate their determination. This stationary phase is especially suited for the analysis of flue gas scrubber solutions in desulfurization plants, where corresponding analytical problems occur. With regard to the pH stability, the same limitations apply for the IonPac AS9 and the Fast-Sep column.

With the introduction of the gradient elution of anions in combination with a subsequent conductivity detection in 1987, a *third* generation of latex-based anion exchangers with 5-µm substrate particles was developed. The small latex beads agglomerated on the substrate surface are barely visible in the scanning electron micrograph shown in Fig. 3-22. As already discussed in connection with the Fast-Sep column, the objective of this development was not a further reduction of the analysis times. The reduction of the particle diameter to 5 µm rather aimed at a further increase in the chromatographic efficiency and, thus, in the separation power. However, this development offers no advantage for the isocratic analysis of inorganic anions. A comparison of the elution profiles of the 5-µm exchanger IonPac AS5A in Fig. 3-23 being developed for conventional anion analysis with that of an AS4A column, shown in Fig. 3-16, reveals that both eluents attain longer retention times at lower resolution between bromide and nitrate employing the AS5A column. This is not surprising, since the 5-µm exchanger can only be operated with a flow rate of 1 mL/min due to the considerably higher back pressure. The lower resolution between bromide and nitrate is caused by

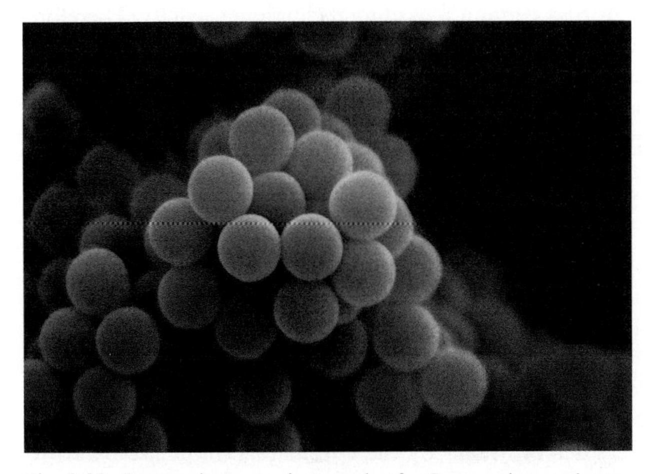

Fig. 3-22. Raster electron micrograph of a 5-µm anion exchange particle.

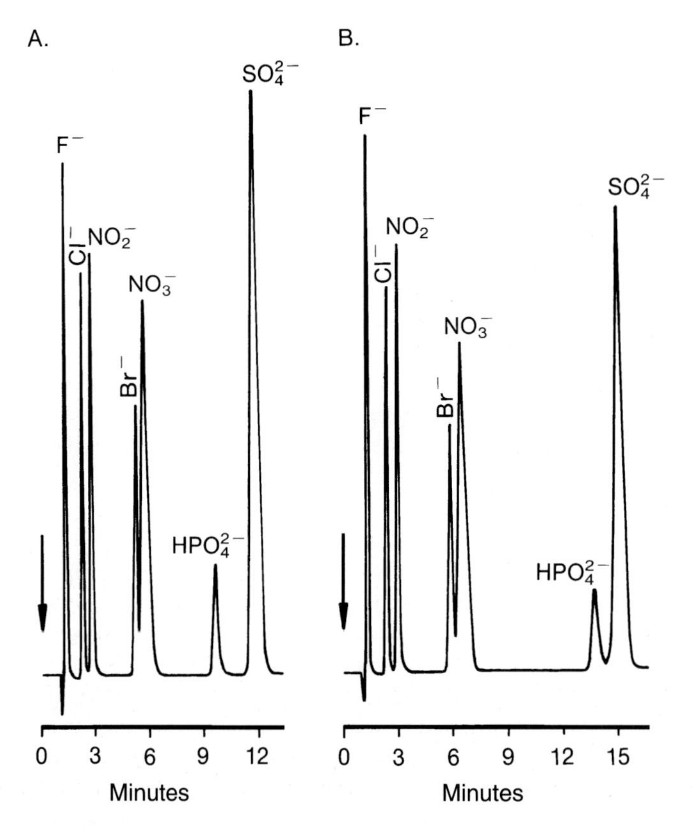

Fig. 3-23. Elution profile of the 5-µm IonPac AS5A anion exchanger under isocratic conditions. – Eluent: (A) 0.0028 mol/L $NaHCO_3$ + 0.0022 mol/L Na_2CO_3, (B) 0.0017 mol/L $NaHCO_3$ + 0.0018 mol/L Na_2CO_3; flow rate: 1 mL/min; all other chromatographic conditions: see Fig. 3-13.

the functionality of the latex material, which is identical with that of an AS5 column. But the degree of crosslink and the size of the latex beads were modified, so that the AS5A column corresponds to an AS4 column regarding its loading capacity. Therefore, the elution profile is very similar to that of the AS5 column displayed in Fig. 3-17. The comparatively better resolution between bromide and nitrate can be attributed to the higher chromatographic efficiency of the AS5A column.

A high separation efficiency is obtained with the AS5A column when it is employed for the gradient elution of anions (see Section 3.3.6). The objective of the gradient elution technique is to analyze anions having a wide range of retention characteristics – e.g., monovalent, divalent, and trivalent anions – within the same chromatographic run via a gradual increase of the ionic strength in the mobile phase. Considering that a suppressor system is essential for the subsequent conductivity detection, a successful gradient elution is only feasible using sodium hydroxide as the eluent. Since sodium hydroxide as a monovalent eluent ion exhibits only a low elution power, the corresponding functionality and dimension of the latex material were adjusted. The high separation efficiency of this column in applying the gradient technique with sodium hydroxide as the mobile phase is impressively demonstrated by the chromatogram depicted in Fig. 3-24.

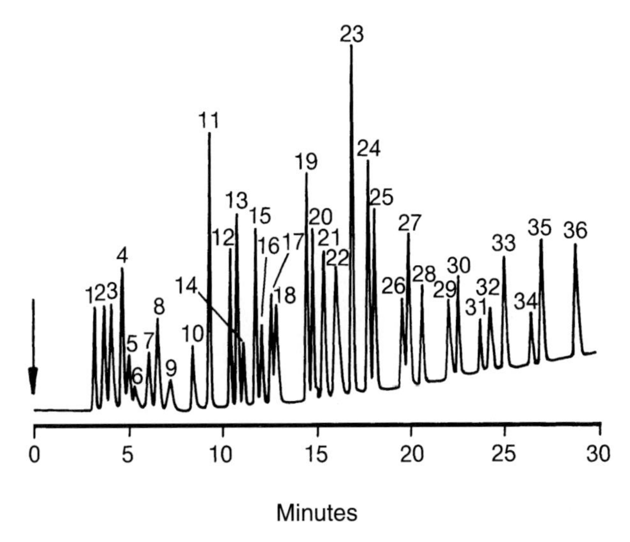

Fig. 3-24. Gradient elution of inorganic and organic anions on IonPac AS5A. – Eluent: (A) 0.00075 mol/L NaOH, (B) 0.1 mol/L NaOH; gradient: 100% A isocratically for 5 min, then linearly to 30% B in 15 min, then linearly to 86% B in 15 min; flow rate: 1 mL/min; detection: suppressed conductivity; injection volume: 50 μL; solute concentrations: 1.5 ppm fluoride (1), 10 ppm α-hydroxybutyrate (2), acetate (3), glycolate (4), butyrate (5), gluconate (6), α-hydroxyvalerate (7), 5 ppm formate (8), 10 ppm valerate (9), pyruvate (10), monochloroacetate (11), bromate (12), 3 ppm chloride (13), 10 ppm galacturonate (14), 5 ppm nitrite (15), 10 ppm glucoronate (16), dichloroacetate (17), trifluoroacetate (18), phosphite (19), selenite (20), bromide (21), nitrate (22), sulfate (23), oxalate (24), selenate (25), α-ketoglutarate (26), fumarate (27), phthalate (28), oxalacetate (29), phosphate (30), arsenate (31), chromate (32), citrate (33), isocitrate (34), *cis*-aconitate (35), and *trans*-aconitate (36).

Another 5-μm anion exchanger, the IonPac AS6A, was developed – but never commercialized – for the gradient elution of carbohydrates. As can be seen in Table 3-4, the degree of crosslinking and the size of the latex beads were only slightly modified. Although high concentrations of sodium hydroxide are also required for the elution of carbohydrates, the functionality of the AS6 column was maintained because of the low affinity of carbohydrates in their ionic form toward the stationary phase of an anion exchanger. A relatively high exchange capacity of the resin is necessary in order to obtain a sufficient separation of the carbohydrates which do not differ significantly in their p*K* values. This is achieved by appropriate dimensioning of the latex beads.

The gradient elution of different sugar alcohols and saccharides, as shown in Fig. 3-25, illustrates the state-of-the-art of modern carbohydrate analysis and, above all, the power of anion exchange chromatography with latex-based anion exchangers of the *third* generation.

3.3.1.3 Silica-Based Anion Exchangers

Parallel to the development of organic polymers as anion exchange substrates, a number of silica-based anion exchangers were introduced over the past years [29,30]. Once again, the development of low capacity exchange materials was in the foreground in order to be able to dispense the suppressor system using eluents with low background conductances.

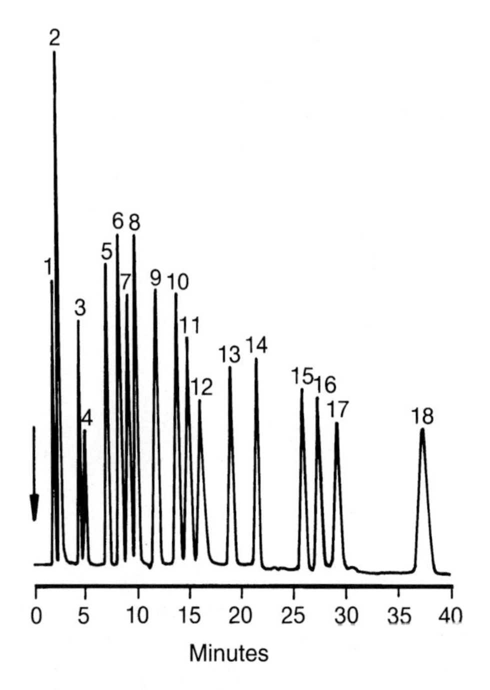

Fig. 3-25. Gradient elution of different sugar alcohols and saccharides. – Separator column: Ion Pac AS6A; eluent: (A) water, (B) 0.05 mol/L NaOH + 0.0015 mol/L acetic acid; gradient: 7% B isocratically for 15 min, then to 100% B in 10 min; flow rate: 0.8 mL/min; detection: pulsed amperometry on a Au working electrode (post-column addition of NaOH); injection volume: 50 µL; solute concentrations: 15 ppm inositol (1), 40 ppm sorbitol (2), 25 ppm fucose (3), deoxyribose (4), 20 ppm deoxyglucose (5), 25 ppm arabinose (6), rhamnose (7), galactose (8), glucose (9), xylose (10), mannose (11), fructose (12), melibiose (13), isomaltose (14), gentiobiose (15), cellobiose (16), 50 ppm turanose (17), and maltose (18).

In contrast to organic polymers, silica-based substrates have the advantages of higher chromatographic efficiency and greater mechanical stability. In general, no swelling and shrinkage problems are encountered, even if the ionic form of the ion exchanger is changed or an organic solvent is added to the eluent. Column temperatures up to 80 °C also have no adverse affect on the stationary phase. Although the chromatographic efficiency of these stationary phases with theoretical plate numbers up to 15,000 to 20,000 is very high [31], they may only be used within a narrow pH range (pH 2 to pH 7). This reduces both the number of possible eluents and the type of samples that can be analyzed using silica-based ion exchangers. Quite remarkable is the behavior of transition and heavy metal ions in ion chromatographic systems without a suppressor system investigated by Jenke and Pagenkopf [32]. Metal cations such as Cu^{2+}, Pb^{2+}, and Zn^{2+} can adversely affect the stability and efficiency of silica-based ion exchangers. The presence of free silanol groups leads to an interaction with these metal ions, which are retained at the stationary phase. A co-elution of lead, zinc, and copper ions with inorganic anions such as chloride, bromide, and nitrate is observed, resulting in misleading data.

In general, silica substrates are grouped according to their particle size. Totally porous substrates with small particle sizes in the range of 3 µm to 10 µm, which are called microparticulate beads, are preferred. Exchange capacities between 0.1 and 0.3 mequiv/g are obtained by adding quaternary ammonium functions via reaction of the free silanol groups with an appropriate chlorosilane. Pellicular materials, on the contrary, consist of much bigger particles with a diameter between 25 µm and 40 µm. Their surface is covered with a polymer such as lauryl methacrylate (Zipax SAX) that carries the actual ion-exchange function. The thickness of the polymer film is kept deliberately low at 1 µm to 3 µm to minimize the contribution σ_s^2 of the peak broadening caused by the stationary phase [see Eq. (22) in Section 2.4]. The exchange capacity of pellicular materials is about 0.01 mequiv/g.

Silica-based anion exchangers are offered by manufacturers such as Wescan (Santa Clara, CA, USA), Separation Group (Hesperia, CA, USA), Toyo Soda (Tokyo, Japan), and Macherey & Nagel (Düren, Germany). Table 3-6 summarizes the characteristic structural and technical properties of the most important column packings available today.

Fig. 3-26 shows a typical example of an inorganic anion separation on a silica-based ion exchanger, Vydac 302 IC 4.6. The stationary phase recently introduced by the same manufacturer under the trade name Vydac 300 IC 405 represents a novelty. Although the substrate of the stationary phase also consists of spherical silica particles, the pH stability of this material was enhanced by polymer coating [33]. According to the manufacturer's specifications, eluents may be used in the pH range between 2 and 10. The standard anion chromatogram shown in Fig. 3-27 is obtained using *o*-phthalic acid as the eluent. While the selectivity of this stationary phase is identical to that of Vydac 302 IC 4.6, the retention times could be reduced by 50% due to the shortening of the column length to 50 mm.

The silica ion exchanger manufactured by Wescan is based on a macroporous substrate with a pore width of 300 Å. The big differences in the retention behavior of monovalent and divalent anions are characteristic for the respective chromatogram depicted in Fig. 3-28. The negative signal appearing at about 20 minutes is the "system peak" (see Section 3.3.4.3), which is inevitable upon employing phthalates as eluents.

The chromatogram shown in Fig. 3-29 exemplifies how strongly the selectivity of a separation system depends on the eluent. A silica ion exchanger, TSK GEL IC-SW from Toyo Soda Company was employed as the stationary phase. The separation of chlorate and nitrate that is achieved using tartaric acid at a concentration of 0.001 mol/L

Table 3-6. Structural and technical properties of various silica-based anion exchangers.

Separator column	Vydac 302 IC 4.6	Vydac 300 IC 405	Wescan 269-001	Nucleosil 10 Anion	TSK Gel IC-SW
Dimensions (length × ID) [mm]	50 × 4.6	250 × 4.6	250 × 4.6	250 × 4.0	50 × 4.6
Manufacturer	Separations Group		Wescan	Macherey & Nagel	Toyo Soda
Capacity [mequiv/g]	ca. 0.1	ca. 0.1	0.08	0.06	0.4
Particle size [µm]	10	15	13	10	5
Type of column packing	Spherical particles with quaternary ammonium function			Spherical particles aminated with trimethyl amine	methyl diethyl amine

as the eluent is remarkable. A corresponding separation on a polymer-based latex-agglomerated anion exchanger at alkaline pH is not possible. However, it should be pointed out that under these chromatographic conditions only monovalent anions can be eluted from the column with acceptable analysis times.

As a comparison, Fig. 3-30 shows the chromatogram of inorganic anions using a silica-based anion exchanger of the type Nucleosil 10 Anion. Orthophosphate elutes as a monovalent ion due to the comparatively low pH value of the mobile phase.

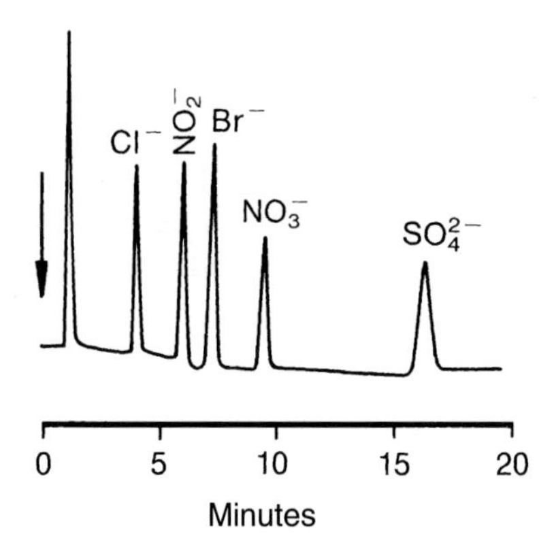

Fig. 3-26. Separation of various inorganic anions on a Vydac 302 IC 4.6 silica-based anion exchanger. – Eluent: 0.002 mol/L *o*-phthalic acid (pH 5.0); flow rate: 2 mL/min; detection: direct conductivity; injection volume: 10 μL; solute concentrations: 100 ppm each; (taken from [25]).

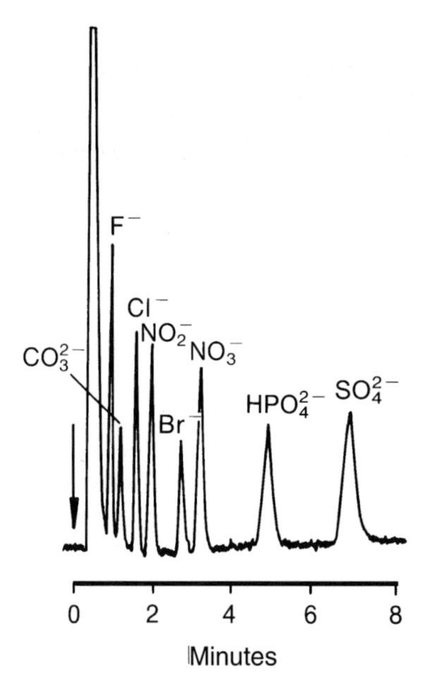

Fig. 3-27. Separation of various inorganic anions on a Vydac 300 IC 405 silica-based anion exchanger. – Eluent: 0.0015 mol/L *o*-phthalic acid (pH 8.9); flow rate: no specification; detection: direct conductivity; solute concentrations: 5 ppm fluoride, 1 ppm chloride, 1.5 ppm nitrite, 3 ppm bromide, 2.5 ppm nitrate, 3 ppm orthophosphate and sulfate.

3.3.1.4 Other Materials for Anion Separations

Crown Ether Phases

Separations of inorganic and organic anions are not only possible by using strongly basic anion exchangers based on organic polymers or silica as substrates. In the mid seventies, Blasius et al. [35] described the separation of ionic species on crosslinked polymers modified with cyclic polyethers. Cyclic polyethers are neutral macrocyclic ligands with more than nine ring atoms, at least three of which act as donor atoms [36]. The ring system most thoroughly investigated and the one most commonly employed is 18-crown-6 (1,4,7,10,13,16-hexaoxacyclooctadecane):

18-Crown-6

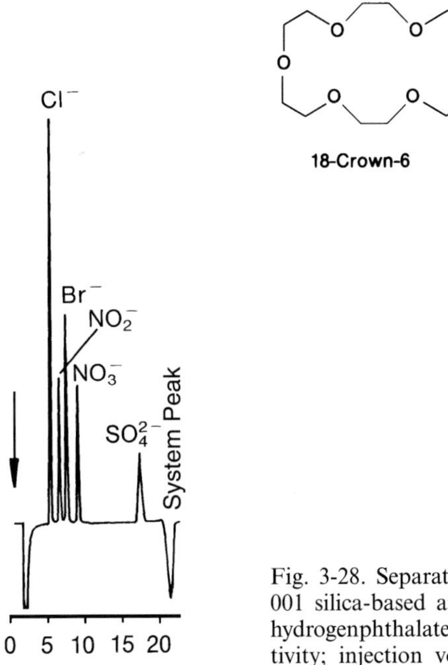

Fig. 3-28. Separation of various inorganic anions on a Wescan 269-001 silica-based anion exchanger. – Eluent: 0.004 mol/L potassium hydrogenphthalate; flow rate: 1.5 mL/min; detection: direct conductivity; injection volume: no specification; solute concentrations: 10 ppm each; (taken from [34]).

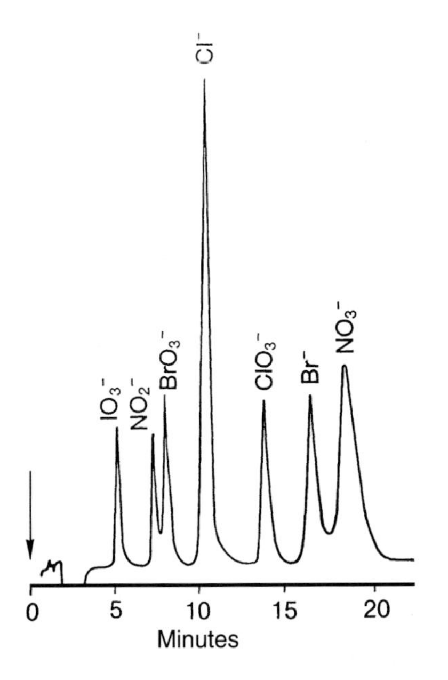

Fig. 3-29. Separation of various inorganic anions on a TSK Gel IC-SW silica-based anion exchanger. – Eluent: 0.001 mol/L tartaric acid (pH 3.2); flow rate: 1.5 mol/L; detection: direct conductivity; injection volume: 100 μL; solute concentrations: 10 ppm each.

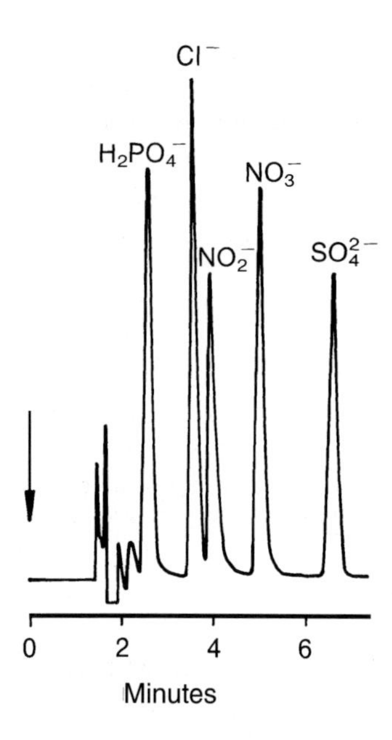

Fig. 3-30. Separation of various inorganic anions on a Nucleosil 10 Anion silica-based anion exchanger. – Eluent: 0.025 mol/L sodium salicylate (pH 4.0 with salicylic acid); flow rate: 1.5 mL/min; detection: refractive index; injection volume: 10 μL; solute concentrations: 1 to 2 g/L of the various anions.

The possibility of complexation of inorganic salts, as well as of ionic and non-ionic organic compounds, can be explained by the dipole effect of the C-O bond in the polyether ring. Other hetero atoms such as nitrogen or sulfur may also act as donor atoms. The selective complexation of a specific cation, which is always associated with an anion because of electroneutrality, depends on the number and position of the hetero atoms.

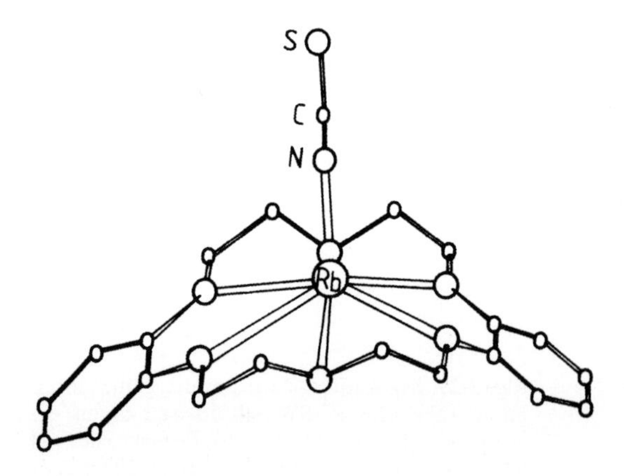

Molecular structure of a RbNCS ion-pair that is coordinated by 18-crown-6

Complexation with organic molecules occurs very often via hydrogen bonding.

Polymeric crown ethers are prepared by condensation, polymerization, or substitution [37]. Above all, condensation polymers based on monobenzo- and dibenzocrown ethers are immobilized on the surface of a solid substrate, and are used in chromatographic applications. They are distinguished by chemical and thermal stability and exhibit a relatively high capacity. Modified and non-modified silicas are used preferentially as support materials.

The immobilization of crown ethers on these support materials may take place in different ways. If silica beads are impregnated with a solution of dibenzo-18-crown-6 in formic acid, a substrate that is stable to hydrolysis is obtained after crosslinking with formaldehyde. Its schematic structure is depicted in Fig. 3-31 (top left). Polymeric crown ethers may be bonded chemically to the silica surface. For that purpose, the support material is modified with 3-(N-methacryloylamino)-propyl groups, and subsequently co-polymerized with (4-methacryloylamino)-benzo-15-crown-5 [38]. Stationary phases, in which the crown ether molecules are chemically bonded to silica via Si-O-C linkages, are obtained with silica treated with thionyl chloride via reaction with 4-hydroxymethyl-benzo-18-crown-6 (Fig. 3-31, top right). Since Si-O-C-bonds are prone to hydrolysis, water-free methanol is used as the eluent for such stationary phases. Much more stable are silicas modified via Si-C-bonds (see Fig. 3-31, bottom), that are prepared by using 4-vinylbenzo-18-crown-6 or 4-butene-18-crown-6 and dimethylchlorosilane in the presence of $H_2[PtCl_6]$. The remaining free silanol groups are reacted with trimethylchlorosilane to prevent interactions with the solutes. The literature calls this process "endcapping".

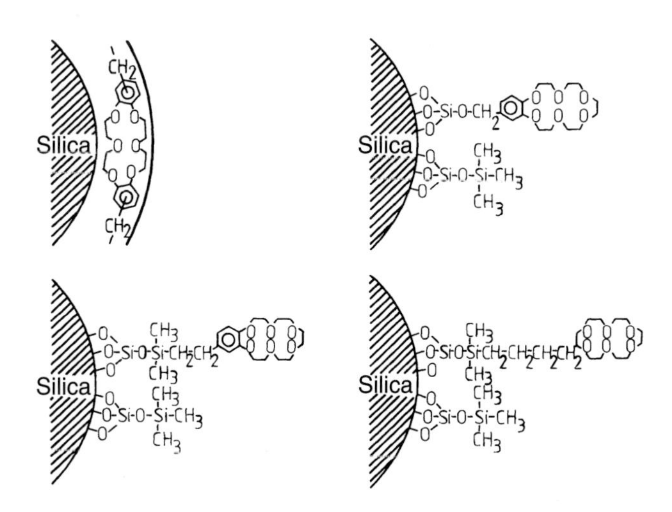

Fig. 3-31. Schematic structures of various silica-based crown ether phases.

As an example, Fig. 3-32 shows the separation of various inorganic anions on a surface-coated silica. As can be seen, the chromatographic efficiency of this stationary phase is very poor compared to modern anion exchangers. However, the possibility to elute anions with pure de-ionized water is advantageous, because in this case the use of a suppressor system necessary for a sensitive conductivity detection is dispensable. Similar points hold for the separation of the sodium salts of various chloro-containing compounds on silica modified with 4-hydroxymethylbenzo-18-crown-6 shown in Fig. 3-33. Pure methanol was used as the mobile phase for this separation.

Also worth mentioning is the polyamide crown ether resin developed by Igawa et al. [39] for the separation of inorganic anions. The monomeric unit has the following structure:

Monomeric unit of a polyamide crown ether resin

This resin is suitable for the coating as well as for the chemical modification of silicas. However, the separation in Fig. 3-34 shows a relatively poor chromatographic efficiency, which is characteristic for crown ether phases but not appropriate for today's requirements.

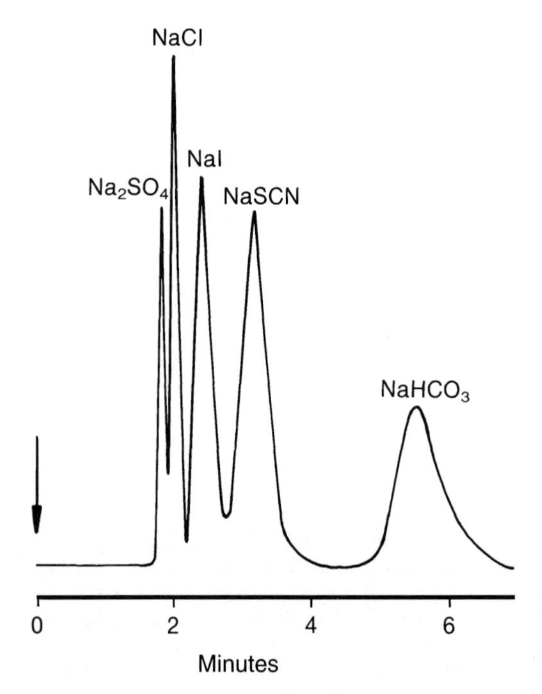

Fig. 3-32. Separation of the sodium salts of various anions on silica coated with crown ether polymers. – Stationary phase: dibenzo-18-crown-6; eluent: water; flow rate: 1 mL/min; detection: direct conductivity; solute concentrations: 0.7 ppm Na_2SO_4, 0.1 ppm NaCl; 1 ppm NaI, 4 ppm NaSCN, and 8 ppm $NaHCO_3$; (taken from [37]).

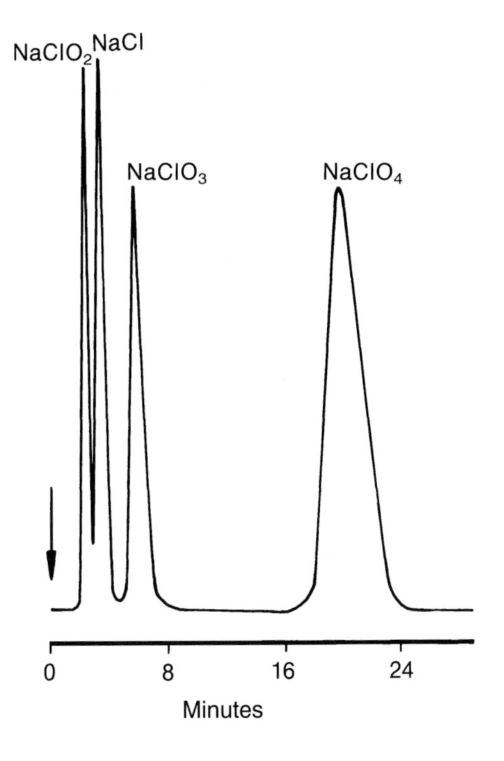

Fig. 3-33. Separation of the sodium salts of various chloro-containing species on silica modified with 4-hydroxymethyl-benzo-18-crown-6. – Eluent: methanol; flow rate: 1.6 mL/min; detection: direct conductivity; solute concentrations: 23 ppm $NaClO_2$, 6 ppm NaCl, 37 ppm $NaClO_3$, and 326 ppm $NaClO_4$; (taken from [37]).

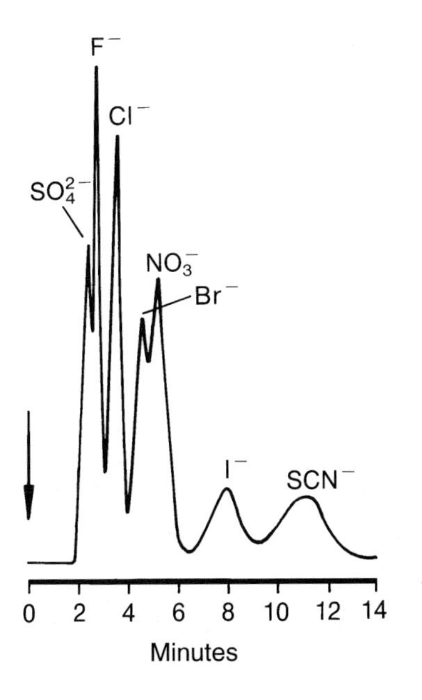

Fig. 3-34. Separation of inorganic anions on silica modified with polyamide crown ether. – Eluent: water; flow rate: 0.6 mL/min; detection: direct conductivity; injection volume: 20 μL; solute concentrations: each 0.1 mol/L of the respective salts; (taken from [39]).

Alumina Phases

In addition to silica, $(SiO_2)_x$, alumina, $(Al_2O_3)_x$, is amongst the most common adsorbents in liquid solid chromatography. Although highly efficient separator columns were developed in the past with spherical beads of small diameter, this separation material is only of minor importance since the introduction of chemically bonded reversed phases on the basis of silica in HPLC.

Like many other metal oxides, alumina exhibits typical ion-exchange properties [40]. It is also mechanically and thermally stable, and exhibits only slight swelling and shrinkage phenomena in aqueous media. Nonetheless, alumina was seldom used in the past as substrate for ion exchangers due to the low ion-exchange capacity and the inadequate stability against strong acids and bases. However, with the introduction of sensitive HPLC detectors, high ion-exchange and loading capacities are no longer necessary because of an intensified interest in alumina phases for the separation of ionic species [41,42].

The first detailed studies of the retention behavior of inorganic anions and cations on alumina were carried out by Schwab et al. [43]. They confirmed that ion-exchange processes are predominantly responsible for the retention of ionic species on alumina. The ion-exchange model is based on the development of an electrical double layer at the alumina surface, which is caused by dissociation of Al-OH groups present at the surface and by abstraction of H^+ and OH^- ions. This process, which is described in a simplified way by the dissociation equilibria (59) and (60),

$$\gt Al-OH \rightleftharpoons \gt Al^+-OH^- \tag{59}$$

$$\gt Al-OH \rightleftharpoons \gt Al^+-O^- \quad H^+ \tag{60}$$

leads to the formation of positive and negative charges at which according to Eqs. (61) and (62), anion or cation exchange occurs with the solute ions X^- and X^+, respectively.

$$\gt Al^+ OH^- + X^- \rightleftharpoons \gt Al^+ X^- + OH^- \tag{61}$$

$$\gt Al-O^- H^+ + X^+ \rightleftharpoons \gt Al-O^- X^+ + H^+ \tag{62}$$

The presence of hydroxide groups at the alumina surface was unequivocally confirmed by IR spectroscopy [44], whereas the number and nature of these groups are determined by the thermal preconditioning of the material. The amphoteric character of alumina and its transformation into an anion or cation exchanger, respectively, via hydration and subsequent treatment with acid or base are illustrated in the following reaction scheme:

$$Al_2O_3 + H_2O \longrightarrow -Al-\bar{O}-Al(\cdots\bar{O}H_2)-\bar{O}-$$

$$\downarrow$$

$$H_2O + -Al(\cdots|\bar{O}|-Na)-\bar{O}-Al-\underline{O}(\cdots H)- \xleftarrow{NaOH} -Al(\cdots|\bar{O}|-H)-\bar{O}-Al-\underline{O}(\cdots H)- \xrightarrow{HCl} -Al(\cdots Cl)-\bar{O}-Al-\underline{O}(\cdots H)-$$

HCl

NaOH

Reaction Scheme

According to this scheme, the anion exchange process occurs between solute anions and OH^- ions attached to the alumina surface, while cation exchange occurs between solute cations and H^+ ions, which are released by dissociation of Al-OH groups.

The retention behavior of inorganic anions on an alumina phase was recently investigated in detail by Schmitt and Pietrzyk [45]. The most important experimental parameters determining retention are the pH value and the ionic strength of the mobile phase. Since alumina is weakly basic, anion exchange can only occur at pH values below the isoelectric point. The latter depends on the type of eluent. In general, an increase of the anion exchange capacity and, thus, the retention, is observed with decreasing pH value of the mobile phase.

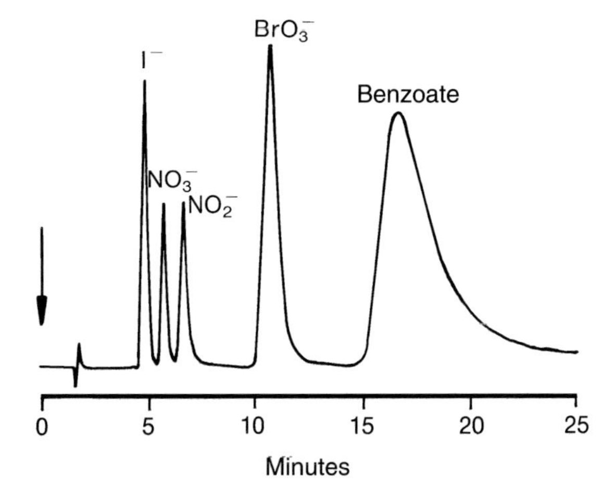

Fig. 3-35. Separation of various anions on an alumina phase. – Separator column: Spherisorb A5Y; eluent: 0.001 mol/L sodium acetate buffer (pH 5.6) + 0.05 mol/L $NaClO_4$; flow rate: 1 mL/min; detection: UV (214 nm); injection volume: 2 µL; solute concentrations: 1 g/L each; (taken from [45]).

Most interesting are the selectivity differences between an alumina phase and a strongly basic anion exchanger with quaternary ammonium groups. In comparison with a conventional anion exchanger, halide ions, for example, are eluted in reverse order: $I^- < Br^- < Cl^- < F^-$. This order corresponds to the formation constants of aluminum halide complexes [46], which suggests an interaction between aluminum and halide ions within the Al_2O_3 structure.

The influence of the ionic strength on retention does not differ from conventional anion exchange chromatography: retention decreases with increasing ionic strength of the mobile phase. However, due to the different selectivity of an alumina phase, other eluent ions must be selected. Fluoride belongs to the strongest eluent ions, but its use is not recommended because the interaction with alumina is too strong. Hydroxide ions also exhibit a strong elution power, since they convert the alumina phase into a cation exchanger at corresponding concentrations. Sodium perchlorate was found to be a suitable eluent for the determination of simple inorganic anions. A typical chromatogram, displayed in Fig. 3-35, reveals that modern alumina phases with spherical 5-μm particles are highly efficient. However, it should be pointed out that the eluents usually employed for the separation of inorganic anions are not compatible with a detection via conductivity measurements.

3.3.2 Eluents for Anion Exchange Chromatography

The kind of eluent that is used for anion exchange chromatography depends mainly on the detection system being employed. Since, in many cases, the detection of inorganic and organic anions occurs via conductivity detection, the eluents used are classified into two groups:

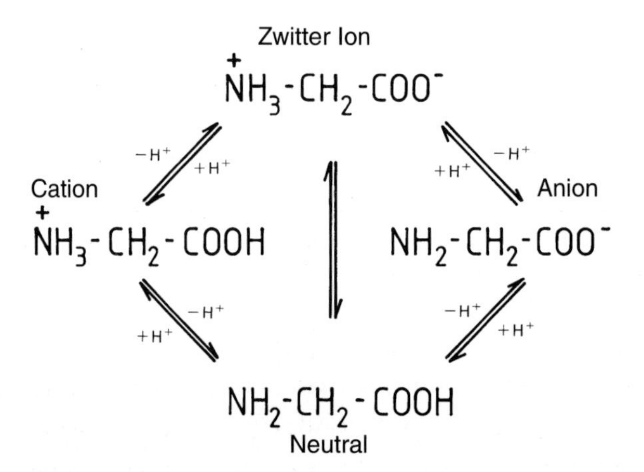

Fig. 3-36. Dissociation behavior of amino acids on the example of glycine.

- Eluents for conductivity detection with *chemical* suppression of the background conductivity
- Eluents for conductivity detection with *electronic* compensation of the background conductivity

Regardless of this, the affinities of eluent ions and solute ions must be comparable; i.e., as a rule of thumb, divalent solute ions may only be eluted with divalent eluent ions.

Eluents of the *first* kind include the salts of weak inorganic acids, which exhibit a low background conductivity after chemical modification within a suppressor system (see Section 3.3.3). In their introductory paper in 1975, Small et al. [3] used sodium phenolate for the separation of non-polarizable inorganic and organic anions. Although good separations were achieved with this eluent, even minute traces of oxygen in the mobile phase lead to oxidation of this compound and, thus, to a gradual poisoning of the stationary phase by the oxidation products. Today, sodium phenolate is no longer used as an eluent because of the toxicity of the phenol resulting from the suppressor reaction.

The versatile mixture of sodium carbonate and sodium bicarbonate, on the other hand, finds widespread application, since the elution power and the selectivity resulting there from are determined over a wide range solely by the concentration ratio of these two compounds. A great variety of inorganic and organic anions can be separated with this eluent combination. As the product of the suppressor reaction, the carbonic acid is only weakly dissociated, so that the background conductivity is very low.

As an alternative to carbonate/bicarbonate systems, amino acids (α-aminocarboxylic acids) may be used as an eluent [47,48]. Their dissociation behavior is depicted in Fig. 3-36. At alkaline pH, amino acids exist in the anionic form due to the dissociation of the carboxyl group and, thus, may act as an eluent ion. The product of the suppressor reaction is the zwitterionic form with a correspondingly low background conductance. This depends on the isoelectric point, pI, of the amino acid.

$$\mathrm{pI} = \frac{\mathrm{p}K_1 + \mathrm{p}K_2}{2} \tag{63}$$

The residual dissociation of the zwitter ion and the background conductance of the eluent is even lower than for the carbonate/bicarbonate system, if amino acids are selec-

Table 3-7. Eluents commonly used for conductivity detection with chemical suppression of the background conductivity.

Eluent	Eluent ion	Suppressor product	Elution strength
$Na_2B_4O_7$	$B_4O_7^{2-}$	H_3BO_3	very weak
NaOH	OH^-	H_2O	weak
$NaHCO_3$	HCO_3^-	$[CO_2 + H_2O]$	weak
$NaHCO_3/Na_2CO_3$	HCO_3^-/CO_3^{2-}	$[CO_2 + H_2O]$	medium strong
$H_2NCH(R)COOH/NaOH$	$H_2NCH(R)COO^-$	$H_3\overset{+}{N}CH(R)COO^-$	medium strong
$RNHCH(R')SO_3H/NaOH$	$RNHCH(R')SO_3^-$	$R\overset{+}{N}H_2CH(R')SO_3^-$	medium strong
Na_2CO_3	CO_3^{2-}	$[CO_2 + H_2O]$	strong

ted with an isoelectric point near the neutral pH value. Similar low background conductances are obtained with N-substituted aminoalkylsulfonic acids, which were recently introduced as eluents by Irgum [48], and which can also be used for gradient elution because of their sufficient purity [49].

The application of pure sodium hydroxide as eluent was regarded to be disadvantageous in the past, although water as the suppressor product produces virtually no background conductance. However, since hydroxide ions exhibit only a small affinity toward the stationary phase, it was necessary to work with relatively high concentrations to elute anions with more than one negative charge. This has an adverse effect on the background conductance because even modern hollow fiber suppressors possess only a limited exchange capacity. Only with the introduction of the micromembrane suppressor systems of high capacity did it become possible to use sodium hydroxide in concentrations of up to 0.1 mol/L. Hence, this eluent is of special significance, since it is possible to elute even polyvalent solute ions at these high concentrations. Sodium hydroxide is therefore perfectly suited for the gradient elution of anions in connection with conductivity detection.

Tetraborate ions exhibit a similar low affinity toward the stationary phase allowing for the separation of fluoride and short-chain aliphatic carboxylic acids. Since boric acid as the suppressor product is only weakly dissociated, sodium tetraborate is also suited for gradient elution to a limited extent.

Aromatic amino acids such as tyrosine, which preferably reduce the retention of polarizable anions at alkaline pH, have a comparatively high elution power. *p*-Cyanophenol acts in a similar way. When it is added to a mixture of sodium carbonate and sodium bicarbonate, a wealth of polarizable anions can be separated at a suitable stationary phase and can be detected via their electrical conductance.

Table 3-7 presents an overview of the various eluents being compatible with suppressor systems and their elution power. With the eluents listed in Table 3-7 and with the aid of organic additives described above, nearly all anions that are detected via conductivity may be analyzed using one of the many available anion exchangers. This multitude of stationary phases, with their different selectivities, enables the vast number of potential eluents to be reduced to a few versatile systems. Therefore, the statement that suppressor systems limit the choice of eluents is not valid.

Table 3-8. Background conductances of the eluents used by Fritz et al. [19] in their introductory paper.

Eluent	Concentration [mol/L]	pH Value	Conductance [μS/cm]
Potassium benzoate	$6.5 \cdot 10^{-4}$	4.6	65.9
Potassium hydrogen phthalate	$5.0 \cdot 10^{-4}$	4.4	74.3
	$5.0 \cdot 10^{-4}$	6.2	112.0
	$6.5 \cdot 10^{-4}$	4.4	90.5
	$6.5 \cdot 10^{-4}$	6.2	158.0
Ammonium *o*-sulfobenzoate	$5.0 \cdot 10^{-4}$	5.8	132.0

Eluents of the *second* kind should have a low background conductivity to enable a sensitive conductivity detection of the anions to be analyzed. The selection of eluents suited for conductivity detection with electronic compensation of the background conductivity was the subject of numerous investigations [16,17,19,50]. Benzoates, phthalates, and *o*-sulfobenzoates are used most frequently, since they exhibit both a sufficient affinity toward the stationary phase of surface-aminated ion exchangers and a relatively low conductivity. When aromatic carboxylic acids are used as eluents, the pH value of the mobile phase must be adjusted between pH 4 and 7, because it affects the degree of dissociation of the organic acid and, thus, determines the retention behavior of the species to be analyzed. In addition, an almost neutralized eluent is necessary to minimize the concentration of oxonium ions with their high equivalent conductivity. Table 3-8 lists the resulting background conductivities of the eluents used by Fritz et al. in their introductory paper [19]. The background conductivities of these eluents are much higher than that of carbonate/bicarbonate-based eluents after passing the suppressor system (ca. 15 to 20 μS/cm). The high background conductivities with values between 60 and 160 μS/cm reduce the detection limit and linear range of the detector. Recent investigations by Gjerde and Fritz [50] show significantly lower background conductivities when adjusting the pH value of the mobile phase to neutral.

Problems are encountered in adjusting the pH value when commercially available pH electrodes are used. They contain, in general, a Ag/AgCl reference electrode with saturated KCl solution as supporting electrolyte, as well as porous frits as salt bridges. Therefore, eluents can be contaminated with chloride, which would result in false data for a subsequent chloride determination [51]. This problem may be circumvented by spatially separating the reference electrode and eluent from each other.

Using aromatic carboxylic acids as the eluent, a separation of the most important inorganic anions such as fluoride, chloride, bromide, iodide, nitrite, nitrate, orthophosphate, sulfate, thiocyanate, and thiosulfate is possible. However, at best, seven anions may be separated within a single run under suitable chromatographic conditions. Sodium and potassium benzoate are primarily used for the separation of monovalent anions. The affinity of benzoate toward the stationary phase is comparable with that of sodium bicarbonate. The corresponding phthalic acid salts exhibiting a higher elution strength have to be used for the elution of monovalent *and* divalent species. Increasing the benzoate concentration is not possible, since the resulting background conductivity becomes too high for a sensitive conductivity detection. Strongly polarizable and polyvalent anions might be eluted employing the salts of trimesic acid that exhibits a strong affinity toward the stationary phase due to its three carboxylate groups.

Potassium hydroxide is a suitable eluent for the determination of inorganic anions with p*K* values above 7. Due to the high pH value in the mobile phase, even weak acids are completely dissociated and, thus, may be detected via their conductivity. However, the conductivity of the species is lower than that of the mobile phase, so that negative signals are observed for these species. This method is called indirect conductivity detection.

The classification of eluents into the two categories mentioned above is only meaningful and necessary in the framework of conductivity detection with its different application requirements. Eluent selection is much easier for applications using spectrophotometric or amperometric detectors, respectively. In photometric detection, both the photometric properties of the eluent ions and their chemical properties have to be taken into account; nevertheless, a large number of eluents are available. The alkali salts of phosphoric acid, sulfuric acid, and perchloric acid have proved successful, because they all feature a good UV transmittance. In the field of amperometric detection, the choice of eluents is still much higher. The electrolyte concentration in the mobile phase must be about 50 to 100 times higher than the concentrations of the analyte ions. The mobile phase acts as a support electrolyte which, via reduction of the mobile phase resistance, R_L, ensures that the voltage drop, $i \cdot R_L$, is kept low. Chlorides, chlorates, and perchlorates of alkali metals and alkali hydroxides and carbonates are suited as supporting electrolytes.

3.3.3 Suppressor Systems in Anion Exchange Chromatography

A "suppressor system" is used for the sensitive detection of ions via their electrical conductance. Its function is to chemically reduce the background conductivity of the electrolyte used as eluent before it enters the conductivity cell. Therefore, the suppressor system may be regarded as part of the detection system.

In its simplest form, a suppressor system consists of a column containing a strongly acidic cation exchange resin in its hydrogen form. The function of this "suppressor column", originally used by Small [3], is illustrated by the analysis of the chloride and bromide anions. After these anions are separated with an eluent such as sodium bicarbonate on one of the anion exchangers described above, the column effluent is passed through the suppressor column prior to entering the conductivity cell, where the following reactions occur:

- Weakly dissociated carbonic acid is formed from the strongly conducting sodium bicarbonate by exchanging the eluent sodium ions with the protons of the cation exchanger:

$$\text{Resin}-SO_3H + NaHCO_3 \longrightarrow \text{Resin}-SO_3Na + [CO_2 + H_2O] \quad (64)$$

- Similarly, sodium chloride and sodium bromide are converted into their corresponding acids:

$$\text{Resin}-SO_3H + NaCl \longrightarrow \text{Resin}-SO_3Na + HCl \quad (65)$$

$$\text{Resin}-SO_3H + NaBr \longrightarrow \text{Resin}-SO_3Na + HBr \quad (66)$$

As the result of the suppressor reaction, strongly conducting mineral acids in the presence of weakly dissociated carbonic acid enter the conductivity cell and are thus easily be detected.

The advantage of the suppressor technique is its higher sensitivity. In addition, the specificity of the method is also increased, since the chemical modification of eluent and sample in the suppressor system converts the conductivity detector from a bulk property detector into a solute specific detector [52]. Thus, exchanging eluent and sample cations with protons means that only the sample anions to be analyzed are detected by the conductivity detector and appear in the resulting chromatogram.

Recently, membrane-based suppressor systems are increasingly used in addition to conventional suppressor columns. The characteristic properties of the various suppressor devices are discussed below.

Suppressor Columns

As already discussed, a suppressor column suitable for anion exchange chromatography contains a strongly acidic cation exchange resin in the hydrogen form. The properties of this resin are extremely important for the quality of the analysis. In addition to the conversion of eluent and sample into their corresponding acids, other phenomena ensue in the suppressor column. Since a regenerated cation exchanger exists in the hydrogen form, it resembles an ICE column. Therefore, weak and strong electrolytes are subject to the Donnan effect (see Chapter 4). While strong acids are excluded from the stationary phase and travel with the mobile phase along the column, weak acids such as nitrite and acetate, that could partly be non-dissociated, may pass through the Donnan layer and may be adsorbed on the surface of the stationary phase. This results in a small increase in the retention times for weak acids, which depends on the pK_a value of the respective species, the total volume of the suppressor column, and its degree of use, respectively. The resulting change in peak height adversely affects the routine analysis. Therefore, gel-type polystyrene/divinylbenzene-based cation exchangers with 8% crosslink, particle diameters between 20 μm and 40 μm, and a low specific surface area are generally used. Although, in principle, cation exchangers with a high specific surface area are suited to be used as suppressor columns because of their fast exchange kinetics, they have to be considered as disadvantageous with respect to the adsorption processes described above.

A major drawback of suppressor columns is the requirement for a periodic regeneration. Depending on the degree of use, the retention time of the negative water dip is shifted. The latter arises due to the fact that the conductance of water as the solvent for the sample is smaller than the background conductance of the eluent after passing through the suppressor column. If pure, de-ionized water is injected into the ion chromatograph, it travels with the mobile phase through the column. The observed negative conductivity change occurs at a time almost simultaneously with the void volume of the column. With the change of the water dip position, the quantification of species eluting near the void, for example, fluoride and chloride, is rendered more difficult.

The particle size of the resin and the void volume of the suppressor column affect the quality of the separation, since both parameters determine the peak broadening. The total volume of the suppressor column should be as small as possible to prevent a mixing of the already separated signals. However, with regard to the resulting exchange capacity, the total volume of the suppressor column should be as large as possible. Both requirements are incompatible, thus the dimensioning of suppressor columns is always a compromise between the exchange capacity and the peak broadening caused by the void volume. It is, therefore, advisable to use suppressor columns only in combination with separator columns packed with materials having a particle diameter of more than 15 μm. Under these conditions, the contribution to the total peak variance is small.

The regeneration of conventional suppressor columns is usually carried out for about 15 minutes using sulfuric acid at a concentration of c = 0.5 mol/L with a flow rate of 4 mL/min. Subsequently, the system is flushed with de-ionized water for some minutes and then it is conditioned with eluent again.

Hollow Fiber Suppressors

Conventional suppressor columns are sometimes still in use, although their importance is more a historical one since the introduction of modern membrane technologies in 1982. The disadvantages of suppressor columns mentioned above could be eliminated with the development of a hollow fiber suppressor. It consists of a semipermeable membrane, which is wrapped around a cylindrically shaped body. This tubular coil is housed in a jacketed vessel. While the column effluent is passed through the interior of the fiber that is packed with inert beads, a dilute sulfuric acid solution flows countercurrent to the eluent, in contact with the exterior of the fiber. The flow rate can easily be controlled by the difference in height between the regenerent reservoir and the suppressor housing. When working in extremely sensitive detector ranges (<1 μS/cm full scale), it is recommended to pump the regenerent solution through the fiber suppressor, since the constantly changing hydrostatic pressure leads to baseline drifts. A much more stable baseline is obtained by pneumatic delivery of the regenerent. For that purpose, the sulfuric acid reservoir is pressurized, resulting in a pulse-free flow.

In contrast to conventional suppressor columns, hollow fiber suppressors are continuously regenerated, and thus do not require an additional pump system. The reactions that occur across the membrane wall are shown in Fig. 3-37. Since fiber suppressors suited for anion exchange chromatography act as cation exchangers, the eluent cations are exchanged with protons in the regenerent solution. The driving force for the diffusion of protons across the membrane is provided by their subsequent reaction with

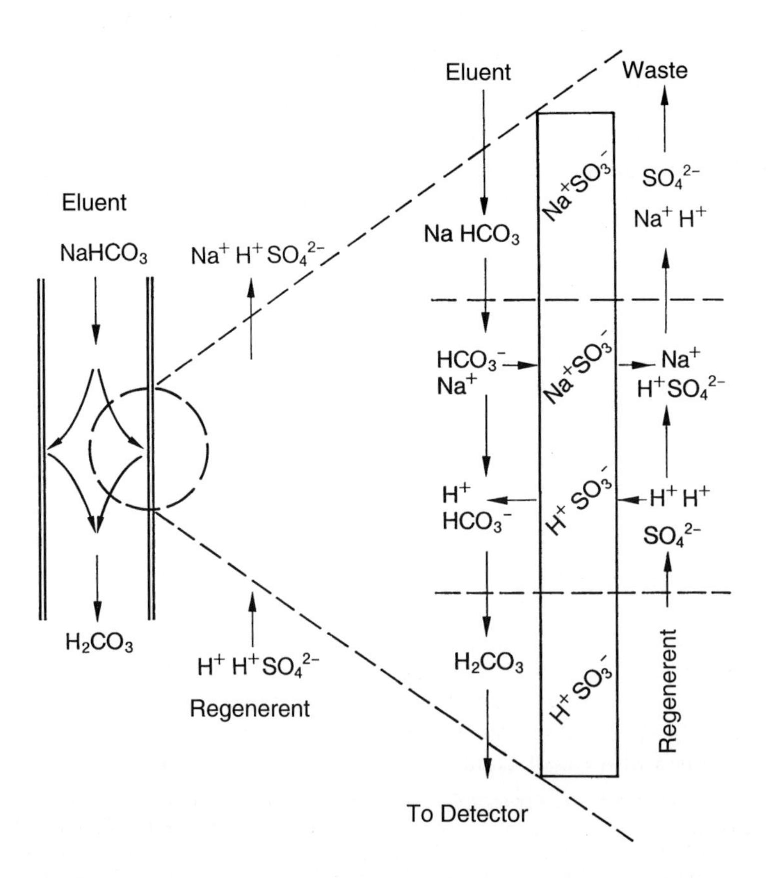

Fig. 3-37. Schematic of a hollow fiber suppressor for anion exchange chromatography.

bicarbonate to form carbonic acid. In order to maintain the ionic balance, the cations diffuse into the regenerent solution. As illustrated in Fig. 3-37, three distinct regions in the fiber can be distinguished:

- An expended region where the eluent enters the fiber suppressor (cation exchanger in the sodium form).
- A region of “dynamic equilibrium” in which the actual suppression reactions occur.
- A regenerated region where the eluent exits the fiber suppressor (cation exchanger in the hydrogen form).

Once established, the dynamic equilibrium is maintained as long as pertinent operating conditions are not altered.

Since the percentage of exchange groups in the sodium form remains constant due to continuous regeneration, weak acids also remain constant and less standardization is required. Additionally, the quantification of fluoride and chloride is simplified, since the water dip immediately precedes the fluoride peak and remains constant in retention time. In comparison with suppressor columns, hollow fiber suppressors have a void volume that is smaller by an order of magnitude. This drastically reduces the mixing and band broadening effects and enhances the sensitivity. Fig. 3-38 displays a comparison of the detection sensitivity when suppressor columns or hollow fiber suppressors are used. Most remarkable is the huge sensitivity increase for nitrite by 300%.

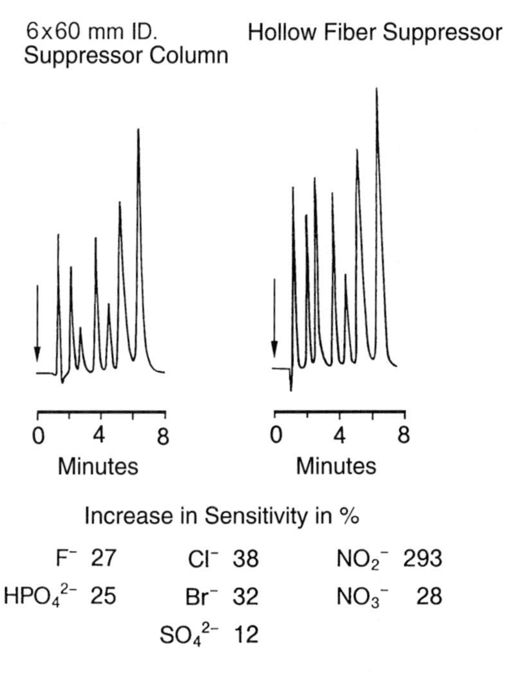

Fig. 3-38. Comparison of detection sensitivities for the application of a conventional suppressor column and a hollow fiber suppressor. – Separator column: IonPac AS4; eluent: 0.0028 mol/L $NaHCO_3$ + 0.0022 mol/L Na_2CO_3; flow rate: 2 mL/min; detection: conductivity with a) ASC-1, and b) hollow fiber suppressor AFS-1; regenerent: 0.0125 mol/L H_2SO_4; injection: 50 µL anion standard.

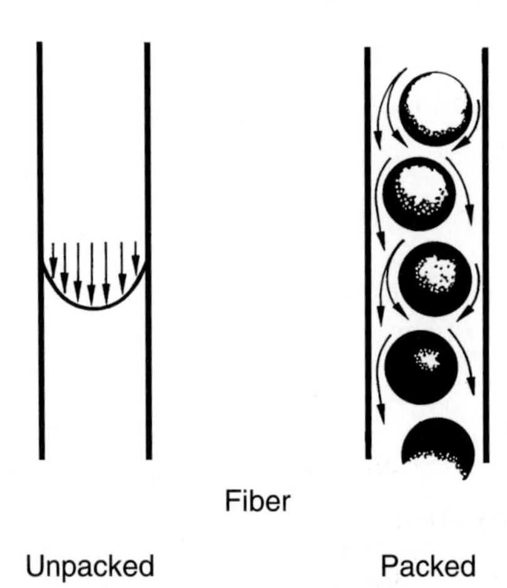

Fig. 3-39. Schematic representation of the flow profile in the interior of a fiber.

Looking at the schematic representation of the flow profile within the fiber shown in Fig. 3-39, it becomes apparent that a further sensitivity increase can be accomplished by changing the parabolic flow profile. When the fiber is packed with inert Nafion beads, the translational diffusion of ions is favored over longitudinal diffusion. This, in turn, improves the mass-transfer across the membrane which leads to a further increase in sensitivity, particularly pronounced in the case of orthophosphate as the salt of a weak acid.

Sulfuric acid at a concentration of c = 0.0125 mol/L is recommended for the continuous regeneration of a hollow fiber suppressor. The flow rate should be adjusted between 2 and 3 mL/min. Under these conditions, a background conductivity of 18 to 25 µS/cm results for the bicarbonate/carbonate eluent (0.0028 mol/L; 0.0022 mol/L) depending on the purity of the water used for eluent preparation. If the ionic strength is lowered, the regenerent concentration should still be maintained, resulting in a decrease of the background conductivity. If more concentrated eluents are required, it is possible to expend the exchange capacity of the hollow fiber suppressor. However, the regenerent concentration can only be increased to a certain extent since, at higher concentrations, the sulfate ions can overcome the Donnan exclusion forces and diffuse into the interior of the fiber. Table 3-9 lists the background conductivities resulting from different concentrations of the regenerent solution. When using more concentrated carbonate-based eluents, outgassing of carbon dioxide may occur within the conductivity cell. This effect may be prevented by applying an adequate backpressure at the outlet capillary of the detector cell. However, the backpressure must not exceed about 13 bar (200 psi) due to the limited pressure stability of the membrane suppressor. If the use of more concentrated carbonate eluents cannot be avoided, a micromembrane suppressor with a much higher exchange capacity has to be employed.

Table 3-9. Dependence of background conductivity on the regenerent concentration.

[H_2SO_4] [mol/L]	Background conductivity [µS/cm]
0.0125	21.7
0.0250	28.2
0.1250	288.0

Micromembrane Suppressors

Although suppressor systems are very effective with respect to a sensitivity increase, its application limits the choice of eluents and their concentrations. Therefore, the objective for developing of a new suppressor generation was to prepare a system that combines the positive properties of packed bed suppressors and hollow fiber suppressors, respectively. This objective was achieved with the introduction of the micromembrane suppressor in 1985 [53], which, like the hollow fiber suppressor, operates by using the principle of continuous regeneration but, in comparison, features a much higher exchange capacity. Its design allows the exchange capacity to be enhanced by more than one order of magnitude, so that, in terms of capacity, a micromembrane suppressor is comparable to a conventional suppressor column. Significant progress was made in further decreasing the void volume of the system, which is only 50 μL for a micromembrane suppressor, a volume that has little effect on the chromatographic separation. (For comparison: the void volume of a conventional suppressor column is about 2000 μL, that of a hollow fiber suppressor is about 200 μL.) Further improvements concerned the handling of the system. While the membrane in a hollow fiber suppressor ages and has therefore to be replaced semi-annually, a micromembrane suppressor is almost maintenance-free when following the operating instructions.

Fig. 3-40 is a schematic illustration of the sandwich structure of a micromembrane suppressor. It consists of a flat, two-part box in which strongly sulfonated ion-exchange screens and thin ion-exchange membranes lie on one another in alternating order. The two parts of the housing keep them together. The ion-exchange screen functions as eluent or regenerent channel, respectively. Gasketing material is attached at the side so that the screen is porous only in the central region. In analogy to a hollow fiber suppressor, the eluent is passed through the eluent chamber located in the middle while regenerent flows countercurrent through the two regenerent chambers. It is clear from Fig. 3-41 that, due to the screen structure, eluent cations are transported much more efficiently to the adjacent membrane wall than in a hollow fiber suppressor by packing the membrane.

The comparatively higher exchange capacity of a micromembrane suppressor is not only caused by the higher diffusion efficiency of eluent cations to the membrane wall, it is based on the ion-exchange characteristics of the screen, the exchange capacity of which is directly proportional to the suppressor capacity. In continuously regenerated suppressors, this is understood as the exchangeable cation concentration per time unit. Therefore, the suppressor capacity is defined for both the hollow fiber suppressor and the micromembrane suppressor according to Eq. (67) as the product of eluent concentration and eluent flow rate:

$$\underset{[\mu\text{equiv/min}]}{\text{Suppressor capacity}} = \underset{[\mu\text{equiv/mL}]}{\text{Eluent concentration}} \times \underset{[\text{mL/min}]}{\text{Eluent flow rate}} \qquad (67)$$

Sodium hydroxide at a maximum concentration of $c = 0.1$ mol/L (100 μequiv/mL) and with a flow rate of 2 mL/min can be used as eluent with a micromembrane suppressor, resulting in a suppressor capacity of about 200 μequiv/min.

Since the exchange capacity is determined by the degree of sulfonation of screen and membrane, the total area finally determines the capacity of the suppressor. As illustrated by the diagram in Fig. 3-42, a linear relationship exists between the suppressor capacity and its length, provided that the width of the eluent chamber is kept constant.

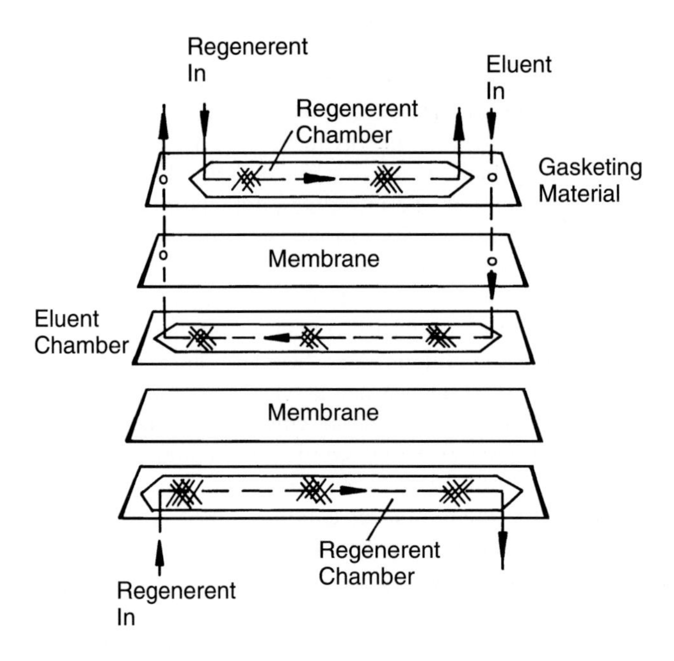

Fig. 3-40. Schematic structure of a micromembrane suppressor.

The high capacity of a micromembrane suppressor and the ability to provide continuous suppression enable the number of possible eluents to be significantly enlarged. In general, any weak acid can be used as eluent as long as it exists in an anionic form above pH 8 and in neutral form between pH 5 and 8. Above all, this includes the

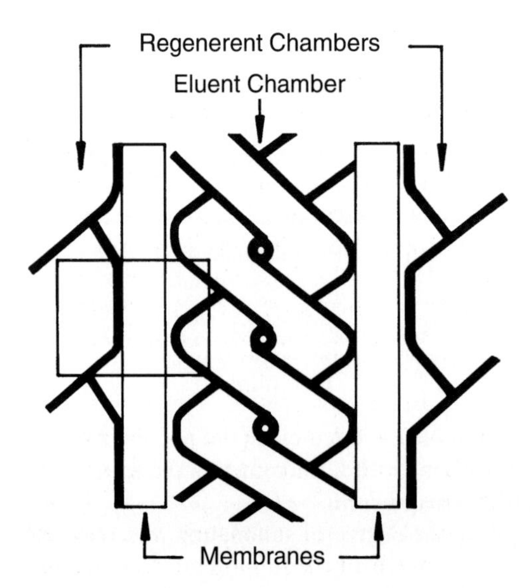

Fig. 3-41. Enlarged view of an eluent chamber in a micromembrane suppressor.

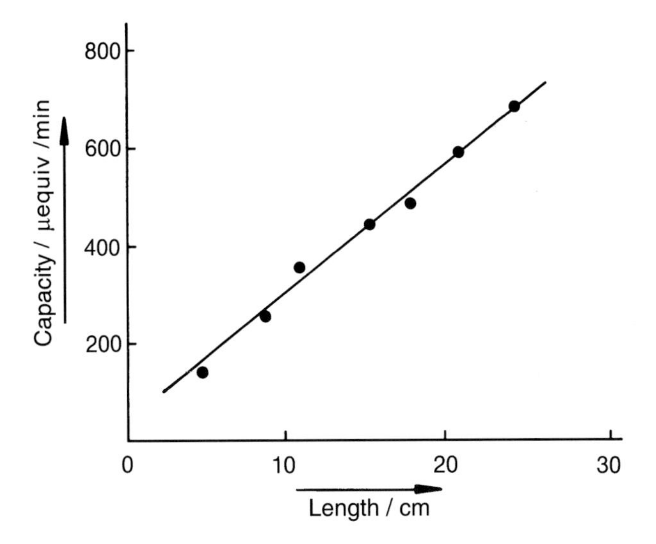

Fig. 3-42. Dependence of the micromembrane suppressor capacity on the length of the eluent chamber at a constant width of 1 cm.

amino acids already mentioned in Section 3.3.2 which, for example, allow the elution of all halide ions within a short time.

The high capacity of a micromembrane suppressor also allows the use of high ionic strength eluents, which significantly reduces the analysis times. The low void volume of the suppressor that hardly affects the separation efficiency also contributes to the sensitivity increase achieved therewith.

The greatest achievement for ion chromatography that was made possible by the development of micromembrane suppressors is the introduction of gradient elution techniques with subsequent conductivity detection. As in conventional HPLC and in anion exchange chromatography, a distinction is made between composition gradients and concentration gradients. Simple composition gradients, such as the stepwise or continuous replacement of bicarbonate ions by carbonate ions, have been successfully employed in the past. Although the background conductivity increases slightly when the eluent ion changes due to increased formation of carbonic acid as the suppressor product, the capacity of conventional hollow fiber suppressors is sufficient for this kind of gradient technique. However, since the ionic strength in the mobile phase remains constant in the case of composition gradients, ions with strongly different retention characteristics such as linear polyphosphates of the type $M_{n+2}P_nO_{3n+1}$ cannot be analyzed within a single run.

This becomes possible by applying a concentration gradient in which the ionic strength in the mobile phase is increased during the gradient run. When sodium hydroxide that is suited for suppression is used as the eluent, often the increase in ionic strength covers more than two orders of magnitude because of the low elution power of hydroxide ions. Without the application of a high capacity suppressor system, the resulting baseline drift would be steep enough to prevent the evaluation of compound signals, even after baseline subtraction by a computer. However, this baseline increase is limited to a few µS/cm when using the micromembrane suppressor. Further details about the handling and applicability of gradient techniques are discussed separately in Section 3.3.6.

The regeneration of a micromembrane suppressor occurs in much the same way as for a hollow fiber suppressor. The regenerent is delivered pneumatically from the reservoir, since the required flow rate cannot be obtained simply via gravity feed. While a sulfuric acid concentration of c = 0.01 mol/L suffices for isocratic operation, a twofold regenerent concentration is recommended for gradient techniques. The flow rate should be adjusted to ensure a sufficiently low background conductivity when the maximum eluent conentration is reached. Maintaining these conditions, one can then switch to the initial eluent concentration.

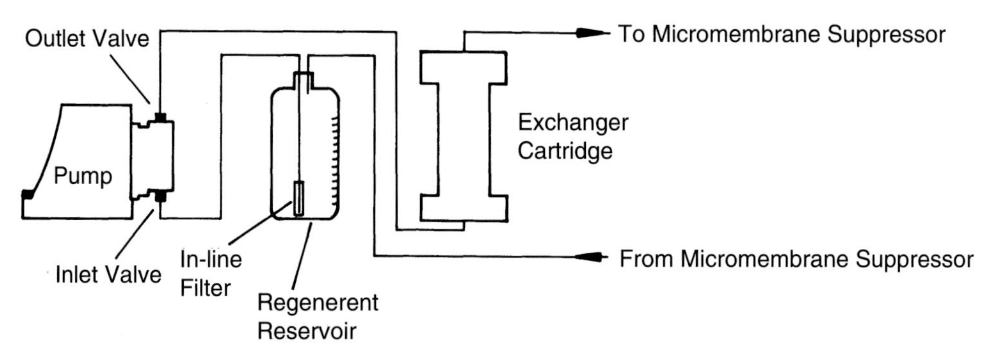

Fig. 3-43. Flow diagram for continuous operation of a micromembrane suppressor.

If the ion chromatograph is to be operated on-line for several days without interruption, the regeneration of the suppressor system must not be interrupted. Thus, the regenerent cannot be delivered pneumatically from a reservoir, because even a relatively large volume of 4 L must be opened and refilled after about 24 h. Such interruptions can only be avoided by closing the regenerent circulation. The regenerent is passed back to the reservoir after flowing through the suppressor system. As shown in the flow diagram in Fig. 3-43, a small pump is used instead of pneumatic delivery. Prior to entering the suppressor system, the regenerent is passed through an exchanger cartridge that contains a cation exchanger in the hydrogen form in order to maintain a constant proton concentration. In that way, the sodium sulfate formed in the suppressor is transformed back into the corresponding acid. The lifetime of such an exchanger cartridge having a fixed capacity of 500,000 μequiv depends both on the concentration and the flow rate of the eluent. It may be calculated according to Eq. (68):

$$\text{Lifetime of cartridge} = \frac{\text{Capacity of cartridge}}{\text{Normality of Eluent} \times \text{Flow rate}} \tag{68}$$

$$[\text{min}] = \frac{[\mu\text{equiv}]}{[\mu\text{equiv/mL}] \times [\text{mL/min}]}$$

When the chromatographic conditions for the analysis of inorganic anions are taken as a basis*), the lifetime of the exchanger cartridge is about 780 h. Even if, as suggested,

*) Eluent: 0.0017 mol/L $NaHCO_3$ + 0.0018 mol/L Na_2CO_3; flow rate: 2 mL/min.

this cartridge is already changed when it is expended to 90%, the suppressor system can be operated for about 30 days without any interruption.

3.3.4 Ion-Exchange Chromatography of Inorganic Anions

Overview

A variety of inorganic anions may be separated using the stationary phases described in Section 3.3.1. They comprise:

Halides:

F^-, Cl^-, Br^-, and I^-

Oxygen-containing halides:

OCl^-, ClO_2^-, ClO_3^-, ClO_4^-, BrO_3^-, and IO_3^-

Oxygen-containing phosphorus compounds:

PO_2^{3-}, PO_3^{3-}, PO_4^{3-}, $P_2O_7^{4-}$, $P_3O_{10}^{5-}$, $P_4O_{13}^{6-}$, and PO_3F^{2-}

Sulfur compounds:

S^{2-}, SO_3^{2-}, SO_4^{2-}, $S_2O_3^{2-}$, and SCN^-

Nitrogen compounds:

CN^-, OCN^-, NO_2^-, NO_3^-, and N_3^-

Silicon compounds:

SiO_3^{2-} and SiF_6^{2-}

Boron compounds:

$B_4O_7^{2-}$ and BF_4^-

Oxy non-metal anions:

AsO_2^-, AsO_4^{3-}, SeO_3^{2-}, and SeO_4^{2-}

Oxy metal anions:

MoO_4^{2-}, WO_4^{2-}, CrO_4^{2-}

When classifying these anions, it must be taken into account that, in contrast to other liquids, water exhibits good solvent properties for salts because of the specific structure of this solvent and the special interaction mechanism between the ion and the water molecule. Hydrogen bonds are broken (cavity effect) and the water structure is destroyed when an ion is solvated by water. The larger the ion, the higher the energy required for the formation of a molecular cavity. On the other hand, electrostatic ion-dipole interactions occur that lead to the formation of a new structure. Thus, the smaller the ion radius and the higher the ion charge, the stronger the effect. As established by the values of the molar hydration enthalpies for halide ions given in Table 3-10, the hydration increases as the size of the ion decreases.

Table 3-10. Molar hydration enthalpies of halide ions [54].

Ion	$\Delta H_{Hydr.}$ [kcal/mol]
F^-	−98
Cl^-	−79
Br^-	−71
I^-	−62

Large ions, for example, iodide, exhibit a very strong affinity toward the stationary phase of an anion exchanger. In these large ions, the hydration enthalpy is partly counterbalanced by the cavity formation energy. Such ions are termed *polarizable* anions in the following; their chromatography is discussed separately.

The polarizibility of an ion is directly connected with the ionic radius in the hydrated state as one of the solute-specific properties that determines the affinity of an ion toward the stationary phase. In general, the retention time increases with increasing ionic radius in the hydrated state and, thus, the stronger polarizability. Accordingly, the halide ions elute in the order of increasing retention times: fluoride < chloride < bromide < iodide. The retention time difference between bromide and iodide is already so large that the set of halide ions can only be analyzed in a single run by using special eluents.

In addition to the ionic radius in the hydrated state, the valency of an ion is another solute-specific property that affects retention. In general, retention is shifted forward with increasing valency. Thus, the monovalent nitrate elutes prior to the divalent sulfate. Exceptions are multivalent ions such as orthophosphate, where the retention depends on the eluent pH due to different dissociation equilibria. However, the size of an ion often influences the retention more strongly than the valency. Hence, the divalent sulfate elutes prior to the monovalent, but strongly polarizable, thiocyanate.

3.3.4.1 General Parameters Affecting Retention

Eluent Flow Rate

The van Deemter plot of a selected anion exchange column of the IonPac AS4 type shown in Fig. 2-7 reveals that, in analogy to HPLC on reversed phases, the height of a theoretical plate changes only at very low flow rates and thereafter approaches a plateau region. Hence, the retention times can be reduced by increasing the flow rate without a significant loss in separation efficiency. An inverse proportionality exists between flow rate and resulting retention time. However, an unrestricted flow rate increase is limited by the maximal operating pressure of the separator column. On the other hand, obtaining a better resolution via a flow rate reduction is only possible to a limited extent. Since the pH value and ionic strength and, thus, the order of elution, remain unaffected by a flow rate change, the control of the retention times via the flow rate is easily accomplished.

Length of the Separator Column

The number of theoretical plates and, thus, the separation efficiency, is determined by the length of the separator column. If two separator columns are used in series, the resulting enhancement of separation efficiency leads to a better resolution between ions with similar retention characteristics, with a corresponding increase in retention times. The separator column length also determines the exchange capacity. An increase of the exchange capacity via elongation of the separator column is recommended in all cases where the ion to be analyzed is present in an excess of another component.

Shortening separator columns is generally not feasible because the alteration process causes the separation efficiency to slowly diminish with time. A resolution between two signals that is too high leads to unnecessarily long retention times. This can be prevented by choosing appropriate eluents and stationary phases, respectively. A drastic reduction in the retention time of an ion may be achieved in special circumstances by a separation on the shorter pre-column.

3.3.4.2 Experimental Parameters Affecting Retention when Applying Suppressor Systems

While the parameters listed in Section 3.3.4.1 that determine the retention are fundamental in character, the parameters concerning the eluent discussed below depend on the detection system being used. This particularly applies to conductivity detection, which is possible directly or in combination with a suppressor system. These two modes of conductivity detection are fundamentally different and require eluents that not only differ with regard to their type but also to their concentrations and pH values, respectively. Therefore, the influence of these parameters on both modes is discussed separately for the most important detection system.

Choice of the Eluent

Table 3-7 lists the selection of eluents and eluent mixtures that are appropriate for applying a suppressor system and which depend on the type of analyte. In general, the eluent ions should exhibit a similar affinity toward the stationary phase as the analyte ion. The mixture of sodium carbonate and sodium bicarbonate has proved successful for the majority of applications, since over a wide range the resulting selectivity is determined solely by the concentration ratio of both components. As can be seen in Fig. 3-44, the most important monovalent and divalent inorganic anions are eluted with this eluent [55]. Under the same chromatographic conditions, it is possible to separate the nitrogen-containing compounds shown in Fig. 3-45. While the application of a suppressor system allows a very sensitive detection of cyanate via conductivity measurement, this method is inappropriate for cyanide, since the corresponding acid, HCN, formed in the suppressor is not dissociated at neutral pH.

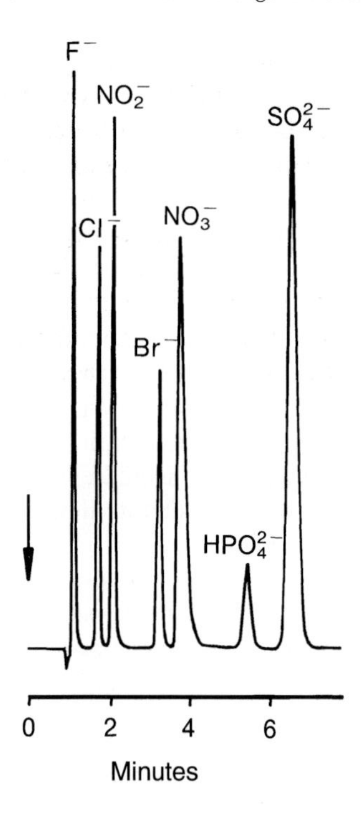

Fig. 3-44. Chromatogram of inorganic anions according to DIN 38405, part 19 [55]. – Separator column: IonPac AS4A; eluent: 0.0017 mol/L $NaHCO_3$ + 0.0018 mol/L Na_2CO_3; flow rate: 2 mL/min; detection: suppressed conductivity; injection: 50 μL; solute concentrations: 3 ppm fluoride, 4 ppm chloride, 10 ppm nitrite, 10 ppm bromide, 20 ppm nitrate, 10 ppm orthophosphate, and 25 ppm sulfate.

A small modification of the concentration ratio between bicarbonate and carbonate leads to a change in pH, thus oxy non-metal anions and simple mineral acids can be analyzed in a single chromatographic run. However, an exception is arsenic(III) in the metaarsenite form. Compared to cyanide, it cannot be detected via conductivity

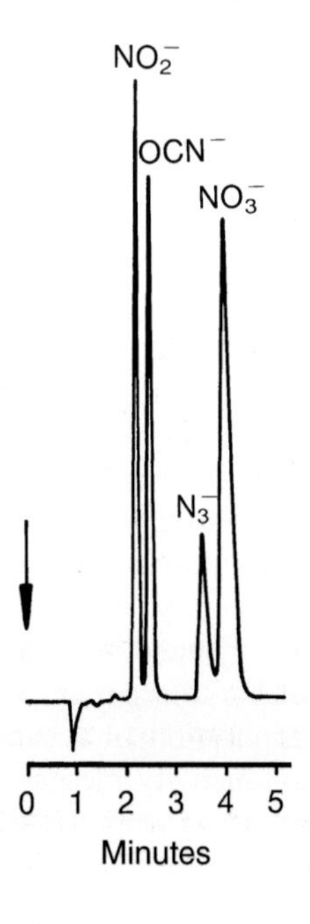

Fig. 3-45. Separation of various nitrogen-containing species. – Chromatographic conditions: see Fig. 3-44; solute concentrations: 10 ppm nitrite, 10 ppm cyanate, 20 ppm azide, and 20 ppm nitrate.

when applying the suppressor technique, since the arsenic acid formed in the suppressor is hardly dissociated. Arsenite, on the other hand, can easily be oxidized at a Pt working electrode. If the effluent of the separator column is passed through an amperometric detection cell before entering the suppressor system, the simultaneous detection of As(III) and As(V) is possible. In this case, the various detection modes contribute to the selectivity of the analytical procedure. The separation of non-metal anions, depicted in Fig. 3-46, that are detectable via conductivity, provides an acceptable alternative to atomic absorption spectroscopy, since it is possible to distinguish between the two most important oxidation states of arsenic and selenium. However, with regard to the detection limit, the AAS hydride system is superior to the ion chromatographic method.

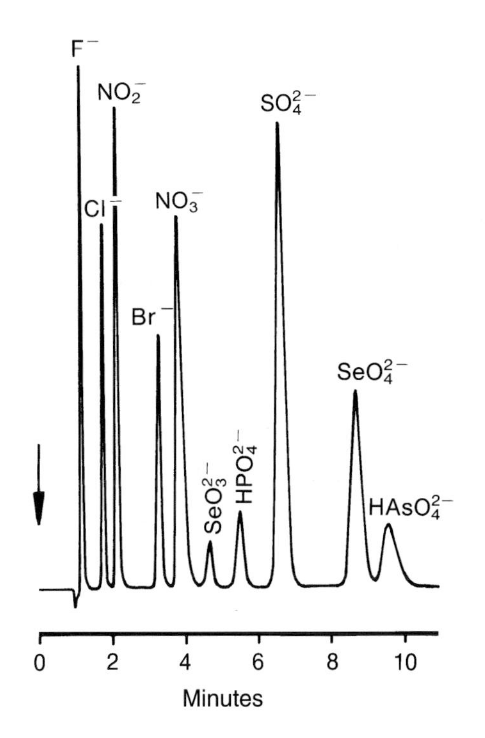

Fig. 3-46. Separation of mineral acids and oxy non-metal anions. – Separator column: Ion Pac AS4A; eluent: 0.00075 mol/L $NaHCO_3$ + 0.002 mol/L Na_2CO_3; flow rate: 2 mL/min; detection: suppressed conductivity; injection: 50 μL; solute concentrations: 3 ppm fluoride, 4 ppm chloride, 10 ppm nitrite, 10 ppm bromide, 20 ppm nitrate, 10 ppm selenite, 10 ppm orthophosphate, 25 ppm sulfate, 20 ppm selenate, and 25 ppm arsenate.

For the detection of mineral acids in the presence of an excessive amount of nitrate, the IonPac AS2 separator column was developed from which bromide and nitrate elute after sulfate. The selectivity of this stationary phase is based on the hydrophobic properties of the exchange groups bound to the latex beads (see Section 3.3.1.2). As shown in Fig. 3-47, small quantities of chloride, orthophosphate, and sulfate can be determined in the presence of high amounts of nitrate. The best separation is obtained with an eluent mixture of sodium carbonate and sodium hydroxide.

Pure sodium carbonate eluents are used for the separation of several phosphorus compounds, such as hypophosphite, orthophosphite, and orthophosphate, which are separated within a single run on an IonPac AS3 separator column and detected via electrical conductivity (Fig. 3-48).

Environmentally important anions such as sulfide and cyanide can be determined very sensitively via amperometric detection. For their separation on a conventional Ion

Pac AS3 anion exchanger (Fig. 3-49), a mixture of sodium carbonate and sodium dihydrogen borate is used as the eluent. A small quantity of ethylenediamine is also added to the mobile phase to allow for the complexation of traces of heavy metal ions, which could be present in the eluent [56,57]. While the detection of these two anions is very easy, the interpretation of experimental results for the investigation of real-world samples is very difficult. These samples normally contain transition and heavy metal ions, in the presence of which sulfide and cyanide are not present or only partly present as free ions. However, only the free ions are detected under the chromatographic conditions listed in Fig. 3-49.

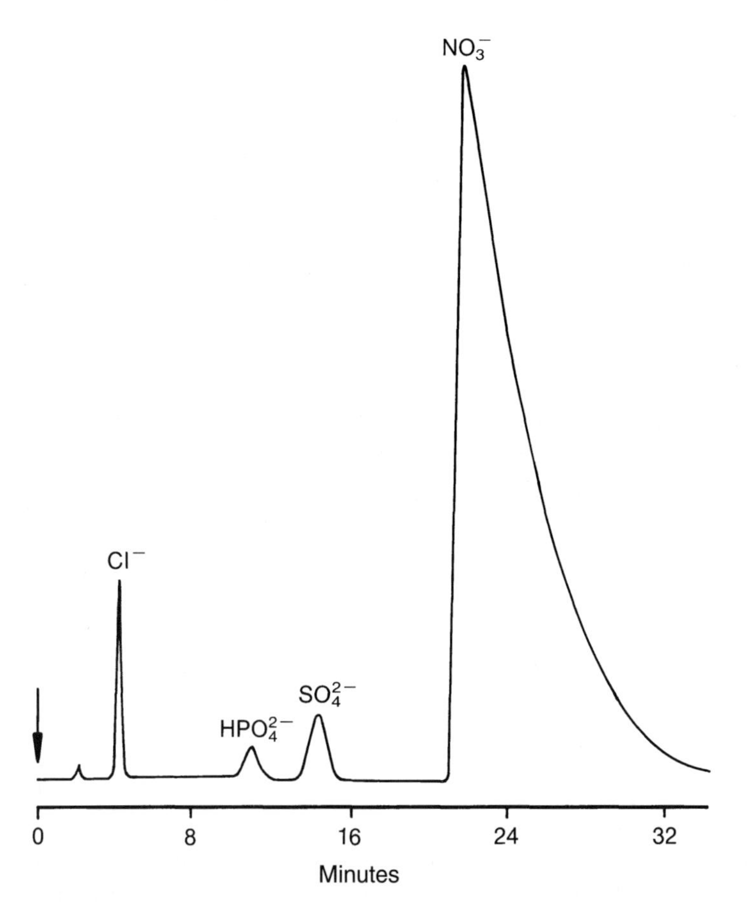

Fig. 3-47. Separation of mineral acids in the presence of large amounts of nitrate. – Separator column: IonPac AS2; eluent: 0.003 mol/L Na_2CO_3 + 0.002 mol/L NaOH; flow rate: 2.3 mL/min; detection: suppressed conductivity; injection: 50 µL; solute concentrations: 5 ppm chloride, 10 ppm orthophosphate, 10 ppm sulfate, and 500 ppm nitrate.

Disposal and drainage waters are among the few matrices in which free sulfide can be determined. Depending on the sample pH value, the transition and heavy metals present in these samples form sparingly soluble precipitates with sulfide. When the pH value of the sample is changed via dilution, the chemical equilibrium in the sample shifts, which could result in an increased amount of free sulfide. However, difficulties in

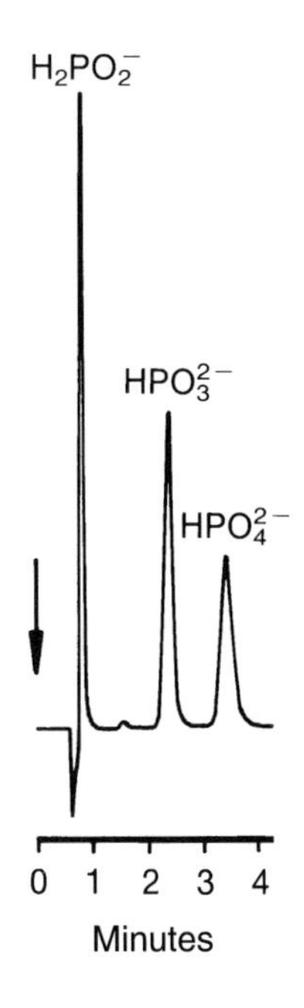

Fig. 3-48. Separation of hypophosphite, orthophosphite, and orthophosphate. – Separator column: IonPac AS3; eluent: 0.003 mol/L Na_2CO_3; flow rate: 2.3 mL/min; detection: suppressed conductivity; injection volume: 50 μL; solute concentrations: 10 ppm hypophosphite, 20 ppm orthophosphite, and 20 ppm orthophosphate.

sulfide determination arise not only because of complex matrices, but also due to the lack of a reference standard. Although sodium sulfide may be obtained in reagent-grade quality, the amount of crystal water cannot be determined unequivocally. Thus, the titer of a freshly prepared sodium sulfide solution must be determined via titration prior to each calibration.

Conversely, the calibration for the cyanide determination poses no problem. Free cyanide is found, for example, in untreated liquors from electroplating processes. Since these samples also contain transition metal ions, part of the cyanide exists in complexed form, whereby the various metal cyano complexes differ significantly in their stability.

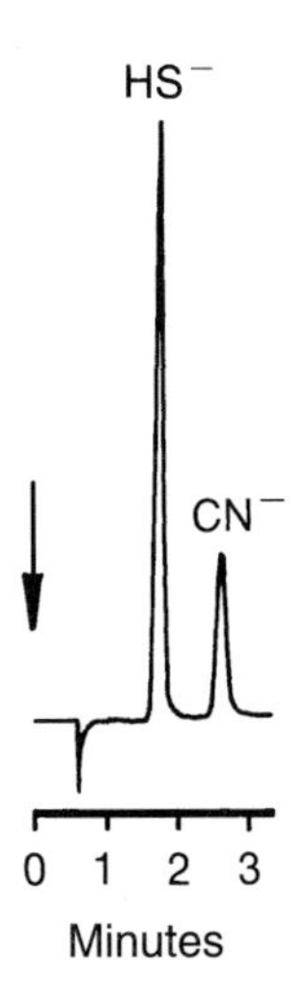

Fig. 3-49. Separation of sulfide and cyanide. – Separator column: IonPac AS3; eluent: 0.001 mol/L Na_2CO_3 + 0.01 mol/L NaH_2BO_3 + 0.015 mol/L ethylenediamine; flow rate: 2.3 mL/min; detection: amperometry on a Ag working electrode; injection volume: 50 μL; solute concentrations: 0.5 ppm sulfide and 1 ppm cyanide.

Table 3-11. Percentage of free cyanide in aqueous solution of various metal cyano complexes during the separation on an anion exchanger.

Complex	Percentage of free cyanide
$Cd(CN)_4^{2-}$	100
$Zn(CN)_4^{2-}$	100
$Ni(CN)_4^{2-}$	86
$Cu(CN)_4^{3-}$	52
$Ag(CN)_2^{-}$	13
$Hg(CN)_4^{2-}$	8
$Au(CN)_2^{-}$	0
$Au(CN)_4^{-}$	0
$Fe(CN)_6^{3-}$	0
$Fe(CN)_6^{4-}$	0
$Co(CN)_6^{3-}$	0

The cyanide signal produced amperometrically represents, therefore, both the free cyanide initially present in the sample and the one that is released in the mobile phase via a shift of the complexation equilibrium according to Eq. (69)

$$M(CN)_x^{n-} \rightleftharpoons M(CN)_{x-1}^{(n-1)-} + CN^- \tag{69}$$

When investigating the content of free cyanide in aqueous solutions of various metal cyano complexes with concentrations $c < 1$ mg/L under the chromatographic conditions listed in Fig. 3-49, the values summarized in Table 3-11 are obtained. Particularly pronounced is the cyanide dissociation from complexes having copper, nickel, zinc, or cadmium as the central atom. Thus, a much higher content of free cyanide is feigned in the respective samples. In contrast, the hexacyano complexes of the metals iron and cobalt are kinetically extremely stable; therefore, the cyanide determination is not affected by their presence. According to Rocklin and Johnson [56], results for free cyanide in real-world samples are as doubtful with respect to their accuracy as results obtained with wet-chemical methods. Recent investigations by Pohlandt [58] showed that the mobile phase can contribute to the decomposition of certain metal cyano complexes. The dissociation of these complexes is caused by ligand-exchange processes that are caused by ehtylenediamine as the eluent component. When the amine concentration in the mobile phase is reduced, a correspondingly lower dissociation of the nickel, copper, silver, and mercury complexes is observed. At an amine concentration of $c = 4.4 \cdot 10^{-4}$ mol/L, this dissociation never occurs. However, the reduction in the amine concentration is accompanied by a significant peak broadening and tailing of the signal for free cyanide, which my be suppressed by addition of small quantities of cyanide (about 50 µg/L) to the mobile phase. Using a modified eluent that consists of

$1.00 \cdot 10^{-3}$ mol/L sodium carbonate +
$1.00 \cdot 10^{-2}$ mol/L sodium dihydrogenborate +
$1.02 \cdot 10^{-6}$ mol/L sodium cyanide +
$4.40 \cdot 10^{-4}$ mol/L ethylenediamine,

hydrogen cyanide (HCN), cyanide ions (CN^-), and metal cyano complexes with formation constants of log $K_n < 20$ are detected ion chromatographically under the general term *free* cyanide. In contrast to titrimetric and spectrophotometric methods, no elaborate sample pretreatment is required. That does not apply to the determination of so-called *easily releasable cyanide*, where sample preparation is indispensable. For the determination of this parameter, which is only defined via this method, an air flow is passed through the sample solution that is adjusted to pH 4. The gaseous hydrocyanic acid released under these conditions is then trapped in a strongly alkaline absorption solution [59]. Usually, sodium hydroxide at a concentration c = 1 mol/L is used for this. Solutions being that basic cannot be analyzed for their cyanide content using the chromatographic conditions given in Fig. 3-49. Due to its limited capacity, the anion exchanger would be totally overloaded by the high hydroxide ion concentration in the sample solution. Significant interferences in the range of the void volume would result. This problem can be circumvented when sodium hydroxide used as the absorption solution is also used as the eluent. An anion exchanger with a relatively high capacity should be employed to allow working with comparatively high concentrations. Therefore, a separator column of the IonPac AS6 (CarboPac) type is suited on which cyanide is strongly retained even in the presence of high hydroxide concentration in the mobile phase. However, the retention time can be significantly reduced by the addition of acetate. The ethylenediamine also added does not interfere in this case, since the sample solution only contains cyanide ions due to the kind of sample preparation. Fig. 3-50 shows the chromatogram of a sulfide and cyanide standard obtained with this method, that was directly injected in a sodium hydroxide matrix at a concentration c = 1 mol/L. As can be seen in the chromatogram, the cyanide signal is not affected by the matrix.

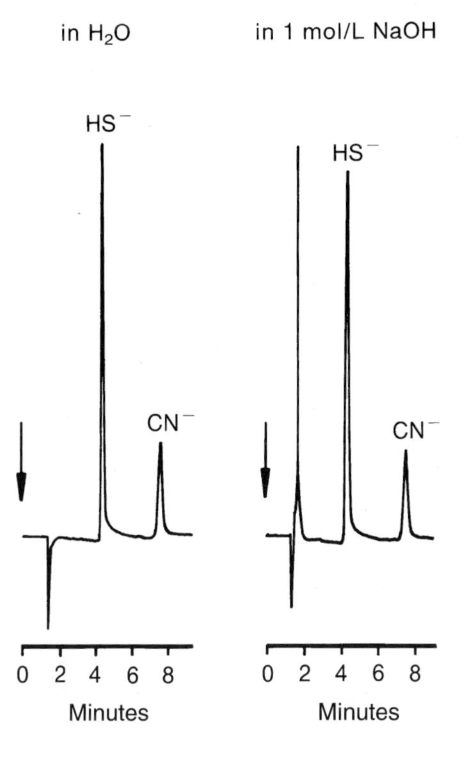

Fig. 3-50. Separation of sulfide and cyanide in a strongly alkaline solution. – Separator column: IonPac AS6 (CarboPac); eluent: 0.1 mol/L NaOH + 0.5 mol/L NaOAc + 0.0075 mol/L ethylenediamine; flow rate: 1 mL/min; other chromatographic conditions: see Fig. 3-49.

Although the sample preparation is the rate-determining step in this analysis, the ion chromatographic method offers the advantage of being less susceptible to interferences than the conventional wet-chemical determinations mentioned above. In addition, it can be easily automated and is applicable to the determination of total cyanide, whereby the cyanide contained in the sample is released in different ways but collected in the same absorption solution [59].

In addition to sulfide and cyanide, bromide and sulfite are eluted under the chromatographic conditions given in Fig. 3-49. They can be detected at a slightly more positive oxidation potential. Of special importance is sulfite, the wet-chemical determination of which is exceptionally problematic. However, caution is also necessary in case of the ion chromatographic determination since, in the presence of oxygen, sulfite is subject to autoxidation to yield sulfate. Therefore, it must be stabilized both in the sample – preferably directly at the time of sampling – and in the standard solution. Of all compounds that have been investigated in the past for their suitability as stabilizing agent for sulfite, formaldehyde has proved to be particularly effective [60,61]. Although the stabilizing effect is not fully understood, a nucleophilic addition of the hydrogensulfite anion with subsequent proton migration under formation of hydroxymethanesulfonic acid is assumed [62]:

$$H_2C{=}O + HSO_3^- \rightleftharpoons \left[H_2C(O^{\ominus})(SO_3H) \longleftrightarrow H_2C(OH)(SO_3^{\ominus}) \right] Na^+ \qquad (70)$$

The latter seems to hydrolyze completely in the alkaline pH region above pH 10, since commercially available hydroxymethanesulfonic acid and sulfite have identical retention times under the chromatographic conditions given in Fig. 3-44 and Fig. 3-49. Only in acid pH, a significantly shorter retention time that is typical for short-chain sulfonic acids is obtained for hydroxymethanesulfonic acid. Formaldehyde seems to have no effect on the chromatographic behavior of sulfite when basic eluents are used. The stabilizing effect is nevertheless very high. Thus, a costly removal of oxygen from sample and eluent is not necessary when using formaldehyde as the stabilizing agent. Even when a stabilized sulfite solution is purged with air for about 15 minutes, no significant loss in the sulfite content is observed [61]. Only small quantities of formaldehyde are required to produce a stabilizing effect. Adding 0.5 mL of a 37% formaldehyde solution per liter sample solution has proved successful. Remarkably, higher amounts lead to a slight reduction in the sulfite retention times, an effect which, at present, is also not understood. In stabilized conditions, sulfite standard solutions with a content of 1000 ppm are stable for about 72 h. Standard solutions with lower sulfite content, however, should be prepared daily from a stock solution. For quantitative sulfite analyses, it should be remembered that even reagent-grade sodium sulfite contains small amounts of sulfate, which have to be taken into account in the calibration of both compounds. The chromatogram in Fig. 3-51 shows that the purity of the reference standard may be easily determined via ion chromatography.

As can be seen in the chromatogram in Fig. 3-44, the most important mineral acids can be eluted under isocratic conditions using a mixture of sodium carbonate and sodium bicarbonate. This does not apply to polarizable anions which, because of adsorp-

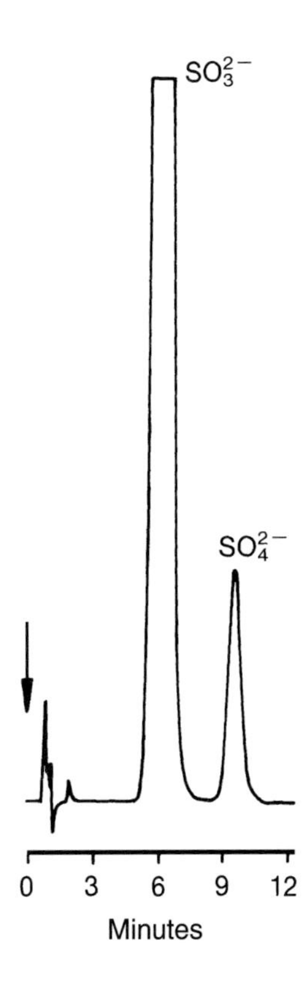

Fig. 3-51. Determination of sulfate in sodium sulfite. – Separator column: IonPac AS3; eluent: 0.0028 mol/L $NaHCO_3$ + 0.0022 mol/L Na_2CO_3; flow rate: 2.3 mL/min; detection: suppressed conductivity; injection: 50 µL, 500 mg/L sulfite.

tion effects (see Section 3.2), are more strongly retained by the stationary phase. However, the retention can be markedly reduced by adding organic additives to the mobile phase. For example, if *p*-cyanophenol, which is especially suited for this purpose, is added to the mixture of sodium carbonate and sodium bicarbonate, certain polarizable anions (e.g., monovalent iodide and tetrafluoroborate) may be detected in the same run together with the anions relevant for conventional water analysis. The resulting chromatogram is shown in Fig. 3-52. At first sight, such a separation seems to be of pure academic interest, but it is applied in the simultaneous determination of boron (as BF_4^-) and phosphorus (as PO_3^{3-} and PO_4^{3-}) in borophosphorosilicate glass films [63,64], which is of great interest for the semiconductor industry. Characteristic for the separation in Fig. 3-52 is the pronounced tailing of the iodide peak, which illustrates the compromise with regard to peak shape and analysis time in the simultaneous determination of anions with strongly different retention characteristics. A separation of bromide and nitrate is impossible under the chromatographic conditions of Fig. 3-52, since this separation is based on the different adsorption behavior of both ions, which does not show to advantage when *p*-cyanophenol is added to the eluent.

The aromatic amino acid tyrosine has proved successful as an eluent for the simultaneous analysis of all halide anions, which also causes a reduction in the iodide retention at alkaline pH, while still allowing for the separation of bromide and nitrate in contrast to *p*-cyanophenol containing eluents. In the respective chromatogram obtained with tyrosine as the eluent, a reversed retention is observed for orthophosphate and sulfate. This is caused by the comparatively high pH value of the mobile phase. The

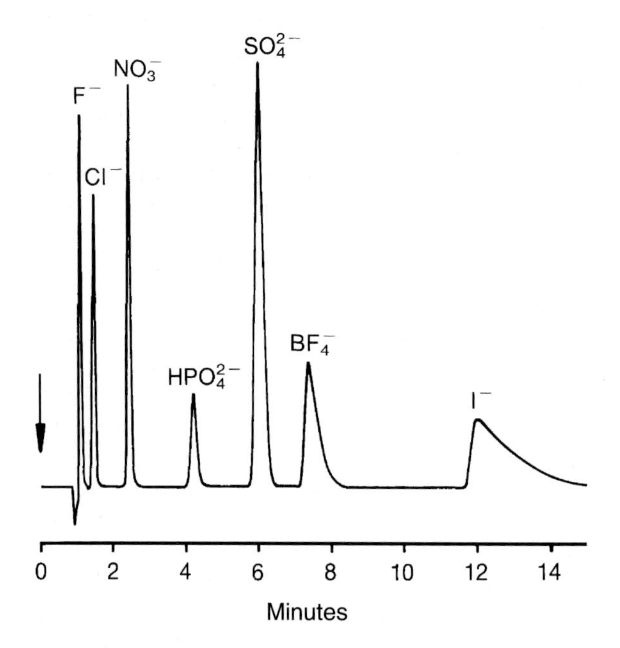

Fig. 3-52. Separation of polarizable and non-polarizable inorganic anions by the addition of *p*-cyanophenol. – Separator column: IonPac AS4A; eluent: 0.0017 mol/L $NaHCO_3$ + 0.0018 mol/L Na_2CO_3 + 100 mg/L *p*-cyanophenol; flow rate: 2 mL/min; detection: suppressed conductivity; injection volume: 50 μL; solute concentrations: 2 ppm fluoride, 4 ppm chloride, 10 ppm nitrate and orthophosphate, 20 ppm sulfate and tetrafluoroborate, 50 ppm iodide.

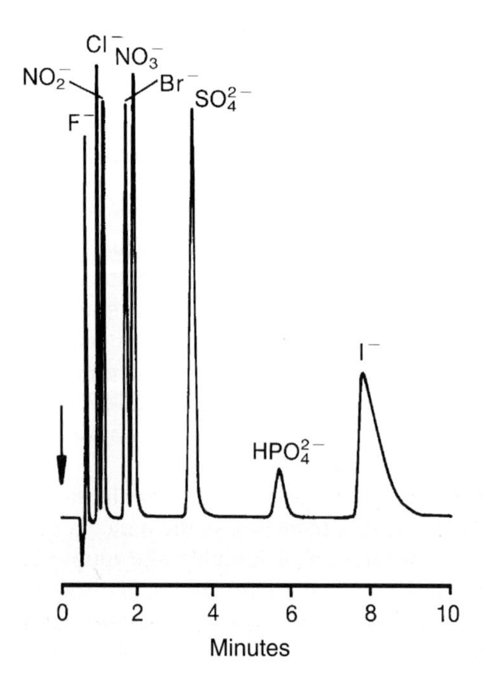

Fig. 3-53. Separation of halide ions using tyrosine as the eluent. – Separator column: IonPac AS4A; eluent: 0.001 mol/L tyrosine + 0.003 mol/L NaOH; flow rate: 2 mL/min; detection: suppressed conductivity; injection volume: 50 μL.

tailing of the iodide signal with tyrosine as the eluent is not completely suppressed, but due to the shorter analysis time is less pronounced than in Fig. 3-52. As shown in Fig. 3-53, α-amino acids may be regarded as alternatives to eluents with other organic additives. However, not all of the potentially suitable compounds are applicable without any restriction, because some of the amino acids do not possess the necessary purity.

A common problem in the isocratic analysis of inorganic anions is the verification of fluoride that elutes near the void volume. Under the chromatographic conditions in Fig. 3-44, monocarboxylic acids are only partly separated or not at all from the fluoride peak. Special caution is necessary in environmental samples (waste waters, ground waters, pulping liquors) in which fluoride can often be detected but, on the other hand, which may contain a variety of organic acids. Since many of these co-elute with fluoride, an interprctation of the signals near the void volume is extremely difficult. Another reason for the appearance of the supposed fluoride signal is, for example, the large amount of bicarbonate ions present in mineral waters. If their concentration is higher than the total carbonate concentration in the mobile phase, a positive signal is obtained within the void volume of the column instead of the negative water dip. This positive peak is hardly distinguishable from a fluoride peak, hence, the fluoride determination is impossible if the negative water dip is missing. Diluting the sample with de-ionized water does not solve this problem, even if no sign of interference can be discerned from the appearance of the negative water dip resulting therefrom. This becomes evident when the peak height is determined in dependence of the bicarbonate concentration in solution. Therefore, increasing amounts of bicarbonate were added to aqueous fluoride standard solutions of a concentration c = 2 mg/L and the resulting fluoride peak heights were measured. The respective results are summarized in Table 3-12. The semi-logarithmic plot of peak height versus bicarbonate concentration in Fig. 3-54 clearly reveals that small amounts of bicarbonate in the sample lead to a significant decrease in the peak height for fluoride. (The same result is obtained by adding carbonate to

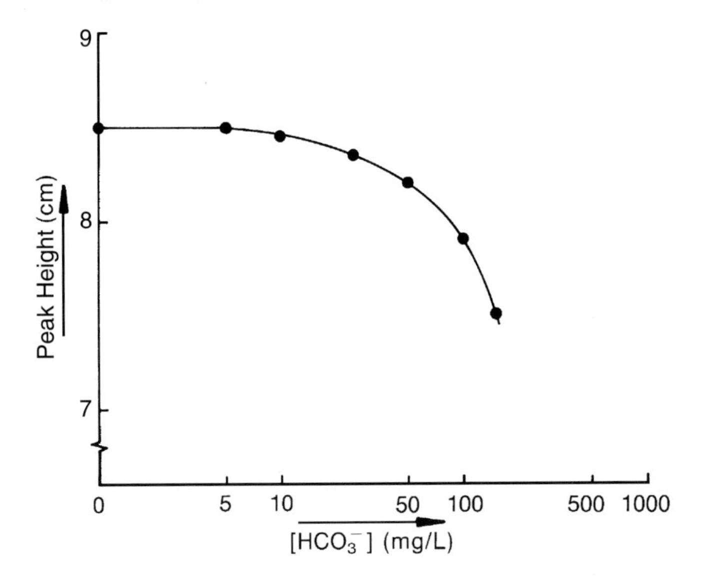

Fig. 3-54. Semi-logarithmic representation of the fluoride peak height versus the bicarbonate concentration being added.

fluoride solutions of defined concentration.) In addition, the percent reduction in peak height depends on the fluoride concentration of the sample. To illustrate this effect, four different fluoride standards were prepared in the concentration range between 0.5 and 3.0 mg/L, and 100 mg/L bicarbonate was added, respectively. The resulting peak heights were then compared with the values obtained with pure aqueous standards. The results in Table 3-13 show that the percent reduction of the fluoride peak height that is caused by the addition of bicarbonate is markedly augmented with increasing fluoride concentration. All these experiments substantiate that the determination of fluoride in real-world samples is problematic, if the fluoride is determined *in the same chromatographic run* together with other mineral acids.

Table 3-12. Fluoride peak height as a function of the carbonate content in solution.

$[CO_3^{2-}]$ [mg/L]	Peak height [cm]
0	8.50
5	8.50
10	8.45
25	8.35
50	8.20
100	7.90
150	7.50

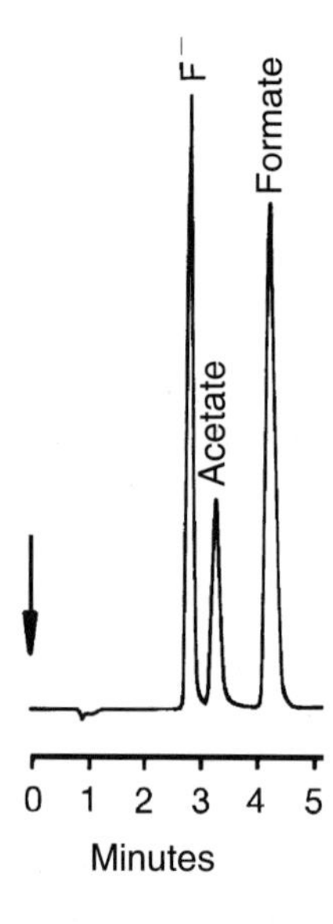

Fig. 3-55. Separation of fluoride, acetate, and formate. – Separator column: IonPac AS4A; eluent: 0.002 mol/L $Na_2B_4O_7$; flow rate: 2 mL/min; detection: suppressed conductivity; injection volume: 50 μL; solute concentrations: 3 ppm fluoride, 20 ppm acetate, and 10 ppm formate.

An exact quantification of fluoride is possible if the advantage of simultaneous analysis is eliminated, and if the chromatographic conditions are changed in the way that fluoride is separated from the carbonate and bicarbonate traveling with the mobile phase. An increase in fluoride retention is achieved by using an eluent of lower elution strength. A dilute solution of sodium tetraborate is particularly suited in this respect.

Table 3-13. Percent reduction of fluoride peak height upon addition of bicarbonate as a function of the fluoride content in solution.

Fluoride content [mg/L]	Percent reduction upon addition of 100 mg/L bicarbonate
0.5	3.9
1.0	3.9
1.5	4.8
2.0	9.8

As can be seen from the chromatogram in Fig. 3-55, a baseline separation between fluoride, acetate, and formate is obtained with this eluent. However, this method has only limited applicability for routine analyses. Ions such as chloride, nitrate, and sulfate have much longer retention times under these chromatographic conditions, which may lead to interferences with subsequent analyses. Therefore, the separator column must be flushed occasionally with carbonate solution of the concentration c = 0.1 mol/L to remove ions that are strongly retained at the stationary phase. However, the relatively long time required for the subsequent reconditioning of the separator column with the tetraborate eluent, which takes at least one to two hours, is a disadvantage.

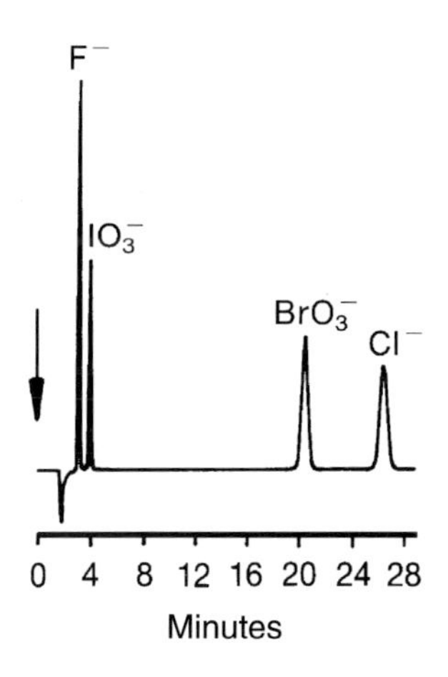

Fig. 3-56. Separation of fluoride, iodate, bromate, and chloride. – Separator column: IonPac AS6 (CarboPac); eluent: 0.0034 mol/L $NaHCO_3$ + 0.0036 mol/L Na_2CO_3; flow rate: 1 mL/min; detection: suppressed conductivity; injection volume: 50 µL; solute concentrations: 2 ppm fluoride, 10 ppm iodate, 20 ppm bromate, and 5 ppm chloride.

A much higher resolution in the retention range between fluoride and chloride is obtained by using a stationary phase of the IonPac AS6 (CarboPac) type. Compared with a conventional anion exchanger such as the AS4A, this separator column, that was initially developed for the analysis of carbohydrates, exhibits a significantly higher capacity. As a result, under the chromatographic conditions in Fig. 3-44 the chloride retention increases to more than 20 minutes resulting in an extremely high resolution between fluoride and chloride. Even the separation between fluoride and iodate and between bromate and chloride are possible under isocratic conditions, something generally considered to be difficult. A corresponding chromatogram is depicted in Fig. 3-56. Since the particle size of the AS6 column is 10 µm, the normal flow rate for AS4A columns of 2 mL/min was halved, while the concentration of the carbonate/bicarbonate mixture was doubled for chromatographic efficiency.

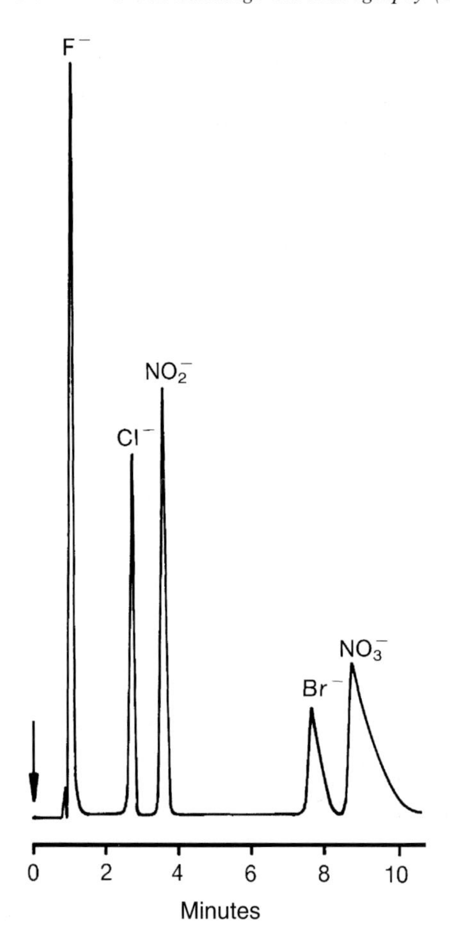

Fig. 3-57. Separation of inorganic anions using NaOH as the eluent. – Separator column: IonPac AS3; eluent: 0.001 mol/L NaOH; flow rate: 2.3 mL/min; detection: suppressed conductivity; injection: 50 μL anion standard solution.

The above examples reveal that ions in comparable as well as in strongly differing concentrations may be determined at the same time by selecting appropriate eluents and detection systems. The applicability of ion chromatographic methods is limited, however, by retention and concentration differences of two signals, as is generally the case for chromatographic techniques. Although anions such as chloride and sulfate, which elute far from each other, may still be separated and determined in a ratio of 100,000:1, the ratio for anions such as chloride and nitrite, which elute very close to each other, especially when conductivity detection is applied, is only 500:1. In those cases, the selectivity of the method may only be enhanced by combining different detection systems. Details on this will be discussed in Section 6.5.

Eluent Concentration and pH Value

In addition to the type of eluent, its concentration is one of the most important parameters affecting retention. Closely connected is the pH value of the mobile phase as a function of the eluent concentration. When a carbonate/bicarbonate mixture is selected as the eluent, parameters such as ionic strength and pH value cannot be considered separately, as the selectivity changes by varying the concentration ratio of the two compounds.

In general, the retention of monovalent and divalent ions shifts forward as the eluent concentration increases. On the other hand, a change in pH primarily affects the retention behavior of multivalent ions, since the valency of such ions depends on the pH value of the mobile phase. This effect is illustrated using orthophosphate as an example of the dissociation which occurs in three steps [65]:

$$H_3PO_4 \rightleftharpoons H^+ + H_2PO_4^- \qquad pK_1 = 2{,}16 \tag{71}$$

$$H_2PO_4^- \rightleftharpoons H^+ + HPO_4^{2-} \qquad pK_2 = 7{,}21 \tag{72}$$

$$HPO_4^{2-} \rightleftharpoons H^+ + PO_4^{3-} \qquad pK_3 = 12{,}33 \tag{73}$$

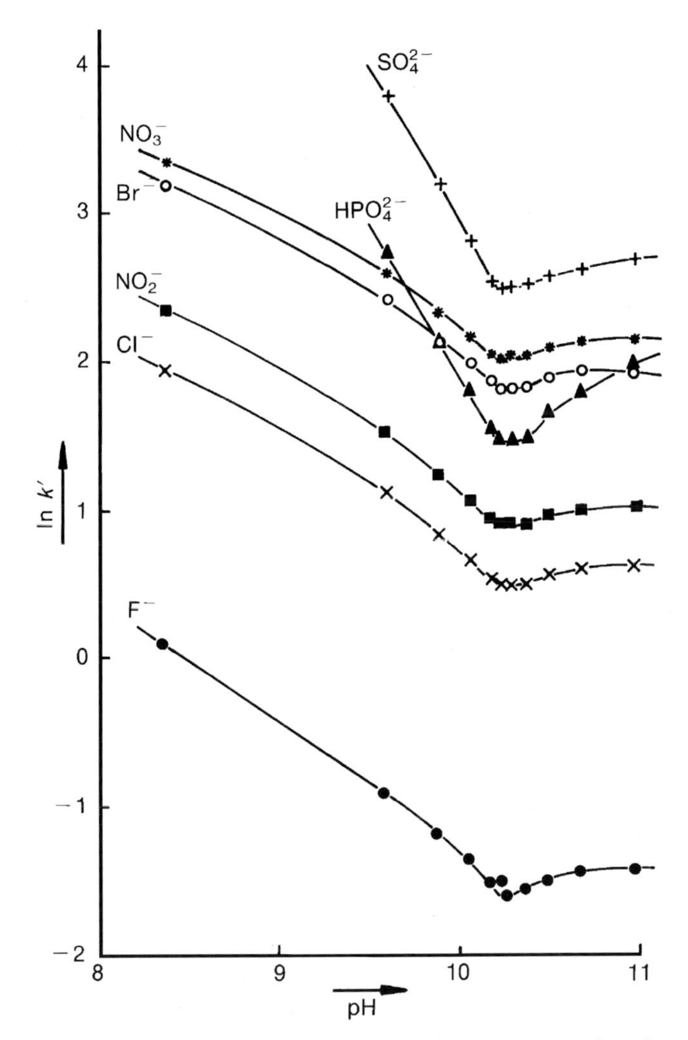

Fig. 3-58. Dependence of the retention of various inorganic anions as a function of eluent pH after adjustment via different carbonate/bicarbonate concentrations. – Separator column: IonPac AS4; flow rate: 2 mL/min; detection: suppressed conductivity; injection: 50 μL anion standard.

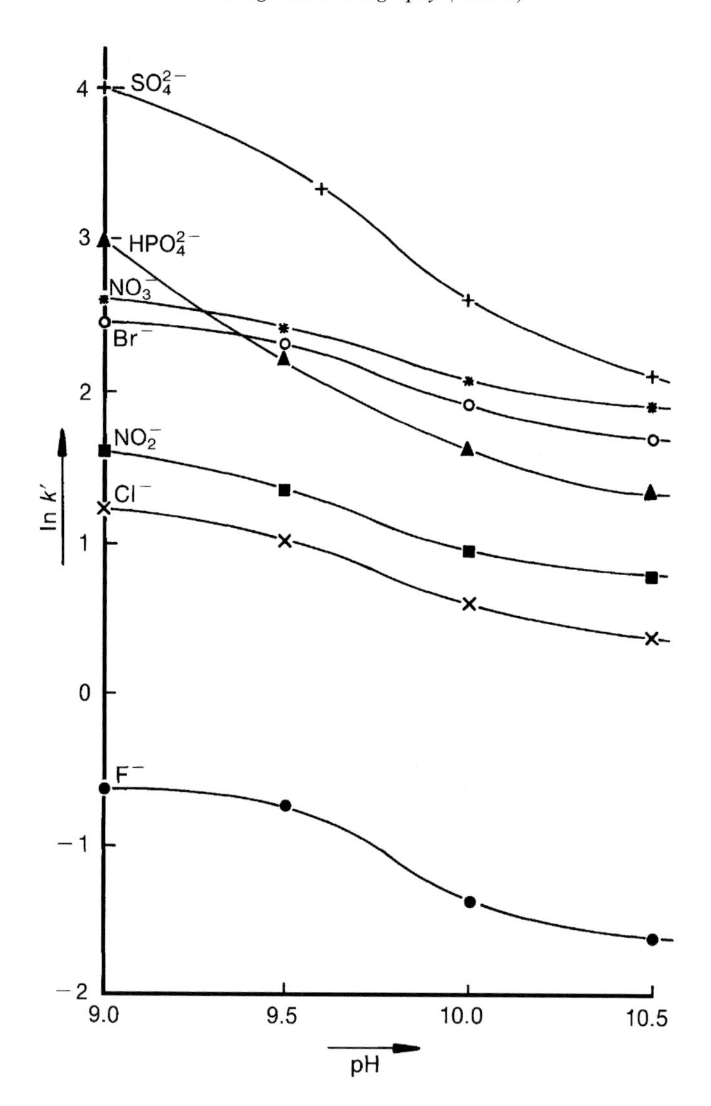

Fig. 3-59. Dependence of the retention of various inorganic anions on the pH value of the buffer mixture at constant ionic strength. – Separator column: IonPac AS4; eluent: 0.0028 mol/L $NaHCO_3$ + 0.0022 mol/L Na_2CO_3, pH adjusted with H_3BO_3/NaOH; flow rate: 2 mL/min; detection: suppressed conductivity; injection: 50 μL anion standard.

An ion chromatographic separation of the three orthophosphate species is impossible because of these p*K* equilibria. Using the weak hydroxide ion as the eluent in the form of NaOH, the pH value increases as the concentration increases. Fig. 3-57 shows a chromatogram of the seven standard anions obtained with an eluent concentration of c = 0.001 mol/L NaOH (pH 11). As can be seen, orthophosphate and sulfate do not elute under these conditions, primarily because the concentration of the monovalent eluent ions is too low for divalent analyte ions to elute. Moreover, at pH 11 about 10% of the orthophosphate exist as PO_4^{3-} ions that have a much longer retention time than HPO_4^{2-} ions. When the sodium hydroxide concentration is increased, the monovalent and divalent ions are shifted forward in retention; orthophosphate, however, elutes later, since the pH value and, thus, the percentage of trivalent orthophosphate ions increases with increasing sodium hydroxide concentration.

A more complex situation is the carbonate/bicarbonate buffer mixture. Due to the dissociation equilibrium,

$$HCO_3^- \rightleftharpoons H^+ + CO_3^{2-} \qquad pK_2 = 10.31 \tag{74}$$

the ratio of carbonate and bicarbonate changes at constant total carbonate concentration, as the pH value of the solution is lowered or raised by adding boric acid or sodium hydroxide, respectively. To investigate these phenomena in more detail – starting from standard conditions (0.0028 mol/L $NaHCO_3$ + 0.0022 mol/L Na_2CO_3) – the carbonate concentration was varied between $c = 0.5$ and $2.22 \cdot 10^{-3}$ mol/L at a constant bicarbonate concentration of c = 0.0028 mol/L. Conversely, the carbonate concentration was kept constant while the bicarbonate concentration was varied accordingly. The capacity factors of the seven standard anions were determined under these chromatographic conditions. Plotting ln k' versus the pH values that are calculated from the carbonate/bicarbonate concentrations according to Eq. (75) gives the correlations depicted in Fig. 3-58.

$$pH = pK_2 + \log \frac{[CO_3^{2-}]}{[HCO_3^-]} \tag{75}$$

Table 3-14. ln k' values of inorganic anions at different pH values of the mobile phase. (See Fig. 3-58 for chromatographic conditions)

c_{eluent}	pH	ln k'						
		F^-	Cl^-	NO_2^-	HPO_4^{2-}	Br^-	NO_3^-	SO_4^{2-}
2.8 mM HCO_3^-	8.35	0.11	1.96	2.35	--	3.19	3.35	--
2.8 mM HCO_3^- 0.5 mM CO_3^{2-}	9.58	−0.89	1.13	1.52	2.72	2.42	2.60	3.79
2.8 mM HCO_3^- 1.0 mM CO_3^{2-}	9.88	−1.17	0.84	1.23	2.13	2.11	2.33	3.19
2.8 mM HCO_3^- 1.5 mM CO_3^{2-}	10.06	−1.35	0.67	1.07	1.79	1.99	2.17	2.82
2.8 mM HCO_3^- 2.0 mM CO_3^{2-}	10.18	−1.51	0.54	0.94	1.53	1.87	2.05	2.55
2.8 mM HCO_3^- 2.2 mM CO_3^{2-}	10.23	−1.51	0.51	0.91	1.47	1.83	2.02	2.48
2.2 mM CO_3^{2-} 0.5 mM HCO_3^-	10.98	−1.43	0.63	1.03	1.94	1.93	2.16	2.69
2.2 mM CO_3^{2-} 1.0 mM HCO_3^-	10.68	−1.51	0.61	1.00	1.78	1.94	2.14	2.62
2.2 mM CO_3^{2-} 1.5 mM HCO_3^-	10.50	−1.56	0.57	0.97	1.64	1.90	2.10	2.58
2.2 mM CO_3^{2-} 2.0 mM HCO_3^-	10.38	−1.56	0.51	0.90	1.48	1.84	2.03	2.44
2.2 mM CO_3^{2-} 2.5 mM HCO_3^-	10.28	−1.61	0.51	0.92	1.48	1.84	2.04	2.48

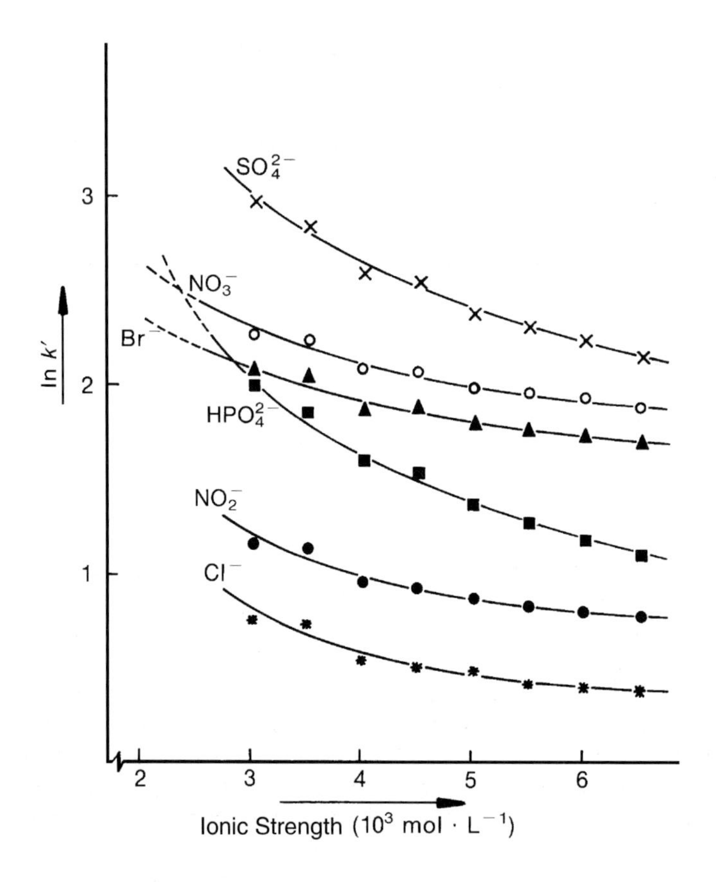

Fig. 3-60. Effect of ionic strength on the retention behavior of various inorganic anions at constant pH value. – Separator column: IonPac AS4; flow rate: 2 mL/min; detection: suppressed conductivity; injection: 50 μL anion standard.

The respective data are summarized in Table 3-14. Clearly, the retention times of all anions investigated decrease with increasing addition of carbonate at a constant bicarbonate content of the solution (pH 8.35 to 10.23). However, when bicarbonate is added at constant carbonate concentrations (pH range 10.98 to 10.28), the retention times hardly decrease. This effect is understandable inasmuch as bicarbonate exhibits only a small elution power. The decisive factor for the retention behavior of multivalent ions is the pH value resulting from the concentration ratio of both eluent components. An inspection of Fig. 3-58 reveals that bromide and nitrate are superimposed by orthophosphate within a very narrow pH range between 9.6 and 10.0. Interferences also occur at pH > 10.8. In both cases, this may be attributed to the dissociation equiblibria of orthophosphoric acid.

Fig. 3-59 shows the retention behavior of inorganic anions changing when the ionic strength is kept constant and the pH value is changed by adding boric acid or sodium hydroxide. The relevant data are summarized in Table 3-15. As expected, the retention is shifted forward with increasing pH value. However, adding boric acid increases the retention times by shifting the dissociation equilibrium of carbonate to the bicarbonate site. Interferences in the determination of bromide and nitrate by orthophosphate are observed between pH 9.0 and 9.5.

Table 3-15. ln k' values of inorganic anions at different pH values but constant ionic strength. (See Fig. 3-59 for chromatographic conditions)

Species	ln k'	ln k'	ln k'	ln k'
F^-	−0.63	−0.73	−1.39	−1.66
Cl^-	1.22	1.00	0.57	0.34
NO_2^-	1.59	1.36	0.92	0.75
HPO_4^{2-}	2.98	2.19	1.59	1.30
Br^-	2.44	2.30	1.87	1.68
NO_3^-	2.61	2.40	2.06	1.87
SO_4^{2-}	4.00	3.33	2.59	2.07
pH	9.0	9.5	10.0	10.5

Figures 3-59 and 3-60 clearly reveal that the influence of the pH value on the retention behavior of inorganic anions can only be investigated indirectly, since changes in the pH value also affect the dissociation equilibrium between carbonate and bicarbonate. Keeping the carbonate/bicarbonate ratio constant allows one to study the retention behavior in dependence of the ionic strength, and to draw conclusions about the pH influence. The total carbonate content was varied between $3 \cdot 10^{-3}$ and $6 \cdot 10^{-3}$ mol/L without changing the carbonate/bicarbonate ratio. Plotting the ln k' values as a function of ionic strength produces the dependences shown in Fig. 3-60. The result of the pH value of 10.28, established by the concentration ratio between carbonate and bicarbonate, is that in addition to divalent sulfate ions, orthophosphate elutes as HPO_4^{2-} ions. As expected, ionic strength exerts a much stronger effect on the retention behavior of the two divalent ions than of the monovalent ions. When the dependences shown in Fig. 3-60 for the ions orthophosphate, bromide, and nitrate are extrapolated to low ionic strength, interferences result for the bromide and nitrate signal due to orthophosphate at an ionic strength between $2 \cdot 10^{-3}$ and $3 \cdot 10^{-3}$ mol/L. At much higher ionic strengths, a reversal in the elution order between nitrate and sulfate may be expected.

Table 3-16. ln k' values of inorganic anions at different ionic strengths but constant pH value. (See Fig. 3-60 for chromatographic conditions)

Ionic strength [10^3 mol/L]	ln k'						
	F^-	Cl^-	NO_2^-	HPO_4^{2-}	Br^-	NO_3^-	SO_4^{2-}
3.02	−1.26	0.76	1.15	2.00	2.07	2.27	2.97
3.52	−1.11	0.74	1.13	1.86	2.05	2.23	2.84
4.02	−1.50	0.55	0.96	1.60	1.89	2.08	2.59
4.52	−1.55	0.51	0.92	1.55	1.88	2.07	2.55
5.02	−1.35	0.49	0.88	1.37	1.80	1.98	2.38
5.52	−1.61	0.42	0.83	1.28	1.77	1.96	2.31
6.02	−1.50	0.40	0.80	1.18	1.73	1.93	2.23
6.52	−1.61	0.38	0.77	1.10	1.70	1.88	2.14

By comparing Figs. 3-59 and 3-60, it is concluded that the retention behavior of inorganic anions is only controlled by ionic strength. Changes in the pH value merely shift the concentration ratio between carbonate and bicarbonate, thus affecting the retention behavior of anions only indirectly.

3.3.4.3 Experimental Parameters Affecting Retention when Applying Direct Conductivity Detection

Choice of Eluent

As already discussed in Section 3.3.2, the choice of eluents suited for the application of direct conductivity detection has been the subject of numerous investigations [16,17,19,50]. Organic acids with an aromatic backbone generally exhibit good elution properties, since they have a high affinity toward the stationary phase of an anion exchanger and, thus, may be used at comparatively low concentrations. Also desirable is a fairly low background conductivity of the eluent which may be ensured by using aromatic carboxylic acids with their low equivalent conductances. In the application of direct conductivity detection, the most important criterion for choosing the eluent is the resulting difference in the conductances between eluent and analyte ions. Thus, the eluent pH value does not have to be close to the neutral point. Strongly alkaline eluents such as sodium hydroxide [23], with a high equivalent conductance, have been successfully employed. However, aqueous solutions of the sodium, potassium, and ammonium

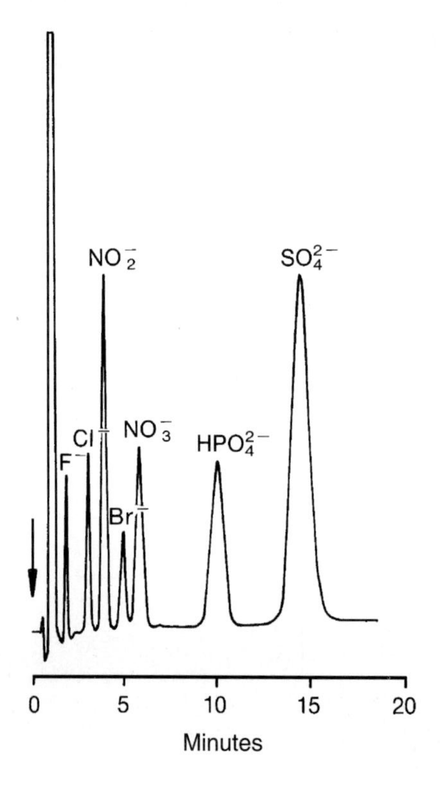

Fig. 3-61. Separation of inorganic anions using a borate/gluconate buffer as the eluent. – Separator column: TSK Gel IC-PW; eluent: 0.0013 mol/L $Na_2B_4O_7$ + 0.0058 mol/L H_3BO_3 + 0.0014 mol/L potassium gluconate/acetonitrile (88:12 v/v), pH 8.5; flow rate: 1.2 mL/min; detection: direct conductivity; injection: 100 µL anion standard (5 to 40 ppm); (taken from [66]).

salts of various organic acids in the concentration range between 10^{-4} and 10^{-3} mol/L are used. The pH value of the mobile phase is adjusted between 4 and 7, since it determines the dissociation of the organic acid and, thus, the retention behavior of the analyte species. In many cases, fully dissociated salts are employed as eluents, as are partially dissociated acids such as benzoic acid and partially neutralized acids such as potassium hydrogenphthalate. Mixtures of eluent ions with different valencies are also common. The pH value of the mobile phase is often adjusted with sodium borate, since the borate ions act as eluent ions. A well-known example is the mixture of gluconate and borate which is used in combination with polymethacrylate-based separator columns. An inorganic complex is formed that has a low background conductivity and that is especially effective as eluent. As shown in Fig. 3-61, the seven most important inorganic anions [55] can be analyzed in a single run. Although the separator column being used is characterized by a high chromatographic efficiency, a long analysis time is required for a baseline-resolved separation of all components. This is primarily due to the positive signal that appears in the void volume. This signal may not be separated from the subsequent fluoride signal without compromizing on the analysis time. Section 3.3.1.1 discussed the reason for the appearance of this signal that is typical for all ion chromatographic systems with direct conductivity detection.

A selectivity comparable to the gluconate/borate mixture is obtained with potassium hydrogenphthalate, the eluent most often used for the application of both direct conductivity detection and indirect UV detection. Depending on the type of stationary phase being used, the pH value of the mobile phase is adjusted between 5.3 and 7.0. In this pH region, phthalate exists mainly as a divalent anion. Fig. 3-62 displays an optimized separation of inorganic anions on Waters IC-PAK A. Potassium hydrogenphthalate is a slightly stronger eluent than the gluconate/borate mixture and, thus, is predestined for the analysis of divalent anions. Polarizable anions such as iodide and thiocyanate may also be separated from simple inorganic anions in the same run by using potassium hydrogenphthalate. Only orthophosphate cannot be determined under the chromatographic conditions given in Fig. 3-62.

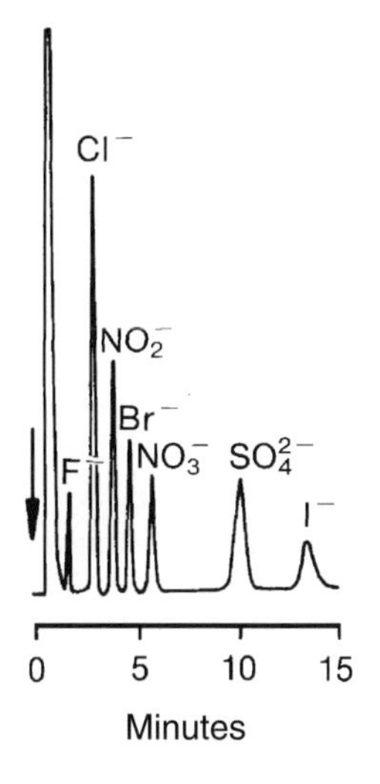

Fig. 3-62. Separation of inorganic anions using potassium hydrogenphthalate as the eluent. – Separator column: Waters IC-PAK Anion; eluent: 0.001 mol/L KHP, pH 7.0; flow rate: 2 mL/min; detection: direct conductivity; injection volume: 10 µL; solute concentrations: 100 ppm each.

The significant advantage of the carbonate/bicarbonate mixture described in Section 3.3.2 to elute monovalent and divalent anions in the same run is obtained upon application of direct conductivity detection by using *p*-hydroxybenzoic acid. When the pH value of the mobile phase is adjusted to 8.5, the carboxylic group is fully dissociated, while the hydroxide group with a p*K* value of 9.3 is only partially dissociated. This eluent also contains a mixture of monovalent and divalent eluent ions, so fluoride and sulfate can be analyzed in a single run within 15 minutes, as shown in Fig. 3-63 using a Wescan 269-029 polymeric stationary phase. The negative signal appearing after 25 minutes is called the *system peak*. The origin and significance of such signals is discussed below in more detail.

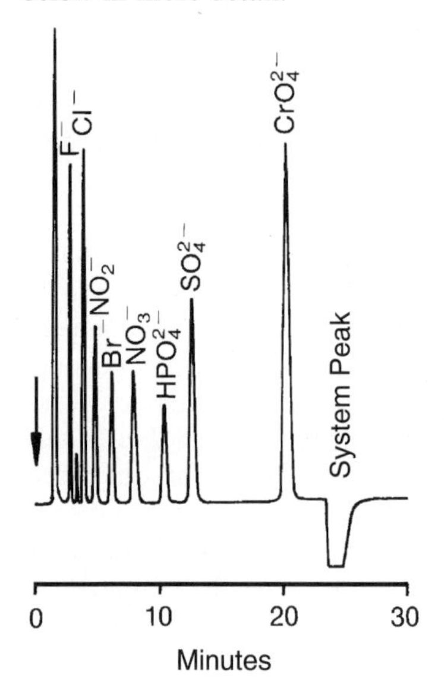

Fig. 3-63. Separation of inorganic anions using *p*-hydroxybenzoic acid as the eluent. – Separator column: Wescan 269-029; eluent: 0.004 mol/L PHBA, pH 8.7; flow rate: 1.5 mL/min; detection: direct conductivity; injection volume: 100 μL; solute concentrations: 5 ppm fluoride, chloride, nitrite, bromide, and nitrate, and 10 ppm orthophosphate and sulfate.

The potassium salt of citric acid has an elution strength comparable to potassium hydrogenphthalate. When fully dissociated, citrate as a trivalent ion exhibits a strong affinity toward the stationary phase and, therefore, may be used as an eluent in relatively low concentrations. However, the elution order of simple inorganic anions that is obtained with potassium citrate depends on the stationary phase being used. When, for example, a polystyrene/divinylbenzene phase is used, the elution order is the same as that obtained with potassium hydrogenphthalate. On the other hand, when using silica-based anion exchangers, Matsushita et al. [31] observed significantly shorter retention times for polarizable anions such as iodide and thiocyanate, which elute prior to the divalent sulfate (Fig. 3-64). The higher affinity of these two anions cannot be attributed to non-ionic interactions, which supersede the actual ion-exchange process.

With regard to the sensitivity of the conductivity detection, phthalate eluents have an advantage over citrate eluents, since the conductivity difference between eluent and analyte ions is much higher upon using salts of aromatic acids as the eluent. However, when the column effluent is passed through a suppressor system before it enters the conductivity cell, a much higher sensitivity is also obtained with potassium citrate eluents because of the incomplete dissociation of the corresponding acid formed in the suppressor (see also Section 3.3.3), which reduces the background conductivity, thus increasing the conductivity difference to the analyte ions correspondingly.

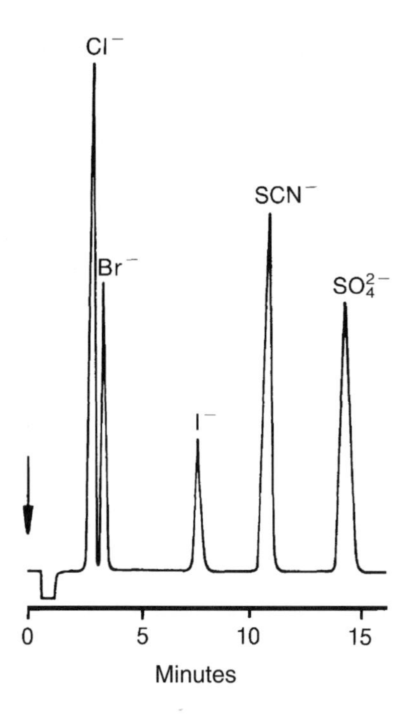

Fig. 3-64. Separation of inorganic anions using potassium citrate as the eluent. – Separator column: TSK Gel IC-SW; eluent: 0.001 mol/L potassium citrate, pH 5.2; flow rate: 1.2 mL/min; detection: direct conductivity; injection volume: 100 µL; solute concentrations: 5 ppm chloride, 10 ppm bromide, iodide, thiocyanate, and sulfate.

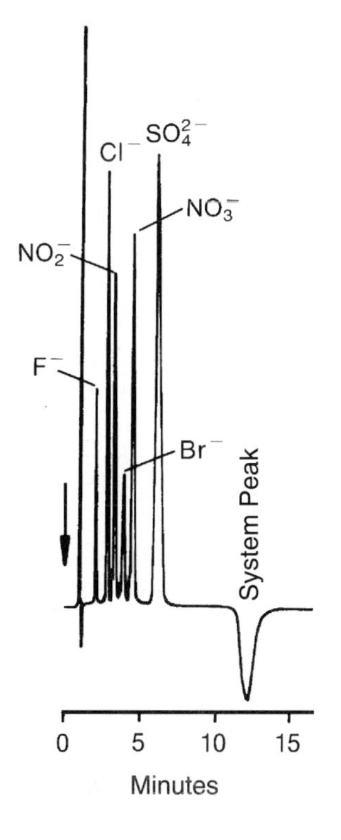

Fig. 3-65. Separation of inorganic anions using phthalic acid as the eluent. – Separator column: Shimpack IC-A1; eluent: 0.0025 mol/L phthalic acid + 0.0024 mol/L tris(hydroxymethyl)aminomethane, pH 4.0; flow rate: 1.5 mL/min; detection: direct conductivity; injection volume: 20 µL; solute concentrations: 5 ppm fluoride, 10 ppm chloride, 15 ppm nitrite, 10 ppm bromide, 30 ppm nitrate, and 40 ppm sulfate.

The above suggests that free carboxylic acids may act directly as eluents. The separation of inorganic anions at a Shimpack IC-A1 separator column (shown in Fig. 3-65) reveals how legitimate this assumption is. As impressively demonstrated in this chromatogram, obtained using free phthalic acid as the eluent, high chromatographic efficiency and relatively short retention times for the individual components are the two most striking features in this separation example. However, the total analysis time increases significantly in this case, since the system peak does not appear until about 12 minutes. Furthermore, the high elution power of the phthalate – either in the form of salt (see also Fig. 3-62) or as free acid – renders impossible the analysis of orthophosphate in the same run.

In some cases, however, the analysis of inorganic anions is of particular interest, especially in the acid pH range. Above all, this holds for anions that are not stable in the alkaline pH range. This includes, for example, the peroxomonosulfate, SO_5^{2-}, that, in contrast to the respective peroxodisulfate, $S_2O_8^{2-}$, decays rapidly at pH values above 5. Another field of application is the determination of fully dissociated anions in the presence of high amounts of weakly dissociated inorganic or organic acids. Since the latter are not dissociated at low pH and therefore travel with the eluent through the column, the separation from the more strongly retained anions is facilitated.

The concept of employing free acids as eluents dates back to the investigations of Fritz et al. [67]. In 1981, they already demonstrated that free benzoic acid may be employed as eluent instead of the sodium or potassium salts that were utilized until then. This is understandable, taking into account that benzoic acid is partially dissociated in aqueous solution. The degree of dissociation may be calculated from the dissociation constant K_a according to the Ostwald dilution law [54]:

$$K_a = \frac{\alpha^2}{1-\alpha} \cdot c_0 \tag{76}$$

c_0 Total acid concentration

The degree of dissociation decreases with increasing electrolyte concentration. With the dissociation constant for benzoic acid, $K_a = 6.25 \cdot 10^{-5}$, the degree of dissociation of a benzoic acid solution is calculated to be $\alpha = 0.2$ for the concentration of $c = 1.25 \cdot 10^{-3}$ mol/L used as an example. Therefore, such a solution contains $2.5 \cdot 10^{-4}$ mol/L of oxonium ions and benzoate ions, respectively. Comparing such a benzoic acid solution with a sodium benzoate solution of the concentration $c = 2.5 \cdot 10^{-4}$ mol/L, an almost identical elution behavior is observed.

With benzoic acid as the eluent, generally higher sensitivities are obtained for the analyte anions in comparison to sodium or potassium benzoate, respectively. This is illustrated in the following reaction scheme:

$$\begin{array}{ccc} \text{Resin}-\text{A}^- + \text{H}^+ + \text{Bz}^- & \rightleftharpoons & \text{Resin}-\text{Bz}^- + \text{H}^+ + \text{A}^- \\ \upharpoonleft\downharpoonright & & \\ \text{HBz} & & \end{array} \tag{77}$$

According to this scheme, analyte ions bound transiently to the fixed stationary charge are exchanged against an equivalent amount of benzoate ions, which leads to a corresponding reduction in the benzoate concentration under the chromatographic signal. However, benzoate ions are reformed from undissociated benzoic acid, since the system is in a state of dynamic equilibrium. With the degree of dissociation of $\alpha = 0.2$ for the benzoic acid concentration of $c = 1.25 \cdot 10^{-3}$ mol/L mentioned above, 80% of the oxonium ions and benzoate ions, respectively, are derived from the molecular benzoic acid, and are subsequently transformed into the fully dissociated species $H^{+}A^{-}$. The much higher equivalent conductance of the oxonium ion as counter ion for A^{-} is the actual reason for the sensitivity increase upon application of free benzoic acid as the eluent.

However, two disadvantages contrast this advantage. Weakly basic anions such as bicarbonate cannot be determined under these chromatographic conditions, since they are hardly dissociated or not dissociated at all at pH 3.6 in the mobile phase. Another limiting factor for the practical application is the long time required for column conditioning, which is due to the tendency of molecular acids to adsorb at the surface of the anion exchanger. Connected with this is the formation of system peaks, whose appearance in the chromatogram may be interpreted as a disturbance of the adsorption equilibrium.

Eluents based on sodium benzoate or benzoic acid, respectively, exhibit about the same elution power as bicarbonate and are thus used for the analysis of monovalent anions. When looking for less strongly adsorbing acids, Fritz, DuVal, and Barron [68] found other compounds that may also be employed as eluents. The properties of these compounds, that are important in this connection, are listed in Table 3-17.

Table 3-17. Properties of various organic acids that may be utilized as eluents [68].

Compound	pK_1	pK_2	pK_3	Degree of dissociation	Elution power
Nicotinic acid	4.87			11	weak
Benzoic acid	4.19			22	strong
Succinic acid	4.16	5.61		25	weak
Citric acid	3.14	4.77	6.39	58	weak
Fumaric acid	3.03	4.44		62	weak
Salicylic acid	2.97	13.40		63	strong

Several factors have to be taken into account when selecting an eluent. The elution power depends upon the acid's degree of dissociation, which determines the concentration of the acid anions acting as eluent ions. Important for the selectivity of the eluent is its affinity toward the stationary phase. Eluent ions with a structure comparable to the matrix of the stationary phase, are characterized by a high affinity. At the same time, they also effect a high selectivity due to intense interactions with the stationary phase. Although the various inorganic analyte ions are eluted by the organic acids listed in Table 3-17 at different speeds, depending on the acid's degree of dissociation, the investigations of Fritz et al. into the relative retention of the solutes [69] reveal only small differences between the various organic acids. However, in the individual case, small differences in the relative retention may be decisive for a successful separation.

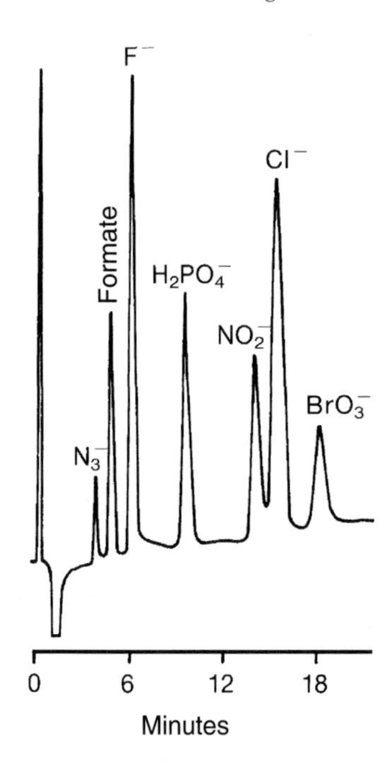

Fig. 3-66. Separation of monovalent anions using nicotinic acid as the eluent. – Separator column: Wescan 269-031; eluent: 0.01 mol/L nicotinic acid; flow rate: 2.7 mL/min; detection: direct conductivity; injection volume: 100 μL; solute concentrations: 20 ppm azide, 10 ppm formate and fluoride, 20 ppm orthophosphate, 10 ppm nitrite and chloride, 15 ppm bromate; (taken from [70]).

The chromatogram shown in Fig. 3-66, obtained using nicotinic acid as the eluent, serves as an example of the variety of organic acids that can be used as eluents. Nicotinic acid is most suitable for the separation of monovalent anions. In contrast to other monocarboxylic acids, it is characterized by a good water solubility and a low equivalent conductance. In addition, its propensity to adsorb at the stationary phase is very low, which mostly prevents the formation of the usual system peak. However, divalent anions are strongly retained under these chromatographic conditions. For their fast separation, eluents based on salicylic acid or trimesic acid are more suitable.

While the eluents described so far are applicable for the separation of a variety of inorganic (and organic) anions, they cannot be employed for the analysis of anions from weak inorganic acids such as sulfide, cyanide, borate, and silicate, since these anions are fully dissociated only in strongly alkaline solution. Therefore, the idea of using a strongly basic eluent for the separation of such anions seems to suggest itself. Corresponding procedures with indirect conductivity detection were introduced by Fritz et al. [71]. Since a sodium hydroxide solution conducts much stronger than the analyte ions, the latter appear in the chromatogram as negative signals. However, the proportionality between peak area and concentration still exists. Suppressor systems cannot be utilized in this case because they would convert the anions mentioned above into the corresponding acids which are not available for conductivity detection owing to their weak dissociation. Although sodium hydroxide fulfills all requirements for an indirect conductivity detection, its elution power frequently is insufficient to elute anions of practical relevance. The retention times are significantly shortened by adding small amounts of sodium benzoate to the sodium hydroxide solution which hardly affect the background conductivity of the eluent. However, the shorter retention times improve the peak shape, thus increasing the sensitivity of the method. Fig. 3-67 shows a corre-

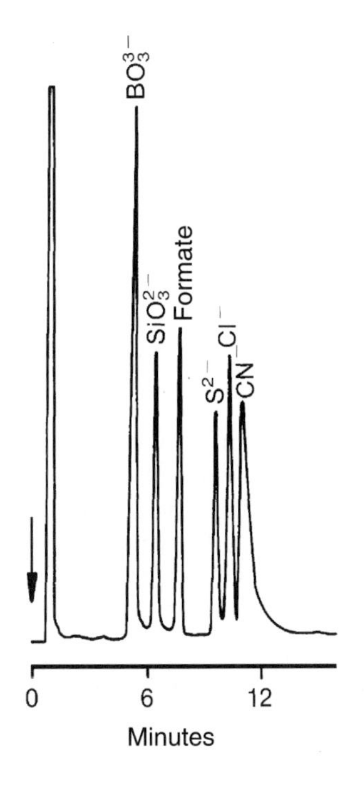

Fig. 3-67. Separation of anions derived from weak inorganic acids using a strongly basic eluent. – Separator column: Wescan 269-029; eluent: 0.004 mol/L NaOH + 0.0005 mol/L sodium benzoate; flow rate: 1.5 mL/min; detection: indirect conductivity; injection volume: 100 μL; solute concentrations: 5 ppm borate (as B), 10 ppm silicate (as SiO_2), 10 ppm formate and sulfide, 20 ppm chloride and cyanide; (system peak appears after 28 min.); (taken from [70]).

sponding separation. Also in this case, the appearance of a system peak after about 28 minutes significantly increases the total analysis time and is therefore disadvantageous. However, this is not the only factor suggesting the limited applicability of this method in routine analysis. In real-world samples, the environmentally relevant anions such as sulfide and cyanide occur only at small concentrations in the presence of high amounts of chloride and sulfate. Since divalent anions are strongly retained under the chromatographic conditions given in Fig. 3-67, a separator column must already be rinsed after only a few injections to avoid overloading the stationary phase by matrix components. Hence, in combination with selective detection modes such as amperometry (see Fig. 3-49), stronger eluents offer significant advantages at least for a routine analysis of such anions. For the analysis of borate and silicate, the reader is referred to ion-exclusion chromatography (see Chapter 4) as an alternative. In ion-exclusion chromatography, the selectivity increase is achieved by a different separation mechanism by which all strongly dissociated anions elute as a single peak within the void volume and, thus represent no interference.

System Peaks

The term system peak refers to signals that may not be attributed to solutes. System peaks are characteristic for ion chromatographic systems that have have no suppressor system when weak organic acids are used as the eluent. Despite numerous publications concerning this subject [72-76], system peaks were often the reason of misinterpretations. However, some facts about the thermodynamic and kinetic processes that occur within the separator column may be inferred [77,78] from their occurrence and help in understanding the chromatographic processes.

The occurrence of system peaks will be discussed, using as an example an anion exchanger that is in equilibrium with the benzoic acid (BA) used as the mobile phase. This equilibrium is maintained as long as the chromatographic system is not disturbed. The equilibrium processes of interest in this connection are illustrated in Fig. 3-68, and comprise:

- Equlibrium between the solute ions in the mobile phase and those that are bound to the fixed stationary charge,
- Dissociation equilibrium of benzoic acid in the mobile phase, and
- Equilibrium between the benzoic acid dissolved in the mobile phase and adsorbed on the hydrophobic surface of the stationary phase.

If this equilibrium is disturbed by a sample injection, a new equilibrium is established via relaxation; i.e., the kind of relaxation process depends on the pH value of the sample injected. If the sample pH is lower than the pH value of the mobile phase, benzoate ions in the mobile phase are protonated due to the sample injection. Thus, the concentration of molecular benzoic acid in the mobile phase increases as does the concentration of the amount adsorbed to the stationary phase. The amount not adsorbed travels through the column and appears as a chromatographic signal: the *system peak*. A qualitatively similar chromatogram is obtained when a sample containing the solute ions and the corresponding eluent component is injected into the system. However, only the position of the system peak is comparable, not its area and direction.

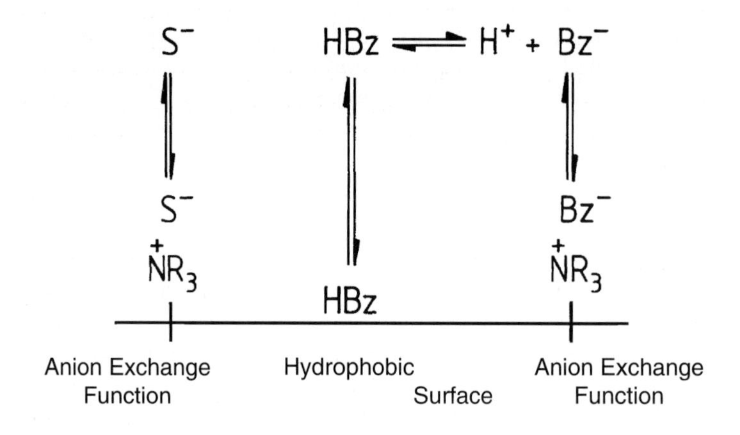

Fig. 3-68. Equilibrium processes between the stationary phase of an anion exchanger and benzoic acid as an example of a weak organic acid that acts as an eluent.

Correspondingly, injecting a sample with a pH higher than the mobile phase pH value shifts the dissociation equilibrium of benzoic acid to the side of the dissociated form. To maintain the equilibrium, molecular benzoic acid is desorbed from the surface of the stationary phase. The resulting deficit in undissociated acid is compensated by the mobile phase. This propagates itself through the column and, finally, a negative peak with a characteristic elution time appears. The migration velocity of the system peak, v_i, through the column is smaller than that of the mobile phase. It is defined according to Eq. (78) as:

$$v_i = R_i\, u \tag{78}$$

with R_i representing the equilibrium amount of the respective eluent component, i, in the mobile phase and u representing the velocity of the mobile phase. In a given chromatographic system, the velocity of the system peak is constant – independent of the kind of sample injected. When the mobile phase contains more than one organic acid, several system peaks are observed.

This explanation is corroborated by the fact that while no system peaks occur when fully dissociated eluent ions are used, their positions can be affected by the addition of organic solvents.

As mentioned, a number of important conclusions can be drawn from the occurrence of system peaks. In this connection, the investigations of Levin and Grushka [77,78] are worth mentioning. From the system peak area the authors derived the amount of eluent component adsorbed at the stationary phase and calculated the capacity ratio, k', from it. Remarkably, this approach allows one to calculate the capacity ratio without prior knowledge of the column dead time, t_{d}, of the separator column being used.

Levin and Grushka performed their studies using an ion-pair chromatographic system. A chemically bonded, reversed octadecyl phase was used as the stationary phase. The aqueous mobile phase contained an acetate buffer, copper ions, and sodium heptanesulfonate as the ion-pair reagent. Thus, three system peaks are obtained in addition to the negative water dip, when pure water is injected. The attribution of each individual eluent component is facilitated because the area of a system peak changes when different concentrations of the respective eluent component are injected as a sample, or when the mobile phase contains different concentrations of it. The desorption of the eluent component that is initially adsorbed at the stationary phase upon injection of pure water occurs in a volume that approximately corresponds to the injection volume. The knowledge of the real volume is insignificant in this respect, since these quantities are put in relation to each other for the calculation of the capacity ratio. However, the transformation of area units into concentration units is of importance for calculating k'. For this, Levin and Grushka prepared a calibration curve by injecting different concentrations of the eluent component at defined concentrations of this component in the mobile phase. When the resulting peak area is plotted versus the injected concentration, a straight line with a positive intercept is obtained. When these measurements are repeated with different concentrations of the eluent component in the mobile phase, straight lines are also obtained and only differ in their intercept. The latter represents the system peak area at injection of pure water. The pertinent concentration of eluent component desorbed from the stationary phase upon water injection is determined via extrapolation of the calibration curve to the abscissa. It corresponds to the absolute value of the negative abscissa intercept. Provided that desorption of adsorbed component is complete, the concentration of the adsorbed component, w_{s}, is thus obtained. Since the concentration of the eluent component in the mobile phase, w_{m}, is known, k' can be calculated according to Eq. (79):

$$k' = \frac{w_{\mathrm{s}}}{w_{\mathrm{m}}} \tag{79}$$

Since the capacity ratio is typically determined according to Eq. (80) from the chromatographic data, inversely, the column dead time may be calculated when k' is known.

$$k' = \frac{t_{ms} - t_d}{t_d} \tag{80}$$

t_d Column dead time
t_m Gross retention time of the system peak

Transforming Eq. (80) to t_d yields

$$t_d = \frac{t_{ms}}{1 + k'} \tag{81}$$

Using this method for the determination of the dead time circumvents the great difficulties that are connected with an exact measurement of the dead volume.

In addition to the usefulness of the information that may be deduced from the system peaks, it is important for the practising analyst to separate these signals from the analyte signals. Some options for optimizing the separation are described below.

Eluent Concentration and pH Value

Theoretical considerations regarding the retention behavior of inorganic anions in ion chromatographic systems without a suppressor device were performed by Jenke and Pagenkopf [79]. Several models were investigated with regard to their suitability for a mathematical description of the retention behavior. Another paper by Jenke and Pagenkopf [80] described options for optimizing the chromatographic conditions by varying the pH value while keeping the eluent concentration constant, and by varying the eluent concentration at constant pH value.

Straight lines are obtained for the various inorganic anions when the eluent concentration is varied at constant pH value and when the logarithm of the net retention volume is plotted versus the logarithm of the eluent concentration. The slope of the straight line depends on the type of eluent and analyte ion, S^{y-}, respectively. Fig. 3-69 illustrates this effect on the eluents and solute ions investigated by Small [81]. The coelution of some anions and the reversal of the elution order under specific chromatographic conditions is remarkable. The retention behavior is described mathematically as:

$$\log V_s \propto -m \cdot \log[E^{x-}] \tag{82}$$

V_s Net retention volume
$[E^{x-}]$ Concentration of eluent ions

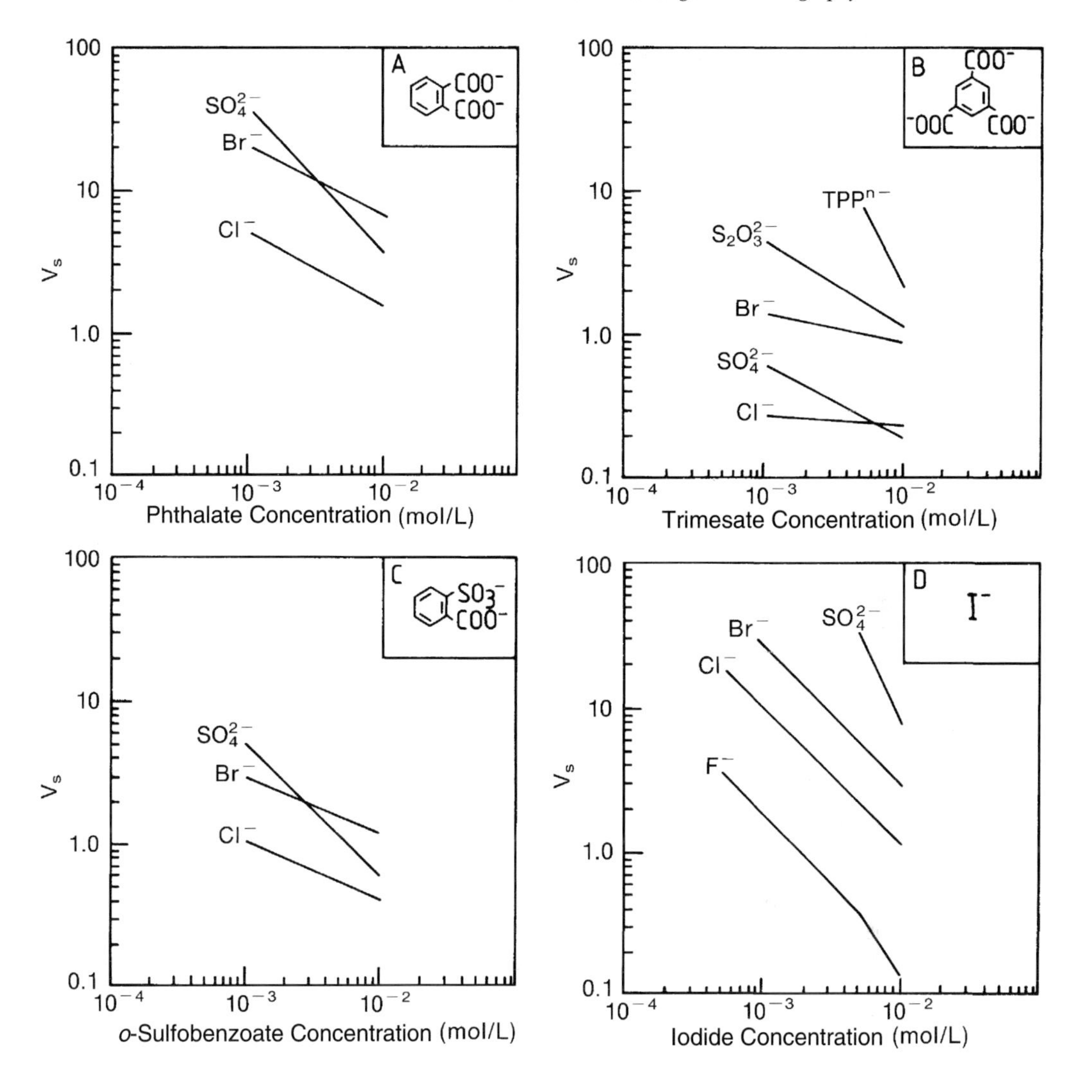

Fig. 3-69. Dependence of the net retention volume of various inorganic anions on the concentration of a) phthalate, b) trimesate, c) *o*-sulfobenzoate, and d) iodide. – Separator column: 250 mm × 2.8 mm I.D. SAR-40-0.6; (taken from [81]).

After transformation [82] one obtains:

$$\log V_s = \text{Konstante} - \frac{y}{x} \cdot \log[E^{x-}] \tag{83}$$

x Charge of the eluent ion
y Charge of the analyte ion

According to Eq. (83), the slope of the straight line is porportional to the quotient of the charge numbers of eluent and analyte ions, respectively. When the calculated quotient y/x is compared with the measured values (Table 3-18), a good agreement is obtained for the eluents and solute ions studied by Small. Provided that the charge numbers of eluent and solute ions are known, it suffices to determine the retention volume of a compound at a single eluent concentration to predict the retention of this species at other concentrations. Furthermore, the dependence of the slope of the straight line explains the observed reversal of the elution order. Based on these findings, co-elutions can be prevented by choosing appropriate chromatographic conditions.

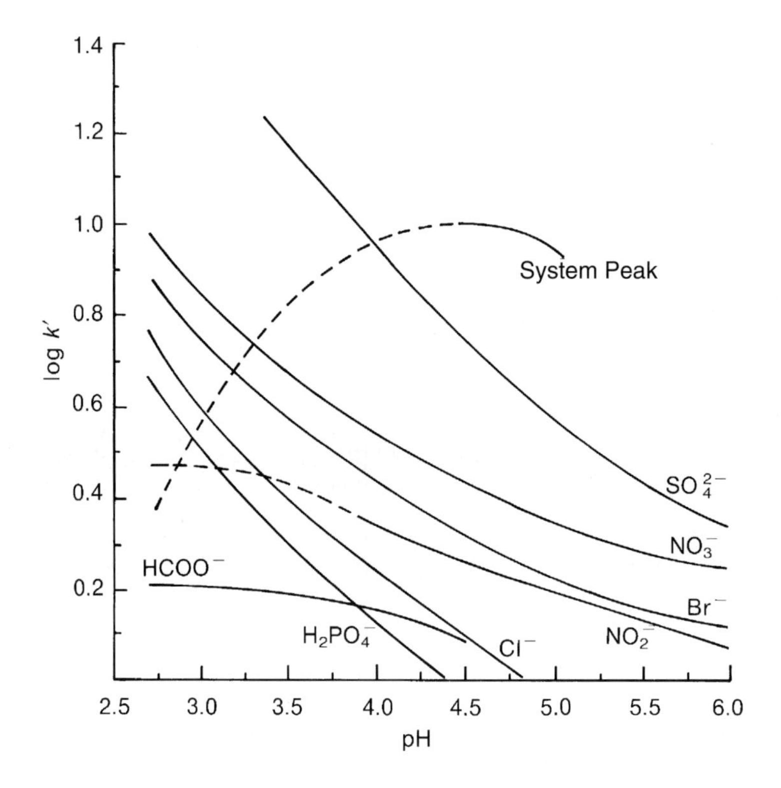

Fig. 3-70. pH dependence of the retention of various inorganic anions with a phthalate eluent of the concentration $c = 0.004$ mol/L; (taken from [70]).

Table 3-18. Comparison of theoretical and measured values for the quotient y/x (taken from [81]).

Eluent ion, E^{x-}	Analyte ion, S^{y-}	$y/x_{theor.}$	$y/x_{meas.}$
I^-	Br^-	1.00	1.00
I^-	SO_4^{2-}	2.00	2.00
Trimesate^{3-}	Br^-	0.33	0.21
Trimesate^{3-}	SO_4^{2-}, $S_2O_3^{2-}$	0.66	0.60
Phthalate^{2-}	Br^-	0.50	0.47
Phthalate^{2-}	SO_4^{2-}	1.00	0.98

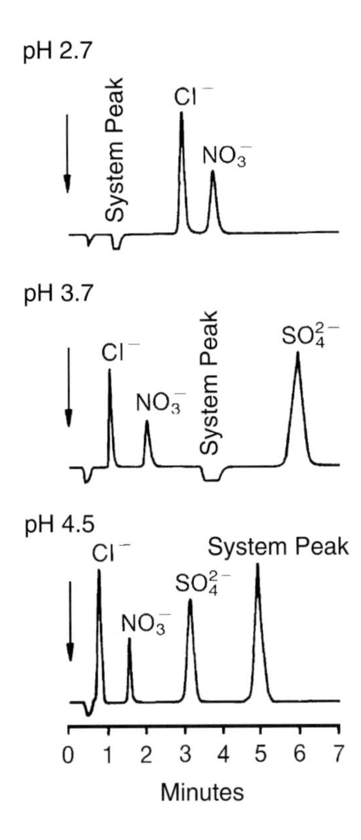

Fig. 3-71. Comparison of the retention of various inorganic anions with a phthalate eluent of the concentration c = 0.004 mol/L at three different pH values; (taken from [70]).

Variation of the pH value at constant eluent concentration primarily affects the dissociation of the organic acid that is employed as the eluent. Therefore, in most cases polybasic acids such as phthalic acid (pK_1 = 2.9; pK_2 = 5.5) are used as the eluent to control the elution power via dissociation to monovalent and divalent anions. Fig. 3-70 shows how the retention of a number of inorganic anions depends on pH using a phthalate eluent at the concentration c = 0.004 mol/L on a commercially available silica-based anion exchanger. The position of the system peak allows the pH value of the mobile phase to be changed only within a limited range. At pH 2.7 – the pK value of phthalic acid itself – the system peak appears behind the water dip (Fig. 3-71). Under these conditions, the elution power of phthalic acid is very low, since only monovalent hydrogenphthalate anions exist as eluent ions at this pH. Thus, divalent analyte ions such as sulfate are strongly retained. Raising the pH value to 3.7, the mobile phase contains divalent phthalate ions in addition to hydrogenphthalate ions. Although monovalent and divalent ions may be analyzed in a single run, unfortunately, the system peak lies between nitrate and sulfate. The retention times decrease as the pH value is further increased to 4.5, since the concentration of divalent eluent ions becomes higher. The system peak elutes after the sulfate signal under these conditions. These three examples illustrate the importance of a correct pH adjustment for positioning of the system peak in the chromatogram.

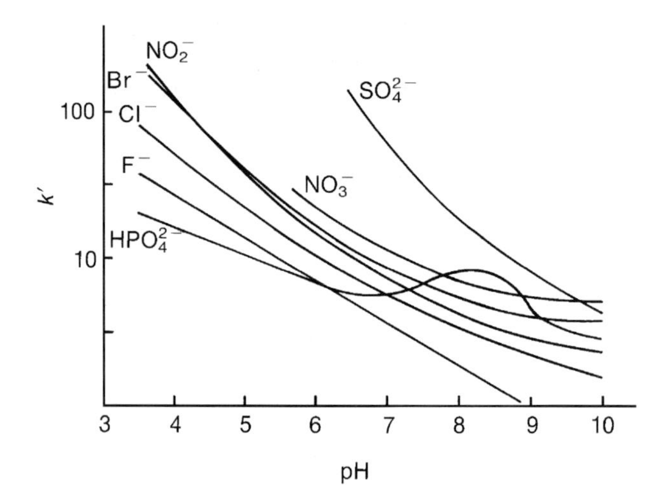

Fig. 3-72. pH Dependence of the retention of various inorganic anions using *p*-hydroxybenzoic acid at a concentration c = 0.005 mol/L as the eluent; (taken from [83]).

The pH value of the mobile phase affects not only the dissociation of the organic acid used as the eluent, but also the valency of some analyte ions. This is demonstrated in Fig. 3-72, where the pH dependence of the retention of inorganic anions using *p*-hydroxybenzoic acid as the eluent is displayed. As expected, the retention decreases as the pH increases. At higher pH not only the carboxyl group (pK_1 = 3.4), but also the hydroxyl group of *p*-hydroxybenzoic acid (pK_2 = 9.4) is dissociated. As is evident from Fig. 3-72, orthophosphate changes its valency at pH 7 and is converted from the monovalent to the divalent form, which increases the retention. Starting at pH 8, divalent eluent ions also exist in the mobile phase, which again decreases the retention of orthophosphate that is predominantly in the divalent form above pH 9. The optimal pH range for the separation of anions specified in DIN 38405 D19 [55] lies between 8.4 and 8.6. Under these chromatographic conditions, orthophosphate is eluted between nitrate and sulfate; thus, the elution order is comparable to that of Fig. 3-16 obtained under application of the bicarbonate/carbonate mixture.

3.3.4.4 Polarizable Anions

The polarizable inorganic anions comprise iodide, thiocyanate, and thiosulfate, as well as oxygen-containing metal anions such as tungstate, molybdate, and chromate. Due to their large radii such ions exhibit a very high affinity toward the stationary phase of anion exchangers. In the past, polarizable anions were separated on conventional anion exchangers such as IonPac AS1 to AS4 with concentrated carbonate eluents (c = 0.008 mol/L Na_2CO_3). Hollow fiber suppressors could not be used to reduce the background conductivity because of the high carbonate concentration. A drawback of conventional packed-bed suppressors, however, is that they require frequent regeneration. Moreover, strong tailing effects attributed to adsorption processes prevented determinations in the sub-ppm range.

The introduction of the IonPac AS5 separator column significantly facilitated the analysis of polarizable anions. Reducing the hydrophobicity of the functional groups bonded to the latex beads makes it possible to elute polarizable anions using a standard mixture of sodium bicarbonate and sodium carbonate. To minimize adsorption phenomena, some *p*-cyanophenol is added to the eluent mixture. The influence of this species on the peak form is evident in Fig. 3-18. The peak broadening could also be greatly reduced because of the compatibility of the eluent with commercial membrane suppressors and the reduction in the void volume.

Figures 3-73 and 3-74 show chromatograms of polarizable anions. The concentration of bicarbonate/carbonate is again the parameter that affects the retention. When this concentration is raised, the retention times are reduced accordingly. As shown in a comparison of both chromatograms, the retention behavior of individual components in both groups of compounds is comparable.

Polarizable anions can also be separated on a conventional exchanger such as the IonPac AS4A, where *p*-cyanophenol is added to the standard buffer prepared from 0.0017 mol/L $NaHCO_3$ and 0.0018 mol/L Na_2CO_3. The respective chromatograms obtained for both groups of compounds under these conditions are shown in Fig. 3-75. No significant advantages are discernable compared with the respective chromatograms in Figs. 3-73 and 3-74. However, when a single-run analysis of iodide, thiocyanate, tungstate, and molybdate is required, the selectivity of the AS4A column seems more suitable. Exceptions to this are thiosulfate and chromate, which are not separated but which cannot exist simultaneously.

An important field of application for the AS5 column is the analysis of polarizable anions in high amounts of sodium chloride. Fig. 3-76 illustrates impressively that the separator column can be loaded with a 1% NaCl-solution due to its high capacity, without affecting the separation of polarizable anions that elute afterwards. The field

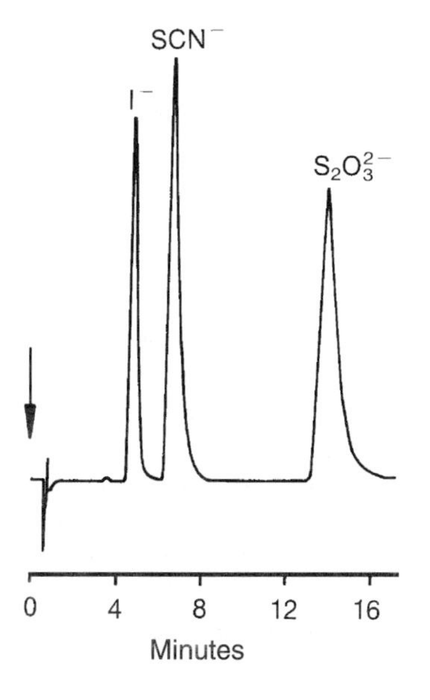

Fig. 3-73. Separation of iodide, thiocyanate, and thiosulfate on an IonPac AS5 separator column. – Eluent: 0.0035 mol/L $NaHCO_3$ + 0.0029 mol/L Na_2CO_3 + 100 mg/L *p*-cyanophenol; flow rate: 2 mL/min; detection: suppressed conductivity; injection volume: 50 µL; solute concentrations: 20 ppm each.

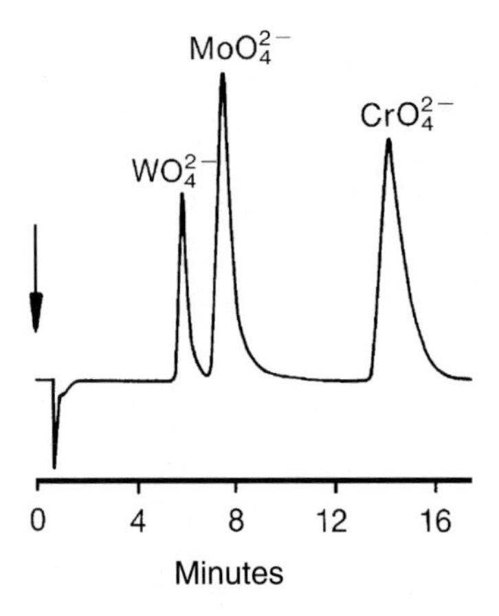

Fig. 3-74. Separation of oxygen-containing metal anions on an IonPac AS5 separator column. – Chromatographic conditions: see Fig. 3-73; solute concentrations: 40 ppm each.

of application for this method is limited to concentration differences of no more than four orders of magnitude when electrical conductivity detection is used. Higher concentration differences are difficult to realize chromatographically, even when applying a more selective detection method such as amperometry at a Ag working electrode, since the processes occurring at the electrode surface are significantly disturbed by high amounts of chloride.

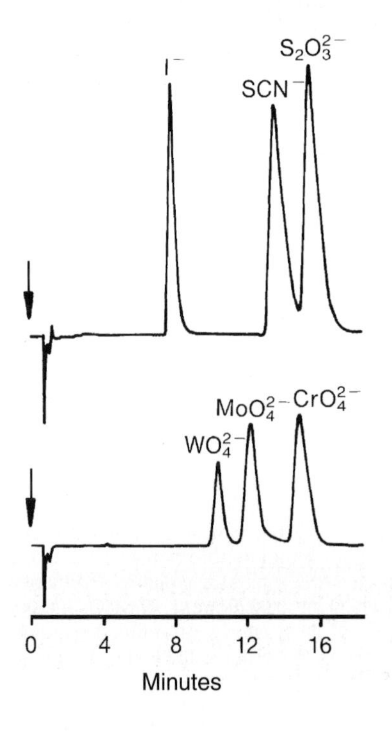

Fig. 3-75. Separation of polarizable anions on an Ion Pac AS4A separator column. – Eluent: 0.0017 mol/L $NaHCO_3$ + 0.0018 mol/L Na_2CO_3 + 100 mg/L *p*-cyanophenol; flow rate: 2 mL/min; detection: suppressed conductivity; injection volume: 50 μL; solute concentrations: a) 20 ppm each of iodide, thiocyanate, and thiosulfate, b) 20 ppm each of tungstate, molybdate, and chromate.

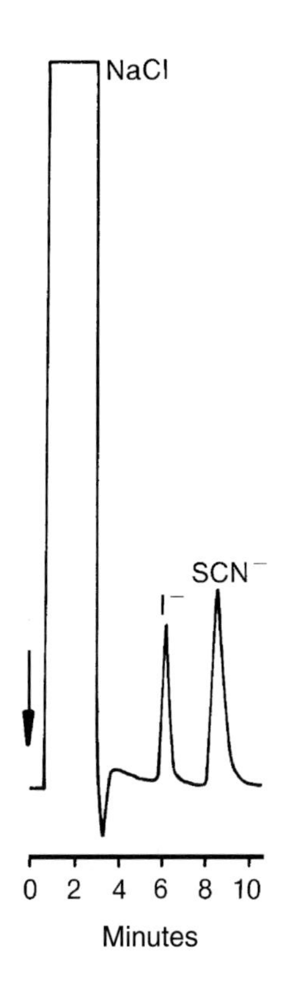

Fig. 3-76. Analysis of iodide and thiocyanate in the presence of high amounts of sodium chloride. – Separator column: IonPac AS5; eluent: 0.0017 mol/L $NaHCO_3$ + 0.0018 mol/L Na_2CO_3 + 100 mg/L *p*-cyanophenol; flow rate: 2 mL/min; detection: suppressed conductivity; injection volume: 50 µL; solute concentrations: 10 ppm iodide and thiocyanate in 1% NaCl.

Potassium hydrogenphthalate is suited for the analysis of iodide, thiocyanate, and thiosulfate in combination with direct conductivity detection. When silica-based exchangers are used, some methanol or 2-propanol is added to the eluent to lessen adsorption effects. Fig. 3-77 shows such a separation, the characteristic of which is the elution of monovalent iodide **prior to** the divalent sulfate. Noticeable is the relatively long time

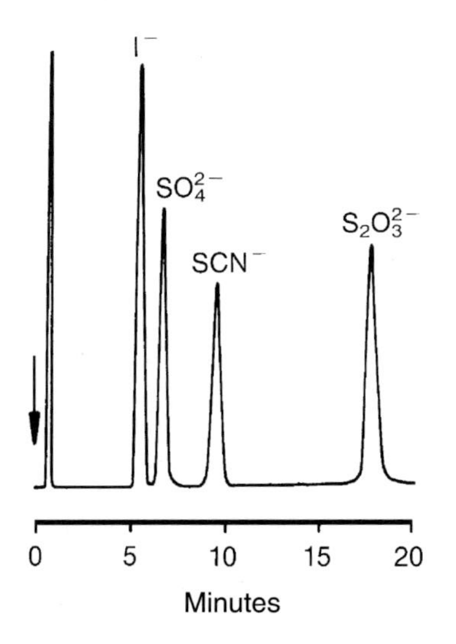

Fig. 3-77. Analysis of iodide, thiocyanate, and thiosulfate on a silica-based ion exchanger Vydac 300 IC 405. – Eluent: 0.002 mol/L KHP – methanol (90:10 v/v), pH 5.0; flow rate: 3 mL/min; detection: direct conductivity; injection volume: 100 µL; solute concentrations: 40 ppm iodide, 20 ppm sulfate and thiocyanate, 40 ppm thiosulfate.

required for the baseline separation of all components, which may only be partly compensated by increasing the flow rate. In comparison with the anion separation obtained on the polymeric phase, it is evident from Fig. 3-73 that the higher chromatographic efficiency of the silica ion exchanger is not exclusively decisive for the optimization of the separation with regard to the speed of the analysis. Much more important is the selectivity of the phase being determined by the type of exchange group that is appropriate for the chemical nature of the analyte species.

Anion exchangers such as the IonPac AS5 are also suited for the analysis of complexed transition metals and heavy metals, which opens another field of application for ion chromatography. Examples of respective chromatograms for the separation of metal-EDTA complexes as well as for metal-chloro complexes are displayed in Figs. 3-78 and 3-79.

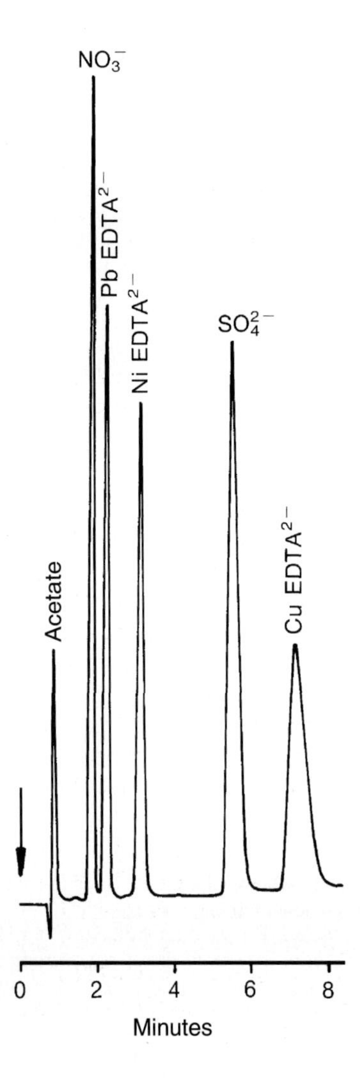

Fig. 3-78. Separation of various metal-EDTA complexes. – Separator column: IonPac AS5; eluent: 0.002 mol/L $NaHCO_3$ + 0.002 mol/L Na_2CO_3; flow rate: 2 mL/min; detection: suppressed conductivity.

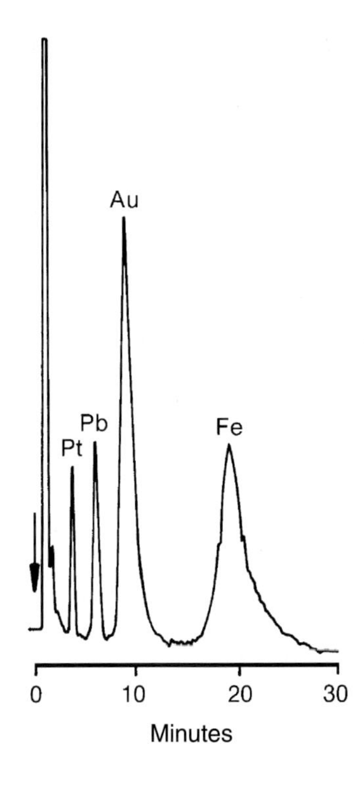

Fig. 3-79. Separation of various metal-chloro complexes. – Separator column: IonPac AS5; eluent: 0.2 mol/L $NaClO_4$ + 0.2 mol/L HCl; flow rate: 2 mL/min; detection: UV (215 nm); injection volume: 50 μL; solute concentrations: 0.2 ppm Pt, 2 ppm Pb, 5 ppm Au, and Fe.

Polyvalent inorganic anions also show a strong affinity toward the stationary phase of anion exchangers. In particular, this includes condensed phosphates such as pyrophosphate and tripolyphosphate. Species of this kind which are of special interest to the detergent industry are separated on a special anion exchanger such as the IonPac AS7. The separation of pyrophosphate and tripolyphosphate depicted in Fig. 3-80 was obtained using relatively concentrated nitric acid eluents. The degree of dissociation of the phosphates was restrained to enable their elution. The detection system used (post-column derivatization with ferric nitrate and UV absorption at 330 nm) allows one to determine other inorganic species such as orthophosphate and sulfate in the same run. However, the determination of the latter two anions is much more sensitive with the aid of conventional ion exchangers and by applying conductivity detection.

Higher polyphosphates, however, cannot be analyzed with this method, because they are retained too strongly by the stationary phase. Application of the phosphorus-specific detection, recently described by Vaeth et al. [84], allows the use of a mixture of potassium chloride and EDTA [85] instead of nitric acid as the eluent. Owing to the formation of iron-chloro complexes, this eluent is not compatible with the detection system described above. While the potassium chloride concentration determines the retention, adding EDTA merely serves to improve peak symmetry. At sufficiently high KCl concentrations, phosphates with a molecular weight of up to 10,000 g/mol can be separated. Fig. 3-81 shows the separation of a "tetrapolyphosphate" by gradient elution. Tetrapolyphosphate is the major component in this sample, but exists in a distribution between P_1 and P_{20}. The chromatogram depicted in Fig. 3-82 is obtained under identical

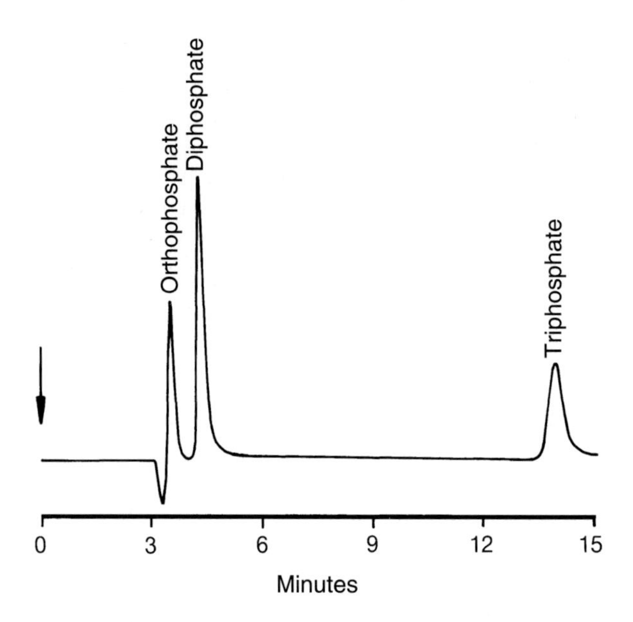

Fig. 3-80. Separation of orthophosphate, pyrophosphate, and tripolyphosphate. – Separator column: IonPac AS7; eluent: 0.07 mol/L HNO_3; flow rate: 0.5 mL/min; detection: photometry at 330 nm after post-column derivatization with ferric nitrate; injection volume: 50 μL; solute concentrations: 100 ppm orthophosphate, 50 ppm pyrophosphate, and 200 ppm tripolyphosphate.

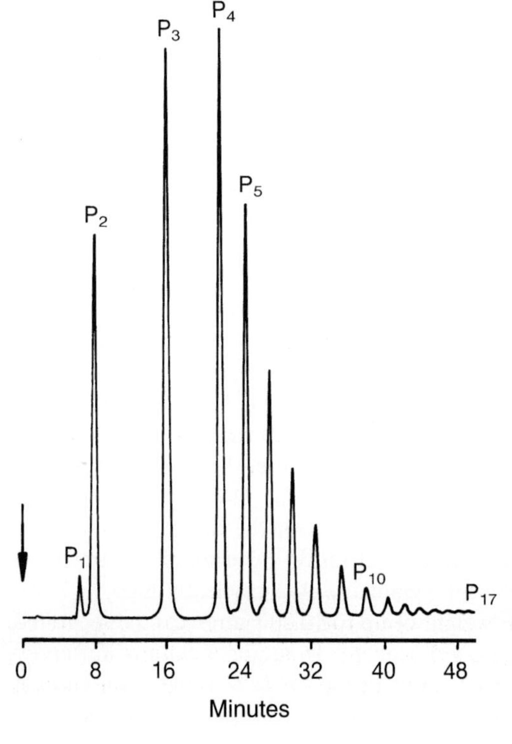

Fig. 3-81. Separation of a tetrapolyphosphate by gradient elution. – Separator column: IonPac AS7; eluent: (A) 0.17 mol/L KCl + 0.0032 mol/L EDTA, pH 5.1, (B) 0.5 mol/L KCl + 0.0032 mol/L EDTA, pH 5.1; gradient: 100% A isocratic for 2 min, then to 30% B in 8 min, then in 60 min to 100% B; flow rate: 0.5 mL/min; detection: phosphorus-specific detection according to Vaeth et al. [84]; injection: 50 μL of a 0.1% solution.

chromatographic conditions. It shows the separation of a polyphosphate that has a chain length of P_{40}, according to the specification of the manufacturer. The slight baseline rise during the chromatogram is explained by the elution of the high molecular weight fraction that is not separated under these conditions. In both cases, the separated compounds were hydrolyzed to orthophosphate using HNO_3 at 110 °C. In a second reaction step, they were reacted with a molybdate/vanadate reagent to yield yellow phosphorvanadomolybdic acid, the light absorption of which is measured at 410 nm. Hence, the calibration can be carried out directly with an orthophosphate standard, since the signal is proportional to the phosphorus content of the eluted component.

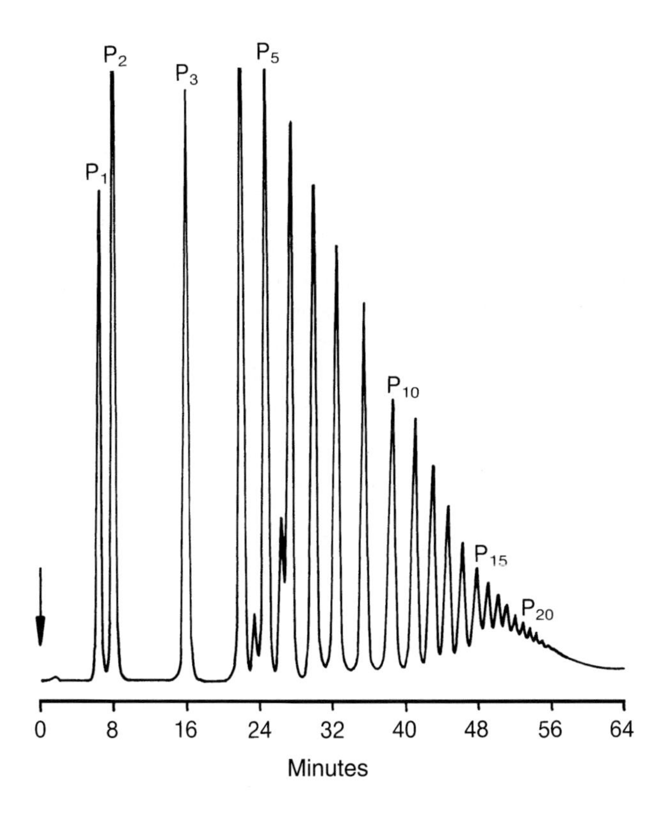

Fig. 3-82. Separation of a polyphosphate by gradient elution. – Chromatographic conditions: see Fig. 3-81; injection: 50 µL of a 0.2% solution.

In addition to polyvalent inorganic anions, numerous organic compounds can be analyzed under similar chromatographic conditions in combination with the various post-column derivatization procedures. A detailed account of this type of analysis is given in Section 3.3.5.2.

3.3.5 Ion-Exchange Chromatography of Organic Anions

3.3.5.1 Organic Acids

Apart from inorganic anions, a number of organic anions can also be separated with a conventional anion exchanger. Interferences are possible because real-world samples often contain both inorganic and organic anions.

Short-chain aliphatic carboxylic acids, for instance, elute near the void volume and render the separation and determination of fluoride more difficult. If an eluent mixture of 0.0017 mol/L $NaHCO_3$ and 0.0018 mol/L Na_2CO_3 is used, which is suitable for the separation of mineral acids by applying the suppressor technique, fluoride and formic acid are barely separated with an IonPac AS4A separator column. If the sample also contains acetic acid or a longer chain carboxylic acid, a separation of fluoride is not possible under these conditions. As already discussed in Section 3.3.4.2, this often leads to a misinterpretation of the fluoride signal. To shed light on the kind of species that elutes near the void volume, it is recommended that dilute bicarbonate solution is used as the eluent. The low elution strength of bicarbonate renders it possible to separate formic and acetic acid from fluoride. (See Fig. 3-83 for a corresponding chromatogram.) However, this procedure is not suitable for routine analysis of fluoride in such complex matrices, since even anions such as chloride have extraordinarily long retention times. Components that elute later, such as nitrate and sulfate, interfere with subsequent analyses and, therefore, have to be frequently removed from the separator column by flushing it with a strong eluent. In individual cases this may be circumvented by employing a step gradient. The application of sodium tetraborate as eluent to obtain a slightly higher resolution between fluoride and acetate and its associated disadvantages was discussed in Section 3.3.4.2.

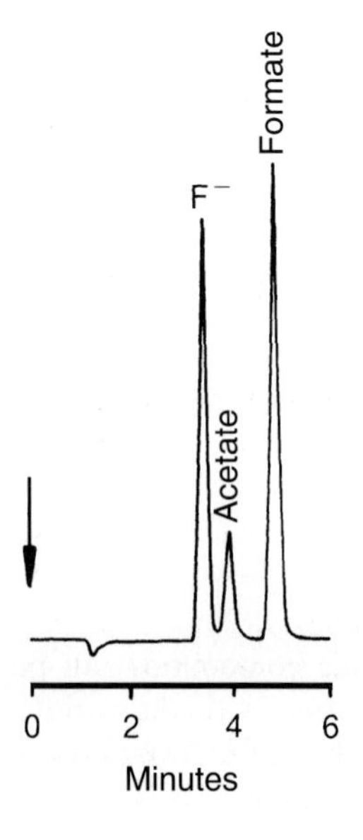

Fig. 3-83. Separation of fluoride, acetate, and formate. – Separator column: IonPac AS4A; eluent: 0.001 mol/L $NaHCO_3$; flow rate: 1.5 mL/min; detection: suppressed conductivity; injection: 50 µL; solute concentrations: 3 ppm fluoride, 20 ppm acetate, and 10 ppm formate.

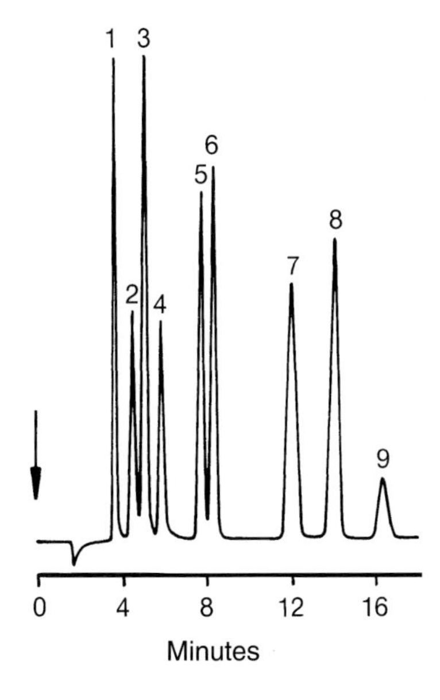

Fig. 3-84. Separation of various monocarboxylic acids at an anion exchanger IonPac AS6 (CarboPac). – Eluent: 0.0017 mol/L $NaHCO_3$ + 0.0018 mol/L Na_2CO_3; flow rate: 1 mL/min; detection: suppressed conductivity; injection: 50 µL; solute concentrations: 3 ppm fluoride (1), 40 ppm acetic acid (2), 20 ppm glycolic acid (3), 10 ppm α-hydroxyisocaproic acid (4), 20 ppm formic acid (5), oxamic acid (6), methanesulfonic acid (7), amidosulfonic acid (8), and α-ketoisocaproic acid (9).

For the analysis of this class of compounds, changing the stationary phase provides an alternative to choosing a different eluent. When the IonPac AS6 (CarboPac) separator column, initially developed for the analysis of carbohydrates, is conditioned with the carbonate/bicarbonate buffer mentioned above, even compounds exhibiting a small affinity are strongly retained due to the comparatively high capacity of this stationary phase. Thus, the retention difference between fluoride and chloride, which is less than 2 minutes under the same conditions at an AS4A column, may be enlarged to more than 20 minutes with an AS6 column. An example of the almost baseline-resolved separation of various monocarboxylic acids is illustrated in Fig. 3-84. All these acids are relevant in current analysis problems, which may be solved elegantly using this procedure. Amidosulfonic acid, for example, is one of the many components in flue gas scrubber solutions of desulfurization devices that have to be analyzed frequently.

Compared to aliphatic monocarboxylic acids, a slightly higher affinity toward the stationary phase is exhibited by aromatic monocarboxylic acids. The simplest congener, benzoic acid, co-elutes at some stationary phases with nitrite. Depending on the separator column the interpretation of the nitrite signal may become a difficult task. Often the peak shape provides the only distinguishing feature. While the nitrite peak is fully symmetrical, the benzoic acid signal shows a slight tailing, which is explained by adsorption at the aromatic backbone of the stationary phase. A separation of benzoic acid and nitrite is possible, for example, in an IonPac AS4A separator column under standard conditions, where benzoic acid elutes immediately after nitrite. An alternative to ion-exchange chromatography is provided by ion-pair chromatography, using a non-polar stationary phase at which both compounds show a markedly different retention behavior.

Under standard conditions (0.0028 mol/L $NaHCO_3$ + 0.0022 mol/L Na_2CO_3) benzoate exhibits only a small retention in an IonPac AS4 separator column. Therefore, in-

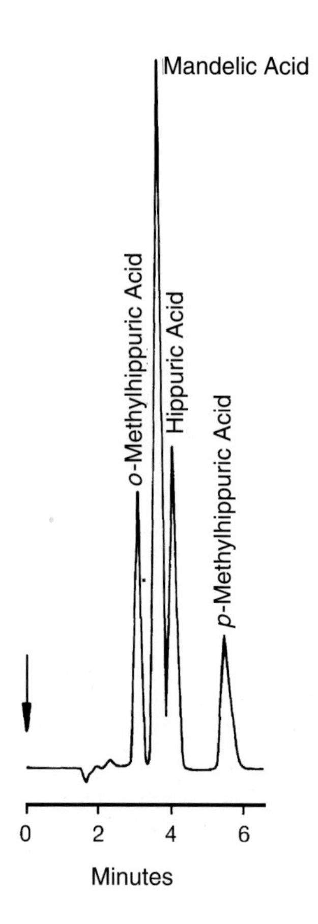

Fig. 3-85. Separation of various aromatic monocarboxylic acids. – Separator columns: 2 IonPac AS4; eluent: 0.0005 mol/L Na_2CO_3 + 0.0005 mol/L NaOH; flow rate: 2 mL/min; detection: suppressed conductivity; injection: 50 μL.

vestigations were performed to determine to what extent other monocarboxylic acids may be separated using anion exchange chromatography. Fig. 3-85 shows a separation of mandelic acid, hippuric acid, and methyl-substituted hippuric acid which can only be obtained by employing two separator columns in series and a modified eluent. The separation between *o*-methylhippuric acid and *p*-methylhippuric acid is remarkably good. It is obvious from Table 3-19 that the elution order may be correlated with the p*K* values of the compounds.

Table 3-19. Elution order and p*K* values [86] of some aromatic monocarboxylic acids.

Elution order	p*K* Value
Mandelic acid	3.36
Hippuric acid	3.64
Benzoic acid	4.20

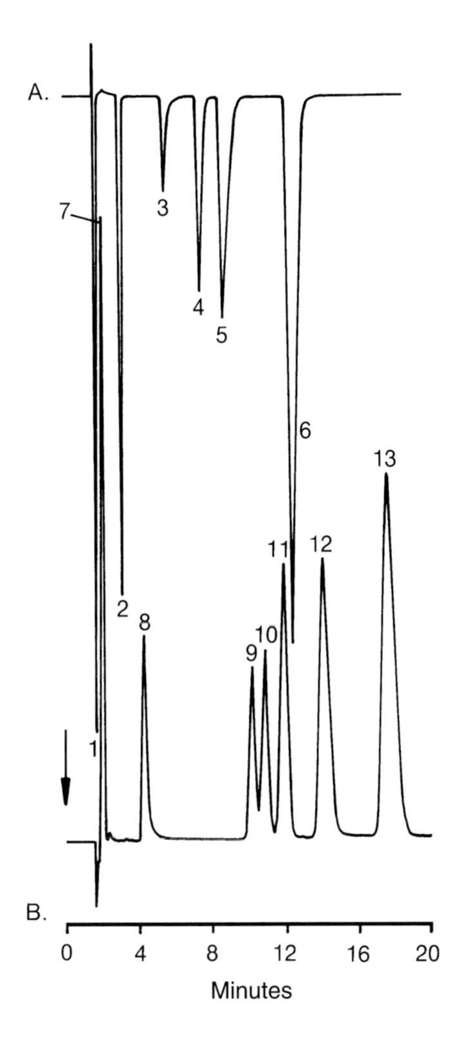

Fig. 3-86. Comparison between the retention behavior of inorganic anions and several organic carboxylic acids, respectively. – Separator column: IonPac AS4; eluent: 0.0028 mol/L $NaHCO_3$ + 0.0022 mol/L Na_2CO_3; flow rate: 1.6 mL/min; detection: suppressed conductivity; injection: 50 μL; solute concentrations: a) 1.5 ppm fluoride (1), 2 ppm chloride (2), 5 ppm orthophosphate (3) and bromide (4), 10 ppm nitrate (5), and 12.5 ppm sulfate (6), b) 5 ppm formic acid (7), 40 ppm benzoic acid (8), 20 ppm succinic acid (9), 10 ppm malonic acid (10), 20 ppm maleic acid (11), tartaric acid (12), and oxalic acid (13).

It is generally observed that the order of elution is aliphatic monocarboxylic acids, followed by aromatic monocarboxylic acids, followed by aliphatic dicarboxylic acids. Fig. 3-86, in which the retention behavior of aliphatic dicarboxylic acids is compared with that of inorganic anions, reveals that compounds such as succinic acid, malonic acid, maleic acid, and tartaric acid elute in the retention volume of nitrate and sulfate.

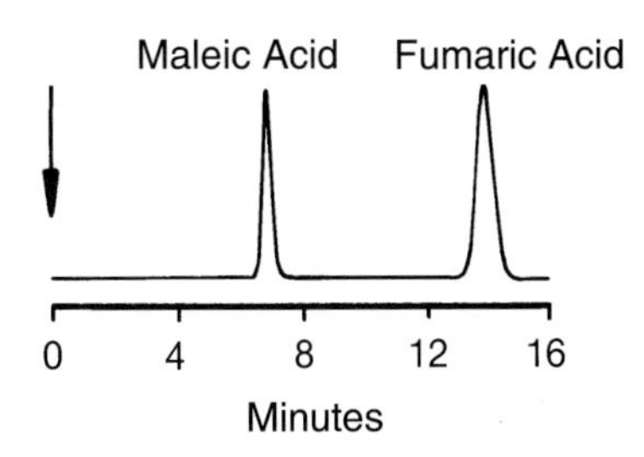

Fig. 3-87. Separation of maleic acid and fumaric acid. – Separator column: IonPac AS4A; eluent: 0.0017 mol/L $NaHCO_3$ + 0.0018 mol/L Na_2CO_3; flow rate: 2.0 mL/min; detection: UV (210 nm); injection: 50 μL; solute concentrations: 2 ppm maleic acid and 5 ppm fumaric acid.

A sufficient separation of all these compounds, therefore, is only achieved by using two AS4 columns in series. However, the separation of the two stereoisomers, maleic acid and fumaric acid, is much easier. It is obtained under standard conditions and is shown in Fig. 3-87. In contrast to monocarboxylic acids, the retention of aliphatic dicarboxylic acids increases with decreasing p*K* value. The corresponding data are summarized in Table 3-20. This finding is explained by the charge-stabilizing effect exerted by the +I-effect of the methylene groups which decreases from succinic acid to oxalic acid:

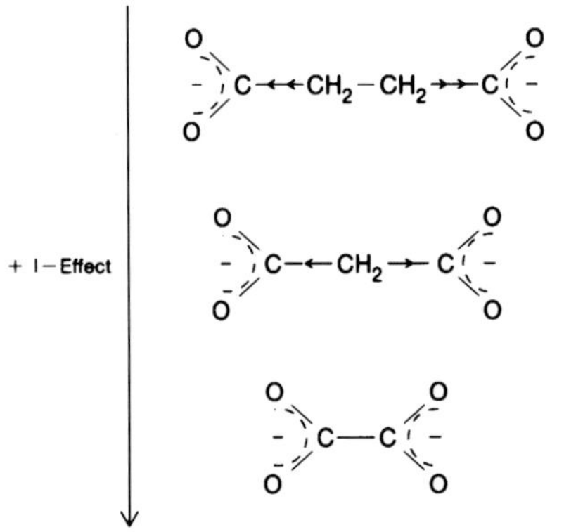

+ I − Effect

$$^{-}OOC \leftarrow\leftarrow CH_2-CH_2 \rightarrow\rightarrow COO^{-}$$

$$^{-}OOC \leftarrow CH_2 \rightarrow COO^{-}$$

$$^{-}OOC-COO^{-}$$

The presence of additional hydroxide groups at the dicarboxylic acid, as in malic and tartaric acid, increases the acidic character and, thus, the retention (Table 3-21). A corresponding chromatogram is shown in Fig. 3-88. However, a separation of malic acid (monohydroxysuccinic acid) and malonic acid is not possible under these conditions, while tartronic acid (hydroxymalonic acid) exhibits a significantly higher retention.

Table 3-20. Elution order and p*K* values [86] of some dicarboxylic acids.

Elution order	p*K* Values	
	pK_1	pK_2
Succinic acid	4.21	5.64
Malonic acid	2.88	5.68
Oxalic acid	1.27	4.29

Table 3-21. Comparison of elution order and p*K* values [86] between hydroxy-substituted and unsubstituted aliphatic dicarboxylic acids.

Elution order	p*K* Values	
	pK_1	pK_2
Succinic acid	4.21	5.64
Malic acid	3.40	5.10
Tartaric acid	3.04	4.37

Aliphatic tricarboxylic acids such as citric acid exhibit a remarkably high affinity toward the stationary phase of an anion exchanger. Hence, low ionic strength bicarbonate/carbonate buffer solutions are not particularly suited as eluents. However, when a sodium hydroxide solution at a comparatively high concentration ($c \approx 0.08$ mol/L) is used, citric acid may be eluted, and may even be separated from its structural isomer isocitric acid. When the detection of these compounds is carried out via electrical conductivity, a suppressor system must be used to reduce the background conductivity. For capacity reasons only a micromembrane suppressor is applicable. Alternatively, a different stationary phase may be used for the separation of organic acids with high affinities. The IonPac AS5 separator column (see Section 3.3.4.4) has also proved to be suited for this field of application. This is illustrated in Fig. 3-89 at the example of 2,6-dihydroxyisonicotinic acid (citrazinic acid), whose elution is obtained using simple carbonate/bicarbonate buffer upon adding some *p*-cyanophenol. A separation of this acid from thiocyanate, which exhibits a similar retention behavior poses no problem under these conditions.

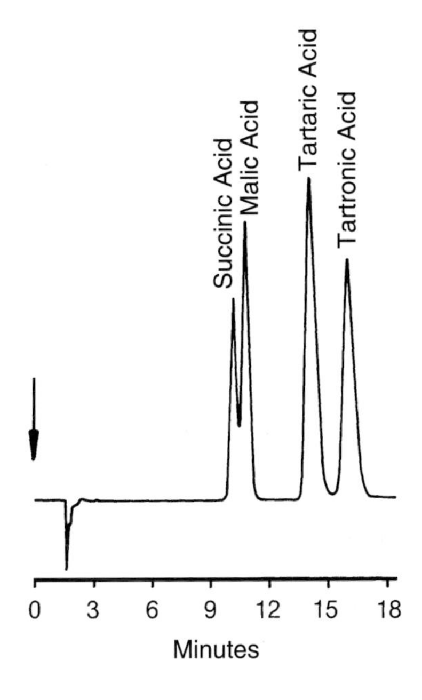

Fig. 3-88. Separation of aliphatic hydroxycarboxylic acids. – Chromatographic conditions: see Fig. 3-86; solute concentrations: 20 ppm each of succinic acid, malic acid, tartaric acid, and tartronic acid.

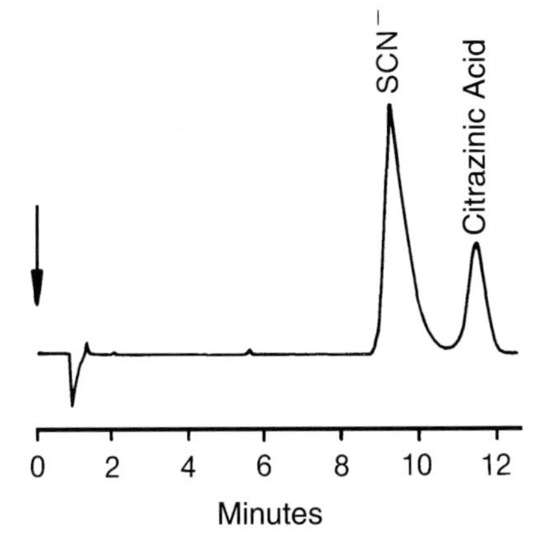

Fig. 3-89. Separation of citrazinic acid. – Separator column: IonPac AS5; eluent: 0.0028 mol/L $NaHCO_3$ + 0.0022 mol/L Na_2CO_3 + 100 mg/L *p*-cyanophenol; flow rate: 2.0 mL/min; detection: suppressed conductivity; injection volume: 50 μL; solute concentrations: 20 ppm each of thiocyanate and citrazinic acid.

An interesting field of application for anion exchange chromatography is the analysis of herbicides based on phenoxycarboxylic acids and their derivatives. Generally, these compounds were separated at chemically modified silicas using a methanol/water mixture as the eluent. Acetic acid was added to the eluent to suppress analyte dissociation. The list of herbicides that may be analyzed with anion exchange chromatography comprises:

- 2,4 D
  2,4-dichlorophenoxyacetic acid
- 2,4,5 T
  2,4,5-trichlorophenoxyacetic acid
- Silvex
  2-(2,4,5-trichlorophenoxy)propionic acid
- Mecoprop
  2-(2-methyl-4-chlorophenoxy)propionic acid

These compounds are also strongly retained due to their aromatic skeleton and the resulting π-π interactions with the stationary phase of an anion exchanger. Since they can be detected sensitively by measuring the UV absorption, strong eluents such as sodium nitrate may be employed, thereby obtaining the chromatogram displayed in Fig. 3-90. For the analysis of herbicides lacking a chromophore such as endothall, 7-oxabicyclo[2,2,1]-heptane-2,3-dicarboxylic acid,

Endothall

anion exchange chromatography with subsequent conductivity detection provides an alternative to RPLC. Endothall can be analyzed under the chromatographic conditions given in Fig. 3-16 in the same run with simple inorganic anions, and is eluted between nitrate and orthophosphate.

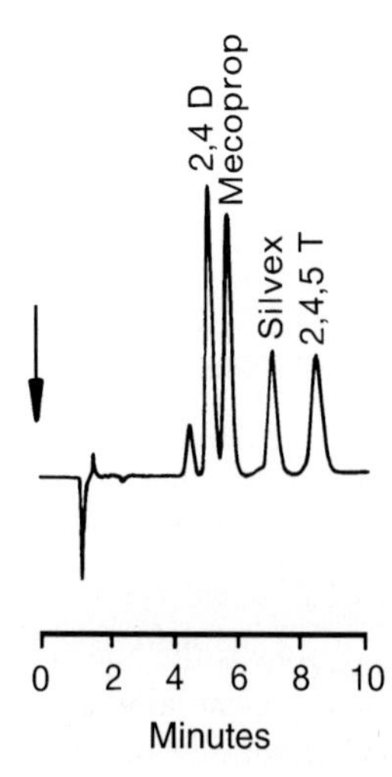

Fig. 3-90. Separation of phenoxycarboxylic acid-based herbicides. – Separator column: IonPac AS4A; eluent: 0.025 mol/L $NaNO_3$ + 0.005 mol/L NaOH; flow rate: 2 mL/min; detection: UV (280 nm); injection volume: 50 μL.

Conventional anion exchangers are also suited for the separation of inorganic and organic phosphates and sulfates. As a rule, the organic species are eluted prior to the inorganic ones. While long-chain alkylphosphates, alkylsulfates, and alkylsulfonates (n_c > 4) can only be separated using ion-pair chromatography, their short-chain homologues exhibit a significant retention. A mixture of sodium carbonate and sodium hydroxide is suited as the eluent for the separation of monobutyl-, dibutyl-, and orthophosphate (see Fig. 3-91). The increase in the pH value with sodium hydroxide is required to improve the separation between inorganic and organic phosphates, respectively.

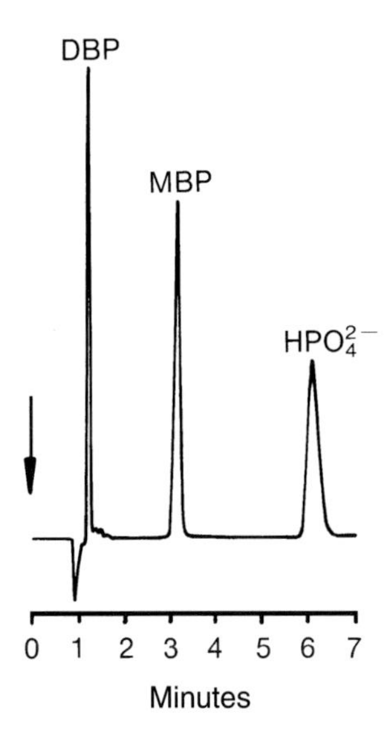

Fig. 3-91. Separation of various alkylphosphates and inorganic phosphates. – Separator column: IonPac AS4A; eluent: 0.003 mol/L Na_2CO_3 + 0.001 mol/L NaOH; flow rate: 2 mL/min; detection: suppressed conductivity; injection volume: 50 μL; solute concentrations: 60 ppm dibutylphosphate (technical grade) and 20 ppm orthophosphate.

In addition to alkyl phosphates, phosphates that are of biological relevance can be separated at a high-performance separator column such as the AS4A. This includes glycerophosphates (see Fig. 3-92), which play a significant role in fat metabolism in that

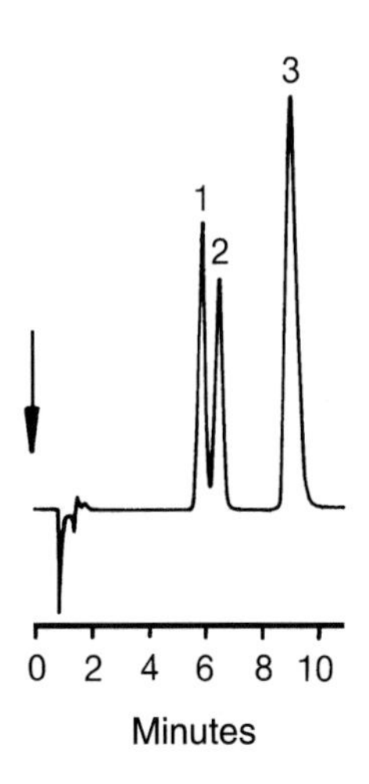

Fig. 3-92. Separation of α- and β-glycerophosphate. – Separator column: IonPac AS4A; eluent: 0.00085 mol/L $NaHCO_3$ + 0.0009 mol/L Na_2CO_3; flow rate: 2 mL/min; detection: suppressed conductivity; injection volume: 50 μL; solute concentrations: 20 ppm each of α-glycerophosphate (1), β-glycerophosphate (2), and orthophosphate (3).

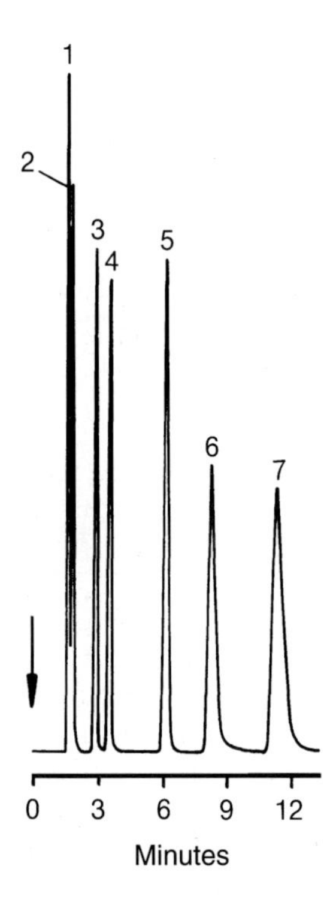

Fig. 3-93. Separation of nucleosides using anion exchange chromatography. – Separator columns: 2 IonPac AS4A; eluent: see Fig. 3-92; flow rate: 1.5 mL/min; detection: suppressed conductivity; injection volume: 50 μL; solute concentrations: 10 ppm cytidine (1), 5 ppm adenosine (2), 10 ppm thymidine (3) and uridine (4), 13 ppm 2′-deoxyguanosine (5), 15 ppm guanosine (6), and inosine (7).

they promote the esterification of fatty acids, which are then stored as glycerides. The separation of α- and β-glycerophosphates shown in Fig. 3-92 is another example of the efficiency of current anion exchangers.

Essential components of the ribonucleic acid and deoxyribonucleic acid (RNA and DNA) such as nucleosides and nucleotides can also be analyzed with anion exchange chromatography.

Nucleosides consist of a purine or pyrimidine base, respectively, to which a sugar component via a βN-glycosidic linkage at the 1′-position is attached. Depending on the type of sugar component – D-ribose or 2′-deoxy-D-ribose – one distinguishes between ribonucleosides and deoxyribonucleosides. Since nucleosides as weak acids exist in an anionic form at basic pH, they can be eluted with a low ionic strength bicarbonate/carbonate buffer. The determination of the UV absorption at 254 nm is suited as the detection method based on the conjugated double bonds in the purine and pyrimidine units. For the separation of the most important nucleosides – the purine derivatives adenosine, guanosine, and inosine, as well as the pyrimidine derivatives cytidine, uridine, and thymidine, two separator columns were used in series in order to obtain the required separation efficiency in the shortest possible analysis time. The comparison between the retention behavior of 2′-deoxyguanosine and guanosine reveals that DNA units elute prior to the RNA units having the same carbon skeleton. In addition to the sugar component, nucleotides contain at the 5′-position of the ribose (or 2′-deoxy-D-ribose) a mono-, di-, or triphosphate group, respectively.

These represent the actual monomeric units of the nucleic acids for which the phosphoric acid groups are always esterified at the 3′- and 5′-position with the hydroxide

NMP Nucleotide monophosphate
NDP Nucleotide diphosphate
NTP Nucleotide triphosphate

group of the pentose in the neighboring nucleotide. Generally, nucleotides are strong acids with a high affinity toward the stationary phase of anion exchangers. A phosphate buffer is typically used for the separation of nucleotides. The pH value of the mobile phase affects retention because it determines both the charge at the purine and pyrimidine bases as well as the degree of dissociation of the phosphate groups. Fig. 3-94 shows a chromatogram with the separation of various nucleotide monophosphates. The phosphate buffer concentration has to be increased for the elution of the respective di- and triphosphates. For the single run determination of the nucleotide monophosphates, diphosphates, and triphosphates, application of the gradient technique is inevitable. Such a chromatogram is shown in Fig. 3-95. The separate determination of ATP, ADP, and AMP with this method is of great practical significance in view of the dominant position of ATP (adenosine triphosphate) in the energy balance of cells.

3.3.5.2 Polyvalent Anions

The ion chromatographic separation and determination of multivalent anions that are derived, for example, from aminopolycarboxylic acids and polyphosphonic acids were two of the most difficult tasks in the past and, therefore, only partially possible

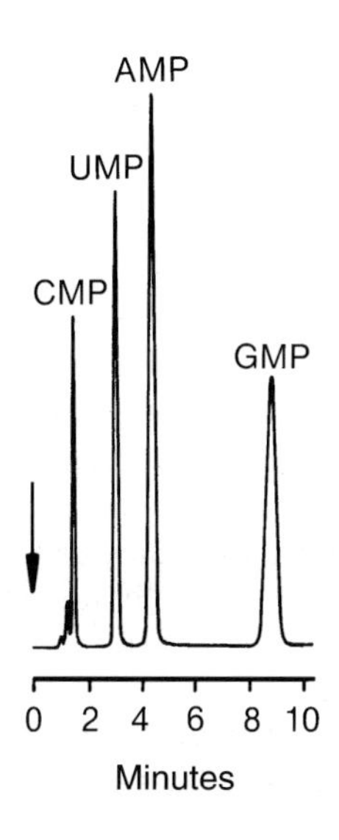

Fig. 3-94. Separation of nucleotide monophosphates. – Separator column: IonPac AS4A; eluent: 0.015 mol/L NaH_2PO_4, pH 3.4 with H_3PO_4; flow rate: 1.5 mL/min; detection: UV (254 nm); injection volume: 50 μL; solute concentrations: 10 ppm each of CMP, UMP, AMP, and GMP.

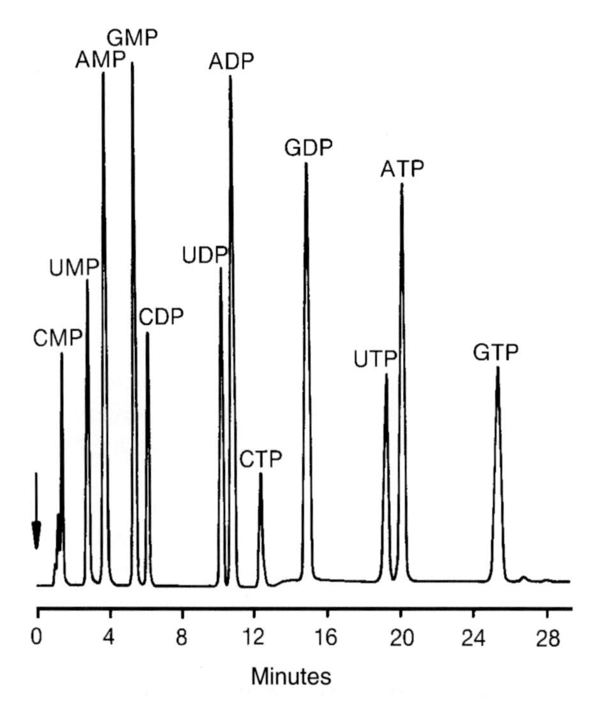

Fig. 3-95. Gradient elution of nucleotide monophosphates, diphosphates, and triphosphates. – Separator column: IonPac AS4; eluent: (A) water, (B) 0.5 mol/L NaH_2PO_4, pH 3.4 with H_3PO_4; gradient: linear, 3% B to 100% B in 40 min; flow rate: 1.5 mL/min; detection: UV (254 nm); injection volume: 50 µL; solute concentrations: 10 ppm each of CMP, UMP, AMP, and GMP, 20 ppm each of CDP, UDP, ADP, GDP, CTP, UTP, ATP, and GTP.

because of the chemical nature of such compounds which, due to their high valencies, exhibit a very high affinity toward the stationary phase of anion exchangers.

The structure and identity of such compounds that are of practical relevance as complexing agents may be elucidated unequivocally by both one-dimensional and two-dimensional nuclear magnetic resonance spectroscopy of the isotopes H-1, C-13, and P-31. Sufficiently high concentrations also render possible their quantitative analysis [87–91]. However, because of the low sensitivity, especially of the phosphorus nucleus, problems are encountered with the limits of detection in practical applications.

The first attempts directed at a chromatographic separation and identification of acidic organophosphates using an autoanalyzer system were presented by Waldhoff and Sladek [92], but typical retention times were in the range of hours. Only with the introduction of a special anion exchanger, the IonPac AS7 (see Section 3.3.1.3), it became possible to separate and to determine a variety of polyvalent anions [93-95]. When nitric acid of the concentration c = 0.03 to 0.05 mol/L is selected as the eluent, the degree of analyte dissociation is pushed back via partial protonation due to the high acid strength in the eluent, reducing the effective charge of the analytes and making their elution possible. In addition, the choice of nitric acid is based on the known high elution strength of the nitrate ion. The nitric acid concentration in the mobile phase is the sole experimental parameter that determines retention.

Conductivity detection of the analytes is not possible owing to the high acid concentration in the mobile phase. Thus, the complexing agents are detected via post-column derivatization with ferric nitrate in acidic solution and measurement of the UV absorption in the wavelength region between 310 nm and 330 nm. The reagent is added from a pressurized vessel to ensure a pulsation-free flow. The reaction with the reagent takes place in a reaction loop of appropriate dimension, that is packed with chemically inert plastic beads. This reduces peak broadening effects and optimizes the mixing.

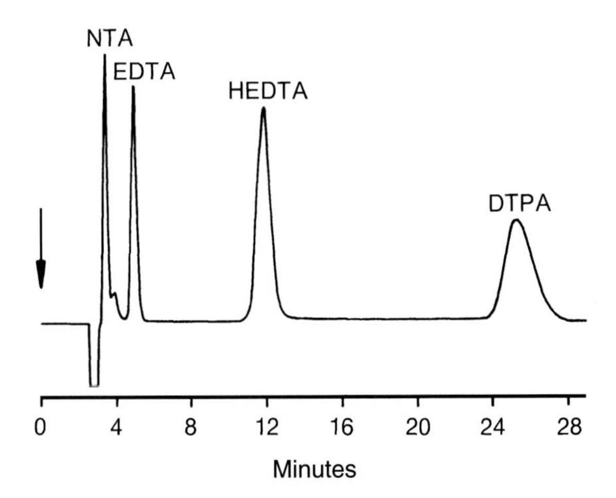

Fig. 3-96. Separation of aminopolycarboxylic acids NTA, EDTA, HEDTA, and DTPA. – Separator column: IonPac AS7; eluent: 0.05 mol/L HNO_3; flow rate: 0.5 mL/min; detection: photometry at 310 nm after reaction with ferric nitrate; injection volume: 50 µL; solute concentrations: 20 ppm NTA and EDTA, 60 ppm HEDTA, and 100 ppm DTPA.

The complexing agents derived from the aminopolycarboxylic acids are one of the classes of compounds that may be analyzed with this procedure.

$$(HOOC-H_2C)_2N-\left[CH_2-CH_2-\underset{\displaystyle CH_2-COOH}{N}\right]_n-CH_2-COOH$$

n = 0: NTA (Trilon A)
n = 1: EDTA (Trilon B)
n = 2: DTPA (Trilon C)

Their separation is shown in Fig. 3-96. As seen in this chromatogram, the retention of these compounds increases with the increasing number of (-CH_2-CH_2-N-CH_2-COOH)-

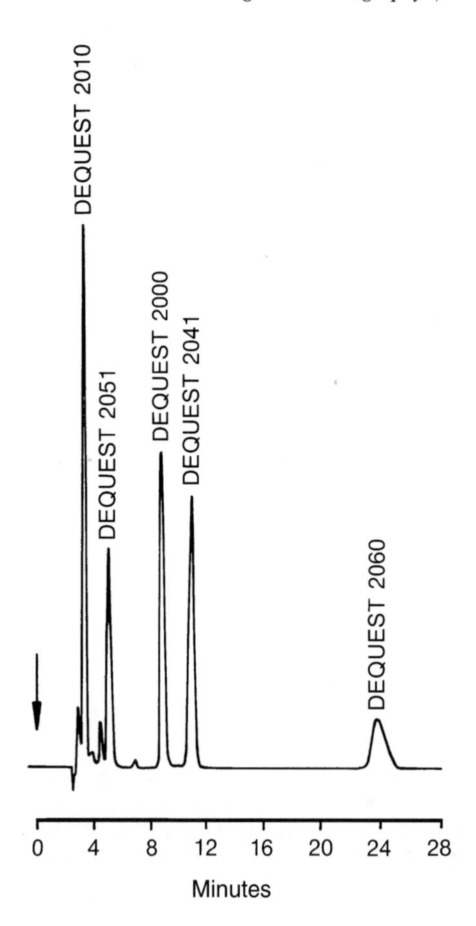

Fig. 3-97. Separation of aminopolyphosphonic acids of the DEQUEST type. – Separator column: IonPac AS7; eluent: 0.03 mol/L HNO_3; flow rate: 0.5 mL/min; detection: photometry at 330 nm after reaction with ferric nitrate; injection volume: 50 μL; solute concentrations: 50 ppm each of DEQUEST 2010, DEQUEST 2051, DEQUEST 2000, DEQUEST 2041, and DEQUEST 2060.

units. The same chromatographic conditions allow the analysis of the hydroxyethylethylenediaminotriacetic acid (HEDTA, Trilon D)

$$(HOOC-H_2C)_2N-CH_2-CH_2-N(CH_2-CH_2-OH)(CH_2-COOH)$$

that is an analogous compound to EDTA. The degradation products from EDTA such as ethylenediaminetriacetic acid and ethylenediaminediacetic acid cannot be determined with this method, since they do not react with ferric nitrate. The analytical procedure for the determination of these compounds is discussed in Section 5.6.

Fig. 3-97 shows the separation of some aminopolyphosphonic acids of the DEQUEST type that can be eluted with nitric acid at the concentration c = 0.03 mol/L. Also the analogous compound to EDTP, hexamethylenediamine-tetramethylenephosphonic acid (DEQUEST 2051)

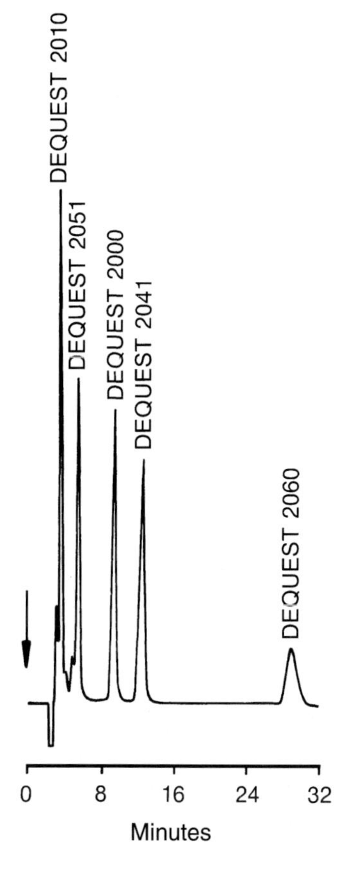

Fig. 3-98. Separation of aminopolyphosphonic acids of the DEQUEST type at a 5-µm anion exchanger. – Separator column: IonPac AS6A; eluent: 0.025 mol/L HNO_3; flow rate: 0.5 mL/min; detection: see Fig. 3-97; injection volume: 50 µL; solute concentrations: see Fig. 3-97.

$$(H_2O_3P{-}H_2C)_2N{-}\left[CH_2{-}CH_2{-}N(CH_2{-}PO_3H_2)\right]_n{-}CH_2{-}PO_3H_2$$

n = 0: NTP (DEQUEST 2000)
n = 1: EDTP (DEQUEST 2041)
n = 2: DTPP (DEQUEST 2060)

$$(H_2O_3P{-}H_2C)_2N{-}(CH_2)_6{-}N(CH_2{-}PO_3H_2)_2$$

DEQUEST 2051

can be analyzed within the same chromatographic run. While the separation displayed in Fig. 3-97 was carried out at a 10-µm IonPac AS7 exchanger, the separation of the same compounds at the IonPac AS6A anion exchanger with a particle size of 5 µm is

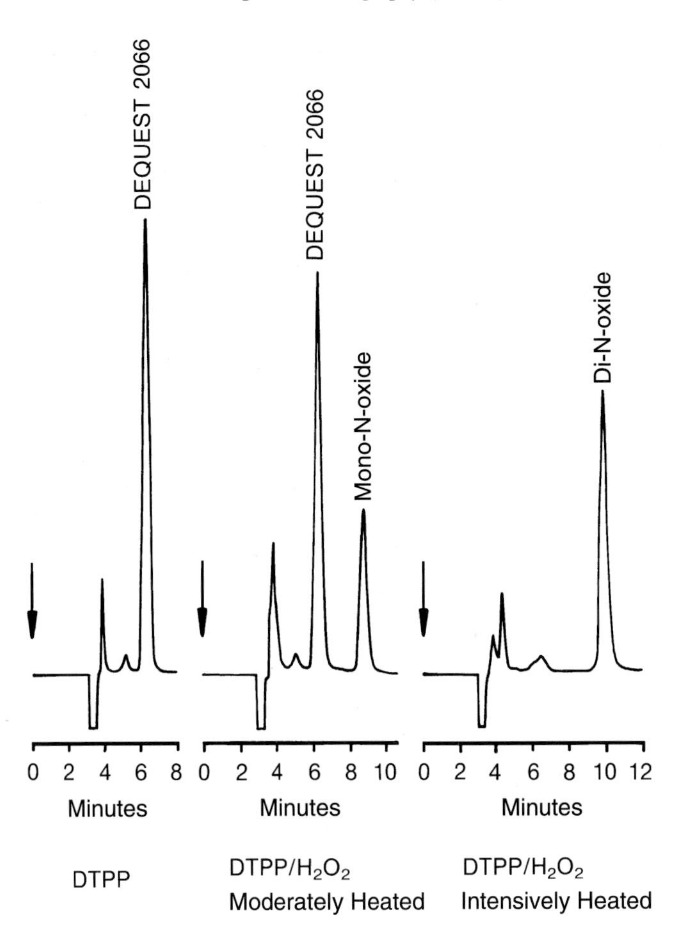

Fig. 3-99. Separation of diethylenetriamine-pentamethylenephosphonic acid and its oxidation products. – Chromatographic conditions: see Fig. 3-96; solute concentrations: 100 ppm DTPA.

shown in Fig. 3-98 for comparison. Although the acid strength was reduced to 0.025 mol/L to obtain comparable retention times, a better separation between DEQUEST 2000 and DEQUEST 2041 is obviously obtained with the AS6A column because of its higher chromatographic efficiency.

Aminopolyphosphonic acids have increasingly been included in detergent formulations. In the peroxide bleach, which is primarily used in Europe, these formulations contain sodium perborate that hydrolyzes in water under formation of hydrogen peroxide. To obtain a good bleaching efficiency at temperatures below 60 °C, a bleaching activator such as tetraacetyl-ethylenediamine (TAED) is added. At pH values between 9 and 12 this compound forms peracetic acid in the presence of hydrogen peroxide. The aminopolyphosphonic acids are oxidized to the stable N-oxides in the presence of these oxidants. The first indications of the chromatographic analysis of such compounds are found in the paper by Vaeth et al. [84], which describes the separation of ethylenediamine-tetramethylenephosphonic acid (EDTP) from its respective di-N-oxide via anion exchange chromatography. The identity of the mono-N-oxide was confirmed and at that time it was postulated as an intermediate in the oxidation reaction. This is shown using diethylenetriamine-pentamethylenephosphonic acid (DTPP) as an example. When a solution of 30% hydrogen peroxide is added to an aqueous DTPP-solution and the mixture is then heated to ca. 90 °C for a certain time, DTPP is quantitatively converted into the respective di-N-oxide. Compared with the starting material, di-N-oxide is retained much stronger by the stationary phase. The resulting two chromatograms are opposed to each other in Fig. 3-99 (left and right side, respectively). If the reaction solution is only heated to 60 °C for a short time, the mono-N-oxide is obtained. The middle chromatogram in Fig. 3-99 reveals that this intermediate can be separated from both the starting material and the di-N-oxide. Thus, ion chromatography can be used to investigate the kinetics of this oxidation.

In addition to aminopolyphosphonic acids, polyphosphonic acids without an amino function may also be analyzed by ion chromatography. Such compounds also serve as complexing agents for alkaline-earth metals and several auxiliary group elements. With respect to monophosphonic acids, the 2-phosphonobutane-1,2,4-tricarboxylic acid is of very high industrial relevance. In diphosphonic acids, one distinguishes between geminal and vicinal structures. Geminally substituted phosphonic acids form stronger metal complexes.

$$\underset{\text{geminal}}{H-\overset{H}{\underset{H}{C}}-\overset{PO_3H_2}{\underset{PO_3H_2}{C}}-H} \qquad \underset{\text{vicinal}}{H-\overset{H}{\underset{H_2O_3P}{C}}-\overset{PO_3H_2}{\underset{H}{C}}-H}$$

A typical compound of the first class is 1-hydroxyethane-1,1-diphosphonic acid (HEDP, DEQUEST 2010), which is often employed in practical applications. As shown in Fig. 3-97, this compound may be analyzed in the same run with other aminopolyphosphonic acids.

Substituting a hydroxide group of the phosphonic acid by an alkyl group or an aryl group, respectively, leads to the compound class of polyphosphinic acids:

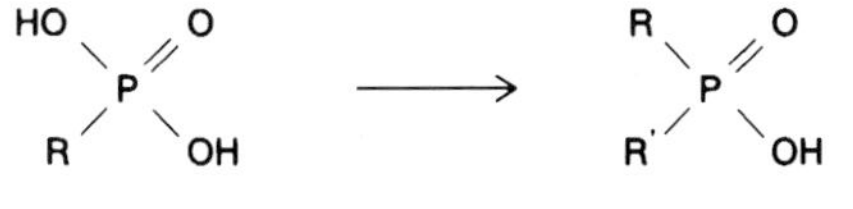

$$(HO)(R)P(=O)(OH) \longrightarrow (R)(R')P(=O)(OH) \qquad (84)$$

Generally, these compounds form complexes with alkaline-earth metals that are less stable than the corresponding complexes of the polyphosphonic acids. Weiß and Hägele [96] investigated the chromatographic behavior of a variety of aliphatic and olefinic polyphosphonic and polyphosphinic acids, respectively, with 1 to 4 phosphorus atoms [87,88], that are applied in medicine and in pharmaceutical chemistry.

1 1,2-Ethanediphosphonic acid

2 Ethane-1,2-bis(P-methylphosphinic acid)

3 Ethane-1,2,2-tris(P-methylphosphinic acid)

4 Ethane-1,1,2,2-tetrakis(P-methylphosphinic acid)

5 1-Phenylethane-1,2-diphosphonic acid

6 1-Phenylethane-1,2,2-triphosphonic acid

7 1-Phenylethane-1,2,2-tris(P-methylphosphinic acid)

8 1-Phenylethene-1-phosphonic acid

9 *trans*-1-Phenylethene-2-phosphonic acid

P*: $-P(O)(OH)_2$
P°: $-CH_3P(O)(OH)$

A comparison of the chromatographic behavior of the polyphosphinic acids **2**, **3**, and **4** (Fig. 3-100) reveals that the elution order depends strictly on the number of phosphinic acid groups. Thus, it follows one of the basic rules in ion chromatography, i.e., that the retention increases with increasing charge of the respective anion. This observation also applies to the polyphosphonic acids **5** and **6**, whose separation is displayed in Fig. 3-101. The example of the compounds **6** and **7** in Fig. 3-102 clearly shows that polyphosphinic acids always elute prior to polyphosphonic acids having the same carbon skeleton.

As revealed by a comparison of the chromatographic behavior of **1** and **5** in Fig. 3-101, the introduction of a phenyl group into the molecule results in an increase in the retention at an identical number of phosphonic acid groups. This finding is explained by π-π interactions between the stationary phase and the aromatic ring system of the analyte.

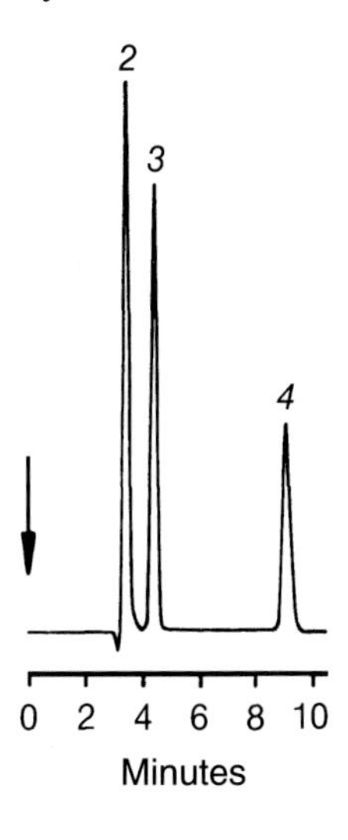

Fig. 3-100. Separation of various aliphatic polyphosphinic acids. – Chromatographic conditions: see Fig. 3-97; solute concentrations; 20 ppm each of **2**, **3**, and **4**.

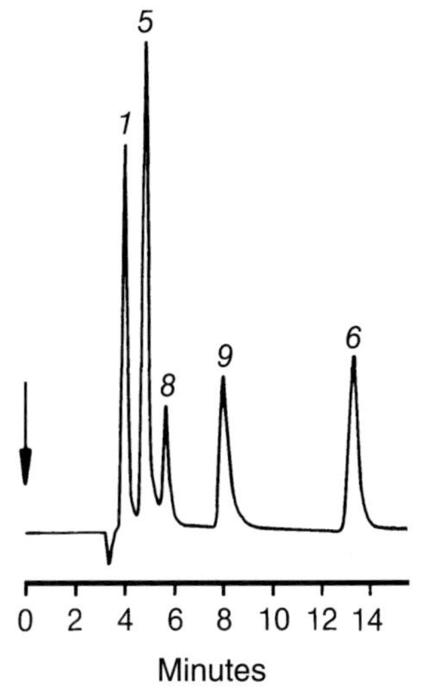

Fig. 3-101. Separation of aliphatic and aromatic polyphosphonic acids. – Separator column: IonPac AS7; eluent: 0.04 mol/L HNO_3; detection and injection volume: see Fig. 3-97; solute concentrations: 100 ppm each of **1**, **5**, **8**, **9**, and **6**.

In general, the chromatographic analysis of polyvalent anions shows an overlapping of ion-exchange phenomena and adsorption phenomena. Above all, the differing adsorption behavior is reponsible for the separation of structural isomeric compounds. This is evident, for example, from the compounds **8** and **9** in Fig. 3-101, which are only distinguished by the position of the phosphonic acid group. Since the dissociation behavior that governs the ion-exchange process is expected to be similar for both compounds, the separation of these compounds is probably based on a different adsorption behavior.

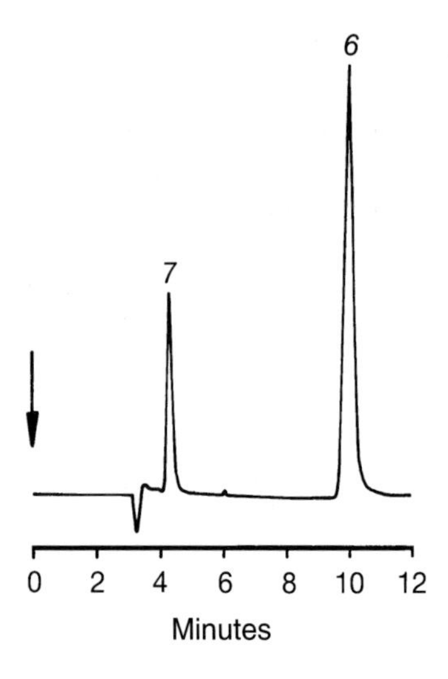

Fig. 3-102. Separation of an aromatic polyphosphinic and polyphosphonic acid of equal carbon backbone. – Chromatographic conditions: see Fig. 3-96; solute concentrations: 200 ppm each of **7** and **6**.

In addition to structural isomers, stereoisomeric and, recently, rotational isomeric polyphosphonic and polyphosphinic acids can be separated. Most intriguing in this connection is the chromatogram in Fig. 3-103 of the sterically-crowded polyphosphonic acid **10**,

10 H–C(tBu)(P*)–C(P*)(P*)–H *tert.*-Butylethane-1,2,2-triphosphonic acid

P*: -P(O)(OH)$_2$

that exists in two rotameric forms. They could be identified unequivocally and quantified using ^{31}P nuclear magnetic resonance spectroscopy. Fig. 3-103 reveals that an ion chromatographic separation of both rotamers is possible, suggesting a slow rotation around the C-C axis labelled in the illustration above. It can be safely assumed that an equilibrium between both rotamers is established in the mobile phase which is characterized by the ratio of their peak areas. The rotation is interrupted only when the compound is retained for a short time at the fixed stationary charge. This is the only way to explain why both rotamer signals are well separated but merge into each other. With respect to the percentage distribution between both rotamers, comparable results are obtained with both analysis methods, i.e., NMR and ion chromatography. Remarkable results were obtained from an investigation into the temperature dependence of the rotamer's retention. As the column temperature decreases, a reduction in the retention difference between both signals is observed, although the opposite effect should be expected. Further studies regarding the dissociation behavior of these compounds, etc., lead one to expect a clarification of this phenomenon in the future.

Recently, another method for the separation and determination of polyphosphonic acids via anion exchange chromatography was described by Vaeth et al. [84]. The objective of this work was to develop a phosphorus-specific detection, because the reaction of polyphosphonic acids with ferric nitrate is not specific, and the determination of these components is impeded by other complexing agents and/or by the presence of high amounts of inorganic anions such as chloride and sulfate in the matrix. In the method developed by Vaeth et al., polyphosphonic acids are hydrolyzed by ammonium peroxodisulfate at high temperature (110 °C) to orthophosphate that, as described in Section 3.3.4.4, is converted in a second reaction step with molybdate/vanadate reagent to yellow phosphorvanadomolybdic acid. The light absorption of that product can be measured at a wavelength of 410 nm. Compared with the detection using ferric nitrate, this method offers the advantage that, even in complex matrices, only the phosphorus compound is detected. Again, a combination of potassium chloride and EDTA is utilized as the eluent that has already proved very worthwhile for the analysis of condensed phosphates. The retention behavior of polyphosphonic acids is affected by the pH value

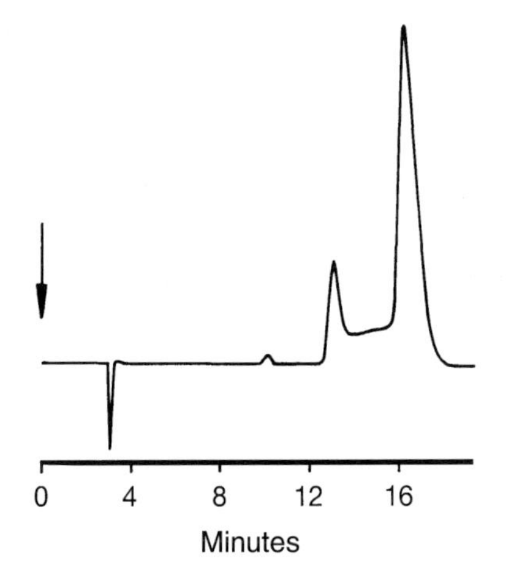

Fig. 3-103. Separation of the two rotamers of *tert.*-butylethane-1,2,2-triphosphonic acid. – Chromatographic conditions: see Fig. 3-97; column temperature: 25 °C; solute concentration: 100 ppm.

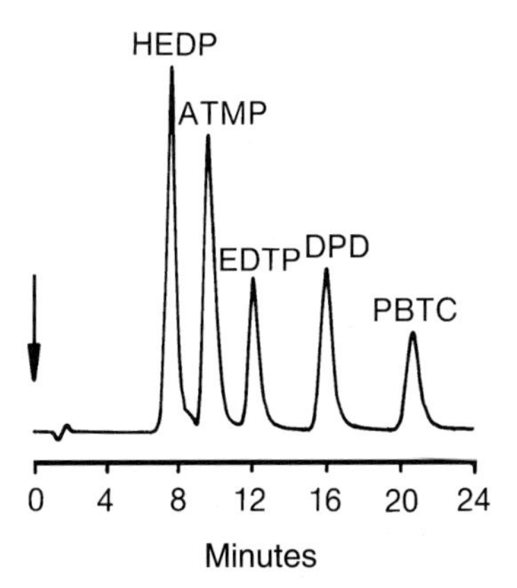

Fig. 3-104. Separation of polyphosphonic acids upon application of phosphorus-specific detection. – Separator column: IonPac AS7; eluent: 0.17 mol/L KCl + 0.0032 mol/L EDTA, pH 5.1; flow rate: 0.5 mL/min; detection: photometry at 410 nm after hydrolysis and derivatization with vanadate/molybdate; injection: 50 μL, 1-hydroxyethane-1,1-diphosphonic acid (HEDP), aminotris-(methylenephosphonic acid) (ATMP), ethylenediamine-tetramethylenephosphonic acid (EDTP), 1,1-diphosphonopropane-2,3-dicarboxylic acid (DPD), and 2-phosphonobutane-1,2,4-tricarboxylic acid (PBTC); (taken from [84]).

of the mobile phase. In general, an increase in the retention is observed with increasing pH value. The high selectivity of this special separation and detection system is exemplified in Fig. 3-104 by the separation of some selected polyphosphonic acids.

3.3.5.3 Carbohydrates

The GC and HPLC methods were developed for the chromatographic separation of carbohydrates. Gas chromatographic methods are very time consuming, because carbohydrates, owing to their non-volatile character, have to be derivatized prior to the determination [97]. On the other hand, the LC method using strong-acid cation exchangers in the calcium form and de-ionized water as the eluent [98] is widespread, but has several drawbacks:

- If the sample to be analyzed contains organic acids, the metal ion fixed at the ion-exchange group can be removed by complexation, which increases the frequency of ion exchanger regeneration.
- High capacity cation exchangers exhibit poor selectivity for higher oligosaccharides. Even the separation of short-chain oligosaccharides requires an ion-exchange resin with a low degree of crosslinking.
- Owing to the resin's compressibility resulting from the crosslinking, only low flow rates can be used which yield long analysis times.
- Optimal separation is achieved only at temperatures around 85 °C, a serious problem for the detection of sugars via measuring the change in refractive index.
- The parameters that affect retention are limited, because only water or mixtures of water and organic solvents can be utilized as eluents.

Another possibility is the separation of carbohydrates after complexation with boric acid [99]. Although such separations are characterized by high selectivities, the kinetics for complex formation are very slow. Thus, low flow rates are required which again result in long analysis times.

Good separations are obtained for mono- and disaccharides using silica-based chemically bonded aminopropyl phases [100, 101]. However, larger oligosaccharides are eluted as broad peaks. Such separator columns have only a limited lifetime, especially when the samples to be analyzed contain aldehydes or ketones. These compounds, that are often ingredients in foods and beverages, react with the amino group to yield a Schiff base (RR′C=NR″, azomethine).

Alternatively, carbohydrates can be separated by anion exchange chromatography [102]. Because the pK values for sugars lie between 12 and 14 [103], they can be converted into their anionic form in a superalkaline environment, and may be separated on a strongly basic anion exchanger in the hydroxide form. Sodium hydroxide in the concentration range between c = 0.001 and 0.15 mol/L is suitable as the eluent. The hydroxide ions have two functions: (1) they act as eluent ions and (2) they determine the pH value of the mobile phase. Thus, a change in the hydroxide ion concentration in the mobile phase has two different effects on the retention behavior of carbohydrates. While the carbohydrate dissociation and, consequently, retention increase with rising pH value, the associated higher eluent ion concentration results in a retention decrease. As long as the carbohydrates are not fully dissociated, the two effects compensate each other. At complete dissociation of the carbohydrates, a further increase in the hydroxide ion concentration merely results in a decreased retention. Carbohydrates with a high affinity toward the stationary phase can be analyzed by adding sodium acetate to the eluent, which is not detected by the pulsed amperometry detection system [104].

In some carbohydrates, such as epimeric compounds, a co-elution is often observed, since the retention behavior of these compounds does not differ very much. With respect to the different dissociation behavior, however, the selectivity of the separation system can be enhanced significantly by lowering the pH of the mobile phase to one that is comparable to the p*K* value of the compound.

In preparing sodium hydroxide-based eluents, care must be taken to ensure that they are carbonate free. Because the carbonate ion exhibits a much higher elution strength than the hydroxide ion, the presence of even small traces of carbonate reduces the resolution. The eluents should, therefore, be prepared from a 50% NaOH concentrate which only contains minute amounts of carbonate. The de-ionized water being used to

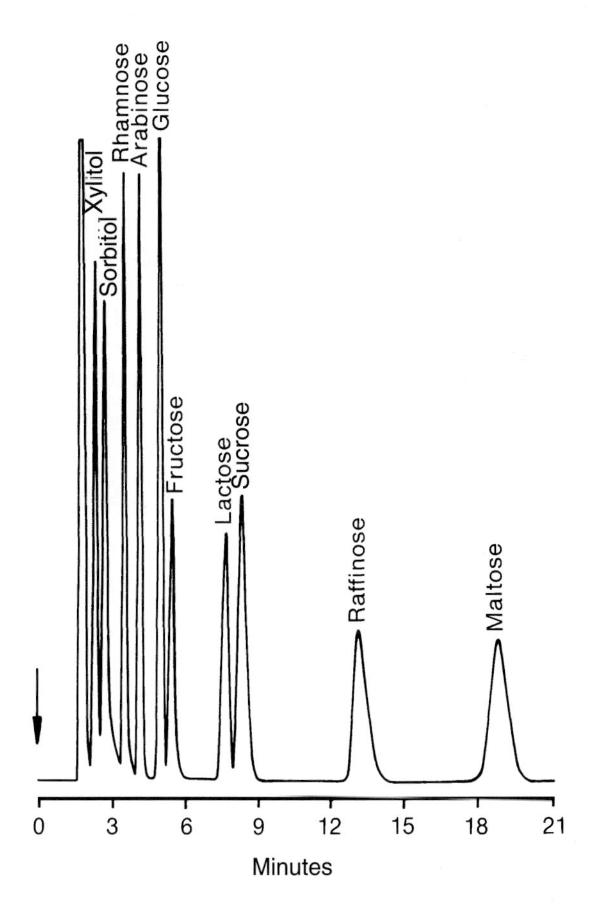

Fig. 3-105. Separation of various sugar alcohols and saccharides. – Separator column: CarboPac PA-1; eluent: 0.15 mol/L NaOH; flow rate: 1 mL/min; detection: pulsed amperometry at a Au working electrode; injection volume: 50 μL; solute concentrations: 10 ppm xylitol, 5 ppm sorbitol, 20 ppm each of rhamnose, arabinose, glucose, fructose, and lactose, 100 ppm sucrose and raffinose, 50 ppm maltose.

make up the mobile phase has to be degassed thoroughly with helium prior to the addition of the concentrate. The eluent itself must be stored under helium.

The anion exchanger CarboPac PA-1 (IonPac AS6) is suitable for the separation of carbohydrates. Fig. 3-105 shows a standard chromatogram with an isocratic separation of various sugar alcohols and saccharides. Although the affinity of carbohydrates toward the stationary phase increases in the following order: sugar alcohols < monosaccharides < oligosaccharides < polysaccharides, overlaps in the elution order are possible in mixtures of different oligosaccharides. Thus, at temperatures slightly above ambient (as seen in Fig. 3-105), the trisaccharide raffinose and the tetrasaccharide stachyose elute ahead of the disaccharide maltose.

A major advantage of using anion exchange chromatography in carbohydrate separations is the possibility to affect both the retention times and the elution order. Fig. 3-106 shows the dependence of the retention on the hydroxide ion concentration exemplified by some selected carbohydrates. This figure reveals that optimal resolution between raffinose, stachyose, and maltose is obtained at a NaOH concentration of c = 0.15 mol/L. A second parameter that affects retention is the column temperature, where a decrease in retention is generally observed with rising temperature. The temperature's influence on the resulting retention time can be correlated with the molecular size of the analyte and decreases in the order: oligosaccharides < sugar alcohols < monosaccharides. The dependence of the retention on the column temperature is displayed in Fig. 3-107. It is noteworthy that the elution order of maltose and stachyose changes with increasing temperature. However, the column temperature should not be raised too much. At temperatures above 45 °C, the column efficiency decreases, and secondary signals and tailing effects are observed, which may be attributed to the Lobry-de-Bruyn-von-Ekenstein-rearrangement. This represents an isomerization occuring predominantly at C-atom 2 of the sugar moiety that leads to a change in configuration via an intermediate endiol. Therefore, the anion exchange column is operated at temperatures between 20 °C and 40 °C.

$$\begin{array}{c} HC{=}O \\ | \\ HC{-}OH \\ | \\ HO{-}CH \\ | \\ HC{-}OH \\ | \\ HC{-}OH \\ | \\ H_2C{-}OH \\ \\ \text{D-Glucose} \end{array} \rightleftharpoons \left[\begin{array}{c} HC{-}OH \\ \| \\ C{-}OH \\ | \\ HO{-}CH \\ | \\ HC{-}OH \\ | \\ HC{-}OH \\ | \\ H_2C{-}OH \end{array} \right]_{\text{Endiol}} \rightleftharpoons \begin{array}{c} H_2C{-}OH \\ | \\ C{=}O \\ | \\ HO{-}CH \\ | \\ HC{-}OH \\ | \\ HC{-}OH \\ | \\ H_2C{-}OH \\ \\ \text{D-Fructose} \end{array} \qquad (85)$$

The conversion of aldoses to ketoses is also effected by dilute sodium hydroxide solutions. However, these reactions occur very slowly and are not observed in the time period necessary for the chromatography.

Monosaccharides

Fig. 3-105 reveals that a variety of sugar alcohols, saccharides, and oligosaccharides can be separated under isocratic conditions using sodium hydroxide at a concentration

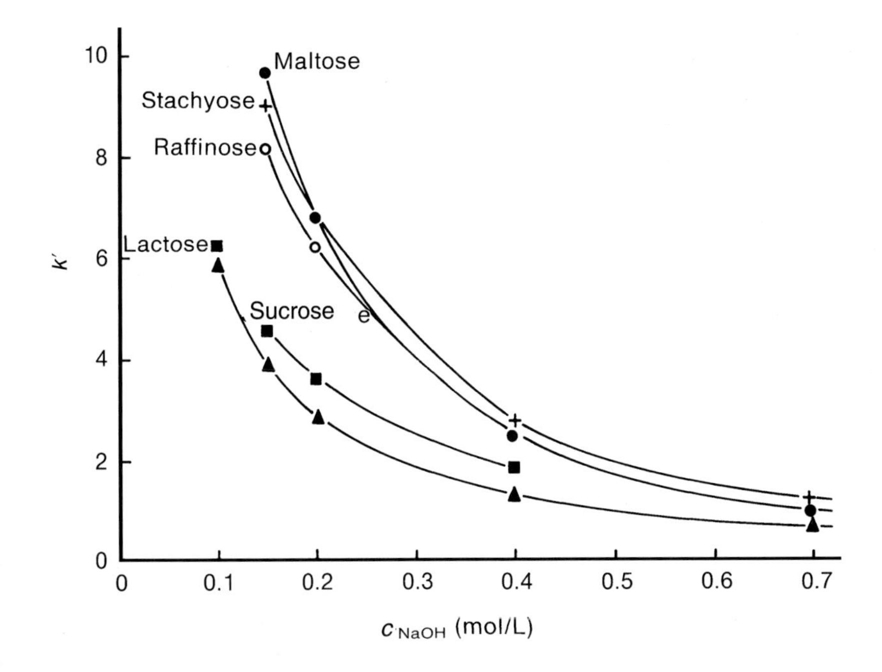

Fig. 106. Dependence of the retention of some selected carbohydrates on the hydroxide ion concentration. – Separator column: CarboPac PA-1; flow rate: 1 mL/min; detection and injection volume: see Fig. 3-105; solute concentrations: 100 ppm each of the respective carbohydrates.

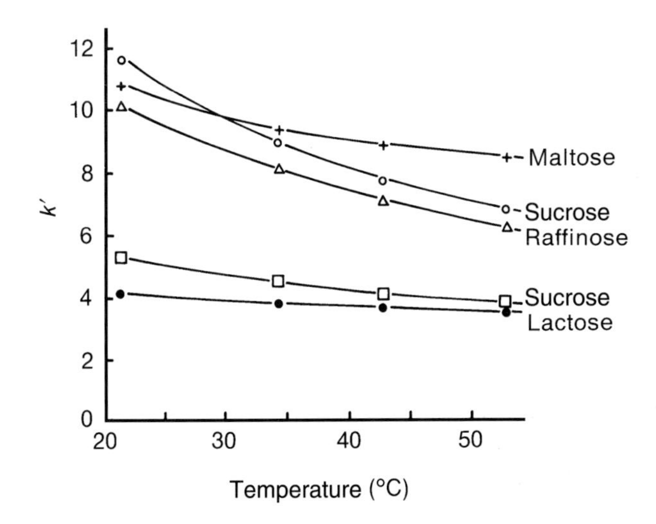

Fig. 3-107. Dependence of the retention of some selected carbohydrates on the column temperature. – Separator column: CarboPac PA-1; eluent: 0.15 mol/L NaOH; flow rate: 1 mL/min; detection and injection volume: see Fig. 3-105; solute concentrations: 100 ppm each of the respective carbohydrates.

c = 0.15 mol/L. The separation of sugar alcohols that exhibit a very low affinity toward the stationary phase of a CarboPac column is extremely difficult. Even lowering the sodium hydroxide concentration to c = 0.001 mol/L and applying two separator columns in series only allows the baseline-resolved separation of sugar alcohols with differing number of C-atoms such as xylitol and sorbitol. However, a complete separation of the most important hexites such as sorbitol, mannitol, and dulcitol is impossible.

Cyclic polyalcohols (or "cyclitols", according to the nomenclature of sugar alcohols) also exhibit a very low affinity. The most important representative of this class of compounds is inositol,

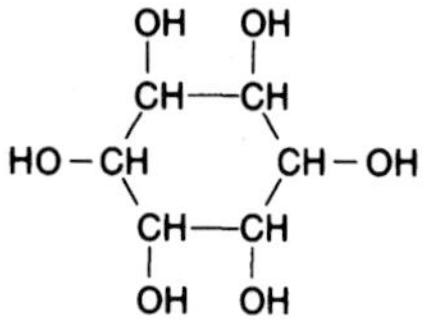

```
        OH   OH
        |    |
        CH---CH
       /       \
  HO-CH         CH-OH
       \       /
        CH---CH
        |    |
        OH   OH
```

Inositol

with a retention behavior comparable to pentitols. In its meso-configuration, inositol can be phosphorylated. Corresponding procedures for the separation and determination of various inositol phosphates were recently introduced by Smith and MacQuarrie [105].

In contrast, the separation of the various monosaccharides is much easier, which is surprising because the affinities of these compounds are not very different. However, when the pH value of the mobile phase is lowered to a value corresponding to the pK value of the respective compounds, even small differences in the dissociation behavior contribute to the separation. Thus, the chromatogram in Fig. 3-108 is obtained using a dilute sodium hydroxide solution of c = 0,001 mol/L. Under these conditions, even epimeric compounds such as glucose, mannose, and galactose are baseline resolved. Two important aldopentoses – arabinose and xylose – can also be determined in the same analysis. Remarkable is the elution order, which cannot be correlated with the carbon chain length of the monosaccharides.

As seen in Fig. 3-108, two other compounds from the group of deoxy sugars are eluted prior to aldopentoses and aldohexoses. While fucose (6-deoxy-L-galactose) and other 6-deoxyhexoses (also called methyloses) are abundant in nature in cardiotoxic glycosides, 2-deoxy-D-ribose, one of the building blocks of deoxyribonucleic acid, is of great biological significance. Rhamnose (6-deoxy-L-mannose), as a fission product of many glycosides, elutes between arabinose and galactose.

By replacing an alcoholic hydoxide group in monosaccharides with an amino group, one obtains the class of amino sugars. The two most important representatives are D-glucosamine and D-galactosamine, which can be separated together with their N-acetylated derivatives via anion exchange chromatography. Usually, sodium hydroxide solution of c = 0.01 mol/L (pH 12) is used as the eluent. If the concentration falls below this value, then concentrated sodium hydroxide (c = 0.3 mol/L) must be added to the column effluent prior to entering the detection cell. This raises the pH value of the mobile phase to pH 13, which is optimal for the detection of amino sugars via pulsed amperometry. The sodium hydroxide concentrate is added pneumatically via a metal-free T-junction connected to a mixing coil of respective dimensions used for mixing the

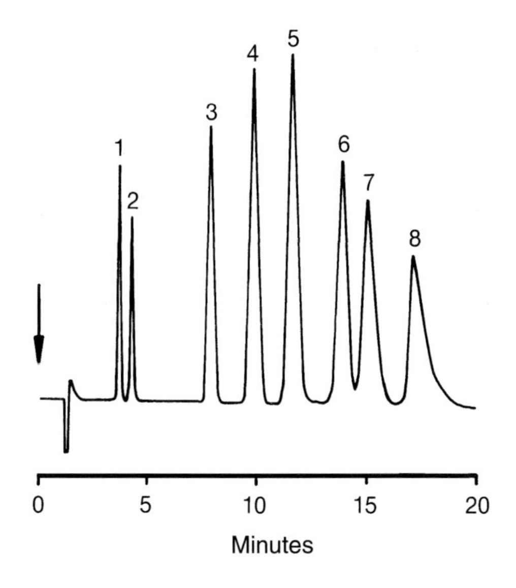

Fig. 3-108. Separation of various monosaccharides under isocratic conditions. – Separator column: CarboPac PA-1; eluent: 0.001 mol/L NaOH; flow rate: 1 mL/min; detection and injection volume: see Fig. 3-105; solute concentrations: 25 ppm fucose (1), deoxyribose (2), arabinose (3), galactose (4), glucose (5), and xylose (6), 50 ppm mannose (7), and fructose (8).

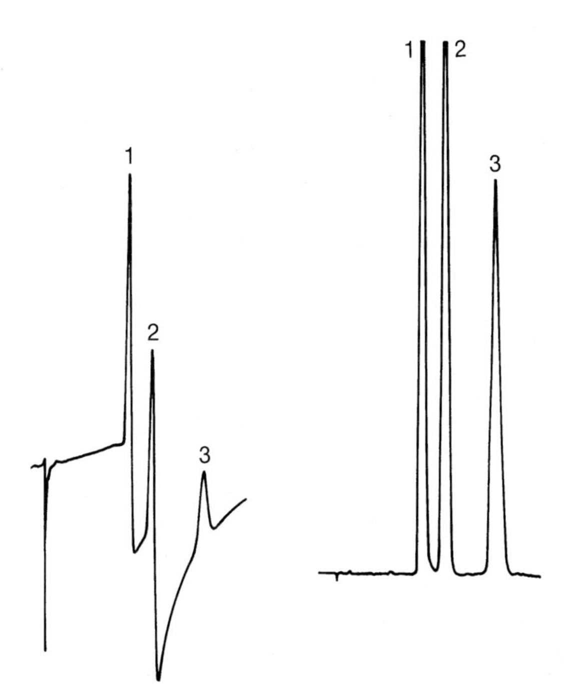

Fig. 3-109. Separation of amino sugars with and without post-column addition of NaOH. – Separator column: CarboPac PA-1; eluent: 0.01 mol/L NaOH; flow rate: 1 mL/min; detection and injection volume: see Fig. 3-105; solute concentrations: 30 ppm galactosamine (1), 60 ppm glucosamine (2), and 30 ppm N-acetylgalactosamine (3).

column effluent and the NaOH concentrate. The importance of adding a NaOH concentrate for the baseline stability is demonstrated in the chromatograms of various amino sugars shown in Fig. 3-109. The left chromatogram clearly reveals that the strong baseline steps renders a reasonable interpretation of the signals normally impossible without post-column addition of a NaOH concentrate.

The post-column addition of concentrated NaOH is significant only for the electrochemical detection of amino sugars. The application of a gradient elution technique allows the analysis of carbohydrates with strongly differing retention behaviors in the same run. As a prerequisite, a constant pH value in the detector cell must be maintained, because in electrochemical detection, pH variations lead to strong baseline drifts. For the gradient elution of mono- and disaccharides, however, the NaOH concentration in the mobile phase must be increased during the run. The resulting pH change requires a post-column addition of NaOH solution in order to ensure baseline stability. The chromatogram of various mono- and disaccharides shown in Fig. 3-110 reveals that, with respect to selectivity and sensitivity, the introduction of the gradient technique with pulsed amperometric detection represents one of the greatest innovations in the field of carbohydrate analysis.

The mono- and diphosphoric acid esters of the sugars are particularly important in enzymatic carbohydrate metabolism. The phosphate group is attached to the glycosidic

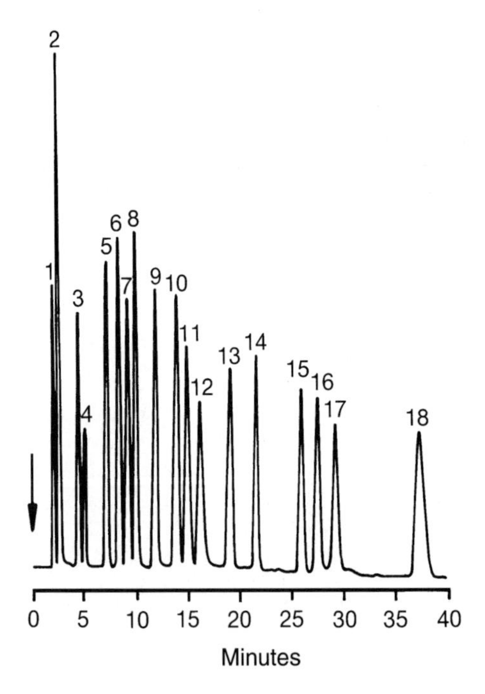

Fig. 3-110. Gradient elution of various mono- and disaccharides. – Separator column: IonPac AS6A; eluent: (A) water, (B) 0.05 mol/L NaOH; gradient: linear, from 7% B to 100% B in 15 min; flow rate: 0.8 mL/min; detection: pulsed amperometry at a Au working electrode with post-column addition of NaOH; injection volume: 50 µL; solute concentrations: 15 ppm inositol (1), 40 ppm sorbitol (2), 25 ppm fucose (3) and deoxyribose (4), 20 ppm deoxyglucose (5), 25 ppm arabinose (6), rhamnose (7), galactose (8), glucose (9), xylose (10), mannose (11), fructose (12), melibiose (13), isomaltose (14), gentiobiose (15), and cellubiose (16), 50 ppm turanose (17), and maltose (18).

hydroxide group or to a primary alcohol group. As primary esters of phosphoric acid, phosphorylated sugars are very resistant to hydrolysis. Biochemically, they are digested only by enzymes. Fig. 3-111 shows a separation of phosphorylated monosaccharides. Again, the simultaneous detection of mono- and diphosphoric acid esters is only possible by application of a gradient elution technique. Because phosphorylated sugars are strongly retained on the stationary phase of an anion exchanger, higher amounts of sodium acetate, which has a much higher elution strength compared to sodium hydroxide, are added to the mobile phase. In this case, sodium hydroxide serves predominantly to adjust the pH required for electrochemical detection. It is particularly important to note that it is possible to distinguish between anomers of phosphorylated monosaccharides with anion exchange chromatography. The comparison of the retention behavior of the two anomers **4** and **6** in Fig. 3-111 reveals that an equatorially bound phosphate group in β-D-glucose-1-phosphate leads to markedly higher retention than the axially bound group in α-D-glucose-1-phosphate. A separation of the glucopyranose anomer mixture present in solution, however, is not possible.

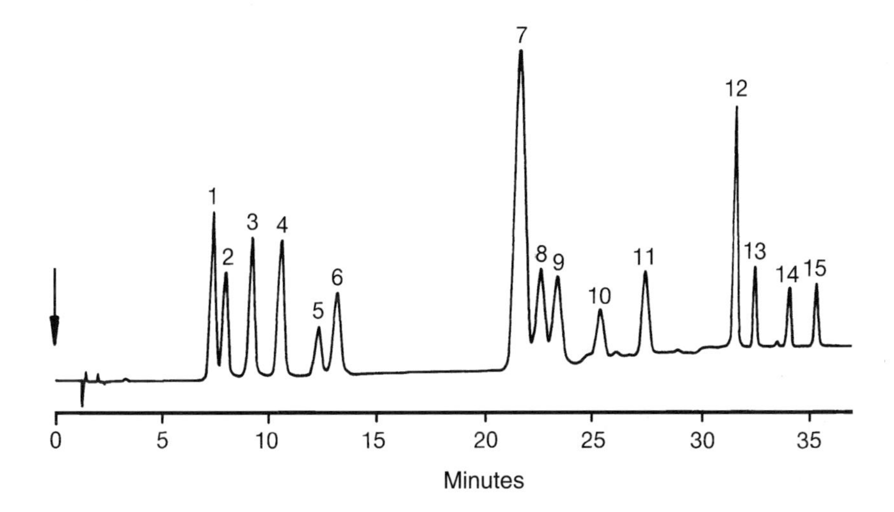

Fig. 3-111. Gradient elution of phosphorylated mono and disaccharides. – Separator column: CarboPac PA-1; eluent: (A) 0.1 mol/L NaOH, (B) 0.1 mol/L NaOH + 1 mol/L NaOAc; gradient: linear, 10% B to 20% B in 20 min, then to 50% B in 10 min; flow rate: 1 mL/min; detection and injection volume: see Fig. 3-105; solute concentrations: 22.6 ppm α-D-galactosamine-1-P (1), 9 ppm α-D-glucosamine-1-P (2), 35 ppm α-D-galactose-1-P (3) and α-D-glucose-1-P (4), 29.2 ppm α-D-ribose-1-P (5), 35 ppm β-D-glucose-1-P (6), 75 ppm D-glucosamine-6-P (7), 40.8 ppm D-galactose-6-P (8), 25 ppm D-glucose-6-P (9), 19.2 ppm D-fructose-1-P (10), 8.4 ppm D-fructose-6-P (11), 61.6 ppm α-D-glucoronic acid-1-P (12), 21.2 ppm α-D-glucose-1,6-diP (13), 18.4 ppm β-D-fructose-2,6-diP (14), and D-fructose-1,6-diP (15).

Oligosaccharides

A disaccharide is formed by coupling the hemiacetalic hydroxide group of a monosaccharide with any hydroxide group of a second sugar molecule. Trisaccharides result from a glycosidic formation between three molecules and, in an analogous manner, tetra-, penta-, and hexasaccharides are formed.

In disaccharides, one distinguishes between reducing and non-reducing compounds, depending on the type of coupling. Non-reducing disaccharides are characterized by different hexoses coupling across the hemiacetalic hydroxide group. The two most important representatives are trehalose and sucrose. In reducing sugars, the hemiacetalic hydroxide group of one sugar molecule is linked via glycosidic coupling to the alcoholic hydroxide group of the other sugar. According to the most important representative, they are called maltose-type disaccharides. In naturally occuring disaccharides of this group, the reducing sugar molecule is coupled to the glycosidic sugar moiety in the 4- or 6-position.

Maltose

Isomaltose

Haworth Projections

*no configuration attributed

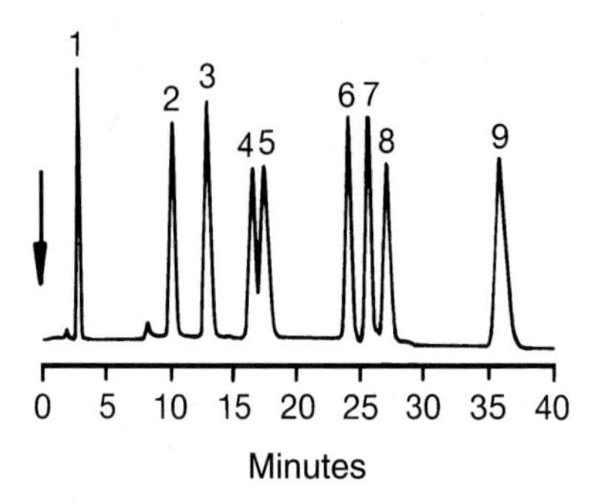

Fig. 3-112. Gradient elution of various disaccharides. – Separator column: IonPac AS6A; eluent: (A) water, (B) 0.05 mol/L NaOH + 0.0015 mol/L acetic acid; gradient: 20% B for 5 min isocratically, then linearly to 100% B in 15 min; flow rate: 0.8 mL/min; detection and injection volume: see Fig. 3-110; solute concentrations: 10 ppm trehalose (1), 25 ppm sucrose (2), lactose (3), isomaltose (4), melibiose (5), gentiobiose (6), cellobiose (7), 50 ppm turanose (8), and maltose (9).

Reducing and non-reducing disaccharides exhibit different retention behaviors, but can be analyzed in a single separation by gradient elution. To reduce the analysis time, small amounts of acetic acid are added to the sodium hydroxide acting as the eluent. Fig. 3-112 shows a separation of different disaccharides on the IonPac AS6A 5-μm anion exchanger.

Tri- and tetrasaccharides such as raffinose and stachyose may be separated with the chromatographic conditions given in Fig. 3-105. When adding sodium acetate to the eluent as is done for the separation of phosphorylated monosaccharides, larger oligosaccharides can be analyzed by anion exchange chromatography. Fig. 3-113 shows the gradient elution of maltose oligomers (DP2 to DP6) as an example of such a separation.

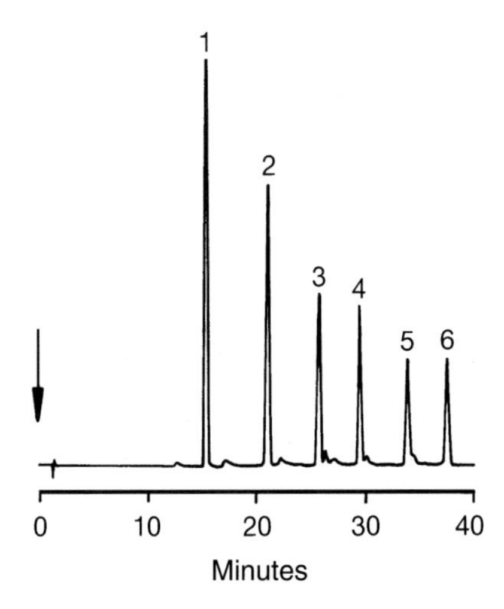

Fig. 3-113. Separation of maltose oligomers. – Separator column: CarboPac PA-1; eluent: (A) 0.1 mol/L NaOH, (B) 0.1 mol/L NaOH + 1 mol/L NaOAc; gradient: 100% A for 2 min isocratically, then linearly to 100% B in 200 min; flow rate: 1 mL/min; detection and injection volume: see Fig. 3-105; solute concentrations: 166 ppm each of maltose (1), maltotriose (2), maltotetraose (3), maltopentaose (4), maltohexaose (5), and maltoheptaose (6).

Polysaccharides

Polysaccharides consist of many monosaccharide residues and occur in the vegetable and animal kingdoms as storage and skeletal carbohydrates. In the past, polysaccharides were separated with chemically bonded octadecyl and aminopropyl phases [106-109] using water or water/acetonitrile mixtures as the eluent. Although these phases exhibit high chromatographic efficiencies, only oligomers with a comparatively low degree of polymerization (< DP15) can be analyzed as a result of the detection system employed. Up to now, carbohydrates have been detected via refractive index measurements. The gradient elution technique, essential for the separation of higher oligo- and polysaccharides, has not been applied.

Using anion exchange chromatography with pulsed amperometric detection, polymers up to DP70 may be analyzed. The necessary gradient elution technique is based on the combination of sodium hydroxide and sodium acetate eluents described above.

An example of the efficiency of anion exchange chromatography is demonstrated by using cyclodextrines [110]. Depending on the number of glucose residues, one distinguishes between α-, β-, and γ-cyclodextrin. Isolated from liquified potato starch after fermentation based on the enzyme system of Bac. macerans, they are constituents of pharmaceutical and cosmetic products. The separation in Fig. 3-114 reveals that the cyclodextrin retention cannot be correlated with its molecular weight.

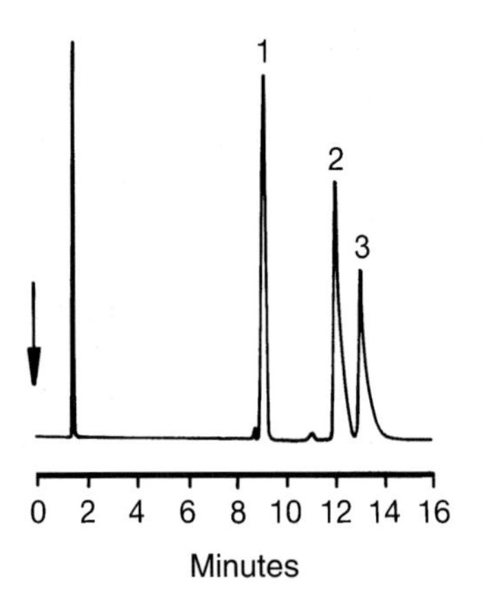

Fig. 3-114. Separation of α-, β-, and γ-cyclodextrin. – Separator column: CarboPac PA-1; eluent: (A) 0.1 mol/L NaOH, (B) 0.1 mol/L NaOH + 0.5 mol/L NaOAc; gradient: linear, 100% A to 100% B in 20 min; flow rate: 1 mL/min; detection and injection volume: see Fig. 3-105; solute concentrations: 500 ppm each of α- (1), β- (2), and γ-cyclodextrin (3).

In dextrans, which are made up of α-D-glucose residues in 1,6-linkage, the retention increases with molecular weight. Dextrans are mainly produced extracellularly by lactic acid bacteria. Sucrose, for example, is transformed into glucose and fructose in a fermentation process by Leuconostoc mesenteroides with transglucosidase. Since only glucose is polymerized, dextrans are homopolymers (α-1,6-glucose). After partial hydrolysis, their molecular weight ranges from 20,000 to 70,000 g/mol. Fig. 3-115 shows the separation of a dextran hydrolysate (dextran 1000). In addition to the polymers up to

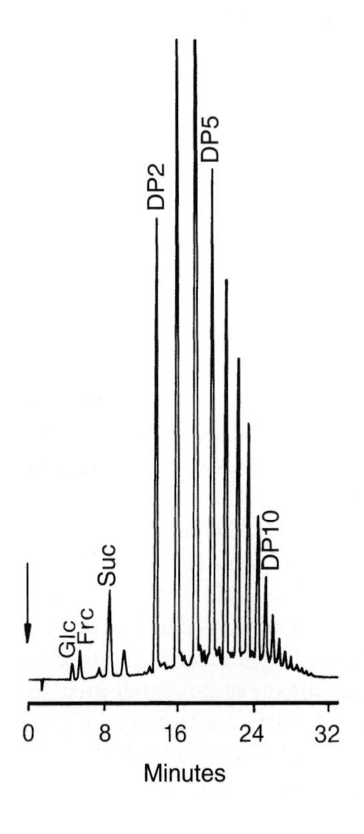

Fig. 3-115. Gradient analysis of a dextran hydrolysate (dextran 1000). – Chromatographic conditions: see Fig. 3-113; injection: 50 µL of a 0.3% solution.

DP 18, small amounts of glucose, fructose, and sucrose from the microbial process were detected. From Fig. 3-115 it appears that a separation of the α- and β-forms (pair of anomers) of the individual polysaccharides is not possible by anion exchange chromatography. The alkalinity of the mobile phase accelerates the transformation of the anomers into each other and, therefore, the sum of the signals is obtained for each anomeric pair. This is in contrast with the double peaks, representing the α- and β-forms, respectively, which are observed when separating compounds on chemically bonded aminopropyl phases. When the gradient profile is increased by adding more sodium acetate per time unit, even higher polymers can be resolved. Fig. 3-116 shows this effect with another dextran hydrolysate separation (dextran 5000), in which polymers up to DP50 can be detected.

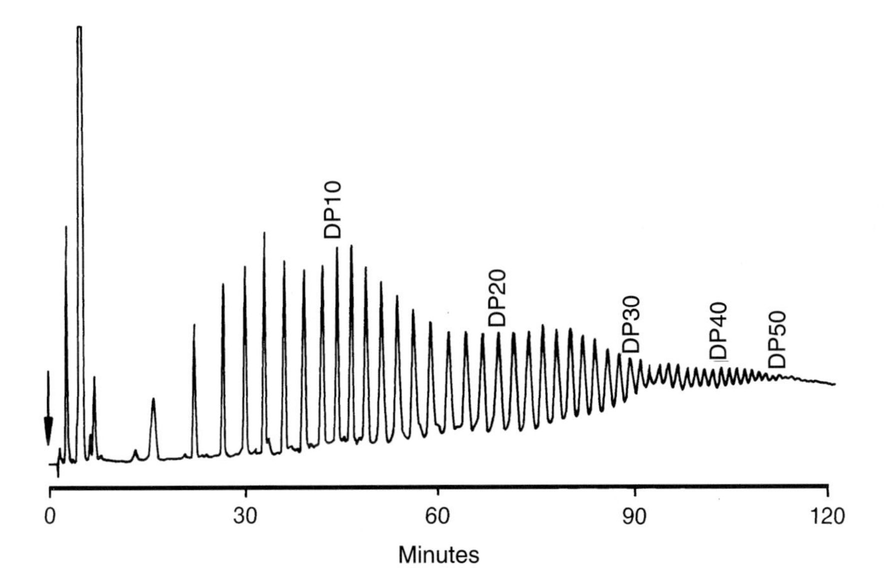

Fig. 3-116. Gradient analysis of a dextran hydrolysate (dextran 5000). – Separator column: CarboPac PA-1; eluent: (A) 0.1 mol/L NaOH, (B) 0.1 mol/L NaOH + 0.5 mol/L NaOAc; gradient: linear, 0% B to 17% B in 40 min, then in 10 min to 100% B; flow rate: 1 mL/min; detection: see Fig. 3-105; injection: 50 μL of a 0.3% solution.

Storage carbohydrates such as polyfructanes may also be analyzed under similar chromatographic conditions. The most important representative of this class of compounds is inulin, which is being increasingly used to diagnose kidney function. It contains about 30 D-fructose residues and a terminal sucrose molecule. The separation of a 0.3% inulin solution in 0.1 mol/L NaOH is displayed in Fig. 3-117. A polymer distribution between DP10 and DP50 is observed. Because the investigated inulin was pretreated with water, polymer residues smaller than DP10 were washed out and, thus, are not detectable.

Still higher polymers were detected after enzymatic degradation of pullulan, a polysaccharide consisting of α-D-glucan and glycosyl residues joined by 1,4- or 1,6-linkages. Pullulan is produced by the fungus Aureobasidium pullulans and can be digested by pullulanase (E.C. 2.2.1.41). This is an enzyme that specifically hydrolyzes α-1,6-linkages. In this reaction, maltotriose residues with a different degree of polymerization are formed, and can be separated chromatographically [110]. The corresponding chromato-

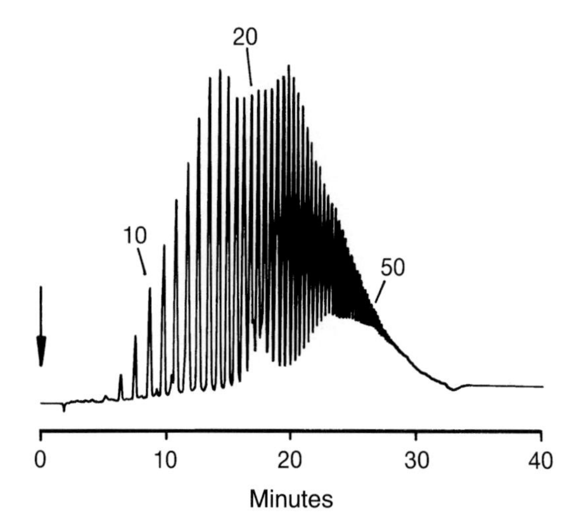

Fig. 3-117. Gradient analysis of an inulin hydrolysate. – Separator column: CarboPac PA-1; eluent: (A) 0.1 mol/L NaOH, (B) 0.1 mol/L NaOH + 1 mol/L NaOAc; gradient: linear, 20% B to 60% B in 40 min; flow rate and detection: see Fig. 3-105; injection: 50 μL of a 0.3% solution.

gram is displayed in Fig. 3-118 and shows the separation of polymers up to DP69. Thus, polysaccharides with a molecular weight of about 10,000 g/mol can be separated and detected via anion exchange chromatography.

3.3.5.4 Oligosaccharides Derived from Glycoproteins

Glycoproteins are extremely interesting biochemical compounds in which the protein is linked to an oligosaccharide via certain amino acids such as serine, threonine, hydroxy-lysine, or asparagine. Glucose, galactose, fucose, mannose, glucosamine, and galactosamine, as well as various acetylated forms of neuraminic acid and its isomers, contribute to the structure of these oligosaccharides. Glycoproteins are present in all eucaryotic cells. While the protein residue is located in the lipid double layer of the cell membrane, the carbohydrate moiety extends into the extracellular space much like an antenna. Specific functions of the carbohydrate result from this arrangement, which plays an important role in cellular and molecular recognition. For example, bacteria that enter cells by specific binding of the proteins, located at the surface of the bacteria, to mannose residues in the glycoprotein.

Many methods are available for the chromatographic separation of proteins, glycoproteins, and their peptide constituents. The majority are RPLC methods which now offer high-performance characteristics and can separate glycoconjugates that differ by only one amino acid unit. In contrast, the separation of glycoconjugates that differ only in their carbohydrate structure poses difficulties. Detailed knowledge of the oligosaccharide portion of a glycoprotein is very important for its complete characterization.

Ion-exchange chromatography combined with pulsed amperometric detection has been successfully employed for the analysis of neutral oligosaccharides (see Section 3.3.5.3). This method can be used for the separation of both oligosaccharides and glycopeptides derived from glycoproteins, especially when they differ only in their oligosaccharide moiety. This method is also suitable for the identification and determination of

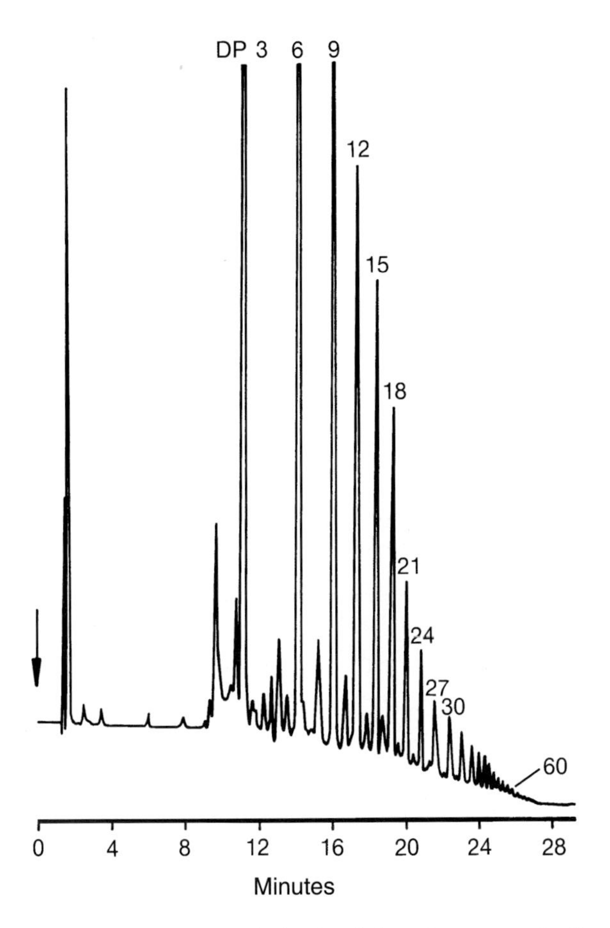

Fig. 3-118. Gradient analysis of pullulan after enzymatic digestion with pullulanase. – Chromatographic conditions: see Fig. 3-115; injection: 50 µL of a 0.2% solution.

the monosaccharide components that make up the carbohydrate structure. They may be isolated from the glycoconjugate for several hours by acid hydrolysis with trifluoroacetic acid at temperatures of 100 °C [111]. A separation of the monosaccharides present in the glycoprotein hydrolysate is shown in Fig. 3-119. It is evident from this figure that all compounds can be separated as sharp and symmetrical signals in less than 15 minutes. To determine the recovery rate after hydrolysis, an internal standard is usually employed. For example, 2-deoxyglucose is particularly useful as the internal standard because it is completely separated from the remaining monosaccharides under the chromatographic conditions given in Fig. 3-119. Compared with other methods, anion exchange chromatography offers the advantage of being able to analyze both free amino sugars and neutral monosaccharides in the same run. However, this does not hold for the N-acetylated derivatives of glucosamine (GlcN) or galactosamine (GalN), which almost co-elute with glucose and mannose. In contrast, interferences by N-acetylated amino sugars do not pose any problems during the analysis of glycoprotein hydrolysates. Under the hydrolysis conditions described above, a complete de-N-acetylation of GlcNAc and GalNAc already occurs after two hours of acid hydrolysis.

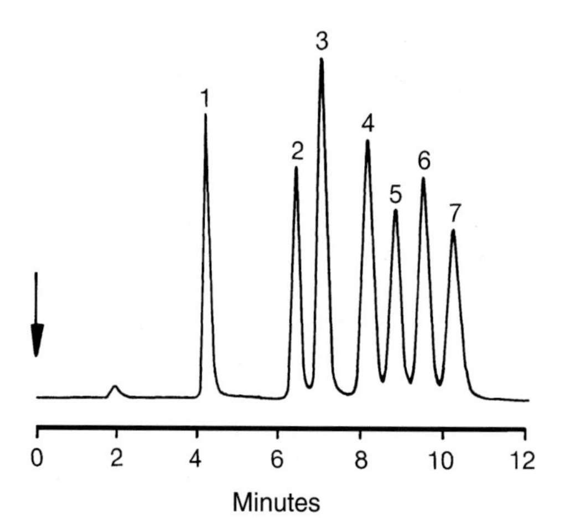

Fig. 3-119. Separation of the monosaccharide components present in glycoprotein hydrolysates. – Separator column: CarboPac PA-1; eluent: 0.022 mol/L NaOH; flow rate: 1 mL/min; detection: pulsed amperometry at a Au working electrode after post-column addition of 0.3 mol/L NaOH; injection volume: 50 μL; solute concentrations: 2.5 nmol each of fucose (1), 2-deoxyglucose (2), galactosamine (3), glucosamine (4), galactose (5), glucose (6), and mannose (7); (taken from [111]).

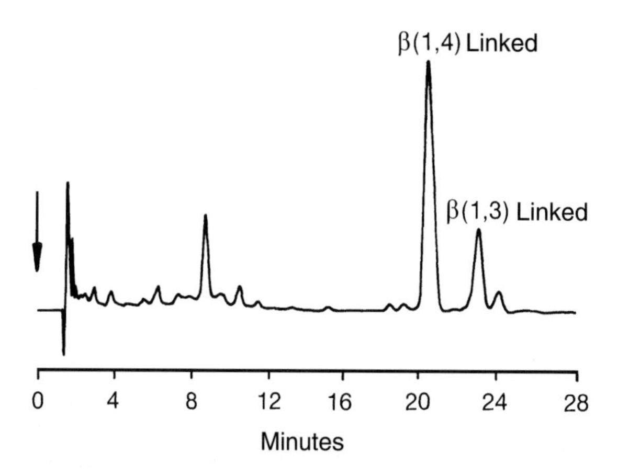

Fig. 3-120. Separation of two structurally similar undecasaccharides from asialo fetuin. – Separator column: CarboPac PA-1; eluent: (A) 0.15 mol/L NaOH, (B) 0.15 mol/L NaOH + 1 mol/L NaOAc; gradient: linear, 50% B to 100% B in 20 min; detection and injection volume: see Fig. 3-105.

As mentioned above, the high selectivity of pellicular anion exchangers allows the separation of structurally isomeric oligosaccharides and glycopeptides derived from glycoproteins. The term "structurally isomeric oligosaccharides" refers to compounds which differ only in the linkage position of a single sugar unit, but not in the number, type, sequence, or anomeric configuration of the monosaccharides used to assemble the oligosaccharide. A separation of such structural isomers is important, for example, in determining the specific action of antibodies [112] and glycosyltransferases [113]. HPLC methods have been developed for the separation of oligosaccharides (trioses to undecaoses) with small structural differences. A survey of these methods can be found in the paper by Honda [114]. Neutral oligosaccharides that differ in their number of sugar units may be easily separated using chemically bonded alkyl [115] or aminopropyl phases [116]. Separations of the 1-6 isomers from their 1-2, 1-3, and 1-4 analogues have also been accomplished for various oligosaccharides with a maximum of 13 monomers [116,117]. In contrast, successful separations of the 1-2, 1-3, and 1-4 isomers from each other are documented only for di-, tri-, and tetrasaccharides [115,118]. Modern latexed anion exchangers are distinguished by a markedly higher selectivity, the most impressive example being the separation of the following two glycopeptides

```
Gal (β1-4) GlcNAc (β1-2) Man (α1-6)                                  Tyr
                               \                                       |
                                Man (β1-4) GlcNAc (β1-4) GlcNAc  --->  Asn
                               /
Gal (β1-4) GlcNAc (β1-2) Man (α1-3)
                           /
   [Gal (β1-4)] GlcNAc (β1-4)

Gal (β1-4) GlcNAc (β1-2) Man (α1-6)                                  Tyr
                               \                                       |
                                Man (β1-4) GlcNAc (β1-4) GlcNAc  --->  Asn
                               /
Gal (β1-4) GlcNAc (β1-2) Man (α1-3)
                           /
   [Gal (β1-3)] GlcNAc (β1-4)
```

that differ merely in the labelled β-1,3 or β-1,4 coupling of galactose and N-acetylglucosamine, respectively [119]. The existence of both isomers, which cannot be separated by conventional RPLC methods, was confirmed by ^{1}H-NMR data. The investigated sample was derived from asialo fetuin; that is, the terminal sialic acid was cleaved using neuraminidase. A prerequisite for the structural analysis of the oligosaccharide moiety of a glycoprotein is the cleavage of the protein moiety via enzymatic hydrolysis. In this respect, the enzyme N-glycanase [peptide-N^{4}-(N-acetyl-β-glucosaminyl)asparagine-amidase] [117] has proved successful in releasing a wide variety of different oligosaccharides from glycoproteins. Because oligosaccharides and glycopeptides can be separated by using the gradient elution technique, as well as detected and quantified by pulsed amperometry, this method is suitable for monitoring the duration of the deglycosylation.

Another example of the high selectivity of a latexed anion exchanger for the analysis of structurally isomeric oligosaccharide side chains of a glycoprotein is the chromatogram in Fig. 3-121. It shows the separation of mannose hexamers with terminal N-

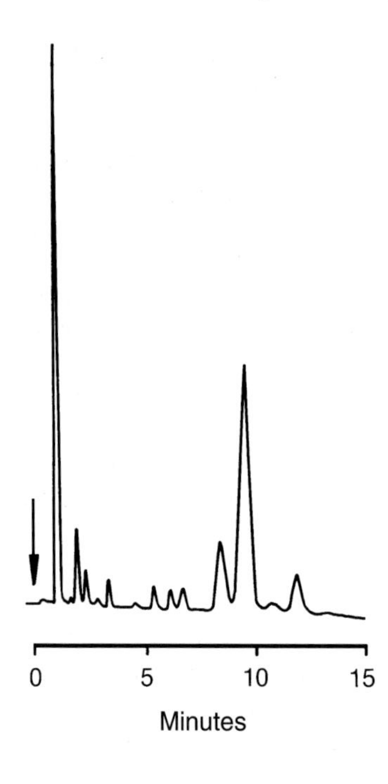

Fig. 3-121. Separation of mannose hexamers from pathological urine. – Separator column: IonPac AS6A; eluent: 0.1 mol/L NaOH + 0.027 mol/L NaOAc; flow rate 0.8 mL/min; detection and injection volume: see Fig. 3-105; solute concentration: 50 ppm.

acetylglucosamine groups which were isolated from pathological urine. The following structures were suggested for the two isomers:

```
Man (α1–6) Man (α1–6) Man (β1–4) GlcNAc
            |          |
       Man (α1–3)     Man (α1–3)

Man (α1–6) Man (α1–6) Man (β1–4) GlcNAc
 |                     |
Man (α1–3)            Man (α1–3)
                       |
                      Man (α1–2)
```

The different linkage of the individual mannose residues allows a baseline-resolved separation of both isomers under isocratic conditions.

The high selectivity of an anion exchanger in its hydroxide form, which is evident from the separation of 1-2, 1-3, and 1-4 isomers, by Hardy and Townsend [120] is attributed to the different acidity of the individual hydroxide groups of a monosaccharide. This, in turn, leads to a different acid strength of oligosaccharides with different linkages of the monomeric residues. The reducing anomeric hydroxide group of a sugar has the highest acidity, followed by the ring hydroxide groups in the order 2-OH >> 6-OH > 3-OH > 4-OH [103]. Correlating the retention behavior of various oligosaccharides with their chemical structure, it becomes evident that the selectivity of the separation is affected by the accessibility of the fixed stationary phase to the oxyanions formed in the mobile phase. Hardy and Townsend investigated a series of synthetic oligosaccharides with chemical structures that resemble N-linked oligosaccharides in glycoproteins. Table 3-22 contains a list of these compounds; their chromatographic separation is shown in Fig. 3-122.

Surprisingly, the retention time does not simply increase with the molecular size. Thus, in comparison to the trisaccharide **1**, only a slightly higher retention is observed for the nonasaccharide **9**. Compared with the non-fucosylated heptasaccharide **7**, the nonasaccharide has a markedly lower retention which can be traced back to the fucosyl substitution at C-3 of the N-acetylglucosamine residues and to steric problems between their acetamido groups and the hydroxide groups at C-2 of the fucose residue.

The influence of the high acidity of an anomeric hydroxide group on retention is demonstrated by the retention behavior of compounds **4** and **5**. The mere conversion of the pentasaccharide **5** to the corresponding sugar alcohol **4** leads to a decrease in retention time of about 15 minutes. This is probably due to a conformational transition of the branching mannose residue from the chair form, which is typical for pyranoses, into an open-chain structure.

A fucose substitution at C-atom 3 of N-acetylglucosamine also shortens retention. This effect is obvious in the branched oligosaccharides **7**, **8**, and **9**. Compound **7**, for example, has a bi-antennaeric structure in which Gal(β1-4) is linked terminally with GlcNAc. When these terminal residues are replaced by fuc (α1-3) (compound **8**), the retention time is significantly reduced, because the high acidity of the hydroxide groups [121] attached to C-atom 3 of N-acetylglucosamine and at C-atom 6 of galactose is not realized.

A similarly strong reduction in retention time is observed when two fuc (α1-3) residues are linked with the galactosylated compound **7**. Although this substitution prevents the formation of oxyanions at the C-atom 3 of the two branching GlcNAc residues, the resulting compound **9** has six additional hydroxide groups. Those located at

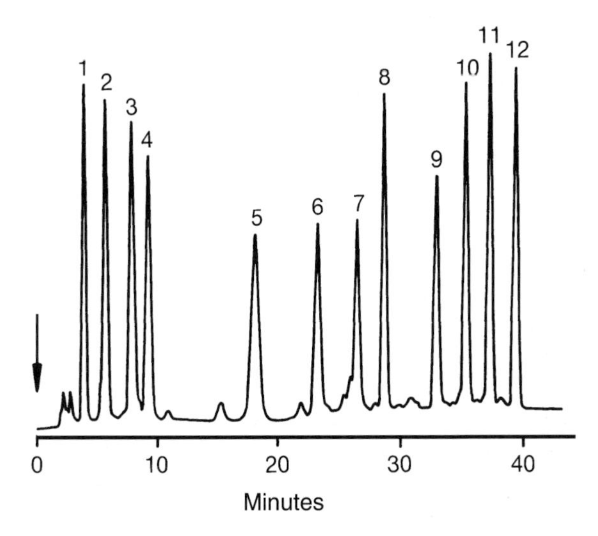

Fig. 3-122. Separation of various neutral oligosaccharides. – Separator column: CarboPac PA-1; eluent: (A) 0.1 mol/L NaOH, (B) 0.1 mol/L NaOH + 0.15 mol/L NaOAc; gradient: 10 min 100% A isocratically, then linearly to 80% B in 60 min; flow rate: 1 mL/min; detection and injection volume: see Fig. 3-105; solute concentrations: 1 nmol each; (taken from [120]).

Table 3-22. Chemical structures of the oligosaccharides separated in Fig. 3-122.

Peak	Structure
1	Fuc (α1−3) GlcNAc (β1−2) Man
2	Gal (β1−4) GlcNAc (β1−3) Gal (β1−4) Glc
3	Gal (β1−3) GlcNAc (β1−3) Gal (β1−4) Glc
4	Gal (β1−4) GlcNAc (β1−2) ⟩Man-OH Gal (β1−4) GlcNAc (β1−4)
5	Gal (β1−4) GlcNAc (β1−2) ⟩Man Gal (β1−4) GlcNAc (β1−4)
6	Gal (β1−4) GlcNAc (β1−6) ⟩Man (α1−2) Man Gal (β1−4) GlcNAc (β1−3)
7	Gal (β1−4) GlcNAc (β1−2) Man (α1−6) ⟩Man Gal (β1−4) GlcNAc (β1−2) Man (α1−3)
8	Fuc (α1−3) GlcNAc (β1−2) Man (α1−6) ⟩Man Fuc (α1−3) GlcNAc (β1−2) Man (α1−3)
9	Fuc (α1−3) Gal (β1−4) GlcNAc (β1−2) Man (α1−6) ⟩Man Gal (β1−4) GlcNAc (β1−2) Man (α1−3) Fuc (α1−3)
10	Gal (β1−4) GlcNAc (β1−2) Man (α1−6) ⟩Man Gal (β1−4) GlcNAc (β1−2) Man (α1−3) Gal (β1−4) GlcNAc (β1−4)
11	Gal (β1−4) GlcNAc(β1−6) Gal (β1−4) GlcNAc (β1−2) Man (α1−6) ⟩Man Gal (β1−4) GlcNAc (β1−2) Man (α1−3)
12	Gal (β1−4) GlcNAc (β1−6) Gal (β1−4) GlcNAc (β1−2) Man (α1−6) ⟩Man Gal (β1−4) GlcNAc (β1−2) Man (α1−3) Gal (β1−4) GlcNAc (β1−4)

C-atom 2 of both fucose residues are particularly acidic. However, the interaction of the oxyanion formed at this position is sterically hindered due to the close proximity of the acetamido group, which explains the distinct retention decrease.

The longer retention time of compound **11** compared with compound **10** is an example for the high selectivity of this method with regard to structural isomers. The acidity of a hydroxide group at C-atom 6 is higher than that of a hydroxide group attached to C-atom 4, hence the nonasaccharide **11** with a (β1-6)-linkage is more strongly retained.

Because certain carbohydrates easily undergo base-catalyzed reactions, the ion-exchange separation of oligosaccharides in a superalkaline eluent may lead to additional peaks in the chromatogram. The most affected are 3-O-substituted compounds for which epimerization and elimination reactions are observed. With the compounds listed in Table 3-22 having terminal mannose and glucose residues, epimerization does not occur during the time required for the chromatographic separation. The small unlabelled signals in Fig. 3-122 are, therefore, of different origin. Oligosaccharides with terminal GlcNAc residues epimerize to a small extent to N-acetylmannosamine at pH 13. This can be easily prevented, without great sacrifices in selectivity, by reducing the NaOH concentration in the mobile phase to c = 0.03 mol/L.

While sensitive detection of oligosaccharides and glycoproteins in the pmol range was only possible by radiolabelling [122], or derivatization with chromophores [123], or fluorophores [117], respectively, the pulsed amperometric detection allows a direct detection of these compounds in this concentration range.

3.3.6 Gradient Elution Techniques in Anion Exchange Chromatography

In conventional HPLC, the gradient elution of compounds is a well-established technique with many diverse applications. For example, RPLC with photometric detection allows the separation of compounds with widely different retention behavior in a single run simply by changing the amount of organic solvent in the mobile phase. In general, a distinction between continuous and discontinuous gradients is made. The term "gradient elution" is reserved for separations obtained with continuous gradients, while "stepwise elution" is attributed to the class of discontinuous gradients. Both chromatographic methods differ in practical application, however, substantial differences are not existent.

The first approach to an application of gradient elution is found in the paper by Mitchell, Gordon, and Haskins [124] published in 1949, although the authors were unaware of the fundamental advantages of the technique. Its unique potential to separate complex mixtures was recognized independently in several laboratories as late as the 1950s. Pioneers in this field were Hagdahl, Williams, and Tiselius [125] and Donaldson, Tulane, and Marshall [126]. Today, the technique of gradient elution is so mature that it has become impossible to completely cover all its applications. In this context, it is worth mentioning the review by Snyder [127] that primarily focuses on theoretical aspects of gradient elution in addition to some ideas concerning the apparatus.

So far, gradient elution techniques have only found limited application in the field of ion chromatography with conductivity detection primarily because the most important inorganic anions such as fluoride, chloride, bromide, nitrite, nitrate, orthophosphate, and sulfate can be separated and determined under isocratic conditions. On the other hand, a variation of the eluent composition is inevitable if inorganic and organic anions or ions with different valencies are to be analyzed in a single run. The resulting change in background conductivity, which manifests itself in a strong baseline drift, is a serious problem for which no solution was offered in the first papers on this subject by Sundén et al. [128] and Tarter [129]. In addition, the application of gradient techniques in ion chromatography is hampered by a second problem that is caused by inorganic impurities in the eluents. These impurities may severely interfere with the analysis, since they are retained at the stationary phase at the beginning of the run. They may also co-elute as peaks with the compounds of interest after the elution power has been raised during the analysis.

These problems have been circumvented by both the introduction of modern micro-membrane suppressors and the development of short clean-up columns for eliminating inorganic impurities in the eluent [130]. Therefore, the gradient elution technique in ion chromatography today is as common as in the area of conventional HPLC.

Theoretical Aspects

A mathematical description of the retention of ions under gradient elution conditions was introduced in 1957 by Schwab et al. [131]. It is based on parameters which are derived from the normal chromatographic elution process for which the eluent composition is kept constant during the separation. Hence, the retention of an ion at isocratic elution may be described according to Eq. (86), taking into account the definitions for the capacity factor, k', and the selectivity coefficient, K, [see Eq. (35) and (36) in Section 3.2]:

$$k' = \frac{V_{ms} - V_d}{V_d} \tag{86}$$

$$= \frac{V_s}{V_d} \cdot K^{1/x} \cdot Q^{y/x} \cdot [E]^{-y/x}$$

V_{ms} Gross retention volume
V_d Void volume
Q Ion-exchange capacity of the resin
$[E]$ Eluent ion concentration
x Charge number of eluent ion
y Charge number of solute ion

With the exception of the eluent ion concentrations, all constants in Eq. (86) can be summarized to a general constant, $Const_i$:

$$k' = Const_i \cdot [E]^{-y/x} \tag{87}$$

A linear relation is obtained by taking the logarithm on both sides of this equation

$$\log k' = -y/x \cdot \log [\mathrm{E}] + \log \mathrm{Const}_i \tag{88}$$

which is described in a slightly different way to Eq. (83) in Section 3.3.4.3. According to Eq. (88) the slope of the straight line is proportional to the quotient of the charge number of both eluent ions and solute ions.

In gradient elution, k' changes as a function of the eluent ion concentration, so that it may not be equated with the ratio $(V_{ms}\text{-}V_d)/V_d$. For a gradient in which the eluent ion concentration starting from zero increases linearly with time, the present eluent ion concentration is calculated by

$$[\mathrm{E}] = R \cdot V \tag{89}$$

V represents the eluent volume delivered since the start of the gradient run and R is the slope of the gradient ramp, which is defined as the ratio of the temporal change in the eluent ion concentration and the flow rate. The corresponding present capacity factor k_a' for the peak maximum is obtained by inserting Eq. (89) into Eq. (87):

$$k_a' = \mathrm{Const}_i \cdot (RV)^{-y/x} \tag{90}$$

where k_a' represents the capacity factor that would result if the eluent ion concentration would be held constant from the moment of injecting a component. It follows that in gradient elution, k_a' must be integrated over this time period to calculate $(V_{ms}\text{-}V_d)/V_d$. Hence, k_a' is replaced by $\mathrm{d}V/\mathrm{d}x$:

$$k_a' = \mathrm{d}V/\mathrm{d}x \tag{91}$$

indicating that with each volume portion $\mathrm{d}V$ the peak maximum travels the portion $\mathrm{d}x$ through the column. Combined with Eq. (90) one obtains:

$$\mathrm{d}x = \frac{\mathrm{d}V}{\mathrm{Const}_i \cdot (RV)^{-y/x}} \tag{92}$$

Integrating both sides yields:

$$\int_0^{V_{ms}-V_d} \mathrm{Const}_i^{-1} \cdot (RV)^{y/x}\, \mathrm{d}V = \int_0^{V_d} \mathrm{d}x \tag{93}$$

$$\frac{V_{ms}-V_d}{V_d} = \left(\frac{x}{y+x}\right)^{-\frac{x}{y+x}} \cdot V_d^{-\frac{y}{y+x}} \cdot \mathrm{Const}_i^{-\frac{x}{y+x}} \cdot R^{-\frac{y}{y+x}} \tag{94}$$

Again, the constants in Eq. (94) may be summarized in a general constant, Const_g:

$$\frac{V_{ms}-V_d}{V_d} = \mathrm{Const}_g \cdot R^{-\frac{y}{y+x}} \tag{95}$$

Taking the logarithm on both sides it follows:

$$\log \frac{V_{ms} - V_d}{V_d} = -\frac{y}{y+x} \cdot \log R + \log \text{Const}_g \tag{96}$$

Eq. (96) derived for gradient elution is very similar to Eq. (88) for the isocratic elution. By plotting $\log (V_{ms}\text{-}V_d)/V_d$ as a function of $\log R$ straight lines with differerent slopes are obtained for each solute ion. However, Eq. (96) only applies to linear gradients with an initial eluent ion concentration of zero.

Choice of Eluent

Above all, the choice of eluent for gradient elution applications depends on the type of gradient. In ion-exchange chromatography two different techniques may be applied. In a *concentration gradient* the concentration of eluent ions is changed during the run, in a *composition gradient* the eluent composition is changed by replacing weakly retained eluent ions by more strongly retained ones. Furthermore, pH gradients are occasionally applied in which an increasing amount of a strong base is added to a weak acid that is present in a defined concentration. Actually, pH gradients are also concentration gradients, since an increase in the concentration of dissociated acid is intended by increasing the pH value during the run.

Composition gradients present a problem in practical applications. Taking into account the large retention differences between weakly and strongly retained solute ions, for example, monovalent acetate and trivalent citrate, respectively, the eluent ions employed must also differ significantly in their affinity toward the stationary phase. This means that the weakly retained eluent ions that predominate at the beginning of a run have to be completely replaced by strongly retained eluent ions to elute the trivalent citrate. The time required for this replacement depends on the ratio between exchange capacity and eluent ion concentration. Furthermore, the reverse operation is necessary to repeat a run, which is only accomplished by flushing the column with a high concentration of the eluent initially being used. However, this increases the equilibration time between two gradient runs. In contrast, such problems are not encountered when applying concentration gradients, because the ionic form of the resin remains unaffected.

As mentioned above, a change in the eluent ion concentration causes a change in the background conductivity. The latter is significant, since the eluent ion concentration has to be increased during a run by one to two orders of magnitude to elute anions with a strongly different retention behavior. Direct conductivity detection is thus impossible because of the possibility of a baseline drift that is much too strong. For successful application of concentration gradients, therefore, the background conductivity has to be markedly reduced. This is only achieved chemically with a micromembrane suppressor (see Section 3.3.3) which reduces the background conductivity to values of less than 10 μS/cm. This limits the background conductivity changes during a gradient run to a few μS/cm.

In principle, salts derived from inorganic or organic acids, with pK values above 7 can be employed as the eluent. Sodium hydroxide has proved to be particularly suitable, because it is converted into water independent of the initial concentration. The resulting background conductivity is hardly affected by the hydroxide ion concentration in the mobile phase as long as the exchange capacity of the suppressor is not exhausted. Examples of gradient elutions using sodium hydroxide as the eluent are displayed in Figs. 3-123 and 3-124. Fig. 3-123 shows the separation of chloride, sulfate, and various phosphorus compounds on IonPac AS5. With a simple linear NaOH gradient, both the monovalent chloride and the hexavalent tetrapolyphosphate can be analyzed in a single run within 15 minutes. Even more impressive is the chromatogram on the right in Fig. 3-124 that was obtained using a modern 5-μm exchanger. It shows the separation of a standard solution containing 36 different inorganic and organic anions with a run time of about 30 minutes. Using a highly diluted sodium hydroxide solution, fluoride and various other monocarboxylic acids that are only weakly retained are separated, while the final sodium hydroxide concentration suffices to elute trivalent anions such as orthophosphate and citrate. The progress made since the introduction of gradient elution is apparent when comparing the chromatogram on right with that on the left-hand side in Fig. 3-124, which was obtained under isocratic conditions. Three different operations were necessary for the complete analysis of this standard [132].

When preparing NaOH eluents care has to be taken that they are carbonate-free! Since carbonate ions exhibit a much higher elution strength than hydroxide ions, even traces of carbonate in the eluent would result in a lower resolution, especially at the beginning of a gradient run. The residual dissociation of the carbonic acid formed in the suppressor would lead to a strong baseline drift during the gradient run. (Incidentally, this is the reason that eluents based on carbonate/bicarbonate are totally unsuitable for a gradient technique.) Therefore, NaOH-based eluents should be prepared from

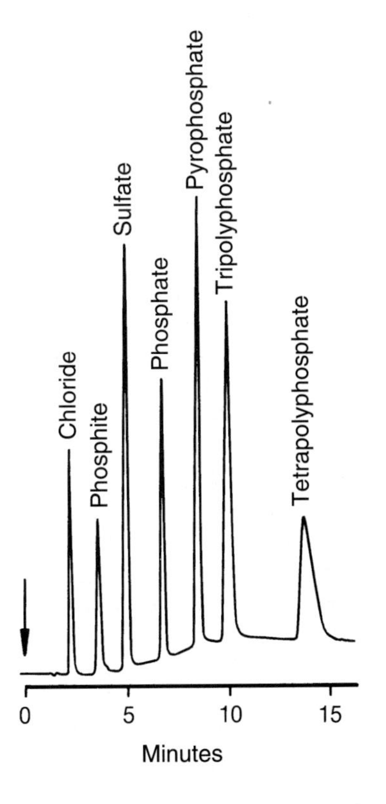

Fig. 3-123. Separation of chloride, sulfate, and various phosphorus compounds using a gradient technique. – Separator column: IonPac AS5; eluent: (A) water, (B) 0.1 mol/L NaOH; gradient: linear, 25% B to 100% B in 5 min; flow rate: 1 mL/min; detection: suppressed conductivity; injection volume: 50 μL.

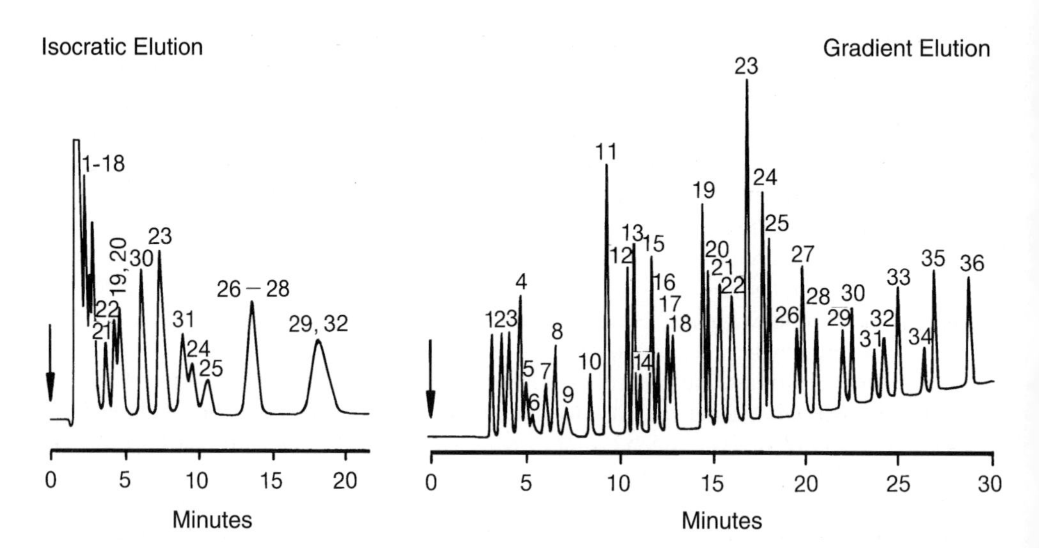

Fig. 3-124. Comparison of the separation of various inorganic and organic anions in a) isocratic elution and b) gradient elution. – Separator column: a) IonPac AS4A, b) IonPac AS5A; eluent: a) 0.0017 mol/L $NaHCO_3$ + 0.0018 mol/L Na_2CO_3, b) see Fig. 3-24; flow rate: a) 2 mL/min, b) 1 mL/min; detection: suppressed conductivity; injection volume: a) 50 μL, b) 10 μL; solute concentration: see Fig. 3-24.

a 50% concentrate in which the carbonate, originally adsorbed on the NaOH pellets, precipitates as a fine slurry. The de-ionized water used for preparing the eluent should have an electrical conductance of 0.05 μS/cm. It should be degassed with helium prior to addition of the NaOH concentrate. After preparation, eluents must be kept under inert gas atmosphere.

By observing these guidelines, the baseline drift for the concentration gradient chosen in Fig. 3-124 does not exceed 3 to 4 μS/cm!

Despite all precautions, impurities from the eluent being used can accumulate on the separator column at the beginning of a gradient run, and as the analysis progresses, they may elute as a peak. To avoid such effects, the eluent is passed through a short anion trap column (ATC) before it enters the injection valve. The anion trap column contains a high-capacity anion exchange resin with a low chromatographic efficiency that completely retains anionic impurities at the beginning of an analysis. As the elution power increases during the gradient run these impurities are released again but do not appear as regular peaks in the chromatogram due to the low efficiency of the material.

Salts of other weak acids may also be employed as the eluent. For example, extensive investigations were performed with p-cyanophenolate ions [130] which, compared to hydroxide ions, exhibit a slightly stronger elution power. A high selectivity for aliphatic dicarboxylic acids is obtained with *p*-cyanophenolate in combination with a CarboPac PA-1 (IonPac AS6) anion exchanger. Nevertheless, this eluent has not gained acceptance because it is not commercially available in the required purity.

A second class of available eluents comprises the family of zwitterionic compounds (see Section 3.3.2), which exist in their anionic form at alkaline pH. The product of the suppressor reaction is the zwitterionic form with a correspondingly low intrinsic conductance. Promising experiments were carried out by Irgum [48] with N-substituted aminoalkylsulfonic acids, which can be employed for both composition gradients as well as concentration gradients. The compounds taurine (2-aminoethanesulfonic acid) and CAPS [3-(N-cyclohexylamino)-1-propanesulfonic acid] used by Irgum are commercially available, but have to be thoroughly purified [133,134] for gradient applications. The extent to which the slow transport of zwitterions through the membrane represents a limitation for the eluent ion concentration that is required for the elution of strongly retained anions is still unresolved.

Possibilities for Optimizing Concentration Gradients

In theory, it is possible to derive the optimal conditions for gradient elution using a sodium hydroxide solution as the eluent from the representation of $\log(V_{ms}\text{-}V_d)/V_d$ as a function of log R. However, as mentioned, this applies only to simple linear gradients with an initial eluent ion concentration of zero, which is rarely used for practical purposes. Much shorter analysis times are obtained when the gradient run starts at a higher eluent ion concentration. Furthermore, gradient programs with different ramps, sometimes combined with isocratic periods, have to be developed to obtain optimal selectivity and speed of analysis. A mathematical description of the retention is impossible

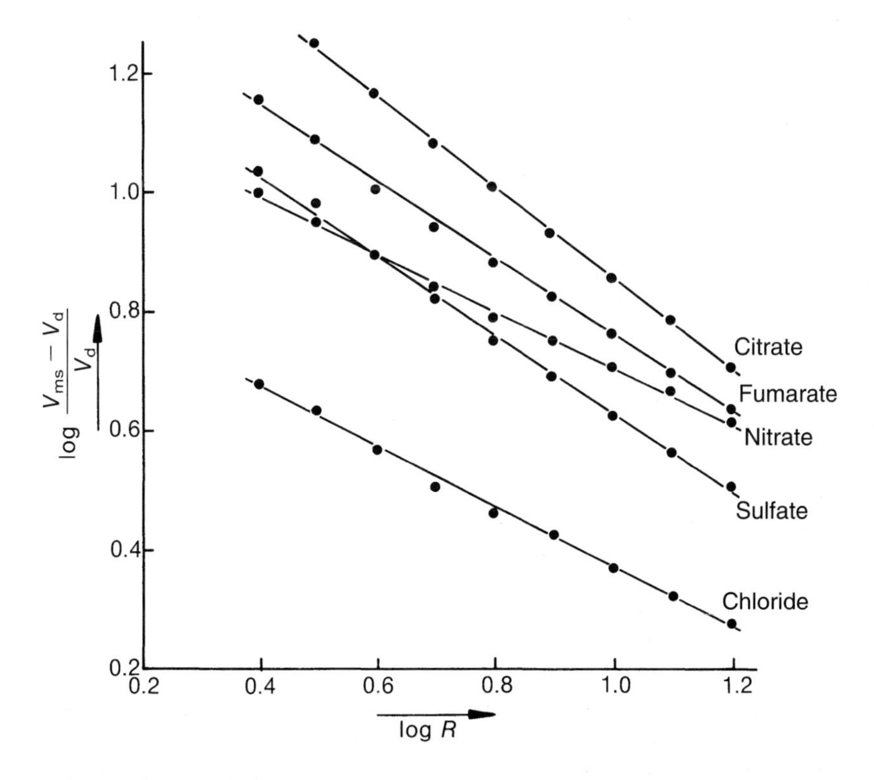

Fig. 3-125. Representation of $\log(V_{ms}\text{-}V_d)/V_d$ as a function of log R for various inorganic and organic anions using an IonPac AS5A separator and NaOH as the eluent. – Flow rate: 1 mL/min; solute concentrations: 3 ppm chloride, 5 ppm nitrate, 10 ppm sulfate, 5 ppm fumarate, and 50 ppm citrate; (taken from [130]).

in all these cases, since the resulting equation for the calculation of the retention volume would be far too complex.

However, Eq. (96) can be employed to predict trends. For this purpose, one should inspect Fig. 3-125 in which $\log(V_{ms}-V_d)/V_d$ is displayed versus $\log R$ for various inorganic and organic anions. It appears from this representation that the elution order of anions with different valencies depends on the steepness of the gradient ramp. When two anions with differing charge number co-elute, a separation of both components can be accomplished by employing a steeper ramp, in which case the ion with the higher charge number elutes first.

Isoconductive Techniques

A completely different approach for the gradient elution of anions was followed by Jandik et al. with the development of "isoconductive eluents" [135]. Intended to be an alternative to the concept of chemical suppression, this technique employs two mobile phases that differ in their elution power but not in the resulting background conductivity. This is achieved by varying the cations present in the eluent as counter ions to the anionic eluent ions. These cations do not contribute to the separation process. The background conductivity remains unaffected, for example, when turning from a cation with high equivalent conductance in the weaker eluent to a cation with lower equivalent conductance in the comparatively stronger eluent, and compensating the resulting conductivity change by a slight increase in the eluent ion concentration. Thus, the gradient technique performed with isoconductive eluents represents a combination of both a concentration gradient and a composition gradient.

A quantitative statement regarding the increase of the elution strength when switching from one eluent to another may be derived from the equation that was proposed by Fritz et al. [69] for the calculation of the background conductivity of a mobile phase with fully dissociated eluent ions:

$$G = \frac{\lambda^+ + \lambda^-}{10^{-3}\,K} \cdot c \tag{97}$$

G Eluent background conductivity
λ^-, λ^+ Equivalent conductances of eluent anions and cations, respectively
K Cell constant
c Eluent concentration

Consider two eluents with different elution power containing the same eluent ions but different cations with their respective equivalent conductivities λ_1^+ and λ_2^+. Under isoconductive conditions, the background conductivity G_1 of the weaker eluent equals that of the stronger eluent G_2. Taking $G_1 = G_2$ Eq. (97) may be rearranged so that the concentrations c_1 and c_2 of both eluents are related to the corresponding equivalent conductivities:

$$\frac{c_2}{c_1} = \frac{\lambda_1^+ + \lambda^-}{\lambda_2^+ + \lambda^-} \tag{98}$$

Eq. (98) facilitates the search for a suitable combination of eluent ions and counter ions, which allows one to maximize the ratio c_2/c_1 as a measure for the increase in elution power. Considerable improvements may only be obtained with eluent ions of low equivalent conductance. Particularly suited is the gluconate/borate mixture used by Jandik et al. [135] in their pioneering paper. The increase in gluconic acid concentration in the presence of a suitable concentration of boric acid was compensated by adjusting with potassium or lithium hydroxide, respectively. Fig. 3-126 illustrates this by comparing the conductivity change that occurs in a pure concentration gradient to the conductivity change resulting from isoconductive conditions. The eluent compositions used for this comparison are summarized in Table 3-23.

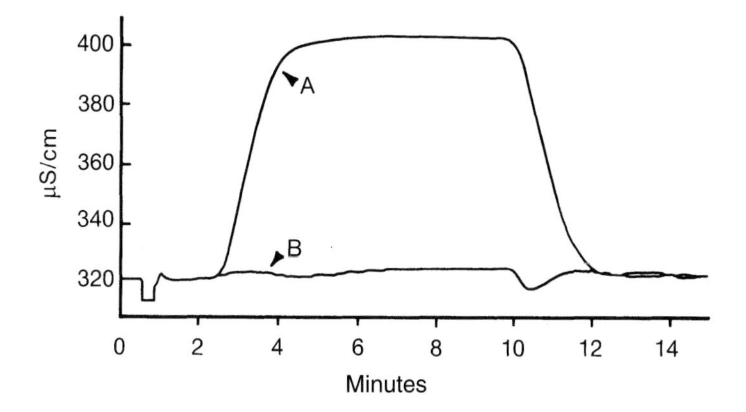

Fig. 3-126. Comparison of the conductivity change resulting from a pure concentration gradient (A) with that obtained under isoconductive conditions (B). – Separator column: Waters IC-Pak Anion; eluent: see Fig. 3-23; (taken from [135]).

In comparing the concentration gradient concept with subsequent chemical suppression, the chromatogram of various inorganic anions in Fig. 3-127 that was obtained with isoconductive eluents clearly shows the limitation of this technique. Although the analysis time for the separation shown is shorter than under isocratic conditions, an improvement in the resolution between the signals, especially in the first part of the

Table 3-23. Eluent composition for a concentration gradient (A) and isoconductive eluent (B) according to Fig. 3-126; (taken from [135]).

Component	A. Eluent 1	Eluent 2	B. Eluent 1	Eluent 2
Boric acid [mol/L]	0.011	0.01375	0.011	0.01375
Gluconic acid [mol/L]	0.00148	0.00185	0.00148	0.00185
Potassium hydroxide [mol/L]	0.00349	0.00436	0.00349	– –
Lithium hydroxide [mol/L]	– –	– –	– –	0.00513
Glycerol [mol/L]	0.00069	0.00081	0.00069	0.00081
Acetonitrile [mL/L]	120	120	120	120

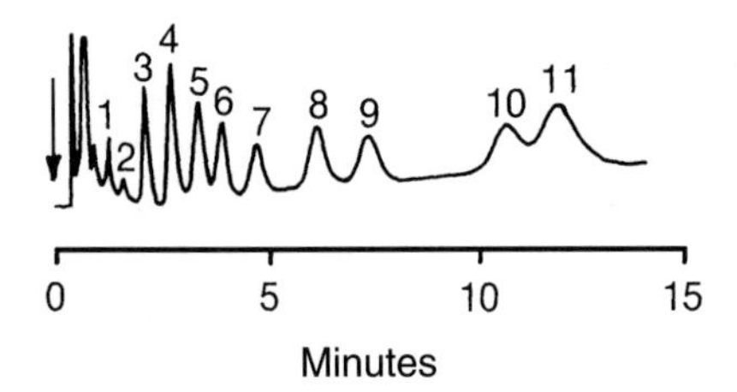

Fig. 3-127. Separation of various inorganic anions with an isoconductive eluent. – Separator column: Waters IC-PAK Anion; eluent: see Table 3-23 (eluent switching at the time of injection); detection: direct conductivity; injection volume: 100 µL; solute concentrations: 1 ppm fluoride (1), 2 ppm carbonate (2) and chloride (3), 4 ppm nitrite (4), bromide (5), and nitrate (6), 6 ppm orthophosphate (7), 4 ppm sulfate (8) and oxalate (9), 10 ppm chromate (10), and molybdate (11); (taken from [135]).

chromatogram, is not achieved. Furthermore, carbonate that often is present in high concentrations in real-world samples elutes between fluoride and chloride which, in practice, may lead to significant interferences. It remains to be seen which development this technique will take in the quest for new eluent compositions and gradient profiles.

3.4 Cation Exchange Chromatography

Small et al. [3] devoted a significant part of their pioneering work in ion chromatography to the separation and determination of cations. The necessary system was equivalent to the system used for anion analysis (see Section 3.3). It comprised a cation exchanger of low capacity and a suppressor column containing a strong-base anion exchange resin in the hydroxide form. With the exception of the suppressor column, which has since been replaced by modern membrane suppressors, the principle setup remains unchanged. As in anion analysis, the application of a suppressor system is not an indispensable prerequisite for the cation determination by means of conductivity detection.

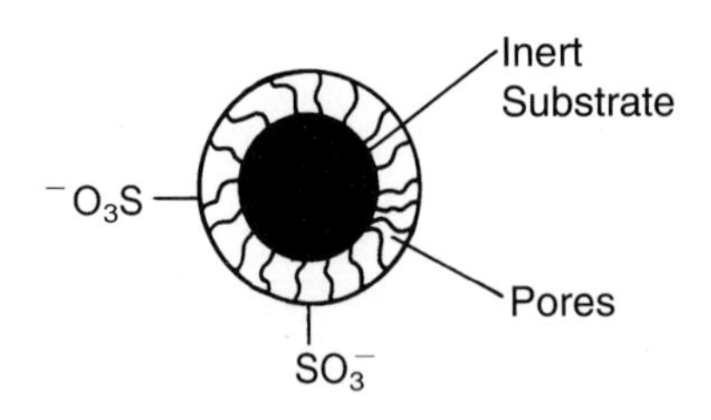

Fig. 3-128. Schematic picture of a surface-sulfonated cation exchanger.

The exchange reaction with a cation M^+ that occurs at the stationary phase of a cation exchanger can be represented as follows:

$$\text{Resin}-\text{SO}_3^-\text{H}^+ + \text{M}^+\text{A}^- \underset{}{\overset{K}{\rightleftharpoons}} \text{Resin}-\text{SO}_3^-\text{M}^+ + \text{H}^+\text{A}^- \qquad (99)$$

The separation of cations is determined by their different affinities toward the stationary phase.

3.4.1 Stationary Phases

Like anion exchangers, cation exchangers are classified according to the type of substrate material. Although organic polymers are predominantly used in the manufacturing of cation exchangers, it is unnecessary. Because dilute acids serve as the eluent for cation separations, the stability over the whole pH range that is typical of organic polymers is not required. Therefore, silica-based cation exchangers which exhibit a significantly higher chromatographic efficiency are now employed.

3.4.1.1 Polymer-Based Cation Exchangers

Styrene/divinylbenzene copolymers

Styrene/divinylbenzene copolymers are the most widely used substrate materials for the manufacture of cation exchangers. The principle properties of this support material were described in detail in Section 3.3.1.1. The material is surface-sulfonated by reaction with concentrated sulfuric acid. The exchange capacity is determined by the degree of sulfonation. The latter depends, essentially, on the reaction time and the temperature program during the reaction [136]. Typical exchange capacities are between 0.005 and 0.1 mequiv/g [137]. A surface-sulfonated cation exchanger is depicted schematically in Fig. 3-128. The diffusion of fully dissociated ions such as Na^+, K^+, or Mg^{2+} into the interior of the stationary phase can be ignored because of its comparatively strong hydrophobic character. This results in short diffusion pathways and, thus, in a higher chromatographic efficiency compared to conventional fully sulfonated cation exchangers.

Table 3-24. Structural and technical properties of surface-sulfonated cation exchangers.

Separator column	Manufacturer	Dimensions [length × I.D.] [mm]	Max. flow rate [mL/min]	Max. operating pressure [MPa]	Solvent stability [%]	Capacity [mequiv/g]	Particle diameter [µm]
IonPac CS1	Dionex	200 x 4.6	3	5	5	0.01....0.05	20
IonPac CS2	Dionex	250 x 4.6	3	10	5	0.01....0.05	15
TSK-Gel IC Cation	Toyo Soda	50 x 4.6	2	5	10	0.012	10
Shimpack IC-C1	Shimadzu	150 x 5	2	5	10	– [a]	10
LCA-K01	Sykam	125 x 4	3	20	5	0.05	10
PRP-X200	Hamilton	250 x 4	8	30	100	0.035	10

[a] no details

Surface-sulfonated styrene/divinylbenzene copolymers are offered by several manufacturers. The structural and technical properties of these columns are summarized in Table 3-24. Because in all strong-acid cation exchangers sulfonate groups represent the ion-exchange function, considerable differences between the different column packings exist only in their chromatographic efficiencies and the resulting column dimensions. The IonPac CS1 separator column, manufactured by Dionex (Sunnyvale, CA, USA), is still commercially available, but is only of historical significance due to the large particle diameter of 20 μm. All other separator columns listed in Table 3-24 are still in use mainly for the analysis of alkali and alkaline-earth metals. As a representative for all separator columns of this type Fig. 3-129 shows a chromatogram with a separation of alkali metals which nowadays requires less than ten minutes.

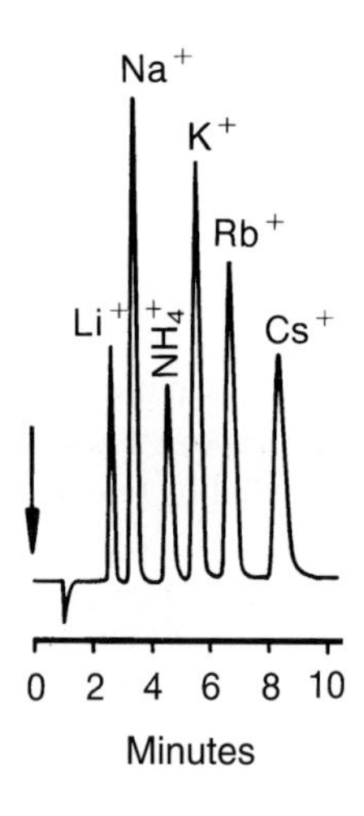

Fig. 3-129. Separation of alkali metals at a surface-sulfonated styrene/divinylbenzene copolymer. – Separator column: IonPac CS2; eluent: 0.03 mol/L HCl; flow rate: 1 mL/min; detection: suppressed conductivity; injection volume: 50 μL; solute concentrations: 5 ppm lithium and sodium, 10 ppm ammonium and potassium, 20 ppm rubidium, and 30 ppm cesium.

Polymethacrylate and Polyvinyl Resins

Polymethacrylate and polyvinyl resins play only a secondary role in the manufacture of cation exchangers. Presently, the only polymethacrylate-based cation exchanger is offered by Sykam (Gauting, Germany) under the trade name LCA K02. This column differs from its PS/DVB analogue (see Table 3-24) only in the particle size (5 μm) and exchange capacity (0.4 mequiv/g). With a tartaric acid eluent, this phase is preferred for the analysis of heavy and transition metals.

In comparison to PS/DVB copolymers, the only polyvinyl-based cation exchanger available (Interaction Chemicals, Mountain View, CA, USA) exhibits poor chromatographic efficiency. The column can be employed for the separation of inorganic and organic cations, and is 50 mm x 3 mm I.D. With a maximum flow rate of 2.5 mL/min, this column may be operated at a maximum pressure of 15 MPa. An exchange capacity of 0.1 mequiv/g is stated for the column packing, the particle size of which is 5 μm. The corresponding chromatogram of an alkali metal separation is shown in Fig. 3-130.

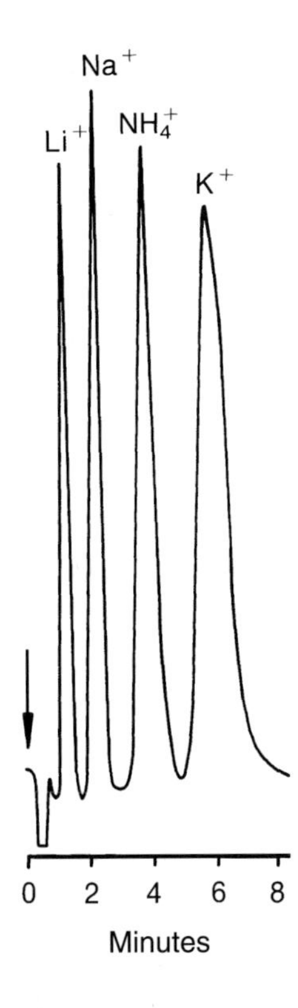

Fig. 3-130. Separation of alkali metals on a surface-sulfonated polyvinyl resin. – Separator column: ION 200; eluent: 0.002 mol/L picolinic acid, pH 2.0; flow rate: 2.6 mL/min; detection: direct conductivity.

3.4.1.2 Latexed Cation Exchangers

While the latex concept for anion exchangers (see Section 3.3.1.2) was realized with the introduction of ion chromatography, it was not until 1986 that its use for cation exchangers was recognized. The reason for the long development time was the lack of suitable surface-aminated substrates. The manufacture of these substrates via direct amination was out of question because of the poor reproducibility of this method. Based on experience gained in the manufacture of latexed anion exchangers, it eventually became possible to cover the anion exchange beads with a second layer of fully sulfonated latex beads. These materials, developed by Dionex Corp., are called latex cation exchangers. Their structure is depicted schematically in Fig. 3-131, which illustrates that they consist of a weakly sulfonated polystyrene/divinylbenzene substrate with a particle size of, for example, 10 μm. Fully aminated latex beads with a much smaller diameter of about 50 nm are agglomerated on its surface by both electrostatic and van-der-Waals interactions. The anion exchange substrate thus produced is covered by a second layer of latex beads which carry the actual exchange function in form of sulfonate groups. In the latex cation exchanger IonPac CS3, which is offered in the standard dimensions of 250 mm x 4.6 mm I.D., the sulfonated latex beads have a diameter of about 250 nm and a degree of crosslinking of 5%. Such a separator column is preferred when extremely high concentration differences, for example, between sodium, ammonium, and potassium, require a maximal resolution between the individual components.

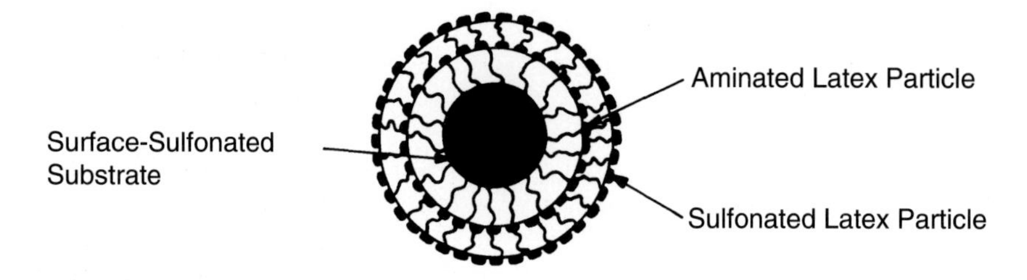

Latex Cation Exchanger Particle

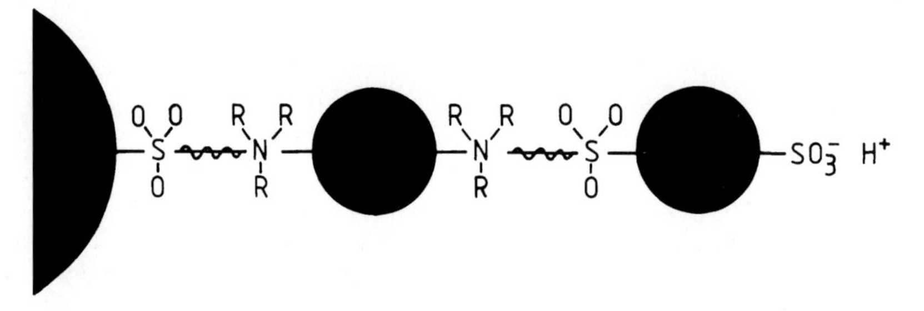

Fig. 3-131. Schematic diagram of a latex cation exchanger.

The separator column introduced as the Fast-Sep-Cation has similar physical properties. This column was developed primarily for the fast separation of alkali or alkaline-earth metals. In combination with a column switching technique, it is also suitable for the simultaneous analysis of the most important cations of both compound classes (see

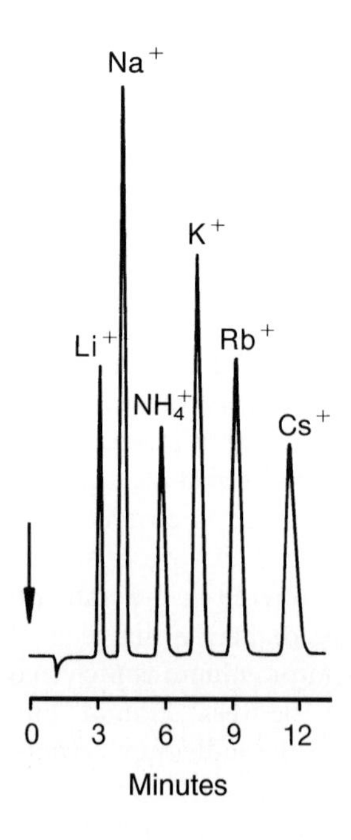

Fig. 3-132. Separation of alkali metals on a latex cation exchanger. – Separator column: Fast-Sep Cation I; chromatographic conditions: see Fig. 3-129.

Section 3.4.4). The dimensions of the Fast-Sep-Cation column are 250 mm x 4 mm I.D. Compared to the CS3 column, the 13-μm particle diameter of the substrate is slightly higher. The sulfonated latex beads are crosslinked with 4% DVB and have a diameter of about 225 nm.

In contrast to surface-sulfonated materials, latex cation exchangers exhibit a significantly higher chromatographic efficiency. This is illustrated by the chromatogram displayed in Fig. 3-132 that was obtained using the chromatographic conditions outlined in Fig. 3-129. Latex cation exchangers may be operated at a flow rate of 2 mL/min without significant loss in separation efficiency. These conditions enable a baseline-resolved separation of sodium, ammonium, and potassium to be carried out within three minutes.

A special latex cation exchanger was recently introduced for the simultaneous analysis of alkali and alkaline-earth metals under isocratic conditions. Its support material is composed of a highly crosslinked ethylvinylbenzene/divinylbenzene copolymer with a particle size of 8 μm. This support material is manufactured as follows: At the time of polymerization, the individual particles are furnished with a reactive surface to which – in contrast to the conventional latex structure according to Fig. 3-131 – a monolayer of a fully aminated colloidal polymer particle is *covalently* bound. However, the fully aminated polymer particles act only as anchor groups for the second layer of latex beads, carrying the actual exchange function in form of sulfonate groups. The covalent bond of the latex beads to the support material provides a high mechanical and chemical stability for this stationary phase. It is offered in the standard dimensions 250 mm x 4.6 mm I.D. under the trade name IonPac CS10. A new feature for latex exchangers is the 100% solvent stability of the IonPac CS10 which results from the high degree of crosslinking of the microporous support. For the first time, this stationary phase can be cleaned using commercial HPLC solvents such as methanol, acetonitrile, and others. However, for the simultaneous analysis of alkali and alkaline-earth metals, an example of which is shown in Fig. 3-133, no organic additives to the mobile phase are required.

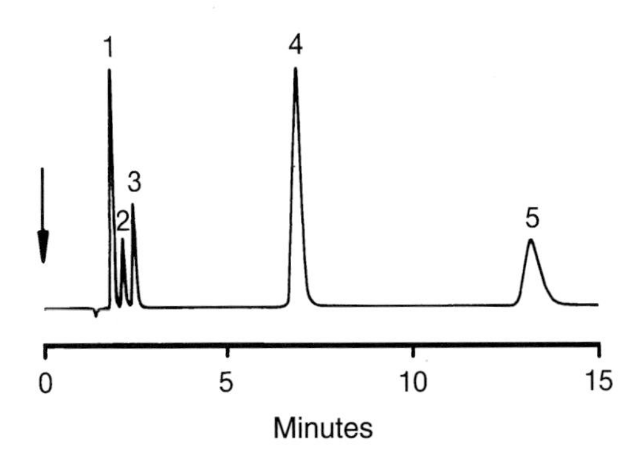

Fig. 3-133. Separation of alkali and alkaline-earth metals at IonPac CS10. – Eluent: 0.04 mol/L HCl + 0.005 mol/L 2,3-diaminopropionic acid; flow rate: 1 mL/min; detection: suppressed conductivity; injection volume: 25 μL; solute concentrations: 5 ppm sodium (1), ammonium (2), and potassium (3), 10 ppm magnesium (4), and calcium (5).

3.4.1.3 Silica-Based Cation Exchangers

Silica-based cation exchangers with sulfonic acid groups as ion-exchange sites, typically produced via silanization of the silica with reagents such as R-Si$(CH_3)_2$-C_3H_6-C_6H_5-SO_3H (R = H, Cl), are not significant despite their comparatively high chromatographic efficiency. The 50 mm x 4.6 mm I.D. Vydac 400 IC 405 is such a stationary phase and is offered by the manufacturer, The Separations Group (Hesperia, CA, USA). It is most suited for the analysis of alkali metals and small aliphatic amines.

In contrast, the two 5-µm-exchangers Nucleosil 5 SA by Macherey&Nagel (Düren, Germany) with the dimensions of 125 mm x 4 mm I.D., and TSK Gel IC Cation SW from Toyo Soda (Tokyo, Japan) with the dimensions 50 mm x 4.6 mm I.D. are designated for the analysis of divalent cations. All three stationary phases exhibit a relatively high exchange capacity of 0.5 mequiv/g and are resistant to all organic solvents that are miscible with water. Fig. 3-134 shows an example of the separation of alkaline-earth metals by Nucleosil 5 SA, which is chararcterized by high resolution and excellent peak symmetry.

A great deal of interest has been shown in a silica-based polymer-coated cation exchange phase recently introduced by Schomburg et al. [138]. It belongs to the class of weak-acid cation exchangers. The coating of the silica is performed with "prepolymers" which are synthesized in a separate step and then applied to the support material and immobilized. For a number of years, this method has been successfully employed for the synthesis of various stationary phases based on silica and alumina, with remarkable selectivities. Further information regarding this subject may be found in the review by Schomburg [139].

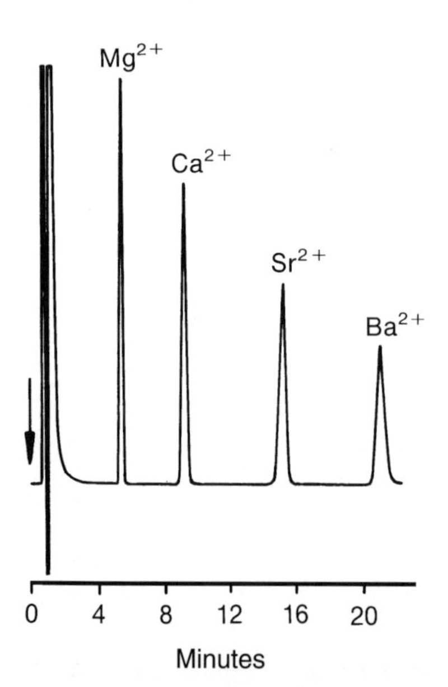

Fig. 3-134. Separation of alkaline-earth metals on a silica-based cation exchanger. – Separator column: Nucleosil 5 SA; eluent: 0.0035 mol/L oxalic acid + 0.0025 mol/L ethylenediamine + 50 mL/L acetone, pH 4.0; flow rate: 1.5 mL/min; detection: direct conductivity; injection volume: 100 µL; solute concentrations: 2.5 ppm magnesium, 5 ppm calcium, 20 ppm strontium, and 40 ppm barium.

The prepolymer used in manufacturing this novel cation exchanger consists of a copolymer that is derived from a mixture of butadiene and maleic acid in equal parts:

Poly(butadiene-maleic acid), PBDMA

The structural formula reveals that this polymer contains two different types of carboxyl group which have different dissociation constants. While the first dissociation step is characterized by a pK value of 3.4, the pK value of the second step is about 7.4. Both pK values were determined via titration of the prepolymer with sodium hydroxide solution. The exchange capacity of the finished stationary phase is directly proportional to its polymer content. It may be calculated in advance, since, owing to the chemical composition, the concentration of the exchange groups in the prepolymer is known.

PBDMA-coated silica is suited for the simultaneous analysis of alkali and alkaline-earth metals. The first coating experiments with 40% PBDMA resulted in a more group-wise separation of both classes of compounds with formic acid as the eluent. However, the separation between the most important components was significantly improved by slightly increasing the thickness of the layer and by changing the eluent. Fig. 3-135 shows a corresponding chromatogram obtained under optimized chromatographic conditions.

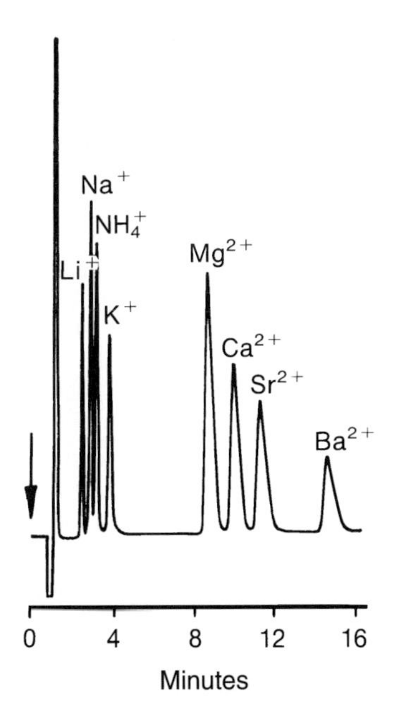

Fig. 3-135. Separation of alkali and alkaline-earth metals on Super-Sep. – Eluent: 0.005 mol/L tartaric acid; flow rate: 1 mL/min; detection: direct conductivity; injection volume: 10 μL; solute concentrations: 1 ppm lithium, 5 ppm sodium, and ammonium, 10 ppm each of potassium, magnesium, and calcium, 20 ppm strontium and barium.

In addition to strong-acid or weak-acid cation exchangers, crosslinked polymers carrying cyclic polyethers as anchor groups have been used for the separation of alkali and alkaline-earth metals, respectively. The structure of these stationary phases that are also based on silica was described in Section 3.3.1.4. A reasonable separation of alkali metals was obtained by Kimura et al. [38] on silica modified with poly(benzo-15-crown-5) using de-ionized water as the eluent (Fig. 3-136). This separation requires a long time; although it can be significantly shortened by coating ODS (octadecyl silica) with lipophilic crown ether derivatives [140]. However, the chromatographic efficiency of these separations does not meet present requirements. The only advantage of this method is the possibility of sensitive detection via electrical conductivity, since the eluent being used does not have an intrisic conductance.

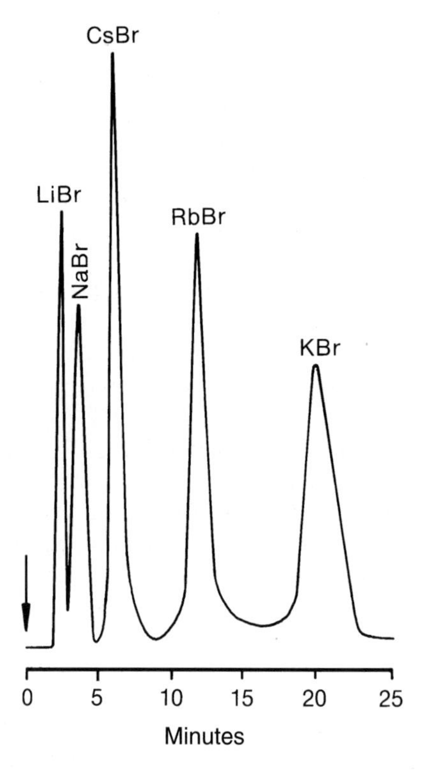

Fig. 3-136. Separation of alkali metal ions on silica modified with poly(benzo-15-crown-5). – Eluent: water; flow rate: 1 mL/min; detection: direct conductivity; injection volume: 1 μL; solute concentrations: 13.6 g/L LiBr, 14.4 g/L NaBr, 26.2 g/L KBr, 36.4 g/L RbBr, and 34 g/L CsBr; (taken from [38]).

In contrast to conventional cation exchangers, a reversed elution order is observed with crown ether phases, which is mainly determined by the size ratio between crown ether ring and alkali metal ion. Due to the high affinity of poly(benzo-15-crown-5) toward potassium and rubidium ions, these are more strongly retained than lithium, sodium, and cesium ions, respectively. However, the complexing properties of crown ethers also depend on the counter ion being employed. Thus, in potassium salts, for example, an increase in retention in the order KCl < KBr < KI is observed with an increasing size of the counter ion.

Alkaline-earth metal ions, on the other hand, elute from a crown ether phase in the normal elution order ($Mg^{2+} < Ca^{2+} < Sr^{2+} < Ba^{2+}$). Such a separation is only of pure academic interest, since the resolution between magnesium and calcium is extraordinarily poor due to the low interactions of both ions with crown ethers.

3.4.2 Eluents in Cation Exchange Chromatography

While the type of available eluent depends upon the detection method being applied in anion exchange chromatography, a corresponding classification is not necessary in cation exchange chromatography. For the separation of alkali metals, ammonium, and small aliphatic amines, mineral acids such as hydrochloric or nitric acid are typically used as eluents, independent of whether the subsequent conductivity detection is performed with or without chemical suppression. The concentration range lies between 0.002 mol/L and 0.04 mol/L. Bächmann et al. [141] employed cerium(III) nitrate in very low concentrations as the eluent for the indirect fluorescence detection of alkali metals.

Organic eluents are not very common in these applications. The respective manufacturer recommends the use of pyridinecarboxylic acids such as picolinic acid or isonicotinic acid only for operating separator columns such as ION-200 and Vydac IC 405.

Divalent cations such as alkaline-earth metals cannot be eluted with dilute mineral acids, since they exhibit a much higher affinity toward the stationary phase of a strong-acid cation exchanger. However, increasing the acid concentration in the mobile phase is impossible for the following reasons:

- In a system *without* chemical suppression, the extremely high background conductance would render impossible the sensitive detection of alkaline-earth metal ions via electrical conductivity.
- In a system *with* chemical suppression, the required reduction of the background conductance cannot be ensured.

In his pioneering paper, Small suggested silver nitrate as the eluent for alkaline-earth metals [3], which also exhibits a high affinity toward the stationary phase. However, this system required a suppressor column in the chloride form to precipitate the silver as insoluble silver nitrate. Application of this eluent quickly proved to be disadvantageous, since peak broadening and high backpressure from the suppressor column is associated with the precipitation.

Also, the mixture of *m*-phenylenediamine dihydrochloride and hydrochloric acid used later has not gained acceptance. The solution turns purple upon light absorption because of a slow dimerization of the amine and, thus, has only a limited stability.

A micromembrane suppressor (see Section 3.4.3) combined with a mixture of 2,3-diaminopropionic acid (DAP) and hydrochloric acid is now used for the separation of alkaline-earth metals. The advantage of this eluent is the possibility to adjust the elution power via the dissociation equilibrium of

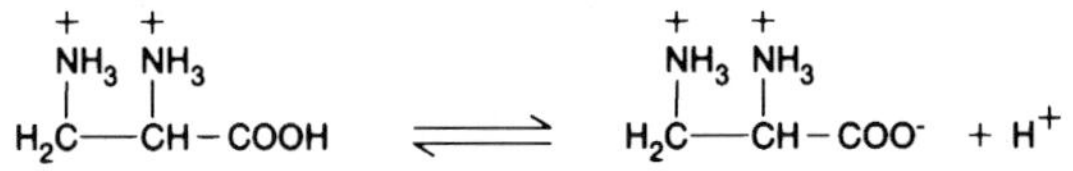

$$H_2C(NH_3^+){-}CH(NH_3^+){-}COOH \rightleftharpoons H_2C(NH_3^+){-}CH(NH_3^+){-}COO^- + H^+ \quad (100)$$

With $pK_a = 1.33$ for the dissociation of the carboxyl group, DAP may exist in the mobile phase as monovalent cation, divalent cation, or mixture of both, respectively. In any case, the product of the suppressor reaction is the zwitterionic form

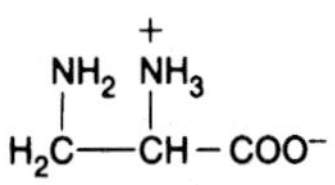

$$H_2C(NH_2){-}CH(NH_3^+){-}COO^-$$

which has no intrinsic conductance. In non-suppressed ion chromatographic systems, mixtures of ethylenediamine and aliphatic dicarboxylic acids such as tartaric acid or oxalic acid, respectively, are typically employed as the eluent for the separation of alkaline-earth metals.

For the elution of heavy and transition metals, weak organic acids are used as complexing agents. Since the metals are separated as anionic or neutral complexes, respectively, type and concentration of the organic acid(s) depend on the separation method. Further details are described in Section 3.4.6.

3.4.3 Suppressor Systems in Cation Exchange Chromatography

Suppressor Columns

In the past, suppressor columns containing an ion-exchange resin were also employed in cation exchange chromatography. The resin was a strong-basic microporous polystyrene/divinylbenzene-based anion exchange resin such as Dowex 1x10. It was functionalized with quaternary ammonium bases and had a relatively high degree of crosslinking of about 10%, which resulted in high mechanical stability, preventing swelling of the resin material in organic solvents miscible with water. Macroporous strong-basic anion exchangers, in contrast, did not gain acceptance as packing material for suppressor columns. The high surface area of the resin (200 to 800 m^2/g) leads to unwelcome adsorption effects with weak elektrolytes.

The function of a suppressor column is described in detail in Section 3.3.3. The disadvantages discussed in this section may be transferred analogously to the cation analysis. Moreover, the exchange function of the suppressor column typically exists in the hydroxide form converting weak-basic cations into the corresponding free bases. The ammonium ion, for example, is converted into the free base NH_3. Since the latter is non-ionic, it is not subject to Donnan exclusion. Weak-basic cations may interact, therefore, with the anion exchange resin of the suppressor column resulting in considerably longer retention times.

In HPLC peak dispersion is significantly affected by the particle size of the resin. Thus, the band broadening in a suppressor column decreases as the particle diameter of the employed resin decreases. Similarly, the void volume of the suppressor column should be as small as possible to reduce band broadening effects. An optimal suppressor column would therefore have a very low volume and would contain an exchange resin with a very small particle diameter. Both requirements are inconsistent with the analytical practice, which calls for a suppressor column that can be employed at least for one working day without regeneration. This requirement can only be met by suppressor columns with a higher volume.

According to the procedure described in Section 3.3.3, the regeneration of conventional suppressor columns such as CSC-1 or CSC-2 is performed using sodium hydroxide solution with a concentration of $c = 0.5$ mol/L.

Hollow Fiber Suppressors

Today, conventional suppressor columns have only historical significance. The above-mentioned disadvantages could be surmounted by the development of the CFS hollow fiber suppressor that was suited for cation exchange chromatography. The schematic of this suppressor corresponds precisely with the respective system for anion exchange chromatography described in Section 3.3.3. Instead of sulfonate groups, the membrane carries quaternary ammonium bases in the hydroxide form to exchange the anions present as counter ions for hydroxide ions. A dynamic equilibrium is established at constant operating conditions so that the detection sensitivity for weak-basic cations remains unaffected. Compared with suppressor columns, hollow fiber suppressors have a void volume more than one order of magnitude smaller. This reduces the peak broadening significantly and, in general, increases detection sensitivity.

For continuous regeneration of the suppressor, either potassium hydroxide with a concentration of c = 0.04 mol/L or tetramethylammonium hydroxide with a concentration of c = 0.02 mol/L may be used. The choice of regenerent depends on the type of analyte. Potassium hydroxide is recommended as a regenerent for the analysis of alkali metals. If ammonium is also to be analyzed, it should be noted that the linear range for the determination of this ion is very small when using potassium hydroxide as the regenerent. This is caused by the equilibrium between NH_4OH as the suppressor reaction product and the free base NH_3.

The regenerent flow rate should be 2 to 3 mL/min. With eluent concentrations of up to c = 0.005 mol/L, background conductivities of 10 μS/cm at maximum are ensured.

Potassium hydroxide solution may also be used as a regenerent for the analysis of alkaline-earth metals, as long as a detection sensitivity in the ppm range is sufficient.

Micromembrane Suppressors

As effective as hollow fiber suppressors are in terms of increased sensitivity, its application is as limiting in cation analysis for the concentration of the acid used as the eluent. As mentioned in Section 3.3.3, a higher ionic strength results in a higher background conductivity, since it is not possible to increase the regenerent concentration accordingly. In such a case, the comparatively small potassium ions would surmount the Donnan exclusion forces and would diffuse into the interior of the membrane. If the application of concentrated eluents is indispensible, as in modern latex-based cation exchangers, a micromembrane suppressor with significantly higher exchange capacity must be employed.

The schematic and operation of the CMMS micromembrane suppressor suitable for cation exchange chromatography corresponds precisely to the AMMS suppressor which was developed for anion exchange chromatography. It is described in Section 3.3.3. The high capacity of this micromembrane suppressor allows the use of eluents with high ionic strengths, which leads to shorter analysis times. The ensuing sensitivity increase contributes to the low void volume of the suppressor, which has little affect on the efficiency of the separation.

The CMMS micromembrane suppressor also allows the application of concentration gradients in combination with conductivity detection that is indispensable for cation detection. A mixture of hydrochloric acid and 2,3-diaminopropionic acid suitable for chemical suppression is used as the eluent. The gradient technique, however, plays a secondary role in cation analysis, since it can only be applied for the analysis of alkali and alkaline-earth metals as well as a number of short-chain aliphatic amines. It is definitely not suitable for the analysis of heavy and transition metals, where different chromatographic conditions are required and conductivity detection has to be ruled out as detection system due to its low selectivity and sensitivity.

The regeneration of a CMMS micromembrane suppressor does not differ fundamentally from that of a corresponding hollow fiber suppressor, except only tetrabutylammonium hydroxide (TBAOH) is used as the regenerent. Depending on the concentration of the acid employed as the eluent, regenerent concentrations from $c = 0.04$ to 0.1 mol/L are believed to be sufficient for isocratic conditions. Upon application of the gradient technique, the flow rate must be adjusted so that a sufficiently low background conductivity is ensured at the maximum eluent concentration being used; thus, the regenerent should be recirculated according to Fig. 3-43 (see Section 3.3.3). Before the regenerent is delivered to the suppressor system via a small pump, it is passed through an ion-exchange cartridge, which contains a strong-basic anion exchanger in the hydroxide form to maintain a constant hydroxide ion concentration. In this way, the tetrabutylammonium chloride formed in the suppressor is converted into the corresponding base. The lifetime of such a cartridge with a stated capacity of about 350,000 μequiv depends, of course, on the concentration and flow rate of the eluent being used. To calculate the cartridge lifetime, use Eq. (68) in Section 3.3.3.

3.4.4 Cation Exchange Chromatography of Alkali Metals, Alkaline-Earth Metals, and Amines

Typically, monovalent cations such as alkali metals are separated using a dilute mineral acid as the eluent. Examples for the separation of alkali metals are displayed in Figs. 3-129, 3-130 (Section 3.4.1.1), and 3-132 (Section 3.4.1.2). These figures reveal that the retention of the alkali metals increases with increasing ionic radius. Compared to conventional instrumental analysis methods, the advantage of ion chromatography is the simultaneousness of the method. Without any doubt, the key ion in this chromatogram is ammonium which elutes between sodium and potassium. Its sensitive detection by other methods is very difficult.

The parameter that determines retention in the separation of alkali metals is solely the concentration of the acid used as the eluent. As seen in Fig. 3-137, a linear relationship is obtained for the different solute ions when the logarithm of the capacity factors is plotted as a function of the ionic strength.

Many aliphatic amines also elute under the chromatographic conditions that are suitable for the separation of alkali metals. This includes hydroxylamine, which can be separated as a cation after protonation in the mobile phase and detected via direct conductivity detection. However, amperometric detection using a Pt working electrode at which hydroxylamine is oxidized at a working potential of $> +0.8$ V is much more selective. The corresponding chromatogram is displayed in Fig. 3-138. Hydroxylammonium ions exhibit about the same retention behavior as ammonium ions. If one component is to be detected in the presence of the other, the necessary selectivity may only be obtained by applying different detection methods, for which the separator column effluent is passed first through the amperometric cell and then through the suppressor system into the conductivity cell. Since hydroxylamine cannot be detected via suppressed conductivity, both detection methods are highly selective. The respective signals may be combined in a dual-channel integrating system. Another way to differentiate between the two is offered by fluorescence detection. While ammonium ions may be detected with a fluorescence detector after derivatization with *o*-phthaldialdehyde/2-mercaptoethanol, hydroxylammonium ions do not undergo this reaction.

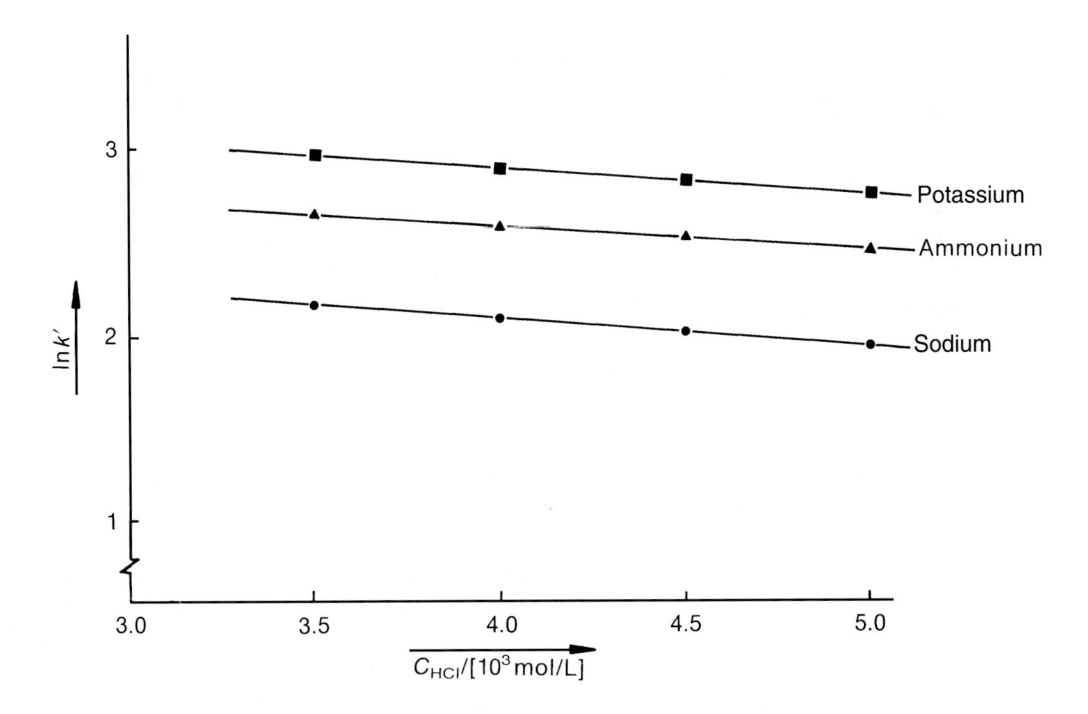

Fig. 3-137. Dependence of the retention of monovalent cations on the ionic strength of the eluent for sodium, ammonium, and potassium. – Separator column: IonPac CS1; eluent: HCl; flow rate: 2.3 mL/min; detection: suppressed conductivity.

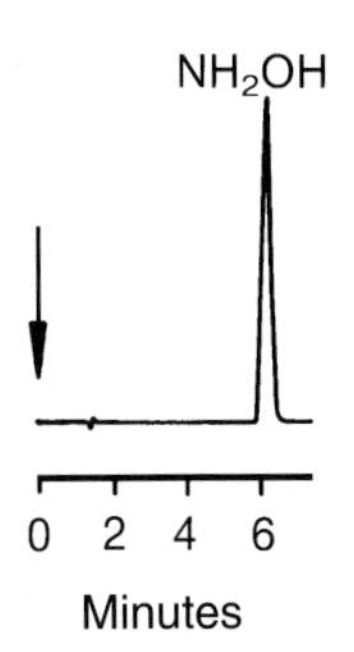

Fig. 3-138. Amperometric detection of hydroxylamine after separation on a cation exchanger. – Separator column: IonPac CS3; eluent: 0.03 mol/L HCl; flow rate: 1 mL/min; detection: amperometry on a Pt working electrode; oxidation potential: +0.85 V; injection volume: 50 µL; solute concentration: 10 ppm.

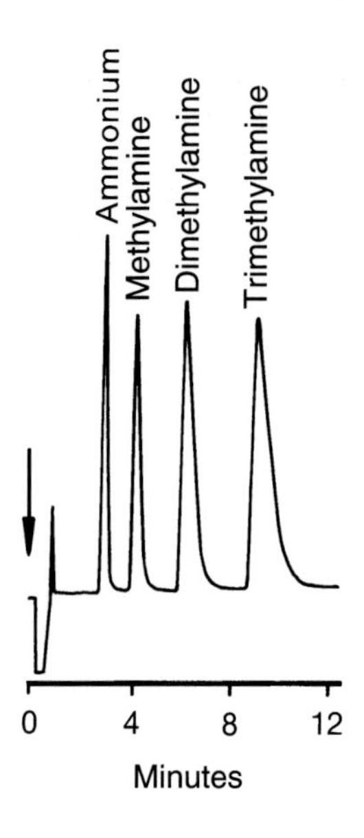

Fig. 3-139. Separation of monomethylamine, dimethylamine, and trimethylamine on Vydac 400 IC 405. – Eluent: 0.005 mol/L isonicotinic acid + 0.0025 mol/L HNO_3 + 100 mL/L methanol; flow rate: 2 mL/min; detection: direct conductivity; injection volume: 100 μL; solute concentrations: 1 ppm ammonium, 2 ppm monomethylamine, 5 ppm dimethylamine, and 10 ppm trimethylamine.

Conductivity detection in *both* of its modes of application is also suitable for the detection of primary, secondary, and tertiary alkylamines. This is shown in Fig. 3-139 using three methylamines which are separated on Vydac 400 IC 405 with isonicotinic acid as the eluent. Interferences are possible if the sample contains potassium, rubidium, and/or cesium in addition to amines. However, a small variation of the solvent content in the mobile phase which mainly affects the retention of alkylsubstituted cations, allows the optimization of the separation. In comparison to Fig. 3-140, the separation of the

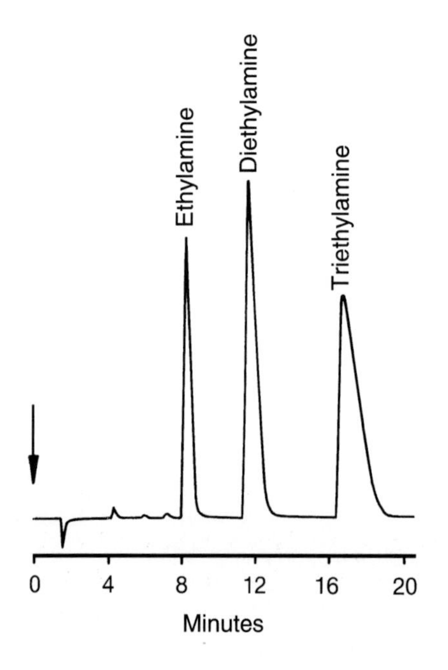

Fig. 3-140. Separation of monoethylamine, diethylamine, and triethylamine on a latex cation exchanger. – Separator column: IonPac CS3; eluent: 0.04 mol/L HCl; flow rate: 1 mL/min; detection: suppressed conductivity; injection volume: 50 μL; solute concentrations: 20 ppm monoethylamine, 50 ppm diethylamine, and 100 ppm triethylamine.

corresponding ethylamines on a latex cation exchanger using hydrochloric acid as the eluent is displayed in Fig. 3-140. If necessary, the higher retention times may be reduced significantly by adding 2,3-diaminopropionic acid.

Furthermore, compounds having several amino functions may be analyzed under the chromatographic conditions given in Fig. 3-140. This includes both the formamidinium and the guanidinium ion:

$$H-C(=\overset{+}{N}H_2)-NH_2 \qquad \overset{+}{H_2N}=C(NH_2)-NH_2$$

Formamidinium Ion **Guanidinium Ion**

The latter is a urea derivative in which the three NH_2 groups are ordered symmetrically and equidistantly to the central C-atom. Both cations, the separation of which is shown in Fig. 3-141, may be detected either with or without chemical suppression.

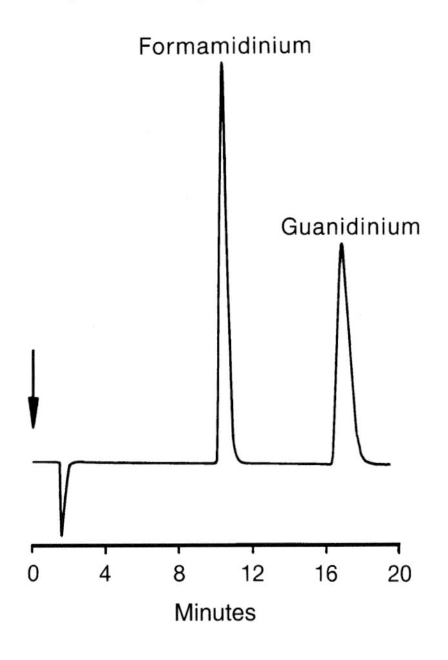

Fig. 3-141. Separation of formamidinium and guanidinium ions. – Chromatographic conditions: see Fig. 3-140; solute concentrations: 50 mg/L each of formamidinium acetate and guanidinium sulfate.

Guanylurea with four amino groups

$$H_2N-\underset{\underset{NH}{\|}}{C}-NH-\underset{\underset{O}{\|}}{C}-NH_2$$

Guanylurea

has an even higher retention time and may be determined either by direct conductivity detection or by measuring the light absorption of the carbonyl group at a wavelength of 215 nm. Fig. 3-142 shows as an example the respective chromatogram obtained upon application of UV detection.

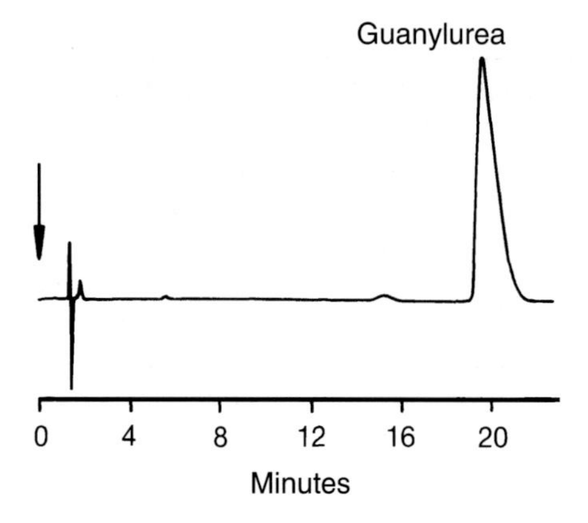

Fig. 3-142. Separation of guanylurea. – Separator column, eluent, and flow rate: see Fig. 3-140; detection: UV (215 nm); injection volume: 50 μL; solute concentration: 100 mg/L guanylurea (sulfate salt).

Alkaline-earth metals, which elute in the order $Mg^{2+} < Ca^{2+} < Sr^{2+} < Ba^{2+}$, can also be determined using both conductivity detection modes. While an eluent mixture of hydrochloric acid and 2,3-diaminopropionic acid is used in the suppressed mode, ethylenediammonium ions are suitable for the direct conductivity detection mode. Fig. 3-143 shows a separation of alkaline-earth metals on Shimpack IC-C1 obtained with this eluent. Because of the high elution power of the mobile phase, all monovalent

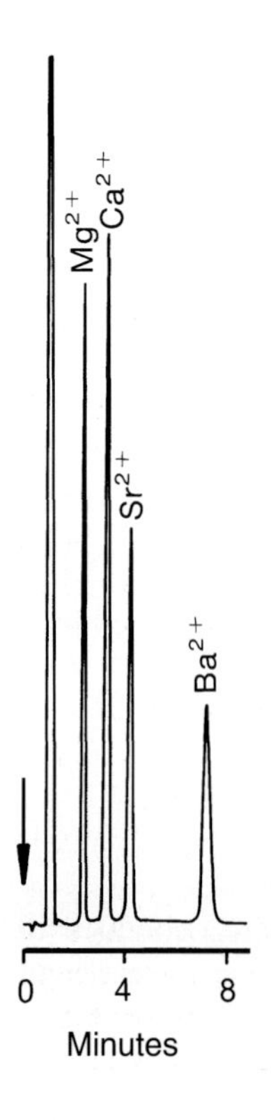

Fig. 3-143. Separation of alkaline-earth metals at Shimpack IC-C1. – Eluent: 0.004 mol/L tartaric acid + 0.002 mol/L ethylenediamine; flow rate: 1.5 mL/min; detection: direct conductivity; injection volume: 20 μL; solute concentrations: 5 ppm magnesium, 10 ppm calcium, 20 ppm strontium, and 36 ppm barium.

cations present in the sample are eluted as one peak within the void volume of the column. As an alternative to strong eluents, shorter separator columns may be employed to reduce the retention of cations with high affinities toward the stationary phase.

The simultaneous analysis of the most important alkali and alkaline-earth metals was once impossible due to their markedly different retention behavior. However, the inorganic chemists have finally realized their dream: this analytical problem no longer poses a problem. One of the two possible solutions is the novel silica-based cation exchanger modified with poly(butadiene-maleic acid), introduced in Section 3.4.1.3. As shown in Fig. 3-135, the most important alkali and alkaline-earth metals and ammonium can be analyzed in a single run via direct conductivity detection using tartaric acid as the eluent. The extremely short time required for such a separation is quite impressive.

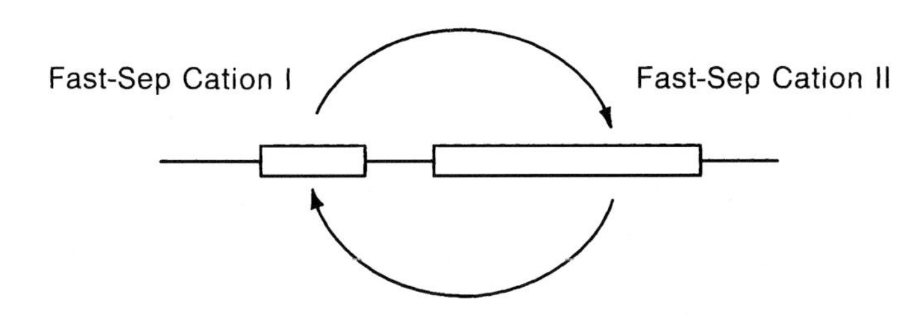

Fig. 3-144. Schematic representation of the column switching for the simultaneous analysis of alkali and alkaline-earth metals.

A comparable result is obtained with the column switching shown schematically in Fig. 3-144. Two latex-based cation exchangers (Fast-Sep Cation I and II) are used; they differ in column length, size of the latex beads, and degree of crosslinking. According to Fig. 3-145, these two columns are connected to the ports 2 and 6 and 4 and 8, respectively, of a 4-way valve which is in the position 0 at the time of injection. The analyte species enter the shorter column, the Fast-Sep Cation I, first. Because of the strongly different affinities of alkali and alkaline-earth metals toward this stationary phase, the alkali metal ions migrate much faster through this column than the alkaline-earth metals. The latter are already separated on this column. The 4-way valve is switched after the alkali metals have passed through the Fast-Sep Cation I column. The alkaline-earth metals only pass through the short column, since the flow direction is maintained in the valve position "1", while the alkali metals have to pass through both columns. Their migration rate is so high that they elute ahead of the alkaline-earth metals. As is evident from Fig. 3-146, a baseline-resolved separation of lithium, sodium, ammonium, potassium, magnesium, and calcium is possible within 10 minutes, when the ion-exchange capacities of both columns are properly dimensioned and the eluent composition optimized, respectively. Although the analysis time is very short and the peak efficiency for monovalent cations is very high, the strong tailing of divalent cations caused by the low chromatographic efficiency of the Fast-Sep Cation I column has to be regarded as a disadvantage. Particularly affected is the calcium signal which can only be reproducibly analyzed with a precise programming of the integration parameters.

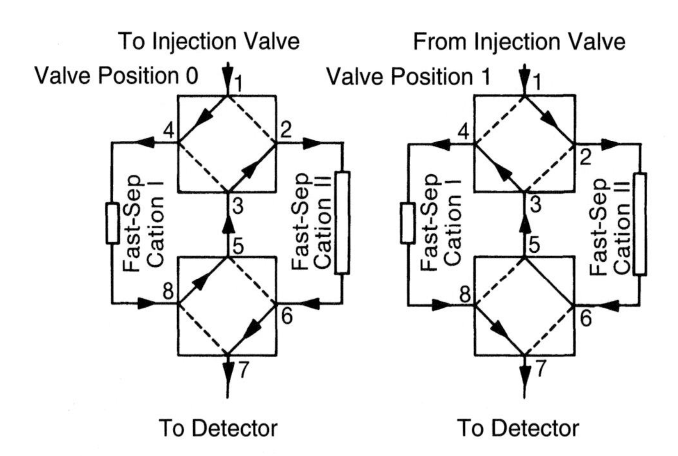

Fig. 3-145. Valve switching for the simultaneous analysis of alkali and alkaline-earth metals.

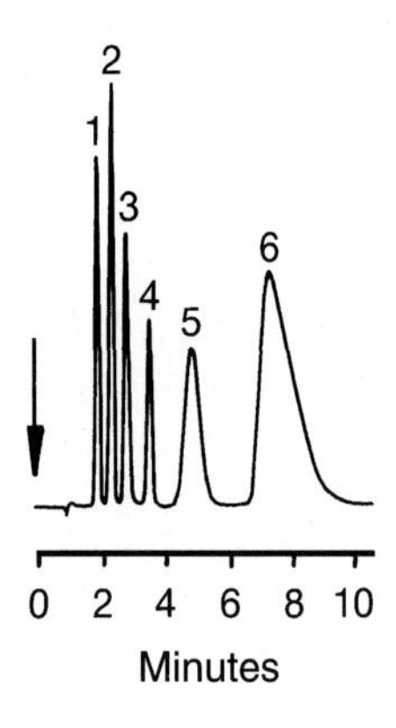

Fig. 3-146. Simultaneous analysis of alkali and alkaline-earth metals by means of column switching. – Separator column: Fast-Sep Cation I and II; switching time: 0.5 min; eluent: 0.02 mol/L HCl + 0.0002 mol/L 2,3-diaminopropionic acid; flow rate: 2 mL/min; detection: suppressed conductivity; injection volume: 50 µL; solute concentrations: 1 ppm lithium (1), 4 ppm sodium (2), 8 ppm ammonium (3), 4 ppm potassium (4), 4 ppm magnesium (5), and 20 ppm calcium (6).

This problem could be resolved by using the IonPac CS10 separator column introduced in Section 3.4.1.2. In this case, the divalent cations also pass through the guard and the separator column, both of which exhibit high chromatographic efficiencies. As is seen in the corresponding chromatogram in Fig. 3-147, all components have an almost symmetrical peak shape. The total analysis time may be adjusted by varying the 2,3-diaminopropionic acid concentration in the mobile phase. The resolution, especially in the front part of the chromatogram, decreases with increasing DAP concentration. It is interesting that the heavy metal manganese can also be determined in the same run. Upon application of suppressed conductivity detection, it is generally assumed that heavy metals precipitate as hydroxides in the suppressor and, thus, may not be detected. However, the kinetics for this reaction are not the same for all metals, especially as a weakly acidic milieu (pH 6) prevails in the suppressor after the suppressor reaction and many heavy metals only precipitate at pH values >7.

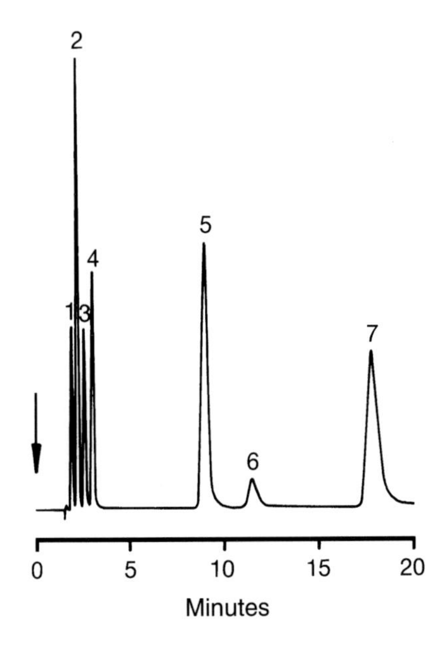

Fig. 3-147. Simultaneous analysis of alkali metals, alkaline-earth metals, and manganese on IonPac CS10. – Eluent: 0.04 mol/L HCl + 0.004 mol/L 2,3-diaminopropionic acid; flow rate: 1 mL/min; detection: suppressed conductivity; injection volume: 20 μL; solute concentrations: 5 ppm lithium (1), sodium (2), ammonium (3), potassium (4), and magnesium (5), 10 ppm manganese (6), and calcium (7).

A simultaneous analysis of alkali and alkaline-earth metals – either via column switching or by using a special cation exchanger – is only possible, however, if the components exist in comparable concentrations. Otherwise, a separate analysis of both classes of compounds is indispensible.

While the separations shown in Figs. 3-146 and 3-147 were obtained under isocratic conditions, Fig. 3-148 displays the chromatogram of a gradient elution of cations.

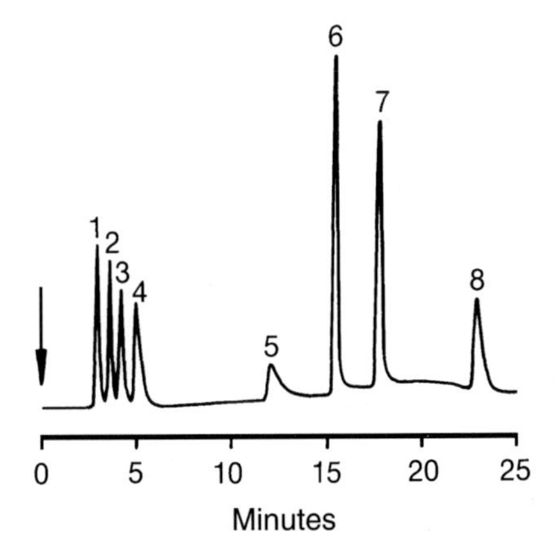

Fig. 3-148. Gradient elution of monovalent and divalent cations. – Separator column: Fast-Sep Cation I and II; eluent: (A) water, (B) 0.04 mol/L HCl + 0.02 mol/L 2,3-diaminopropionic acid; gradient: 5 min 7% B isocratically, then linearly to 100% B in 10 min; flow rate: 1 mL/min; detection: suppressed conductivity; injection volume: 50 μL; solute concentrations: 2 ppm lithium (1), 5 ppm sodium (2), 10 ppm ammonium (3), and potassium (4), 20 ppm tetrabutylammonium (5), 10 ppm magnesium (6) and calcium (7), 20 ppm ethylenediamine (8).

Again, a mixture of hydrochloric acid and DAP was employed as the eluent. Since the concentration of the eluent components is increased by about one order of magnitude during the run, their purity is of essential importance. In the field of cation analysis, the gradient technique is used predominantly for screening purposes. In comparison with isocratic techniques, a significantly better peak shape is observed for strongly retained cations such as ethylenediamine.

3.4.5 Analysis of Heavy and Transition Metals

As early as 1939, Samuelson [142] reported the successful separation of heavy and transition metals by means of ion-exchange chromatography. This method utilized an ion-exchange resin to separate the metal ions. Quantitation was achieved via manual spectrophotometry of the individual fractions. The introduction of highly efficient low capacity ion exchangers by Fritz et al. [143] in 1974 and a continuous solute-specific photometric detection after derivatization of the column effluent significantly improved the ion chromatographic determination of heavy and transition metals. In 1979, Cassidy et al. [144] successfully employed ion chromatography for the analysis of lanthanides. The commercialization of this method since 1982, however, is attributed to the work of Riviello et al. [145]. By introducing membrane reactors, he succeeded in significantly improving the reagent delivery for post-column derivatization. Following the introduction of highly selective pellicular ion exchangers with defined anion and cation exchange capacity [146], ion chromatography gained general acceptance as a fast and sensitive multi-element detection method for the determination of heavy metals, allowing detection limits in the lower ppb range and analysis times of about 15 minutes.

3.4.5.1 Basic Theory

The separation of heavy and transition metals with ion exchangers requires a complexation of the metal ions in the mobile phase to reduce their effective charge density. Monovalent cations such as Na^+ or H^+ are unsuitable as eluents. Because the selectivity coefficients for transition metals of the same charge number hardly differ, a selectivity change is obtained only by the introduction of a secondary equilibrium, such as a complexation equilibrium, which is established by adding suitable complexing agents to the mobile phase. Weak organic acids such as citric acid, oxalic acid, tartaric acid, or pyridine-2,6-dicarboxylic acid form, preferably, anionic or neutral complexes with metal ions. Such complexes may be separated on anion or cation exchangers. Nowadays, stationary phases having both anion and cation exchange capacities are used. The separations obtained with aliphatic or aromatic carboxylic acids are clearly superior to those obtained with ammonia or ethylenediamine exclusively forming neutral complexes. Two different organic acids are frequently used as complexing agents to optimize certain separations. For simplicity, however, only a single ligand is considered in the following kinetic treatment.

Consider a solution containing a multiply charged transition metal ion M^{n+} in low concentration and a monovalent cation A^+ (Na^+ or H^+) in a significantly higher concentration. When this solution is in contact with a cation exchanger in the A^+-form, the following equilibrium is established:

$$M^{n+} + n \cdot RA \rightleftarrows R_nM + n \cdot A^+ \tag{101}$$

R Resin

With the law of mass action the equilibrium constant or selectivity coefficient is expressed as:

$$K_{MA} = \frac{[R_nM] \cdot [A^+]^n}{[RA]^n \cdot [M^{n+}]} \tag{102}$$

If we are only dealing with an ion-exchange mechanism, the distribution coefficient D_M is given by the ratio of the analyte concentration M^{n+} in the stationary and in the mobile phase at equilibrium:

$$D_M = \frac{[R_nM]}{[M^{n+}]} \tag{103}$$

$$= K_{MA} \cdot \frac{[RA]^n}{[A^+]^n}$$

The metal ion concentration is very low compared with the eluent concentration, hence:

$$[M^{n+}] \ll [A^+] \quad \text{and} \quad [R_nM] \ll [RA] \tag{104}$$

It follows that K_{MA} is constant for any given eluent concentration. Since, in good approximation, [RA] is also constant and represents the exchange capacity C of the resin, the distribution coefficient may be expressed as:

$$D_M \sim \frac{K_{MA} \cdot C^n}{[A^+]^n} \tag{105}$$

In logarithmic form one obtains:

$$\log D_M \sim \log K_{MA} + n \cdot \log C - n \cdot \log[A^+] \tag{106}$$

In general, however, the capacity factor k' according to Eq. (7) in Section 2.1 is measured instead of the distribution coeffcient D_M.

From Eq. (106) it follows that the distribution coefficient D_M depends linearly on the cation concentration $[A^+]$ in the mobile phase. Plotting log D_M versus log $[A^+]$ yields a straight line with the slope n, which corresponds to the charge number of the ion.

The selectivity for separating metal ions having the same charge number is exclusively determined by their K_M values. The concentration of eluent cations $[A^+]$ does not contribute to the selectivity increase, since the differences in the logarithms of the distribution coefficients remain unchanged with different eluent concentrations. Adequate separations are only obtained if the selectivity coefficients of two transition metal ions

M_1^{n+} and M_2^{n+} differ significantly. The only exception are metal ions of differing charge number whose separation depends on the eluent concentration. If the eluent contains complexing agents such as citric acid or oxalic acid, then the equation for the distribution coefficient of the metal ion M^{n+} must also take into account the competing equilibrium of the complex formation. The complex formation coefficient may be expressed as follows:

$$a_{M(L)} = 1 + [L]\, B_{ML} + [L]^2\, B_{ML2} + \cdots \tag{107}$$

$a_{M(L)}$ Complex formation coefficient
[L] Concentration of anionic ligand
B_{ML} Formation constant for equilibrium
$M^{n+} + L^{x-} \rightleftharpoons ML^{n-x}$
B_{ML2} Formation constant for equilibrium
$M^{n+} + 2\,L^{x-} \rightleftharpoons ML_2^{n-2x}$

If the complexing form of the ligand is a sufficiently strong conjugated base, protonation of the non-complexing ligands must be taken into account for the calculation of $a_{M(L)}$. Under normal chromatographic conditions, the concentration of the complexing form of the ligand, L^{x-}, is much higher than that of the metal ion. It follows that the complex formation equilibrium is hardly affected by a partial protonation of the ligand. Thus, the distribution coefficient of a metal ion in the presence of complexing agents can be expressed as follows:

$$\log D_M = \log K_{MA} - \log a_{M(L)} + n \log C - n \log [A^+] \tag{108}$$

The separation of the various transition metals may be optimized by varying the pH value. If weak organic acids are used as complexing agents, lowering the pH leads to a decrease in the effective ligand concentration. Therefore, an increase in the proton concentration results in a longer retention time. This is easily recognized when including the dissociation equilibrium for [L] in Eq. (107) and combining it with Eq. (108). The effective ligand concentration can also be raised by a higher concentration of the organic acid, although this leads also to a higher concentration of the respective counter ion (Na^+ or H^+). This accelerates the displacement of metal ions from the cation exchange function and results in a severe loss in separation efficiency.

With the use of low-capacity cation exchangers, the exchange capacity depends on the kind of counter ion. Accordingly, lithium hydroxide is recommended for adjusting the pH value, because the lithium ion exhibits a low affinity toward the stationary phase.

The ability of these metals to form anionic complexes is used for the separation of transition metals or heavy metals on anion exchangers. Such separations are based on (1) the significant differences in the stability of these complexes with inorganic complexing agents and (2) the different affinity of anionic complexes toward the stationary phase of anion exchangers. The most important parameter determining retention is the valency of the complex, which, in turn, depends on the charge numbers of the central metal ion and the ligand, as well as on the coordination number of the complex.

The distribution coefficient of a metal ion between the mobile and the stationary phase with a monovalent anionic ligand is given by:

$$D_M = \frac{[R_{n-x} \cdot ML_x]}{[M^{n+}]} \tag{109}$$

$$= K_M \cdot \Phi_x \cdot \left(\frac{[RL]}{[L^-]}\right)^{n-x}$$

K_M is called the selectivity coefficient for the anions ML^{n-x} and L^-. It is described as:

$$K_M = \frac{[R_{n-x} \cdot ML_x] \cdot [L^-]^{n-x}}{[RL]^{n-x} \cdot [ML_x^{n-x}]} \tag{110}$$

Φ_x is the molar portion of the complex ML_x^{n-x} in solution, which may be calculated via Eq. (111):

$$\Phi_x = \frac{[L^-] \cdot \beta_{ML_x}}{1+[L^-] \cdot \beta_{ML}+[L^-]^2 \cdot \beta_{ML_2} + \cdots} \tag{111}$$

Under normal chromatographic conditions, however, it holds

$$[M^{n+}] \ll [L^-]$$

so that the concentration of the free ligand ions $[L^-]$ may be substituted by the total ligand concentration $[L_t]$. When [RL] is expressed in terms of the exchange capacity C, one obtains for D_M

$$D_M = K_M \, \Phi_x \, C^{n-x} \, [L^-]^{n-x} \tag{112}$$

or in a logarithmic form

$$\log D_M = \log K_M + \log \Phi_x + (n-x) \log C + (n-x) \log [L^-] \tag{113}$$

In cation exchange chromatography of transition metals, an increase in the concentration of the complexing ligand reduces the resolution. In contrast, in anion exchange chromatography optimal separation is achieved at a ligand concentration where predominantly neutral complexes prevail. If the concentration is further raised, the retention also increases because of the increased formation of anionic complexes. This effect is partly compensated for, however, since the free anionic ligands are responsible for the elution of inorganic metal complexes. Excellent separations were accomplished on pellicular anion exchangers using oxalic acid as the eluent [147].

When selecting a ligand for the separation and elution of transition metal ions from cation exchangers the following guidelines should be taken into account:

- Metal ion and ligand must form neutral or anionic complexes.
- Different complex formation constants for the various metals increase the selectivity.

- The transition metal complexes being formed should be thermodynamically stable and kinetically labile; that is, the complex should have a high energetic and entropic formation tendency and the thermodynamic equilibrium should be established quickly and be unrestricted. This requirement shall be discussed in the following examples:
  a) If a transition metal complex has a high formation constant but is kinetically stable, then the cation exchange process is impossible; i.e., the anionic complex is not retained due to Donnan exclusion.
  b) If the formation constant is small but the complex is kinetically unstable, the retention mechanism is dominated by cation exchange. Long retention times could result.
  c) If the formation constant is small and the complex is kinetically stable, the chromatographic signals are often tainted with a strong tailing.
- If the transition metal complexes are detected by post-column derivatization with a suitable metallochromic indicator, the formation constant of the reaction should be high. Also, the indicator complex, MeIn, should be kinetically stable.

$$MeL_x + In \rightleftharpoons MeIn + x \cdot L \tag{114}$$

Me Metal
L Ligand
In Indicator

This also applies to the separation of anionic transition metal complexes on *anion exchangers.* Here, the ligand molecules should be relatively small so that the resulting complex has a similar size to the hydrated metal ion. In metal ions of the same valency and that share the same coordination number and geometry with any given ligand, large ligands cause a loss in efficiency if the formation constants of these complexes do not differ significantly.

3.4.5.2 Analysis of Heavy and Transition Metals by Direct Conductivity Detection

Both conductivity detection modes are suitable for the detection of alkali and alkaline-earth metals. With the analysis of heavy and transition metals, however, only direct conductivity detection may be employed. Upon application of the suppressor technique heavy and transition metals would be transformed by the suppressor reaction into the mostly insoluble hydroxides, but the direct conductivity detection is also impeded by the presence of complexing agents in the mobile phase that are required for separating heavy metals. In 1983, Sevenich and Fritz [148] found an eluent mixture suited for this detection mode. It is comprised of ethylenediamine and tartaric acid. Ethylenediamine serves as the eluent ion since it is already fully protonated (EnH_2^{2+}) at pH values ≤ 5. The selection of tartrate as the complexing anion is based on the resulting degree of complexation for the metal ions to be analyzed, which should be only partly complexed by the selected ligand because complete complexation results in a loss of selectivity. If both eluent components are employed in almost equimolar concentrations, the separation of various divalent metal ions on a surface-sulfonated 20-μm cation exchanger (Fig. 3-149) is obtained. Again, the pH value of the mobile phase is the experimental parameter determining retention. It primariliy affects the complexing properties of tartrate. Its complexing ability increases with increasing pH, so that a decrease in retention, which is caused by a shift of the equilibrium

$$\text{Resin}-M^{2+} + EnH_2^{2+} \rightleftharpoons \text{Resin}-EnH_2^{2+} + M^{2+} \quad \xrightleftharpoons{L^{2-}} \quad ML \tag{115}$$

to the right-hand side, is observed.

In addition to tartrate, α-hydroxyisobutyrate can bc utilized as the complexing anion. It was employed by Cassidy et al. [144] for the separation of lanthanides. In comparison to tartrate, no particular advantages are revealed in the separation of divalent metal ions. Fig. 3-150 shows a separation of the seven heaviest lanthanides with an eluent mixture of ethylenediamine and α-hydroxybutyric acid. The pronounced tailing effects observed under isocratic conditions for late eluting ions are unsatisfactory.

In most applications, the method of cation exchange with direct conductivity detection does not have the required specificity for the analysis of heavy and transition metals. Ionic matrix components which are often present in high excess are also detected.

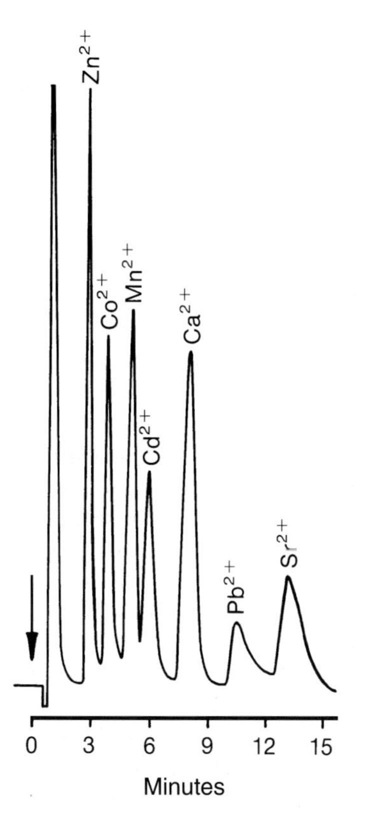

Fig. 3-149. Separation of divalent cations with direct conductivity detection. – Separator column: surface-sulfonated cation exchanger (Benson Co., Reno, USA); eluent: 0.0015 mol/L ethylenediamine + 0.002 mol/L tartaric acid, pH 4.0; flow rate: 0.85 mL/min; injection volume: 100 µL; solute concentrations: 10.3 ppm Zn^{2+}, 9.1 ppm Co^{2+}, 16 ppm Mn^{2+}, 16.1 ppm Cd^{2+}, 17.1 ppm Ca^{2+}, 16 ppm Pb^{2+}, and 20.3 ppm Sr^{2+}; (taken from [148]).

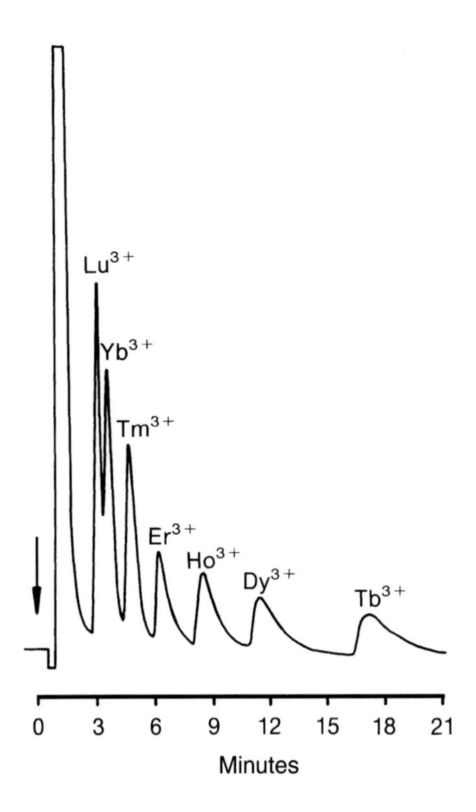

Fig. 3-150. Separation of the seven heaviest lanthanides with an eluent mixture of ethylenediamine and α-hydroxyisobutyric acid. – Separator column: see Fig. 3-149; eluent: 0.004 mol/L ethylenediamine + 0.003 mol/L α-hydroxybutyric acid, pH 4.5; flow rate: 0.85 mL/min; detection: direct conductivity; injection volume: 100 μL; solute concentrations: 10 ppm each; (taken from [148]).

3.4.5.3 Analysis of Heavy and Transition Metals with Spectrophotometric Detection

According to Fig. 3-151, both cation and anion exchange processes contribute to the separation of heavy and transition metals with complexing agents in the mobile phase. Hence, before the introduction of ion exchangers with defined anion and cation exchange capacities, surface-sulfonated cation exchangers as well as latexed cation exchangers were used, resulting in totally different selectivities [149]. Fig. 3-152 shows the separation of various heavy metals on an IonPac CS2 surface-sulfonated cation exchanger. Although satisfactory separation results can also be obtained with oxalic acid as the sole complexing agent, citric acid was added to the eluent to increase the selectivity. Nevertheless, the resolution between iron(III), copper, and nickel in the front of the chromatogram is not optimal. The complete exclusion of iron(III) is attributed to the formation of $Fe(Ox)_3^{3-}$, a very kinetically-stable species. This trivalent complex does not interact with the cationic stationary phase because of its high stability ($\log K = 18.5$ [150]), thus it elutes within the void volume. In addition, the sample pH adjustment to pH 1.5 was of crucial importance in achieving an acceptable separation of copper and nickel. Cadmium and manganese are strongly retained under the chromatographic conditions given in Fig. 3-152, but may be eluted with a mixture of citric acid and tartaric acid. The detection involves derivatization of the column effluent with 4-(2-pyridylazo)-resorcinol (PAR) and subsequent photometric determination of the chelate complexes being formed at a wavelength of 520 nm (see Section 6.2.1).

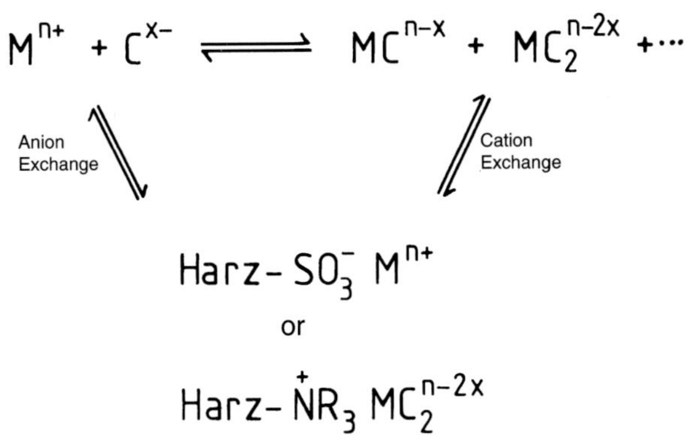

$$M^{n+} + C^{x-} \rightleftharpoons MC^{n-x} + MC_2^{n-2x} + \cdots$$

Anion Exchange

Cation Exchange

$$\text{Harz-}SO_3^- \, M^{n+}$$

or

$$\text{Harz-}\overset{+}{N}R_3 \, MC_2^{n-2x}$$

Fig. 3-151. Overview of the equilibria involved in the separation of heavy and transition metals.

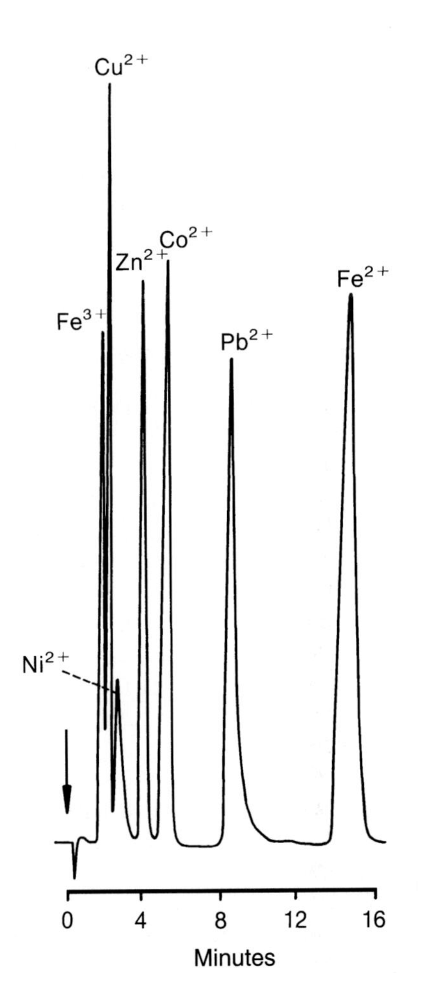

Fig. 3-152. Separation of heavy and transition metals on a surface-sulfonated cation exchanger. – Separator column: IonPac CS2; eluent: 0.01 mol/L oxalic acid + 0.0075 mol/L citric acid, pH 4.2; flow rate: 1 mL/min; detection: photometry at 520 nm after reaction with PAR; injection volume: 50 μL; solute concentrations: 5 ppm Fe^{3+}, 0.5 ppm Cu^{2+}, Ni^{2+}, and Zn^{2+}, 1 ppm Co^{2+}, 10 ppm Pb^{2+}, and 5 ppm Fe^{2+}.

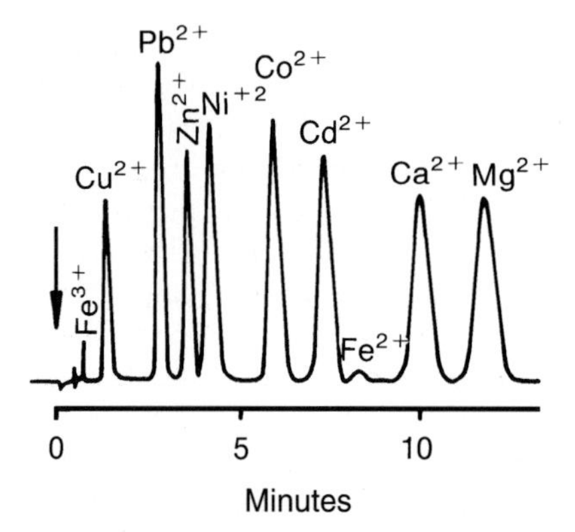

Fig. 3-153. Separation of heavy and transition metals on a polymethacrylate-based cation exchanger. – Separator column: Sykam LCA A02; eluent: 0.1 mol/L tartaric acid, pH 2.95 with NaOH; flow rate: 2 mL/min; detection: photometry at 500 nm after reaction with PAR and ZnEDTA; injection volume: 100 μL; solute concentrations: 2 ppm Fe^{3+} and Cu^{2+}, 4 ppm Pb^{2+}, 1 ppm Zn^{2+}, 2 ppm Ni^{2+} and Co^{2+}, 4 ppm Cd^{2+}, 1.8 ppm Fe^{2+}, 1 ppm Ca^{2+} and Mg^{2+}.

A significantly better separation between iron(III), copper, and nickel is accomplished on a polymethacrylate-based cation exchanger using pure tartaric acid as the eluent. As the respective chromatogram in Fig. 3-153 reveals, iron(III) still elutes near the void volume, which renders the quantitation of this signal more difficult. The addition of ZnEDTA to the PAR reagent is very advantageous, because it enables the simultaneous detection of alkaline-earth metals.

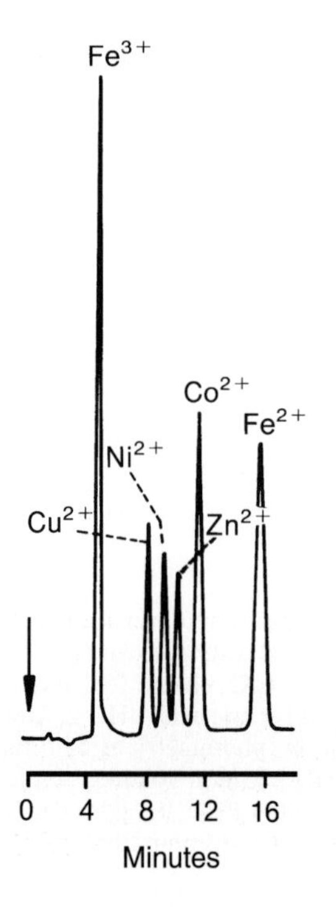

Fig. 3-154. Separation of heavy and transition metals on an ion exchanger with both anion and cation exchange capacity. – Separator column: IonPac CS5; eluent: 0.006 mol/L pyridine-2,6-dicarboxylic acid, pH 4.8 with LiOH; flow rate: 1 mL/min; detection: see Fig. 3-152; injection volume: 50 μL; solute concentration: 1 ppm Fe^{3+} and Cu^{2+}, 3 ppm Ni^{2+}, 4 ppm Zn^{2+}, 2 ppm Co^{2+}, and 3 ppm Fe^{2+}.

The introduction of the IonPac CS5 separator column with mixed anion and cation exchange capacity [146] finally remedied the problem with iron(III)-determination. The CS5 column represents a latexed anion exchanger with dimensions 250 mm x 4 mm I.D., and a PS/DVB substrate particle size of 13 μm. The latex bead diameter is 150 nm and has a degree of crosslinking of about 2%. As revealed by the chromatogram in Fig. 3-154, iron(III) is more strongly retained on this stationary phase. The prerequisite for this is the use of pyridine-2,6-dicarboxylic acid (PDCA) as the complexing agent with a respective different behavior for complexing iron(III). Anion exchange is the dominating separation mechanism, since the stability constants of the metal-PDCA complexes are very high. Lead elutes under these chromatographic conditions, but it cannot be detected because the lead-PDCA complex is more stable than the corresponding chelate complex with the PAR reagent. This stationary phase exhibits a higher chromatographic efficiency than the CS2 column, which can be attributed to the smaller particle diameter of the substrate.

A reversal of elution order for the ions Pb^{2+}, Co^{2+}, Zn^{2+}, and Ni^{2+} is obtained with the use of oxalic acid as the sole complexing agent. Such a separation is displayed in Fig. 3-155. Under these conditions, the metal separation is controlled by anion and cation exchange processes. The degree to which both mechanisms contribute to the separation process is different for each metal ion. The anion exchange mechanism predominates where stable anionic oxalate complexes are formed. In metal ions which do not form stable oxalate complexes, such as Cu^{2+}, the cation exchange mechanism is

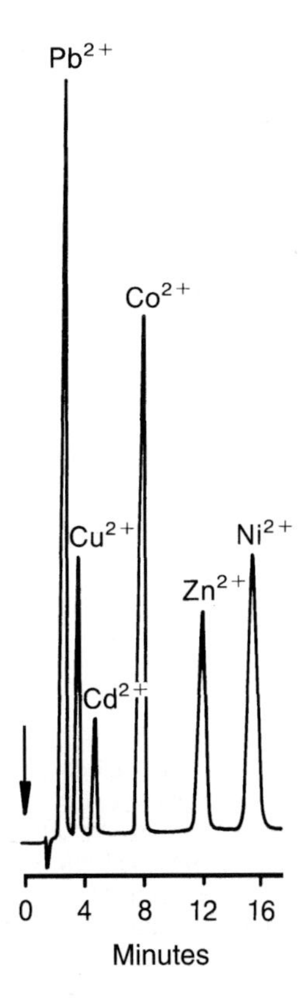

Fig. 3-155. Separation of some selected heavy metals with oxalic acid as the complexing agent. – Separator column: IonPac CS5; eluent: 0.05 mol/L oxalic acid, pH 4.8 with LiOH; flow rate, detection, and injection volume: see Fig. 3-152; solute concentration: 4 ppm Pb^{2+}, 0.5 ppm Cu^{2+}, 4 ppm Cd^{2+}, 2 ppm Co^{2+} and Zn^{2+}, 4 ppm Ni^{2+}.

favored. Under the chromatographic conditions given in Fig. 3-155, iron(II) and iron(III) cannot be eluted. The high affinity of iron(III) toward the stationary phase is explained by the formation of a stable $Fe(Ox)_3^{3-}$-complex which is not eluted by the divalent oxalate. However, this explanation does not apply to the unusual retention behavior of iron(II), which at present, cannot be explained conclusively.

For sometime, the post-column derivatization with PAR was limited to the analysis of iron, cobalt, nickel, copper, cadmium, manganese, zinc, lead, uranium, and lanthanides, primarily due to the kinetics of complex formation with PAR, which is hindered by the complexing agents present in the mobile phase. New investigations reveal that the ligand exchange between the transport complex and PAR is aided by the addition of ligands such as phosphate, carbonate, or alkylamines to the eluent. In this way, heavy metals that are also masked by the presence of complexing agents in the eluent can be detected sensitively with PAR.

Another limiting factor of complexation with PAR is the pH value of the reagent. As the pH value in the reagent increases, so does the PAR dissociation, which is necessary for the complexation of the metals. At higher pH values in the reagent, however, metal ions hydrolyze, thus inhibiting their complexation with PAR. By lowering the pH value in the PAR reagent, heavy metals can be detected that are sensitive to hydrolysis. On the basis of these new insights, the composition of the mobile phase was modified. It is now possible to determine nine different metals within about 15 minutes (Fig. 3-156).

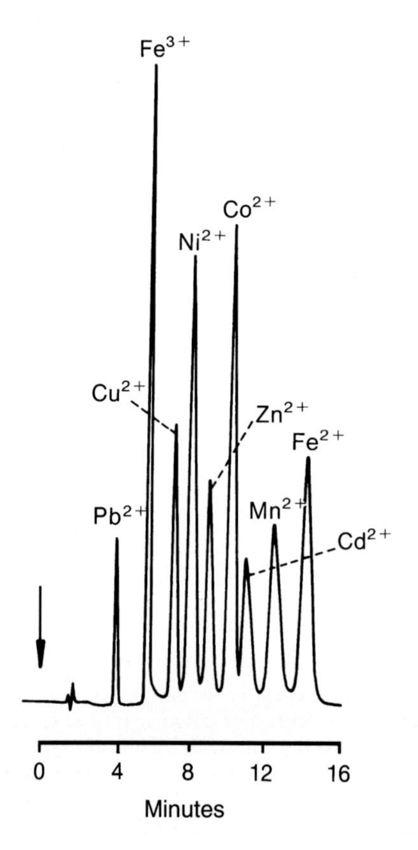

Fig. 3-156. Simultaneous analysis of nine different heavy metals. – Separator column: IonPac CS5; eluent: 0.004 mol/L pyridine-2,6-dicarboxylic acid + 0.002 mol/L Na_2SO_4 + 0.015 mol/L NaCl, pH 4.8 with LiOH; flow rate, detection, and injection volume: see Fig. 3-152; solute concentrations: 10 ppm Pb^{2+}, 1 ppm Fe^{3+}, Cu^{2+}, Ni^{2+}, Zn^{2+}, and Co^{2+}, 3 ppm Cd^{2+}, 1 ppm Mn^{2+} and Fe^{2+}.

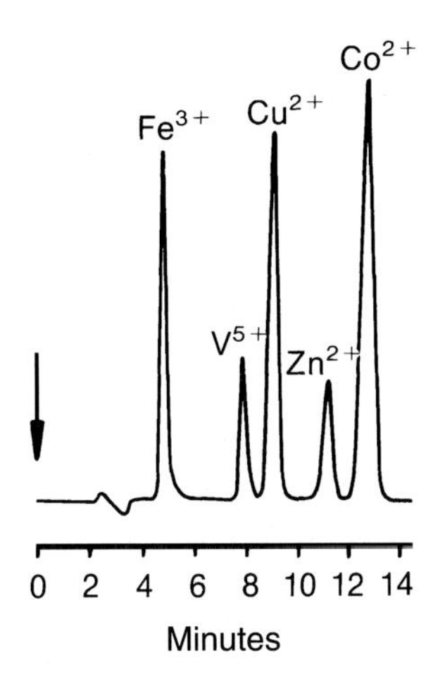

Fig. 3-157. Analysis of vanadium(V). – Separator column: Ion Pac CS5; eluent: see Fig. 3-154; flow rate: 1 mL/min; detection: photometry at 520 nm after reaction with PAR in a phosphate buffer; injection volume: 50 μL; solute concentrations: 0.5 ppm Fe^{3+}, 10 ppm V^{5+} (as VO_3^-), 2 ppm Cu^{2+}, Zn^{2+}, and Co^{2+}.

If the pH value in the PAR reagent is lowered to 8.8 by adding a phosphate buffer, gallium(III), vanadium(IV)/(V), and mercury (II) can be detected and may be separated from the other heavy metals with PDCA as the eluent [151]. Fig. 3-157 shows the separation of vanadium(V) that was applied as ammonium(meta)vanadate, NH_4VO_3. Under these conditions, vanadium(IV) elutes after about 17 minutes; the two most important oxidation states of vanadium being easily distinguished. Gallium(III), of particular importance for the semiconductor industry, elutes near the void volume.

The problem of simultaneously determining chromium(III)/(VI) was also resolved by ion chromatography [152]. The procedure involves the photometric detection of chromium(III) as PDCA-complex and chromium(VI) via a post-column reaction with 1,5-diphenylcarbazide (DPC) both at 520 nm. In this reaction, diphenylcarbazide is oxidized to diphenylcarbazone, while at the same time chromium(VI) is reduced to chromium(III), forming a chelate complex that absorbs at 520 nm. Both chromium species are separated by anion exchange chromatography: chromium(III) as a monovalent $Cr(PDCA)_2^-$ complex and chromium(VI) as a chromate ion, CrO_4^{2-}. Because of the slow ligand exchange kinetics for chromium(III) a pre-column derivatization with PDCA is performed to form the chromium(III)-PDCA complex. The pH value of the

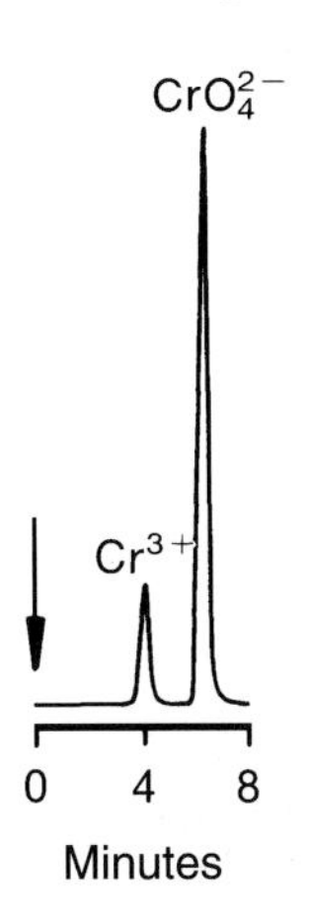

Fig. 3-158. Simultaneous determination of chromium(III) and chromium(VI). – Separator column: IonPac CS5; eluent: 0.002 mol/L pyridine-2,6-dicarboxylic acid + 0.002 mol/L Na_2HPO_4 + 0.01 mol/L NaI + 0.05 mol/L NH_4OAc + 0.0028 mol/L LiOH; flow rate: 1 mL/min; detection: photometry at 520 nm after reaction with 1,5-DPC; injection volume: 50 μL; solute concentrations: 10 ppm chromium(III) and 0.5 ppm chromium(VI).

eluent has to be close to neutral to ensure that chromium(VI) exists in the chromate form. Fig. 3-158 shows the corresponding chromatogram from which the sensitivity of this method may be inferred. Thus, it is possible to determine 10 ppb chromium(VI) with an injection volume of 50 μL.

Since the chromium(III)-PDCA complex can be measured directly at 520 nm, this method allows the determination of chromium(III) in an excess of chromium(VI) utilizing direct UV/Vis detection. Conversely, it is possible to determine traces of chromium(VI) in pure chromium(III) by applying post-column derivatization with 1,5-DPC and abstaining from pre-column derivatization with PDCA.

For the simultaneous detection of chromium(III) and (VI) after both species have been separated on an ion exchanger, Somerset [153] suggested an oxidation of chromium(III) to chromium(VI). The latter can be determined directly via its absorption at 365 nm. Such an oxidation is possible by using potassium peroxodisulfate with a silver catalyst. A complete conversion within a short time is ensured by raising the reaction temperature to about 80 °C.

In addition, cation exchange chromatography in combination with a post-column derivatization is also suitable for analyzing the uranyl cation UO_2^{2+}. Since this cation is hardly retained on a conventional cation exchanger, a buffer mixture of ammonium sulfate and sulfuric acid is used as the eluent. Fig. 3-159 shows a corresponding chromatogram with the separation of UO_2^{2+} that is detected similarly to heavy and transition metal ions via derivatization with PAR.

A buffer mixture of ammonium sulfate and sulfuric acid can also be employed for the elution of aluminum. Its specific detection is carried out via derivatization with Tiron (4,5-dihydroxy-1,3-benzenedisulfonic acid disodium salt) with subsequent photometry of the complex formed at a wavelength of 313 nm (Fig. 3-160). In an acidic sulfate medium aluminum ions exist as $AlSO_4^+$ ions depending on the hydrogen ion and sulfate ion concentration. The parameter determining retention is, in this case, the ammonium sulfate concentration.

Heavy and transition metals may also be separated via ion-pair chromatography on macroporous PS/DVB-resins or chemically bonded silica phases, respectively [154]. The mobile phase contains complexing agents and a respective ion-pair reagent. If these columns are equilibrated with a surface-active acid such as octanesulfonic acid, metal ions such as Cu^{2+}, Ni^{2+}, Zn^{2+}, and Co^{2+} elute in the same order as on surface-sulfon-

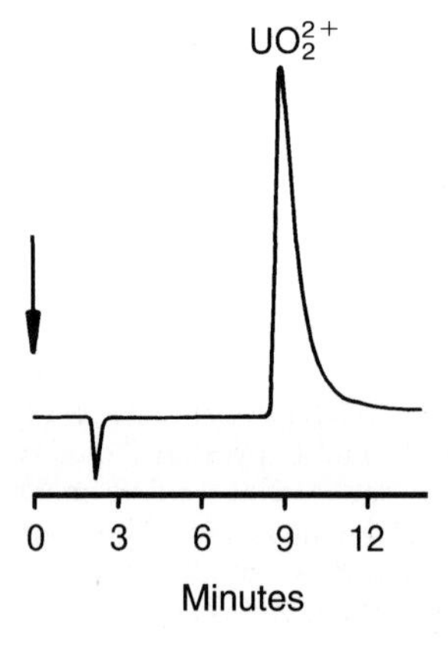

Fig. 3-159. Separation of the uranyl cation. – Separator column: IonPac CS3; eluent: 0.2 mol/L $(NH_4)_2SO_4$ + 0.05 mol/L H_2SO_4; flow rate: 1 mL/min; detection: see Fig. 3-152; injection volume: 50 μL; solute concentration: 5 ppm UO_2^{2+}.

Fig. 3-160. Separation of aluminum. – Separator column: IonPac CS3; eluent: 0.2 mol/L $(NH_4)_2SO_4$ + 0.01 mol/L H_2SO_4; flow rate: 1 mL/min; detection: photometry at 313 nm after reaction with Tiron; injection volume: 50 µL; solute concentration: 5 ppm Al^{3+}.

ated cation exchangers. This suggests that when compared to ion-pairing, ion-exchange is the predominating separation mechanism. The separation results obtainable with MPIC phases are clearly inferior to those of modern ion exchangers.

The method of anion exchange chromatography for heavy metal analysis may also be applied to ion-pair chromatography. With oxalic acid as the complexing agent and tetrabutylammonium hydroxide as the ion-pair reagent, an elution order opposite to cation analysis is obtained. The separation of anionic metal complexes on MPIC phases

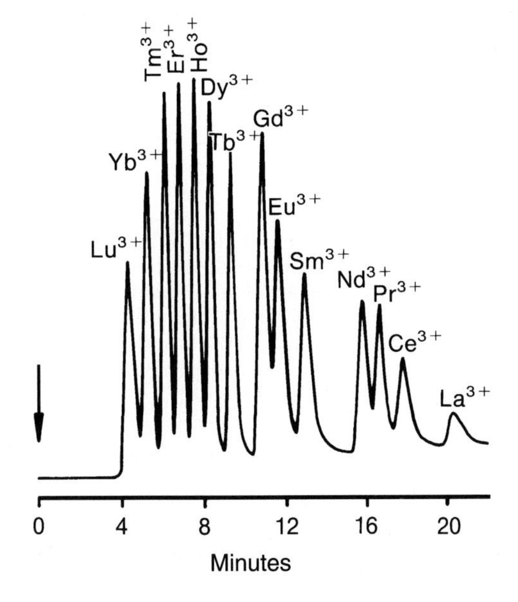

Fig. 3-161. Separation of lanthanides via cation exchange processes. – Separator column: IonPac CS3; eluent: (A) water, (B) 0.4 mol/L α-hydroxyisobutyric acid; gradient: linear, 14% B to 70% B in 18 min; flow rate: 1 mL/min; detection: see Fig. 3-152; injection volume: 50 µL; solute concentrations: 10 ppm each.

occurs exclusively via anion exchange processes. In contrast, the IonPac CS5 ion exchanger features anion exchange as well as cation exchange capacities because of the manufacturing process. Due to the separation mechanisms being involved, this characteristic is of decisive importance for the selective separation of heavy and transition metals.

Analysis of Lanthanides

The separation of lanthanides may be accomplished by anion as well as cation exchange chromatography. In 1979, Cassidy et al. [144] presented in their pioneering paper a successful separation on silica-based cation exchangers. In analogy to heavy and transition metal analysis, they used an organic compexing agent such as α-hydroxyisobutyric acid (HIBA) as the eluent to enhance the selectivity of the separation. As trivalent cations, lanthanides are strongly hydrated in aqueous solution and differ very little in their physical-chemical properties that are of importance for the separation. Conventional methods of cation exchange chromatography, therefore, do not furnish satisfactory results. The various lanthanides differ, however, in their complexing behavior. The use of the gradient technique is indispensable for a fast and efficient separation. Analysis times of about 20 minutes may be realized, for example, with a modern IonPac CS3 latexed cation exchanger by using a linear concentration gradient on the basis of α-hydroxyisobutyric acid [155]. As seen in the corresponding chromatogram in Fig. 3-161 the elution order starts with lutetium and ends with lanthanum. By means of photometric detection after reaction with PAR detection limits in the lower ppb range are realized. Similarly good separations were accomplished by Knight and Cassidy [156] on dynamically coated ODS phases. They employed this procedure for the analysis of rare-earth elements in burned-up nuclear fuel rods.

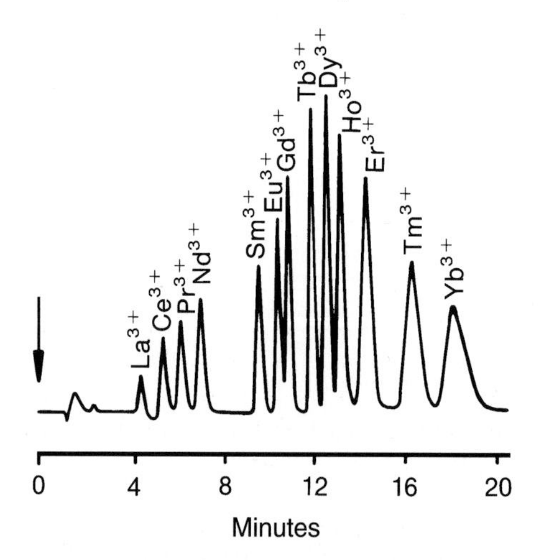

Fig. 3-162. Separation of lanthanides by anion exchange. – Separator column: Ion Pac CS5; eluent: (A) 0.1 mol/L oxalic acid, (B) 0.1 mol/L diglycolic acid, (C) water; gradient: linear, 80% A in 8 min to 26% A and 23% B; all other chromatographic conditions: see Fig. 3-161.

The lanthanides may be eluted in reverse order on a latexed anion exchanger. Oxalic acid and diglycolic acid have proved to be suitable as organic complexing agents. In this case, a linear composition gradient is applied. Such a chromatogram is shown in Fig. 3-162. Under the selected chromatographic conditions, however, lutetium and ytterbium are not completely separated.

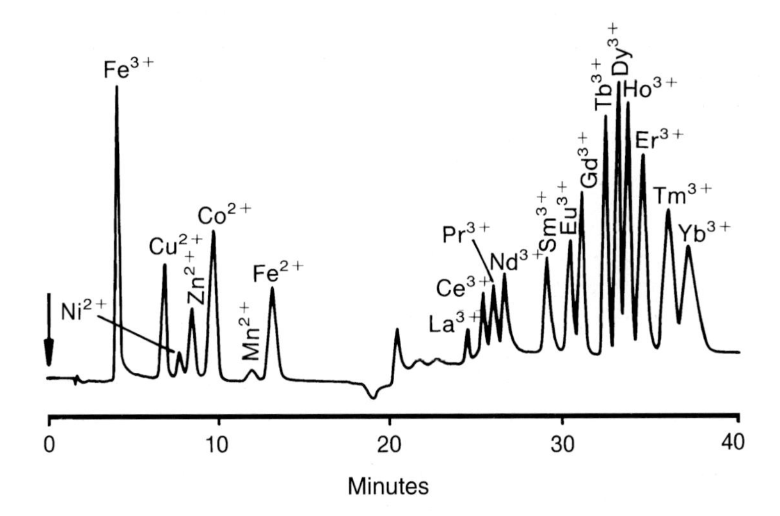

Fig. 3-163. Simultaneous analysis of transition metals and lanthanides. – Separator column: Ion Pac CS5; eluent: (A) water, (B) 0.006 mol/L PDCA + 0.05 mol/L NaOAc + 0.05 mol/L HOAc, (C) 0.1 mol/L oxalic aicd + 0.19 mol/L LiOH, (D) 0.1 mol/L diglycolic acid + 0.19 mol/L LiOH; flow rate: 1 mL/min; detection: see Fig. 3-150; injection volume: 50 µL; solute concentrations: 2 ppm Fe^{3+}, 1 ppm Cu^{2+}, 3 ppm Ni^{2+}, 4 ppm Zn^{2+}, 2 ppm Co^{2+}, 1 ppm Mn^{2+}, 3 ppm Fe^{2+}, and 7 ppm of each lanthanide; gradient program:

Time [min]	A [%]	B [%]	C [%]	D [%]
0	0	100	0	0
12	0	100	0	0
12.1	100	0	0	0
17	100	0	0	0
17.1	40	0	60	0
21	40	0	60	0
21.1	20	0	80	0
30	51	0	26	23

For the determination of lanthanides in geological samples, a problem arises regarding the co-elution with excess amounts of heavy and transition metals. For an improved separation between both classes of compounds, one capitalizes on the different complexing behavior of heavy metals and rare-earth elements with pyridine-2,6-dicarboxylic acid. Heavy and transition metals form stable monovalent and divalent complexes with PDCA, while the complexes of lanthanides with PDCA are trivalent:

$$M^{3+} + 2\,PDCA^{2-} \longrightarrow [M(PDCA)_2]^{-} \qquad (116)$$

M^{3+}: Fc^{3+}, Ga^{3+}, Cr^{3+}

$$M^{2+} + 2\,PDCA^{2-} \longrightarrow [M(PDCA)_2]^{2-} \qquad (117)$$

M^{2+}: Cu^{2+}, Ni^{2+}, Zn^{2+}

$$L^{3+} + 3\,PDCA^{2-} \longrightarrow [M(PDCA)_3]^{3-} \qquad (118)$$

L^{3+}: $La,^{3+}$, Ce^{3+}, Pr^{3+}

On the basis of this charge difference transition metals can be eluted with PDCA, while lanthanides are retained at the beginning of the column. After the transition metals have been completely eluted, one switches in a second step (as described above) to the mixture of oxalic and diglycolic acid which elutes the rare-earth elements. It is possible to analyze transition metals and lanthanides in the same run by optimizing this technique, as shown in Fig. 3-163.

3.4.6 Analysis of Polyamines

In the past, the separation of polyamines was a big problem. These compounds exhibit a very high affinity toward conventional cation exchange resins. While an eluent mixture of hydrochloric acid and 2,3-diaminopropionic acid suffices to elute ethylenediamine, buffer solutions with very high ionic strengths are required for the separation of polyamines. Therefore, only pellicular cation exchangers can be utilized, since fully sulfonated materials are compressed by the high ionic strength, leading to a drastic loss in separation efficiency.

The first successful polyamine separations were accomplished in the mid '70s after the introduction of surface-sulfonated cation exchangers. Fig. 3-164 displays a standard chromatogram with the separation of putrescine (1,4-diaminobutane) and cadaverine (1,5-diaminopentane) as well as spermidine (*N*-(3-aminopropyl-1,4-diaminobutane) and

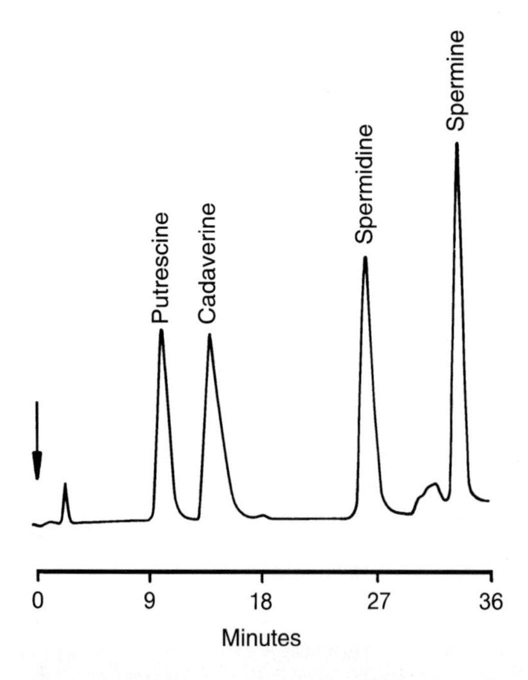

Fig. 3-164. Separation of various polyamines. – Separator column: IonPac CS1; eluent: see Table 3-25; flow rate: 0.6 mL/min; detection: fluorescence after reaction with *o*-phthaldialdehyde; injection volume: 20 µL; solute concentrations: 4.4 ppm putrescine, 5.1 ppm cadaverine, 7.3 ppm spermidine, and 10.1 ppm spermine.

Table 3-25. Eluent composition for the separation of polyamines on a surface-sulfonated cation exchanger.

Composition	Buffer type A	B	C	Regenerent
pH Value	5.80	5.55	5.55	–
Trisodium citrate [g/L]	0.98	1.97	7.84	–
Sodium chloride [g/L]	16.94	42.63	83.00	5.84
Phenol [g/L]	1.0	1.0	1.0	–
EDTA [g/L]	–	–	–	0.25
Sodium hydroxide [g/L]	–	–	–	4.01

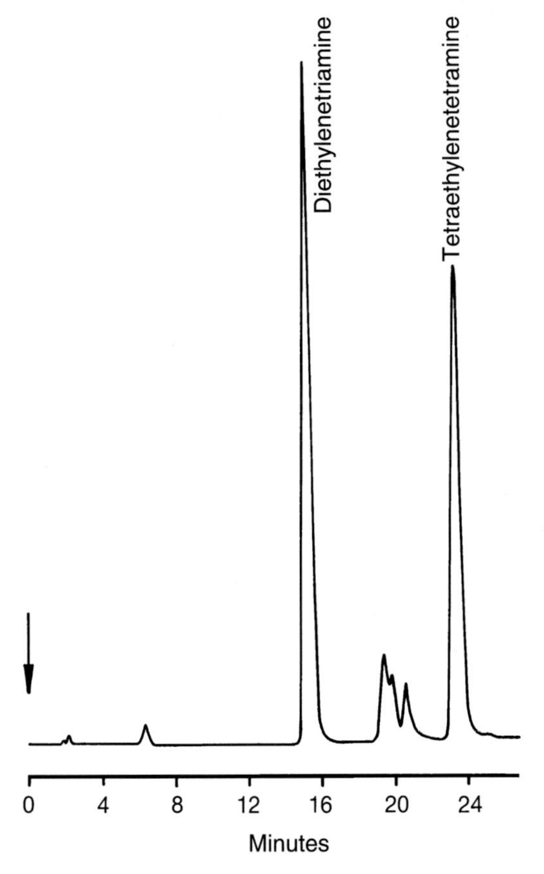

Fig. 3-165. Separation of diethylenetriamine and triethylenetetramine. – Separator column: IonPac CS3; eluent: (A) 0.2 mol/L KCl + 0.0032 mol/L EDTA, pH 5.1, (B) 1 mol/L KCl + 0.0032 mol/L EDTA, pH 5.1; gradient: linear, 100% A for 2 min isocratically, then to 100% B in 20 min; flow rate: 1 mL/min; detection: see Fig. 3-164; injection volume: 50 μL; solute concentrations: 10 ppm each.

spermine (*N,N′*-Bis-(3-aminopropyl)-1,4-diaminobutane). The required eluent was comprised of several buffer solutions. Their composition is listed in Table 3-25. Owing to the high ionic strength of these buffer solutions detection cannot be carried out by means of conductivity measurements. Polyamines carry a terminal NH_2-group, thus, fluorescence detection after reaction with *o*-phthaldialdehyde represents a suitable and very sensitive detection method.

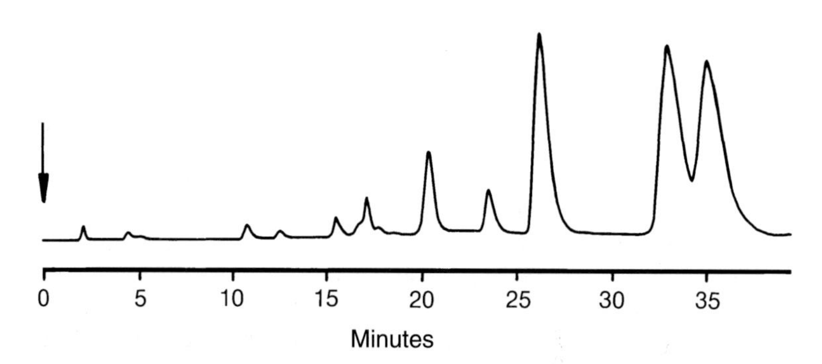

Fig. 3-166. Separation of tetraethylenepentamine (technical grade). – Chromatographic conditions: see Fig. 3-165; gradient: linear, 30% B for 2 min isocratically, then to 100% B in 15 min; injection volume: 50 µL; solute concentration: 20 ppm.

Following the introduction of highly efficient latex-based cation exchangers such as the IonPac CS3, it was possible for the eluent composition to be simplified considerably. With the eluent mixture of potassium chloride and EDTA that was developed for the polyphosphate analysis (see Section 3.3.4.4) and application of a linear concentration gradient, excellent polyamine separations are achieved with remarkably good peak symmetry. This is illustrated in Fig. 3-165 with the separation of diethylenetriamine and triethylenetetramine. While these two amines were available as relatively pure reference compounds, an unequivocal assignment of the signals in the chromatogram of the next higher tetraethylenepentamine shown in Fig. 3-166 is not possible. This product could only be obtained in technical grade quality, thus it contains more than one major component. Even more difficult is the interpretation of chromatograms of higher polyamines. For their elution, the potassium chloride concentration of the mobile phase has to be increased to c = 2 mol/L.

4 Ion-Exclusion Chromatography (HPICE)

The introduction of ion-exclusion chromatography is attributed to Wheaton and Bauman [1]. It serves, above all, for the separation of weak inorganic and organic acids. In addition, ion-exclusion chromatography can be utilized for the separation of alcohols, aldehydes, amino acids, and carbohydrates. Due to Donnan exclusion, fully dissociated acids are not retained at the stationary phase, eluting therefore within the void volume as a single peak. Undissociated compounds, however, can diffuse into the pores of the resin, since they are not subject to Donnan exclusion. In this case, separations are based on non-ionic interactions between the solute and the stationary phase.

In combination with ion-exchange chromatography (HPICE/HPIC coupling), a wealth of inorganic and organic anions can be separated within 30 minutes in a single run.

Detection is usually carried out by measuring the electrical conductivity. When combined with a suppressor system, this detection method is superior to all other detection methods such as, for example, refractive index or UV detection at low wavelengths with regard to specificity and sensitivity.

4.1 The Ion-Exclusion Process

Typically, HPICE separator columns contain a totally sulfonated high-capacity cation exchange resin. The separation mechanism occuring at this stationary phase is based on threc phenomena:

- Donnan exclusion
- Steric exclusion
- Adsorption

Fig. 4-1 represents a schematic picture of the separation process on a HPICE column. It shows the surface of the resin with its bonded sulfonic acid groups. If pure water is passed through the separator column, a hydration shell is formed around the sulfonic

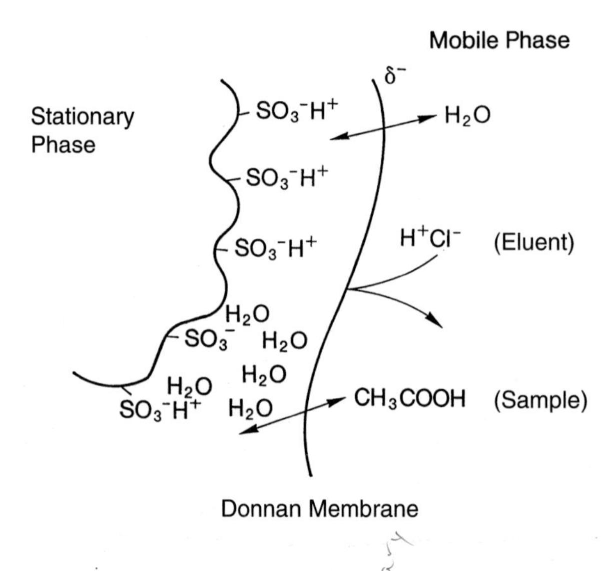

Fig. 4-1. Schematic representation of the separation process on a HPICE column.

acid groups. In a way, some of the water molecules are therefore in a higher state of order compared with the water molecules in the bulk mobile phase.

In this retention model, a negatively charged layer analogous to the Donnan membrane characterizes the interface between the hydration shell, which is only permeable for undissociated compounds, and the bulk mobile phase. Fully dissociated acids such as hydrochloric acid that acts as the eluent cannot penetrate this layer because of the negative charge of the chloride ion. Thus, such ions are excluded from the stationary phase. Their retention volume is called the *exclusion volume* V_e. On the other hand, neutral water molecules may diffuse into the pores of the resin and back into the mobile phase. The volume corresponding to the retention time of water is called the *total permeated volume* V_p. Depending on the pH of the eluent, a weak organic acid (e.g., acetic acid) is present after injection in partly undissociated form and is not subject to Donnan exclusion. Although both acetic acid as well as water may interact with the stationary phase, a retention volume is observed for acetic acid that is higher than V_p. This phenomenon can only be explained by adsorption occuring on the surface of the stationary phase. The separation mechanism in case of aliphatic monocarboxylic acids, therefore, is determined by Donnan exclusion and adsorption. The retention time increases with increasing length of the acid's alkyl side chain. By adding organic solvents such as acetonitrile or 2-propanol, the retention of aliphatic monocarboxylic acids may be reduced. This is due to: (1) adsorption sites being blocked by solvent molecules and (2) the solubility in the eluent being enhanced.

Di- and tricarboxylic acids such as oxalic and citric acid elute between the excluded and the total permeated volume. Apart from Donnan exclusion, the predominating separation mechanism is, in this case, mainly steric exclusion. The retention is determined by the size of the sample molecule. Since the pore volume of the resin is established by its degree of crosslinking, the resolution can only be improved by applying another or by coupling with another separator column, respectively.

4.2 Stationary Phases

The selection of stationary phases for ion-exclusion chromatography is relatively limited. Typically, totally sulfonated polystyrene/divinylbenzene-based cation exchangers in the hydrogen form are employed. The percentage of divinylbenzene – expressed as degree of crosslinking – is particular important for the retention behavior of organic acids. Depending on the degree of crosslinking, these acids may diffuse into the stationary phase to a greater or lesser degree, resulting in a change of retention. In their investigations into the retention behavior of various inorganic and organic acids, Harlow and Morman [2] found that resins with a relatively high degree of crosslinking of 12% are suitable for the separation of weakly dissociated organic acids. On the other hand, more strongly dissociated acids are best separated on resins with a low degree of crosslinking of 2%. Most of the materials offered today have a degree of crosslinking of 8%. This compromise takes into account the different retention behavior of strong and weak acids. Apart from organic polymers, silica-based cation exchangers are used for ion-exclusion chromatography [3]. These materials, however, are only of secondary importance due to their lack of pH stability.

Characteristic for the stationary phases that are used today in ion-exclusion chromatography are the comparably small particle diameters between 5 µm and 15 µm which enable fast diffusion processes. The structural-technical properties of these phases are listed in Table 4-1.

Dionex offers two ion-exclusion columns: IonPac ICE-AS1 and AS5. The former is a totally sulfonated 7-µm cation exchanger used primarily for the separation of aliphatic monocarboxylic acids. Difficulties are encountered, however, in the separation of aliphatic di- and tricarboxylic acids. They elute from such stationary phases within the total permeated volume. The selectivity of the separation in this retention range is usually very poor. It cannot be improved significantly, even when altering the chromatographic conditions. The problem was solved with the introduction of the 6-µm IonPac ICE-AS5 separator column. With this stationary phase, the selectivity of the separation is significantly improved by hydrogen bonding between the resin and the acid to be analyzed by incorporating methacrylates in the substrate material. This stationary phase is also very well suited for the analysis of aliphatic hydroxycarboxylic acids, which are not well separated on a conventional ion-exclusion column. Because non-ionic adsorption interactions also contribute to retention, mono- and dicarboxylic acids with hydrophobic segments such as propionic and butyric acid, as well as fumaric and succinic acid, are retained very strongly.

Polystyrene/divinylbenzene-based ion-exclusion columns are also offered by Hamilton Co. (Reno, NV, USA) under the trade name PRP-X300. This is a 10-µm material with an exchange capacity of 0.2 mequiv/g [4]. It is obtained by sulfonation of PRP-1, a macroporous PS/DVB polymer with reversed-phase properties. Fig. 4-2 shows the separation of various organic acids on this stationary phase. Dilute sulfuric acid was used as the eluent. The much higher retention of succinic acid compared to acetic acid reveals that the retention of organic acids is chararcterized, apart from reversed-phase effects, by the formation of hydrogen bonds.

Interaction Chemicals also offers two ion-exclusion columns for the separation of organic acids under the trade names ORH-801 and ION-300. Both stationary phases

Table 4-1. Structural-technical properties of various polystyrene/divinylbenzene-based ion-exclusion columns.

Separator column	Manufacturer	Dimensions [length × I.D.] [mm]	Particle diameter [μm]	Purpose	Max. flow rate [mL/min]
IonPac ICE-AS1	Dionex	250 × 9	7	high-efficiency separator column for various organic acids	1.5
IonPac ICE-AS5	Dionex	250 × 4	6	separator column for hydroxycarboxylic acids	1.0
PRP-X300	Hamilton	250 × 4.1	10	high-efficiency separator column for various organic acids	8.0
ORH-801	Interaction Chemicals	300 × 6.5	8	high-efficiency separator column for various organic acids	1.5
ION-300	Interaction Chemicals	300 × 7.8	8	separator column for organic acids of the Krebs cycle	1.0

contain a 8-μm substrate which was optimized for its respective purpose by slight modifications. The column ORH-801 is very similar to the Dionex IonPac ICE-AS1 column with regard to its separation properties. The ION-300 column, on the other hand, was especially developed for the analysis of the Krebs-cycle acids. Citric and pyruvic acid,

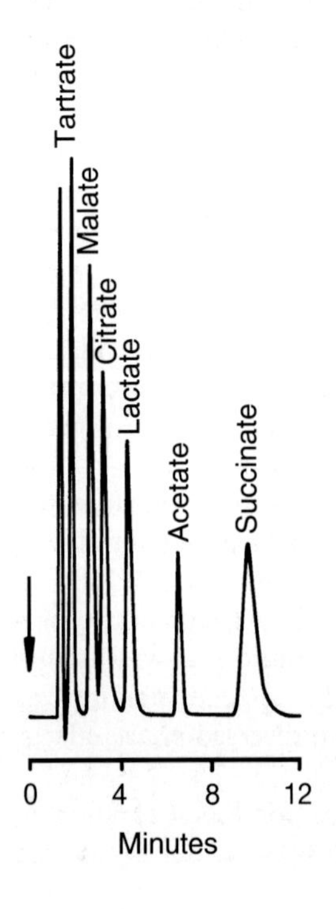

Fig. 4-2. Separation of organic acids on PRP-X300. – Eluent: 0.0005 mol/L H_2SO_4; flow rate: 1 mL/min; detection: direct conductivity; injection volume: 100 μL; solute concentrations: 4 ppm tartaric acid, 7.5 ppm malic acid and citric acid, 10 ppm lactic acid, 25 ppm acetic acid, and 40 ppm succinic acid.

succinic and lactic acid, as well as oxalacetic and α-ketoglutaric acid may also be successfully separated on this stationary phase. Such separations are of importance for the analysis of fruit juices, foods, and physiological samples. Fig. 4-3 displays an example with the separation of a standard mixture of organic acids, carbohydrates, and alcohols, which resembles the matrix of a variety of beverages.

A common characteristic of all polystyrene/divinylbenzene-based ion-exclusion phases is the high retention of aromatic carboxylic acids. This is due to π-π interactions between the aromatic ring systems of the polymer and the solute. The separation of aromatic carboxylic acids, therefore, is more elegantly accomplished by the various procedures of reversed-phase chromatography.

4.3 Eluents for Ion-Exclusion Chromatography

The selection of eluents in ion-exclusion chromatography is very limited. In the most simple case, pure de-ionized water can be utilized. However, the work of Turkelson and Richard [5] on the separation of the organic acids occuring in the citric acid cycle on an Aminex 50W-X4 cation exchanger (30 µm to 35 µm), has shown that the peak form of the acids is characterized by a large half width and a strong tailing when pure water is used. This phenomenon probably occurs because organic acids exist in their molecular as well as in their anionic form. Acidifying the mobile phase suppresses the dissociation, thus improving the peak shape significantly. Nowadays, pure de-ionized water is recommended only for the analysis of carbonate which can be easily determined in this way.

For the separation of organic acids, mineral acids are usually employed as the eluent. If a cation exchanger in the silver form is operated as the suppressor column, hydrochloric acid is the only possible eluent. Applying direct UV detection sulfuric acid would be the best eluent [6,7]. In combination with the introduction of novel AFS-2 and AMMS-ICE membrane suppressors long-chain aliphatic carboxylic acids are also recommended as eluents. The hydrochloric acid used so far is compatible with membrane suppressors. The high equivalent conductance of the chloride ion results, however, in an unneccessarily high background conductivity. Good separation results are obtained with tridecafluoroheptanoic acid (perfluoroheptanoic acid), which is used like all other eluents in the concentration range between c = 0.0005 mol/L and 0.01 mol/L. Octanesulfonic acid exhibits similar elution properties. It is also suited for the analysis of borate and carbonate. Since boric acid is only weakly dissociated, a mixture of octanesulfonic acid and mannitol should be utilized as the eluent for the sensitive detection of this compound by measuring the electrical conductivity. The sugar alcohol complexes boric acid, which leads to a significantly higher conductivity. For separator columns such as the IonPac ICE-AS5, on the other hand, short-chain perfluorated fatty acids such as heptafluoropropanoic acid (perfluorobutyric acid) are recommended as the eluent.

The high retention of aliphatic monocarboxylic acids (n_c > 4) and aromatic carboxylic acids can be reduced by adding small amounts of an organic solvent (10 to 30 mL/L) to the eluent. This blocks adsorption sites on the surface of the stationary phase [8]. Particularly suited are acetonitrile, 2-propanol, or ethanol. If possible, methanol

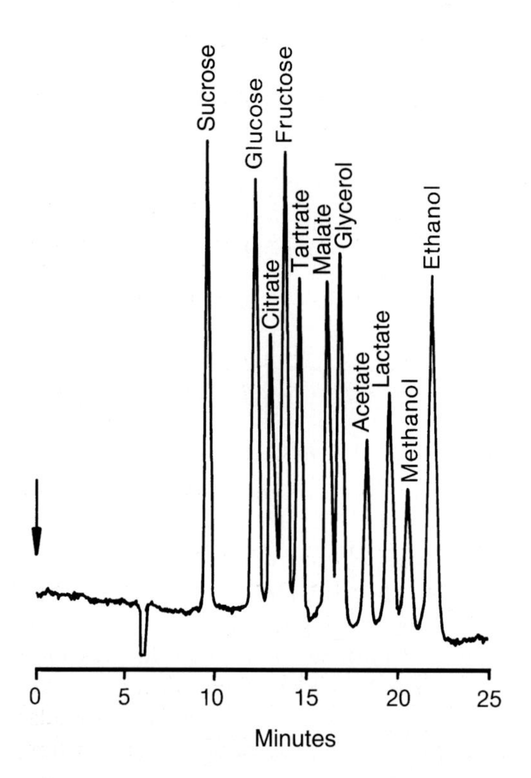

Fig. 4-3. Separation of a standard mixture of organic acids, carbohydrates, and alcohols on ION-300. – Column temperature: 30 °C; eluent: 0.005 mol/L H_2SO_4; flow rate: 0.5 mL/min; detection: RI.

should not be used, since the stationary phase is subject to strong volume changes when using this solvent.

According to Tanaka and Fritz [9], comparably low background conductivities were obtained with benzoic acid as the eluent, because benzoic acid is only partly dissociated. However, the acid strength of a dilute benzoic acid solution is sufficient to elute aliphatic carboxylic acids with a good peak shape.

4.4 Suppressor Systems in Ion-Exclusion Chromatography

The detection of aliphatic carboxylic acids usually involves the measurement of the electrical conductivity. In ion-exclusion chromatography, therefore, suppressor systems are used to chemically reduce the background conductivity of the acid that acts as the eluent. Due to a lack of modern membrane suppressors in the past, suppressor columns were used. They contained a cation exchange resin in the silver form. With dilute hydrochloric acid as the eluent, the suppressor reaction is described as follows:

$$\text{Resin}-SO_3^- \, Ag^+ + H^+Cl^- \longrightarrow \text{Resin}-SO_3^- \, H^+ + AgCl\downarrow \qquad (119)$$

The organic acids to be analyzed do not precipitate with silver and therefore reach the conductivity cell unchanged. This type of suppression has a number of disadvantages:

- With increasing formation of silver chloride in the suppressor column a pressure increase and a significant peak broadening were observed. The signals could only be evaluated by comparing the *peak areas* with that of reference compounds of known concentrations.
- In theory, a periodical regeneration of the suppressor column is possible with a concentrated ammonium hydroxide solutions. However, it does not make any sense since it is very time-consuming.
- A continuous regeneration of the suppressor column – a standard feature in anion exchange chromatography for years – was impossible with the suppressor columns used for ion-exclusion chromatography.

The only possibility to limit the steadily increasing back pressure caused by the continuing precipitation of silver chloride in the suppressor column was to cut off the exhausted part of the cartridge after some days of operation.

Table 4-2. Eluents and regenerents in ion-exclusion chromatography.

Eluent	Concentration [10^3 mol/L]	Regenerent[a] [mol/L]	Background conductivity[b] [µS/cm]
HCl	0.5 1		50 1000
Perfluoroheptanoic acid	0.5 1	TBAOH 0.005....0.01	20 40
Octanesulfonic acid	0.5 1		25 45

[a] The regenerent concentration should be about 10 times as high as the eluent concentration, based on a flow rate of 2 mL/min.

[b] The values indicated for the resulting background conductivity refer to an eluent flow rate of 0.8 mL/min.

A cation exchange membrane was developed to overcome these disadvantages. This membrane allows continuous regeneration and consists of a sulfonated polyethylene derivative. It is resistant to water-miscible organic solvents and exhibits a high permeability for quaternary ammonium bases such as tetrabutylammonium hydroxide. Hence, the suppression mechanism differs substantially from the process for anion exchange chromatography described in Section 3.3.3.

In the regenerated state, the membrane exists in the tetrabutylammonium form. The oxonium ions of the organic acids that flow through the interior of the membrane are replaced by tetrabutylammonium ions in the suppressor reaction. Because of the different equivalent conductances of oxonium and tetrabutylammonium ions, the salt formed as the suppressor product has a markedly lower conductivity than the corresponding acid form. In the zone of dynamic equilibrium the neutralization reaction between H^+ and OH^- ions to form water as the suppressor product takes place which is discharged from the suppressor together with an excess of tetrabutylammonium hydroxide. The higher background conductivity resulting from the use of hydrochloric acid as the eluent is due to the higher equivalent conductance of the chloride ion in comparison with the octanesulfonate ion.

Eluents and regenerents suitable for the analysis of organic acids when applying an AFS-2 hollow fiber membrane suppressor are listed in Table 4-2. For the analysis of borate and carbonate with octanesulfonic acid as the eluent, an ammonium hydroxide solution with a concentration c = 0.01 mol/L can also be used as the regenerent.

A micromembrane suppressor under the trade name AMMS-ICE has also been introduced for ion-exclusion chromatography. Its structure corresponds to the systems developed for anion and cation exchange chromatography (see Sections 3.3.3 and 3.4.3). However, in its mode of operation, it corresponds to the AFS-2 hollow fiber suppressor. An AMMS-ICE micromembrane suppressor also contains membranes that are stable against water-miscible organic solvents. Therefore, it is used for the analysis of long-chain fatty acids, which are separated on an unpolar stationary phase in a weakly acidic medium with methanol or acetonitrile as mobile phase components. In this case, dilute potassium hydroxide solution is utilized as the regenerent. With respect to the exchange capacity no significant differences exist between both types of membrane suppressors. The distinctly lower dead volume of the micromembrane suppressor, however, contributes to the sensitivity increase. The regeneration of an AMMS-ICE micromembrane suppressor is no different than that of the respective hollow fiber suppressor. It is usually carried out with tetrabutylammonium hydroxide of the concentration c = 0.01 mol/L (see Table 4-2).

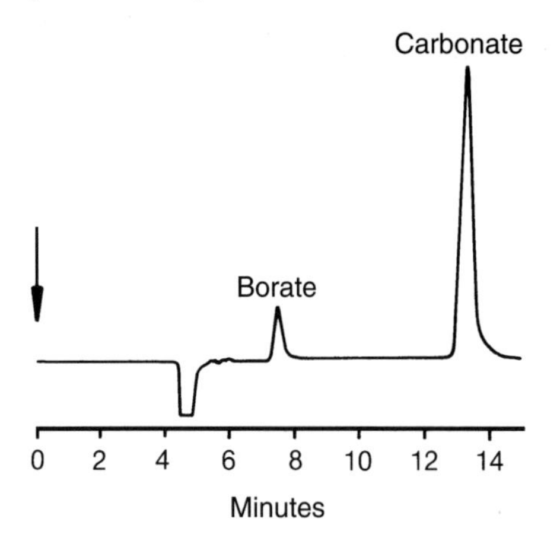

Fig. 4-4. Analysis of borate and carbonate. – Separator column: IonPac ICE-AS1; eluent: 0.001 mol/L octanesulfonic acid; flow rate: 1 mL/min; detection: suppressed conductivity; injection volume: 50 μL; solute concentrations: 10 ppm borate and 50 ppm carbonate.

4.5 Analysis of Inorganic Acids

The separator columns listed in Table 4-1 (see Section 4.2) allow the separation of a variety of inorganic acids. While strong inorganic acids elute within the void volume because of Donnan exclusion, weak inorganic acids are more strongly retained. This includes the hardly dissociated boric acid (pK_a = 9.23), which is detected by measuring

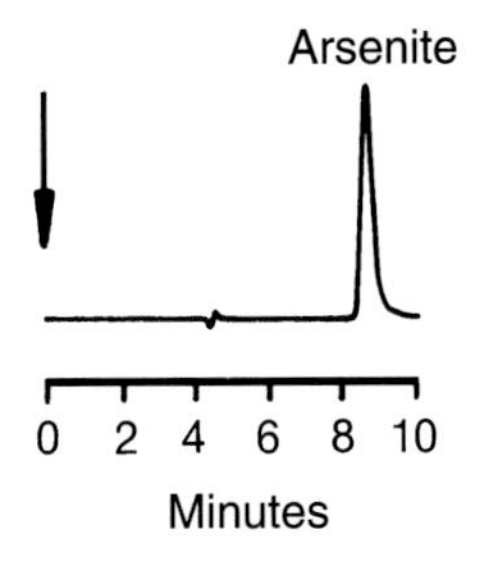

Fig. 4-5. Analysis of arsenite with amperometric detection. – Separator column: IonPac ICE-AS1; eluent: 0.0005 mol/L HCl; flow rate: 1 mL/min; detection: D.C. amperometry on a Pt working electrode; oxidation potential: +0.95 V; injection volume: 50 µL; solute concentration: 10 ppm arsenite.

the electrical conductivity. For sensitive borate detection, more than ten times the amount of mannitol should be added to the eluent. This increases the borate conductance by complexation with the boric acid. The application of indirect conductivity detection usually involves dilute sulfuric acid as the eluent. Fig. 4-4 displays the chromatogram of a 10-ppm borate standard with pure octanesulfonic acid as the eluent. The detection limit for the borate ion is about 500 ppb. Fig. 4-4 shows that the more strongly retained carbonate may also be detected under the same chromatographic conditions. The detection limit for carbonate is about 1 ppm. Apart from borate and carbonate, fluoride can also be analyzed by ion-exclusion chromatography. Again, both conductivity detection modes may be employed. Since fluoride is clearly separated from the strong mineral acids that elute within the void volume, this method for analyzing fluoride is much more selective than ion-exchange chromatography. The only disadvantage is the poor sensitivity caused by the low equivalent conductance of tetrabutylammonium fluoride formed in the suppressor.

Fig. 4-6. Analysis of silicate after derivatization with sodium molybdate. – Separator column: IonPac ICE-AS1; eluent: 0.0005 mol/L HCl; flow rate: 1 mL/min; detection: photometry at 410 nm after reaction with sodium molybdate; injection volume: 50 µL; solute concentration: 20 ppm Si (as $SiCl_4$).

In combination with an amperometric detection, ion-exclusion chromatography is suitable for the determination of sulfite and arsenite. Both ions can be oxidized at a Pt working electrode and, thus, can be selectively and very sensitively detected. Fig. 4-5 displays the chromatogram of a 10-ppm arsenite standard that was obtained with an oxidation potential of +0.95 V.

High selectivities are also obtained by combining ion-exclusion chromatography with photometric detection after derivatization of the column effluent with suitable reagents. As a typical example, Fig. 4-6 shows the chromatogram of a 20-ppm silicate standard obtained by derivatization with sodium molybdate in acid solution and subsequent photometric detection at 410 nm. Orthophosphate may also be detected under these conditions. Because of its different acid strength it elutes *prior* to silicate, so that this method is very selective for the determination of both anions.

Table 4-3. Elution order of some organic acids which can be analyzed by ion-exclusion chromatography on a totally sulfonated polystyrene/divinylbenzene-based cation exchanger.

Maleic acid	↓	Adipic acid
Oxalic acid		Fumaric acid
Citric acid		Acetic acid
Pyruvic acid		Propionic acid
Malonic acid		Acrylic acid
α-Ketobutyric acid		Isobutyric acid
α-Ketovaleric acid		Butyric acid
Succinic acid		Mandelic acid
Glycolic acid		Pivalic acid
Lactic acid		α-Hydroxybutyric acid
Formic acid		and others

4.6 Analysis of Organic Acids

Apart from a couple of inorganic acids, a wealth of organic acids may be separated by using the separator columns listed in Table 4-1. A selection of the analyzable compounds and their elution order are listed in Table 4-3.

As mentioned in Section 4.5, fully dissociated acids elute within the void volume due to Donnan exclusion. The retention behavior of weak organic acids, on the other hand, may be predicted from the following criteria:

- Members of a homologous series are eluted in order of decreasing acid strength and decreasing water solubility, that is, increasing solvophobicity.
  Taking aliphatic monocarboxylic acids as an example, the elution order, according to Fig. 4-7, is formic acid > acetic acid > propionic acid and so forth. Such separations have characteristics similar to that of reversed-phase chromatography. Van-der-Waals forces between the solute and the polymeric resin material (mainly benzene rings) as well as the decrease in solubility of the solutes in the eluent influence their distribution between the stationary and mobile phases.

- Di-basic acids are eluted prior to the corresponding mono-basic acids because of their higher solubility in polar eluents. Thus, it is observed that oxalic acid elutes prior to acetic acid, and malonic acid prior to propionic acid.
- Organic acids with a branched carbon skeleton such as isobutyric acid are generally eluted prior to the non-branched analogues such as *n*-butyric acid. Again, this corresponds to the retention behavior in RPLC.
- A double-bond in the carbon skeleton leads to a significantly higher retention because of π-π interactions with the aromatic rings of the polymer: acrylic acid is eluted *after* propionic acid.
- Aromatic acids are strongly retained because of the interactions described above. Thus, they should not be analyzed by ion-exclusion chromatography. The high retention of unsaturated and aromatic moieties on ICE stationary phases is unlike the behavior on ODS phases, where π-π interactions are not possible and where only the enhanced solubility in the eluent comes to fruition.

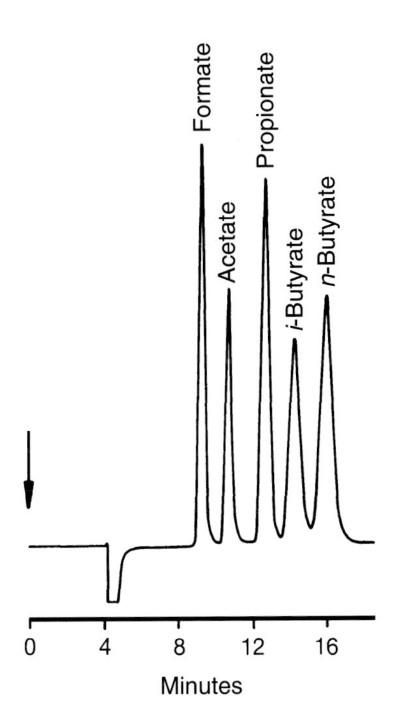

Fig. 4-7. Separation of aliphatic monocarboxylic acids. – Separator column: IonPac ICE-AS1; eluent: 0.001 mol/L octanesulfonic acid; flow rate: 1 mL/min; detection: suppressed conductivity; injection volume: 50 μL; solute concentrations: 20 ppm formic acid and acetic acid, 40 ppm propionic acid, 40 ppm *iso*-butyric acid, and 60 ppm *n*-butyric acid.

Fig. 4-8 shows a chromatogram of various organic acids obtained by using a modern membrane-based suppressor system. The acid concentration in the mobile phase determines the retention in the analysis of di-basic and poly-basic acids. It affects the degree of dissociation and, thus, the retention time of the carboxylic acid to be analyzed. In general, a higher resolution is observed with increasing acid concentration. This effect

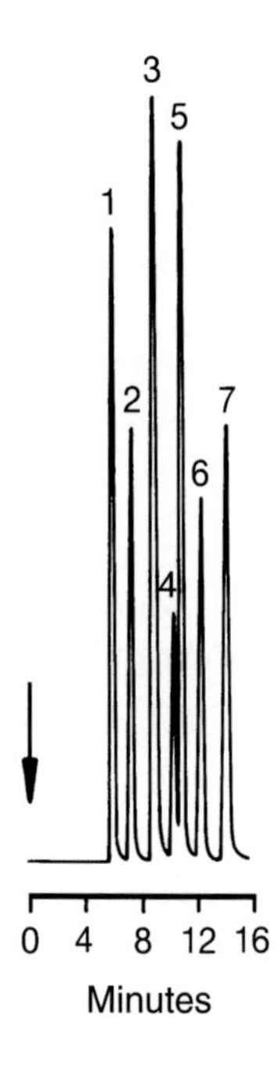

Fig. 4-8. Separation of organic acids upon application of a membrane-based suppressor system. – Separator column: IonPac ICE-AS1; eluent: 0.001 mol/L octanesulfonic acid; flow rate: 1 mL/min; detection: suppressed conductivity; injection volume: 50 μL; solute concentrations: 50 ppm oxalic acid (1), 50 ppm tartaric acid (2), 25 ppm fluoride (3), 50 ppm lactic acid (4), 50 ppm formic acid (5), 50 ppm acetic acid (6), and 100 ppm propionic acid (7).

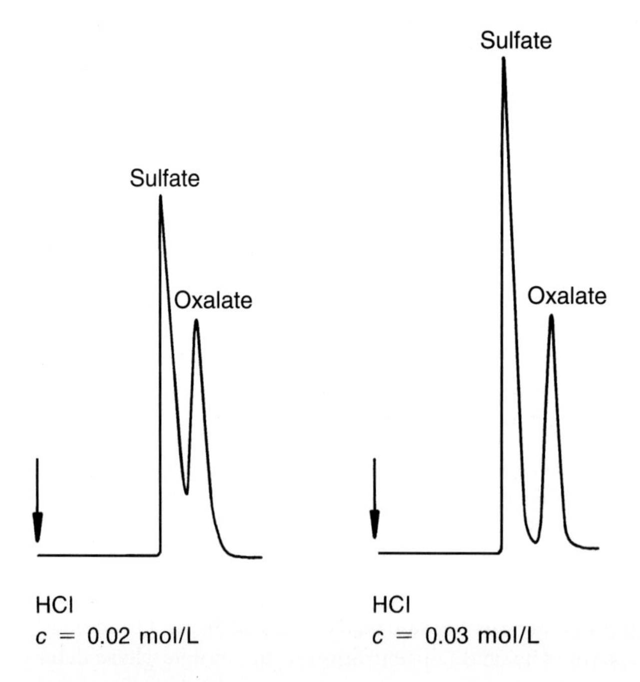

Fig. 4-9. Illustration of the influence of the acid concentration in the mobile phase on the separation of sulfate and oxalate. – Separator column: IonPac ICE-AS1; eluent: HCl; flow rate: 0.8 mL/min; detection: suppressed conductivity.

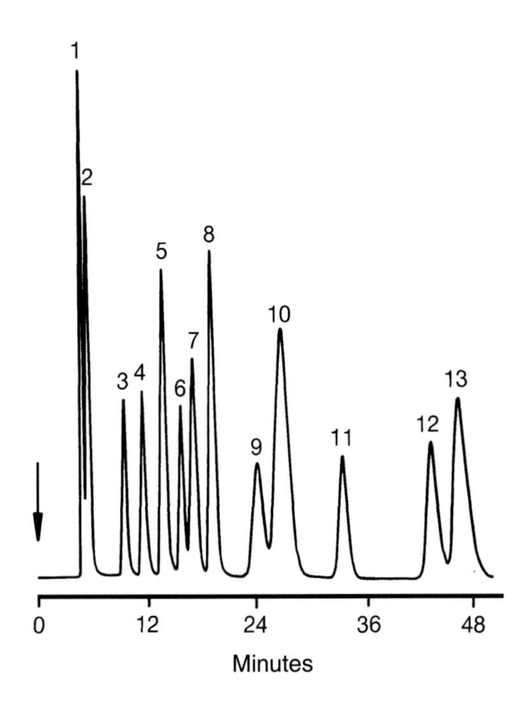

Fig. 4-10. Separation of organic acids on IonPac ICE-AS5. – Eluent: 0.0016 mol/L perfluorobutyric acid; flow rate: 0.3 mL/min; detection: suppressed conductivity; injection volume: 50 µL; solute concentrations: fully dissociated compounds (1), 10 ppm oxalic acid (2), 25 ppm pyruvic acid (3), and tartaric acid (4), 30 ppm malonic acid (5), lactic acid (6), malic acid (7), and acetic acid (8), 20 ppm isocitric acid (9), 30 ppm citric acid (10), 40 ppm β-hydroxybutyric acid (11), succinic acid (12), and propionic acid (13).

Table 4-4. pH Dependence of the retention of various aliphatic carboxylic acids on IonPac ICE-AS5.

Solute	pK Values	k' [a]			
		pH 2.2	pH 2.6	pH 2.8	pH 3.8
Tartaric acid	2.96 4.24	2.04	1.50	1.24	1.15
Formic acid	3.76	2.04	1.50	1.38	1.38
Lactic acid	3.86	2.42	2.33	2.25	2.17
Malic acid	3.40 5.05	3.17	2.83	2.63	2.38
Acetic acid	4.76	3.17	3.08	3.00	2.96
Citric acid	3.13 4.75 5.40	7.33	5.83	5.13	4.88

[a] Chromatographic conditions: see Fig. 4-10.

is illustrated in Fig. 4-9, which shows the separation of sulfate and oxalate at two different acid concentrations. The retention of short-chain fatty acids, on the other hand, is hardly affected by changing the pH value in the mobile phase. These acids are mainly retained by non-ionic interactions resembling reversed-phase effects. It follows that the retention times can only be reduced by adding small amounts of 2-propanol or acetonitrile (10 mL/L up to 30 mL/L).

The IonPac ICE-AS5 separator column has proved the best for the separation of aliphatic hydroxycarboxylic acids. Fig. 4-10 shows the chromatogram of a standard mixture containing various organic acids. Perfluorobutyric acid was employed as the eluent. In this stationary phase, the acid concentration in the eluent is one of the most important parameters determining retention. This is reflected in Table 4-4, using as an example the difficult separation between tartaric acid/formic acid and lactic acid/malic acid in the pH range between 2.2 and 3.8. As shown, the resolution is optimal at pH 2.8 for the investigated acids. Furthermore, the selectivity of the IonPac ICE-AS5 separator is affected by the column temperature. Hydrogen bonding contributes to the solute retention at this stationary phase. Therefore, apart from a general retention decrease with raising column temperature according to Eq. (40) (Section 3.2), an additional loss in selectivity is observed by the breaking of hydrogen bonds at elevated column temperature. Compounds with similar retention behavior such as malic acid and lactic acid, therefore, can only be completely separated at ambient temperature (23 °C to 25 °C).

4.7 HPICE/HPIC-Coupling

The concept of combining ion-exclusion and ion-exchange chromatography was introduced by Rich et al. [10]. It serves to improve the selectivity of the chromatographic separation of inorganic and organic acids in complex matrices. Its mode of action is schematically depicted in Fig. 4-11.

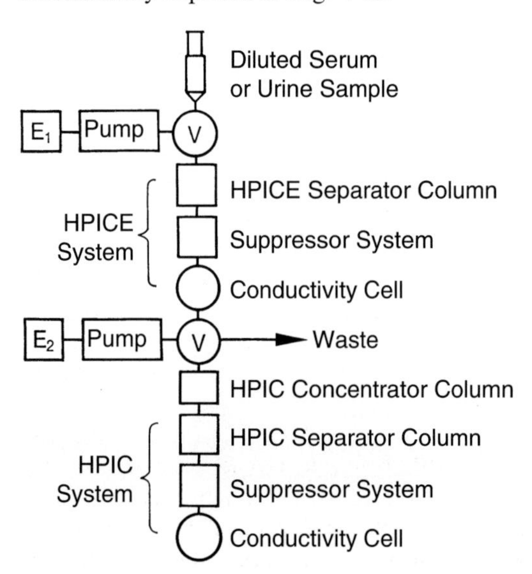

Fig. 4-11. Schematic representation of the HPICE/HPIC-coupling.

After its injection, the sample to be analyzed reaches an ion-exclusion column. All mineral acids are separated as a one peak from weak organic acids. The column effluent is directed into the suppressor system and passed on to the conductivity cell. A column switching technique connects the cell outlet to a concentrator column that contains a low-capacity anion exchange resin. This concentrator column collects one or several of the peaks separated by HPICE. They are separated on a subsequent latex anion exchanger with a carbonate/bicarbonate buffer into the respective components. By selecting eluents of suitable ionic strengths a wealth of inorganic and organic anions can be determined with a minimum of sample preparation.

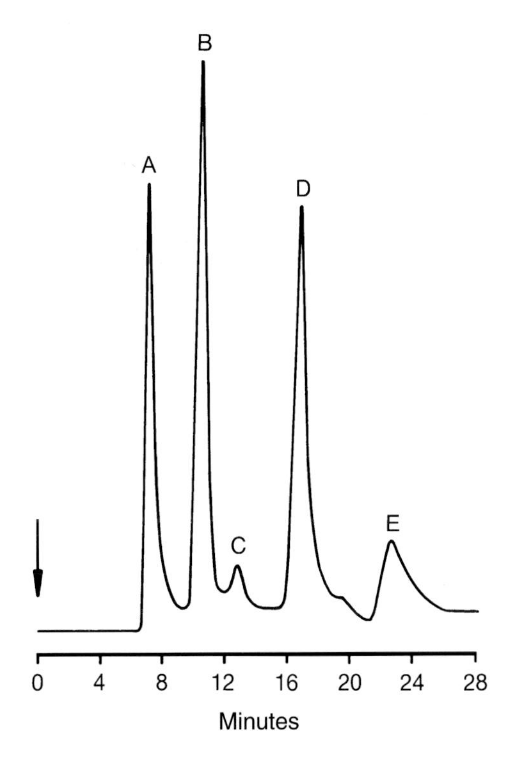

Fig. 4-12. HPICE chromatogram of a human serum. – Separator column: IonPac ICE-AS2 (precursor of IonPac ICE-AS1); eluent: 0.01 mol/L HCl; flow rate: 0.8 mL/min; detection: suppressed conductivity (cation exchanger in Ag form); (A) strong inorganic acids, (B) phosphoric acid group, (C) pyruvic acid group, (D) lactic acid group, (E) hydroxybutyric acid group.

The combination of HPICE/HPIC was successfully applied by Rich et al. [11] to the separation of pyruvic and lactic acid in blood sera. At that time, a separation of organic acids in biological liquids could not be accomplished solely by ion-exchange chromatography because of the similarity of the retention behavior of the organic acids of interest. The application of HPICE/HPIC solved that problem. In additon, organic acids with identical retention on HPICE phases are often well separated on anion exchangers. Characteristic examples are α-ketoisovaleric and pyruvic acid, which cannot be separated on a conventional ion-exclusion resin, while their capacity factors differ significantly upon application of an anion exchanger.

With the introduction of gradient techniques in anion exchange chromatography (see Section 3.3.6), applications involving HPICE/HPIC play only a secondary role. Today, most if not all of these problems can be solved by gradient elution applications using sodium hydroxide as the eluent. Thus, the HPICE chromatogram of a human serum depicted in Fig. 4-12 is of purely historical interest. It was obtained using a suppressor column in the silver form that was commonly employed in the beginning of the 1980s. The individual signals that are marked by letters represent sum peaks of organic acids with similar p*K* values and comparable hydrophobicity. The chromatogram in Fig. 4-13 shows the separation of the peak marked “A” in Fig. 4-12 into its components nitrate and sulfate. This separation was obtained with the IonPac AS1 anion exchanger used at that time, after concentrating the HPICE effluent on a corresponding pre-column.

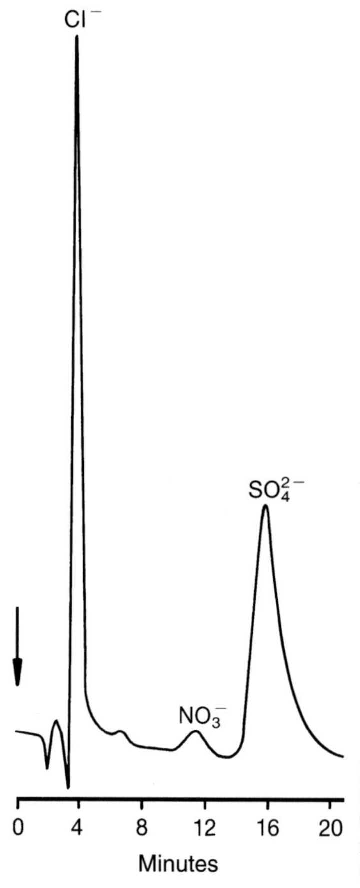

Fig. 4-13. HPIC chromatogram of the peak marked “A” in Fig. 4-12. – Separator column: IonPac AS1; eluent: 0.003 mol/L $NaHCO_3$ + 0.0024 mol/L Na_2CO_3; flow rate: 2.3 mL/min; detection: suppressed conductivity (packed-bed suppressor ASC-2).

4.8 Analysis of Alcohols and Aldehydes

Totally sulfonated cation exchangers are suitable not only for the separation of weak inorganic and organic acids, but they can also be used for the analysis of monohydric and polyhydric alcohols and aldehydes with a small number of C-atoms. For the separation of monohydric alcohols, Jupille et al. [12] used Aminex HPX-85H (Bio-Rad, Richmond, VA, USA) as the stationary phase and dilute sulfuric acid as the eluent. The corresponding chromatogram is displayed in Fig. 4-14. As on a chemically bonded reversed phase, the investigated compounds are eluted in the order of increasing sorption area which points to a similar retention mechanism. Detection was carried out in this case by measuring the change in refractive index.

In comparison, Fig. 4-15 shows the simultaneous separation of alcohols and aldehydes on IonPac ICE-AS1 using pure water as the mobile phase.

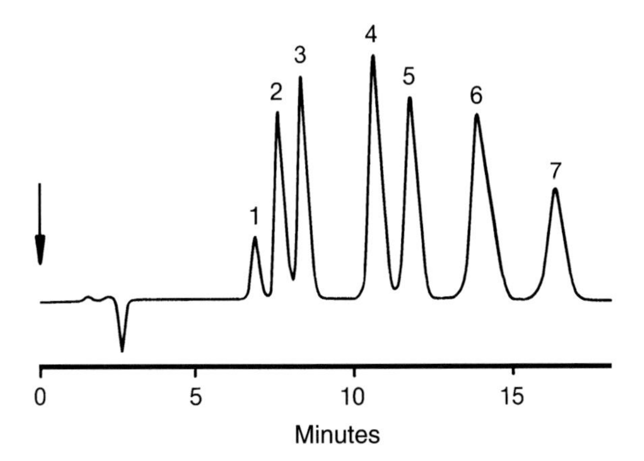

Fig. 4-14. Separation of monohydric aliphatic alcohols on Aminex HPX-85H. – Eluent: 0.005 mol/L H_2SO_4; flow rate: 0.4 mL/min; detection: RI; injection volume: 20 µL; solute concentrations: 0.5% (w/w) each of methanol (1), ethanol (2), 2-propanol (3), 2-methylpropanol-1 (4), 1-butanol (5), 3-methylbutanol-1 (6), and 1-pentanol (7); (taken from [12]).

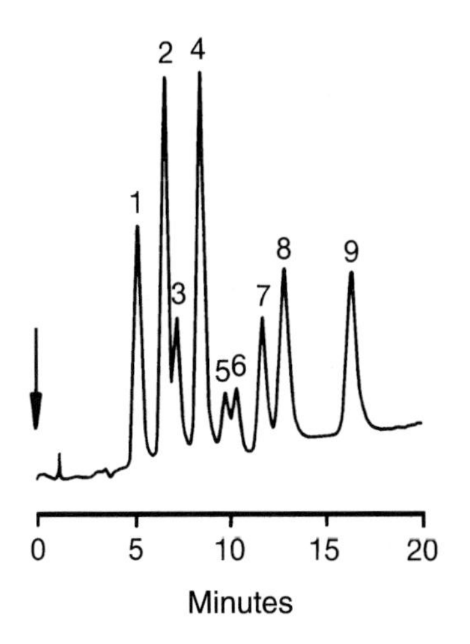

Fig. 4-15. Simultaneous separation of alcohols and aldehydes on IonPac ICE-AS1. – Eluent: water ; flow rate: 1 mL/min; detection: RI; injection volume: 50 µL; solute concentrations: 100 ppm each of glyoxal (1), glycerol (2), formaldehyde (3), ethylene glycol (4), glutaric dialdehyde (5), methanol (6), ethanol (7), 2-propanol (8), and 1-propanol (9).

Pulsed amperometry on a Pt working electrode provides an alternative to RI detection for the detection of alcohols and results in a higher sensitivity and selectivity. Fig. 4-16 shows a standard chromatogram with a simultaneous separation of monohydric and dihydric alcohols. Clearly, glycerol is the key substance in this representation, since it cannot be detected by gas chromatography because of its high boiling point. Although corresponding RPLC methods are available for the analysis of monohydric alcohols, ion chromatography offers the advantage of the simultaneous detection of polyhydric alcohols.

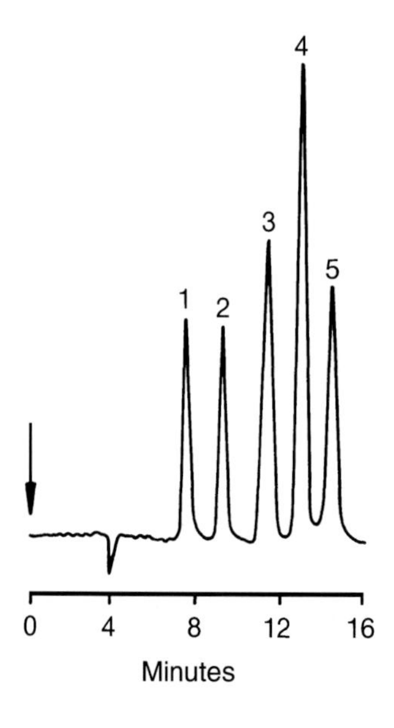

Fig. 4-16. Analysis of monohydric and polyhydric alcohols with pulsed amperometric detection. – Separator column: IonPac ICE-AS1; eluent: 0.1 mol/L $HClO_4$; detection: pulsed amperometric detection on a Pt working electrode; injection volume: 50 μL; solute concentrations: 2.5 ppm glycerol (1) and ethylene glycol (2), 1.75 ppm methanol (3), 10 ppm ethanol (4), and 15 ppm 2-propanol (5).

4.9 Analysis of Amino Acids

The separation of amino acids on totally sulfonated cation exchangers is still one of the most commonly used methods for amino acid analysis. A milestone in the development of this method is the paper by Spackman, Stein, and Moore [13], published in 1958, wherein an automated quantitative detection of physiological amino acids is described for the first time. This method utilized two ion-exchange columns; acidic and neutral amino acids were separated in one column, while the other column was used for basic amino acids. The separated compounds were then reacted post-chromatographically with ninhydrin. The resulting derivatives were detected by measuring their light absorption. At that time, an analysis took about two days and required 4 mL sample of plasma. Today, about a 50-μL sample is sufficient for a single run with a commercial amino acid analyzer based on an ion-exchange separation with subsequent post-column derivatization with ninhydrin. Analysis times are about 30 minutes for protein hydrolysates and 2 hours for physiological samples.

Although conventional amino acid analysis is extremely reliable, it does not satisfy today's demands for both higher sensitivity (fmol range) and shorter analysis times (10 to 15 min) while maintaining the same high resolution. These requirements can only be met by HPLC methods that have since been developed. Applications involve a precolumn derivatization with either *o*-phthaldialdehyde (OPA) [14,15], fluorenylmethyloxycarbonylchloride (FMOC) [16,17], phenylthiohydantoin (PTH) [18-20], phenylisothiocyanate (PITC) [21,22], or dansyl chloride (Dns) [23,24]. The methodological variety, which characterizes amino acid analysis today, appears confusing but should not be regarded as a disadvantage. None of these methods is capable of analyzing all of the more than 300 amino acids and their metabolites known today. A review of the various methods has been published by Ogden [25].

4.9.1 Separation of Amino Acids

Amino acids are amphoteric compounds; that is, they may exist in anionic, cationic, and zwitterionic form, depending on the pH:

$$H_3\overset{+}{N}-\underset{R}{\overset{H}{C}}-COO^- \longleftrightarrow H_2N-\underset{R}{\overset{H}{C}}-COO^- \longleftrightarrow H_3\overset{+}{N}-\underset{R}{\overset{H}{C}}-COOH$$

According to the structure of the rest R, one distinguishes between "acidic", "neutral", and "basic" amino acids. In buffer solutions of low pH the equilibrium is shifted to the right-hand side.

$$H_2N-\underset{R}{\overset{H}{C}}-COOH + H^+ \rightleftharpoons H_3\overset{+}{N}-\underset{R}{\overset{H}{C}}-COOH \qquad (120)$$

Amino acids may interact in their cationic form with the negatively charged sulfonate groups of a cation exchanger. At higher proton concentrations in the eluent the equilibrium represented by Eq. (120) is shifted to the right-hand side. This also increases the concentration of exchangeable cations. Therefore, amino acids are strongly retained on the stationary phase at very low pH. If the pH of the eluent is raised, the equilibrium is shifted more and more to the left-hand side. Amino acid molecules without any positive charge are formed. They are not retained on the stationary phase and, thus, elute very quickly.

Stationary Phases

Totally sulfonated polystyrene/divinylbenzene-based cation exchangers are generally used as stationary phases. The technical specifications of some selected materials are listed in Table 4-5. Totally sulfonated cation exchangers are mechanically not very rugged and have low permeabilities. Hence, at high back pressure of the separator column, only low flow rates can be selected. If a flow rate is accidentally adjusted too high, the quality of the packing material deteriorates, resulting in a loss in separation efficiency. Shorter analysis times can only be realized by selecting suitable eluents. A serious disadvantage of totally sulfonated cation exchangers is the slow kinetic process because of the diffusion of sample molecules into the interior of the stationary phase.

Table 4-5. Characteristic structural-technical properties of some selected fully sulfonated cation exchangers for amino acid separations.

Designation	Manufacturer	Particle diameter [μm]	Degree of cross-linking [%]
DC 6A	Dionex	11	8
DC 4A	Dionex[a)]	8	8
BTC 2710	Mitsubishi[b)]	7	10

[a)] Can also be purchased from Pharmazia/LKB.
[b)] Can also be purchased from Biotronik.

Thus, the development of pellicular ion exchangers was pushed ahead. Two important objectives could be realized:

a) With a pellicular anion exchanger, analysis times for hydrolysates could be reduced to 30 to 40 minutes.
b) At a flow rate of 1 mL/min, the pressure drop along the separator column does not exceed 6.5 MPa (1000 psi).

To realize these objectives, two different methods were applied: the separation of amino acids by both anion *and* cation exchange. The basic principle of both column types is the latex configuration, which has already been described for anion and cation exchangers in Sections 3.3.1.2 and 3.4.1.2, respectively.

The latex cation exchanger, introduced under the trade name AminoPac PC-1, consists of a polystyrene/divinylbenzene substrate with a particle diameter of 10 μm. In contrast to the columns employed in conventional cation analysis, these latex beads are not constructed of the same material but of methylmethacrylate and styrene. In comparison to totally sulfonated cation exchangers, this newly developed latex exchanger is characterized by a significantly lower capacity. This is determined by the size of the latex beads, which is about 200 nm. The chromatogram shown in Fig. 4-17 with the separation of amino acids eluting early on an AminoPac PC-1 latex anion exchanger impressively demonstrates the very high selectivity of this separator column, which is mainly due to the polymer that is employed for synthesizing the latex beads. In addition to the high chromatographic efficiency that is typical for latex exchangers, the isocratic mode of operation at ambient temperature is a characteristic feature of this separation. The composition of the eluent is remarkably simple. With dilute nitric acid (pH 3.2) a complete separation of all signals is achieved in less than 14 minutes. For comparison: using conventional exchangers this run requires almost 30 minutes. Retention is determined in this separation by the proton concentration in the mobile phase, which mostly affects the retention behavior of glutamic acid. This is attributed to the low isoelectric point of this compound.

The corresponding AminoPac PA-1 latex anion exchanger consists of a polystyrene/divinylbenzene substrate with a particle diameter of 10 μm. The synthesis of the latex beads is carried out, however, using dimethylamine instead of divinylbenzene for crosslinking the polymer. Until now, no methods have been available for the separation

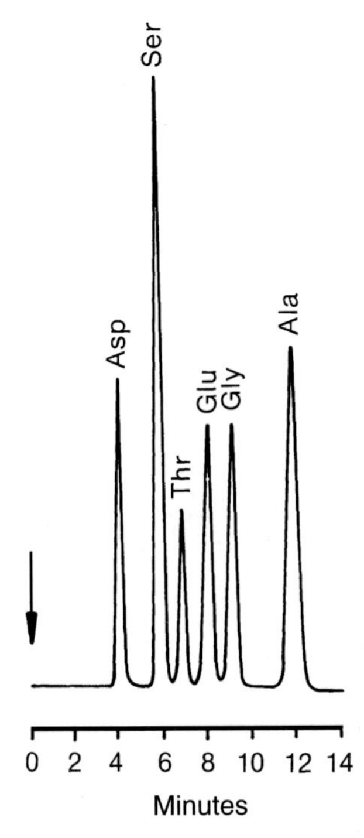

Fig. 4-17. Separation of some early eluting amino acids on a latex cation exchanger such as AminoPac PC-1. – Eluent: 0.00025 mol/L HNO_3; flow rate: 1 mL/min; detection: fluorescence after reaction with OPA; injection volume: 50 μL; solute concentrations: 125 pmol each of aspartic acid, serine, threonine, glutamic acid, glycine, and alanine.

of amino acids by anion exchange, primarily because of the difficulties in manufacturing *surface-aminated* materials with good chromatographic properties. Thus, latex-based anion exchangers offer themselves as an alternative. Fig. 4-18 reveals that the elution order is reversed in comparison to cation exchange: arginine and lysine elute first and threonine and serine are separated much better. The baseline-resolved separation of 17 amino acids contained in a hydrolysate standard requires about 40 minutes, which is significantly longer than with a latex cation exchanger.

Eluents

A series of buffer solutions with increasing cation concentration and increasing pH value are generally used for the separation of amino acids. Dus et al. [26], for example, employed four different citrate buffer solutions with pH values between 3.2 and 10.9 and sodium concentrations between 0.2 mol/L and 1.4 mol/L. Hare [27] developed an elution procedure based on a buffer system of constant cation concentration ($c(Na^+)$ = 0.2 mol/L). The pH value is varied between 3.25 and 10.1. Finally, in his Pico-Buffer® system, Benson [28] used a series of buffer solutions having similar pH values but different cation concentrations. Compared with systems involving pH gradients, this method, originally introduced by Piez and Morris [29], offers the advantage of a more stable baseline.

The Hi-Phi buffer solutions recommended by Dionex for separations on totally sulfonated cation exchangers are similar in their composition to those employed by Dus et al. Their composition is listed in Table 4-6. The sodium-based eluents are intended

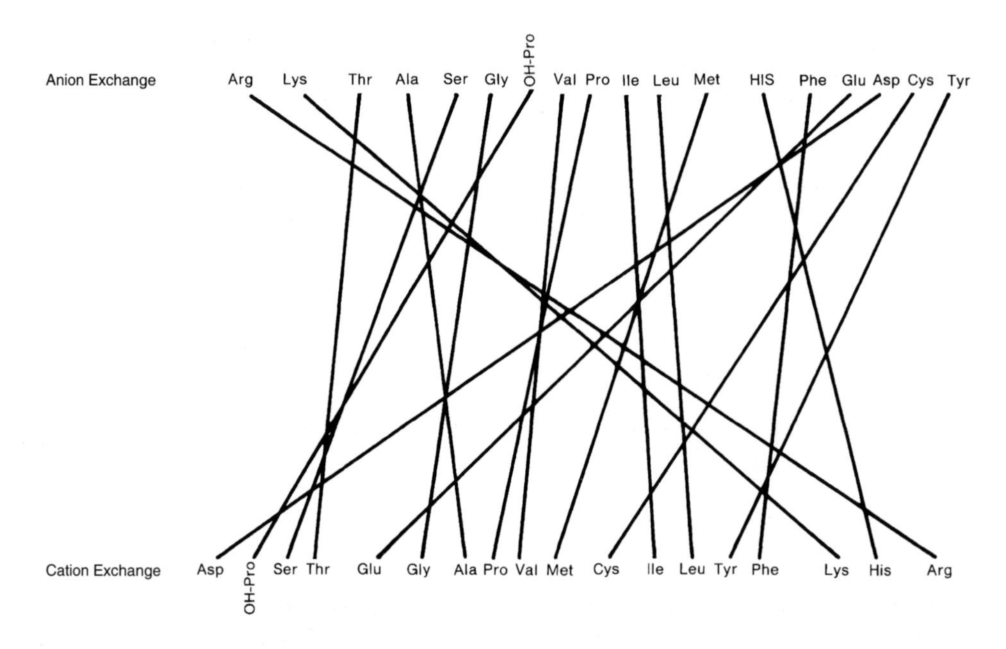

Fig. 4-18. Comparison of relative retentions of amino acids on anion and cation exchangers.

for the separation of hydrolysates. On the other hand, the separation of asparagine, glutamine, and glutamic acid, important for physiological samples, can only be accomplished with buffer solutions in the lithium form.

The chemicals from which the various buffer solutions are prepared should be of high purity. Particularly disturbing are ammonia and other amines present in laboratory air which dissolve very well in the buffer solutions. Ammonia is eluted only very slowly as ammonium ion. If the sample to be analyzed contains ammonia in appreciable quantities, it elutes as a sharp signal in the range of strong basic amino acids. As a contaminant of the buffer solution, in contrast, ammonium ions are continuously bound to the stationary phase. Ammonium ions exhibit a very high affinity toward the stationary phase. They pass very slowly through the column thereby leading to a broad and diffuse signal which disturbs the baseline. For these reasons, the eluents required for amino acid analysis should be stored under inert gas. To prevent the growth of diverse micro organisms, a very common problem, a concentration c = 1 g/L phenol is added to the Hi-Phi buffer solution. Fig. 4-19 displays the chromatogram of a hydrolysate standard obtained with citrate buffers on a totally sulfonated AminoPac Na-1 cation exchanger. Under the same chromatographic conditions, hydrolysates from connective tissue samples can also be analyzed which additionally contain hydroxyproline and hydroxylysine. The only disadvantage is the long analysis time of about 70 minutes, which is mainly due to the low number of buffer solutions utilized for elution. When the number of citrate/borate buffers is increased from three to five, slightly shorter analysis times with significantly improved peak shapes are obtained. This is illustrated in Fig. 4-20, which shows the separation of a hydrolysate standard on a totally sulfonated BTC 2710 cation exchanger (Biotronik, Maintal, Germany).

Table 4-6. Composition of the buffer solutions recommended by Dionex for the separation of amino acids on totally sulfonated cation exchangers.

Eluent type	pH Value	Cation concentration [mol/L]
Na-A	3.25	0.2
Na-B	4.25	0.2
Na-C	7.40	1.0
Li-A	2.75	0.238
Li-B	2.65	0.238
Li-C	3.60	0.340
Li-D	4.17	0.667
Li-E	5.30	0.643

The eluents suitable for the separation of amino acids on latex cation exchangers do not comprise the classical citrate/borate buffers but mixtures of nitric acid and potassium oxalate. In comparison to buffers composed of sodium citrate and borate, these components may be obtained at much higher purity. The retention of the amino acids to be analyzed, however, is possibly affected by the sample pH due to the limited buffer capacity of the eluents that are based on nitric acid and potassium oxalate. Fig. 4-21 shows the separation of a calibration standard for collagen hydrolysates on an Amino Pac PA-1 latex cation exchanger at ambient temperature. The advantage is the short

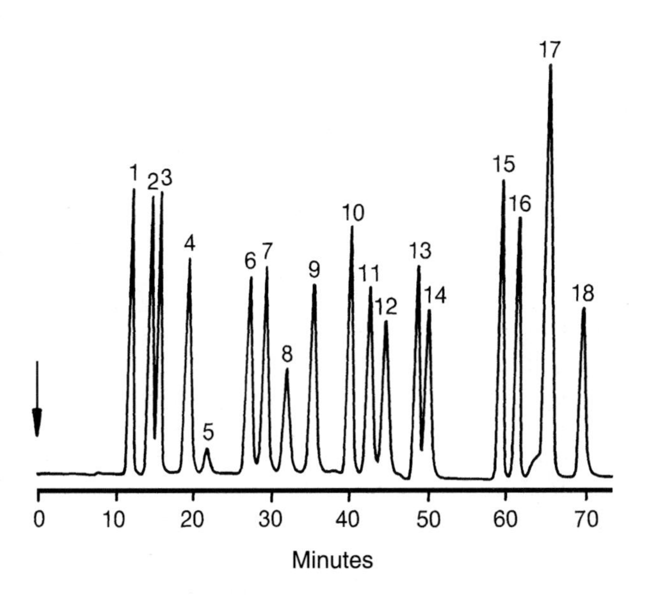

Fig. 4-19. Separation of a hydrolysate standard on a totally sulfonated cation exchanger AminoPac Na-1. – Column temperature: 50 °C; eluent: Hi-Phi buffer A, B and C; flow rate: 0.3 mL/min; detection: photometry with NIN-filter after reaction with ninhydrin at 130 °C; injection volume: 50 μL; solute concentrations: 10 nmol each of ASP (1), THR (2), SER (3), GLU (4), PRO (5), GLY (6), ALA (7), CYS (8), VAL (9), MET (10), ILE (11), LEU (12), TYR (13), PHE (14), HIS (15), LYS (16), NH_3 (17), and ARG (18).

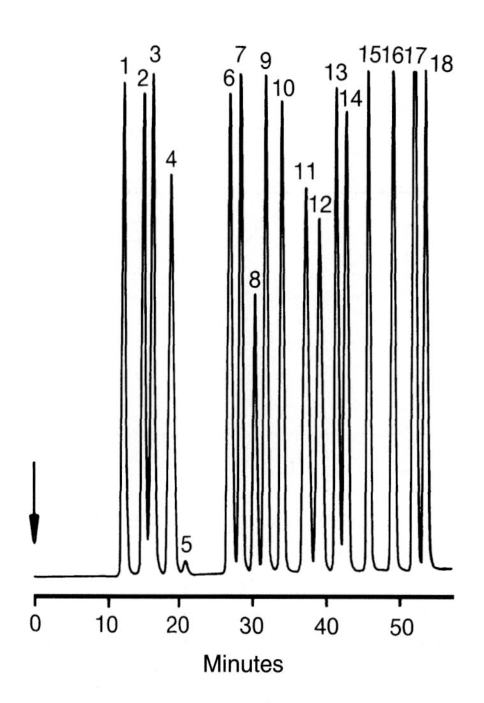

Fig. 4-20. Separation of a hydrolysate standard on a totally sulfonated cation exchanger BTC 2710 (Biotronik, Maintal, Germany). – Column temperature: four-step temperature program from 48 °C to 70 °C; eluent: 5 sodium citrate/borate buffers; flow rate: 0.3 mL/min; for all other chromatographic conditions and elution order: see Fig. 4-19.

analysis time of about 30 minutes, which is further reduced by applying a *five-step* gradient program with graduated nitric acid and potassium oxalate concentrations. Apart from the unneccessarily high resolution between histidine and arginine, such a fine tuning of pH value and cation concentration also results in reduced tailing of the isoleucine and leucine signals.

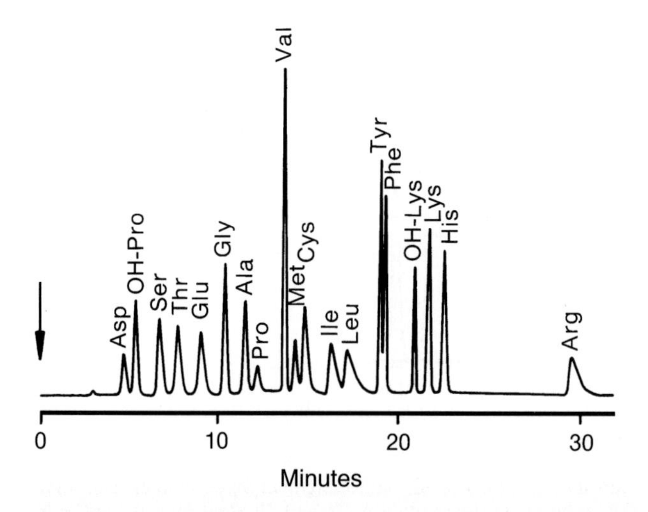

Fig. 4-21. Separation of a calibration standard for collagen hydrolysates on a latex cation exchanger AminoPac PC-1. – Eluent: (A) 0.00025 mol/L HNO_3; (B) 0.006 mol/L $K_2C_2O_4$ + 0.0056 mol/L HNO_3, (C) 0.009 mol/L $K_2C_2O_4$, regenerent: 0.005 mol/L HNO_3; flow rate: 1.2 mL/min; detection: see Fig. 4-19.

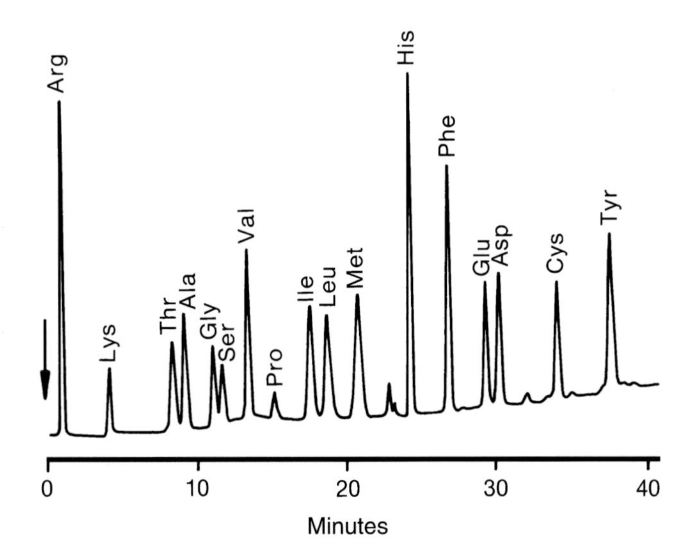

Fig. 4-22. Separation of a hydrolysate standard on a latex anion exchanger AminoPac PA-1. – Eluent: (A) 0.028 mol/L NaOH + 0.008 mol/L $Na_2B_4O_7$, (B) 0.07 mol/L NaOH + 0.02 mol/L $Na_2B_4O_7$, (C) 0.16 mol/L NaOAc, (D) 0.32 mol/L NaOAc, regenerent: 0.56 mol/L NaOH + 0.64 mol/L H_3BO_3; flow rate: 1 mL/min; detection: see Fig. 4-19; injection: 1 nmol of each of the various amino acids.

A completely different selectivity is obtained when amino acids are separated on a latex anion exchanger (see Fig. 4-18). A four-step gradient program is applied with which the analysis of a hydrolysate can be carried out in 35 minutes. The program is started with two mixtures of sodium tetraborate and sodium hydroxide followed by two sodium acetate solutions of different concentration. The corresponding chromatogram in Fig. 4-22 reveals that the anion exchange method allows a complete separation of phenylalanine and tyrosine, which is not possible on any cation exchangers. O-Phosphorylated amino acids such as P-Thr, P-Ser, and P-Tyr are hardly retained on cation exchangers and, therefore, are only poorly separated. These compounds elute from a latex anion exchanger at the end of the hydrolysate program in the retention range of tyrosine and may thus be separated. Furthermore, anion exchange chromatography is suited for the separation of arogenic acid,

NH_2
$CH_2-C-COOH$
H
OH

Arogenic acid

an intermediate in the biosynthesis of phenylalanine and tyrosine. Arogenic acid is not stable at acid pH, hence, it cannot be analyzed by cation exchange chromatography. However, Fig. 4-23 shows the separation of this compound from the amino acids phenylalanine and tyrosine accomplished on AminoPac PA-1 with alkaline eluents.

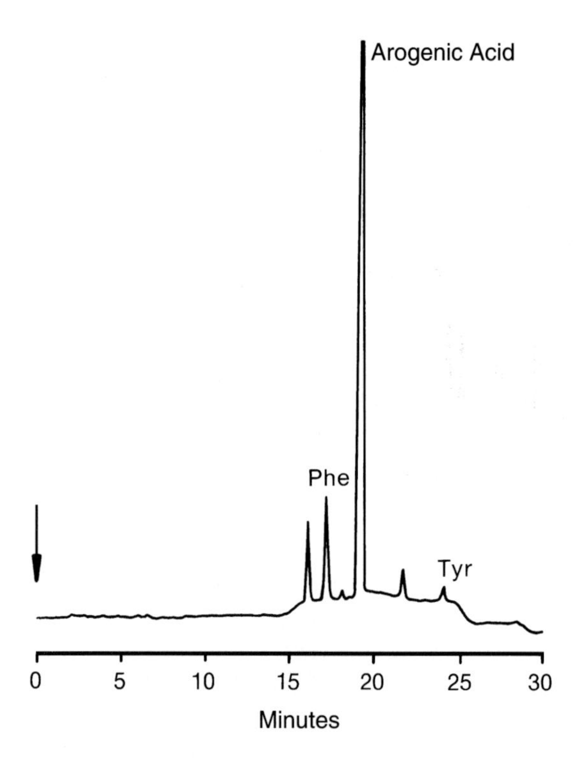

Fig. 4-23. Separation of arogenic acid on a latex anion exchanger. – Chromatographic conditions: see Fig. 4-22.

The determination of amino acids in physiological samples is one of the most demanding tasks in the field of amino acid analysis. Often more than 40 components have to be quantified in matrices as complex as serum or urine. For such separations, lithium citrate buffers are still used as the eluent. The separation of asparagine, glutamic acid, and glutamine, important for physiological samples, is only possible on totally sulfonated cation exchangers in the lithium form. Fig. 4-24 shows the optimized separation of 40 physiologically relevent amino acids on BTC 2710 as a representative example of this technique. The separation has been obtained with five different lithium citrate buffers and a four-step temperature program. The analysis time is more than two hours, but may be reduced significantly if one can dispense with the separation of sarcosine, homocystine, ethanolamine, hydroxylysine, and anserine. For the analysis of free amino acids in fruit juices, beer, wine, and milk the gradient program must be adapted accordingly.

4.9.2 Detection of Amino Acids

Ninhydrin

The post-column derivatization with ninhydrin introduced by Spackman, Stein, and Moore [13] represents even today the most customary detection method for quantitative amino acid analysis. As a strong oxidant, at temperatures around 130 °C ninhydrin reacts with the α-amino groups of eluting amino acids, according to Eq. (121), releasing ammonia and carbon dioxide.

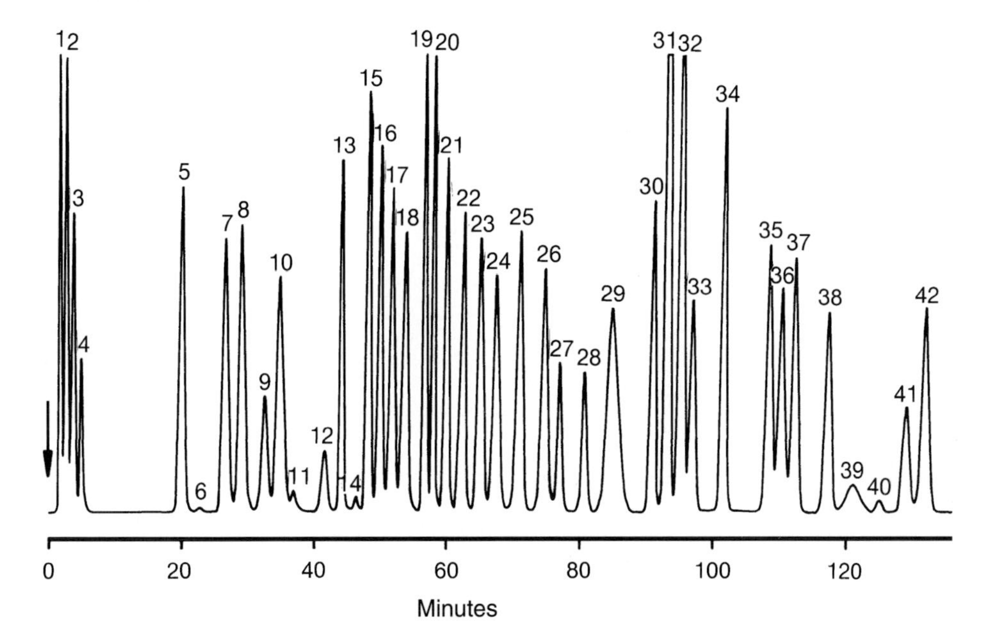

Fig. 4-24. Separation of a standard of physiologically relevant amino acids on a totally sulfonated cation exchanger BTC 2710. – Column temperature: four-step temperature program from 48 °C to 70 °C; eluent: 5 lithium citrate buffer; flow rate: 0.3 mL/min; detection: see Fig. 4-19; solute concentrations: 10 nmol each of PSER (1), TAU (2), PETA (3), UREA (4), ASP (5), OH-PRO (6), THR (7), SER (8), ASN (9), GLU (10), GLN (11), SARC (12), AAAA (13), PRO (14), GLY (15), ALA (16), CITR (17), AABA (18), VAL (19), CYS (20), MET (21), CYST (22), ILE (23), LEU (24), TYR (25), PHE (26), B-ALA (27), B-AIBA (28), HOMCYS (29), GABA (30), ETA (31), NH_3 (32), OH-LYS (33), ORN (34), LYS (35), 1-METH (36), HIS (37), 3-METH (38), TRP (39), ANSER (40), CARN (41), and ARG (42).

Ninhydrin + Amino acid ($R-CH(NH_2)-COOH$) → Hydrindantin + $R-CHO$ + NH_3 + CO_2

Hydrindantin → (– H_2O; + NH_3, + Ninhydrin) → Ruhemann's Purple (121)

An aldehyde shorter by one C-atom and hydrindantin, the reduced form of ninhydrin, are formed in this reaction. Hydrindantin reacts with ammonia and a second molecule of ninhydrin to form a red dye called Ruhemann's purple. This dye has an absorption maximum at 570 nm (see Fig. 4-25).

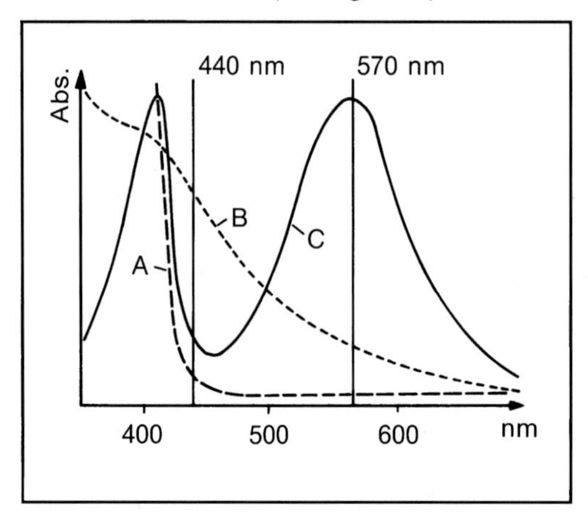

Fig. 4-25. Absorption spectrum of ninhydrin and its reaction products: (A) ninhydrin, (B) derivatized imino acid, (C) derivatized amino acid.

Secondary amino acids, "imino acids", such as proline and hydroxyproline, do not possess an α-amino group, and react with ninhydrin to form a yellow product which is usually detected at 440 nm. Therefore, amino acid analyzers are equipped with a photometer capable of measuring at two different wavelengths (570 nm and 440 nm). The sensitivity of this detection method is about 200 pmol.

o-Phthaldialdehyde

o-Phthaldialdehyde (OPA) is the reagent most widely used to convert primary amino acids into fluorescing derivatives. The reaction with OPA, first described by Roth [30] in 1971, quickly proved to be a more sensitive alternative to the ninhydrin method [31-33]. Initially, OPA was used exclusively for post-column derivatization. Today, it is increasingly employed for pre-column derivatizations. In an alkaline medium, *o*-phthaldialdehyde reacts with primary amino acids, according to Eq. (122), to form a strongly blue-fluorescing adduct, which can be detected in the lower pmol range. The reaction requires only a few seconds, even at room temperature.

$$C_6H_4(\overset{O}{\overset{\|}{C}}-H)_2 + R-\underset{H}{\underset{|}{\overset{NH_2}{\overset{|}{C}}}}-COOH + HS-CH_2-CH_2-OH \tag{122}$$

$$\longrightarrow \text{(isoindole)}\,(S-CH_2-CH_2-OH)\,N-\underset{R}{\underset{|}{\overset{H}{\overset{|}{C}}}}-COOH + H_2O$$

postulated Adduct

The subsequent fluorescence detection is carried out at an excitation wavelength of 340 nm, the emission is measured at 455 nm. The addition of thiols such as 2-mercaptoethanol increases the fluorescence yield. However, OPA only reacts with primary amino acids. Secondary amino acids can only be detected after their oxidation (for example with hypochlorite or chloramine T) [34]. In practical applications, this results in significant difficulties.

Other Detection Methods

An interesting derivatization method for cysteine was recently described by Jenke and Brown [35]. They mixed the column effluent with a buffered solution of 5,5′-dithiobis(2-nitrobenzoic acid) (DTNB), yielding a strongly yellow-colored chromophore, which can be detected photometrically at 412 nm:

Cysteine + DTNB $\xrightarrow[OH^-]{pH\ 6}$ Mixed disulfide + Chromophore (123)

As an alternative to derivatization techniques, amino acids can also be detected directly via pulsed amperometry (see Section 6.3.2.2). This method was first described in 1983 by Johnson et al. [36]. Amino acids are anodically oxidized in an alkaline medium at a gold working electrode. The resulting oxidation products remain on the electrode surface, but may be electrochemically removed by quickly repeating a potential sequence comprised of three different potentials. With an injection volume of 50 μL, this method also enables detection limits for amino acids in the lower ng range. The same potential sequence may also be used for the detection of carbohydrates (see Section 3.3.5.3) using a gold working electrode.

4.9.3 Sample Preparation

Sample preparation is the key step in an accurate and reproducible amino acid analysis. In principle, physiological samples should be processed as quickly as possible to avoid contamination by sample handling or by the reagents being used. In addition, metabolic activities may lead to false results.

Investigations into urine and plasma samples must consider that certain compounds such as antibiotics are not metabolized after its application. They pass through the body without changing their structure and, finally appear in the urine or plasma and interfere with the amino acid analysis via reaction with ninhydrin. Perry et al. [37] reported this problem for the first time. In urine samples oxidized with hydrogen peroxide they detected a signal in the retention range of cysteic acid and homocystine, which they attributed to D-penicillaminesulfonic acid. This compound was formed by oxidation of the antibiotic D-penicillamine that was administered to the patient whose urine was investigated.

If free amino acids are to be analyzed in foods or feeds, it is advisable to remove matrix components such as lipids, carbohydrates, and nucleic acids from the sample. Of course, the necessary extraction steps are time-consuming, but higher concentrations of such compounds may severely interfere with the amino acid analysis. After hydrolysis of protein-free mixtures of nucleosides, nucleotides, and nucleic acids, Paddock et al. [38] even observed the formation of amino acids, predominantly glycine.

To analyze free amino acids in plasma or tissue homogenates, it is necessary to remove proteins and peptides present in solution. The most widely used deproteinization method is precipitation with 5-sulfosalicylic acid followed by centrifugation for separating the precipitate. In comparison to other precipitation agents such as trichloroacetic acid, perchloric acid, picrinic acid, or acetonitrile, the best results with respect to completeness of precipitation are obtained with 5-sulfosalicylic acid [39]. Other deproteinization methods comprise ultrafiltration and ultracentrifugation [40], which have only recently been considered as sample preparation methods for amino acid analysis.

Cleavage of the proteins into their amino acid moieties is typicallly accomplished via acid hydrolysis at 110 °C for 24 hours with hydrochloric acid in a concentration c = 6 mol/L. This is a compromise, since none of the existing hydrolysis methods provide satisfactory results for all amino acids. Sulfur-containing amino acids, for example, have to be oxidized prior to hydrolysis; while isoleucine, valine, threonine, and serine require differently long hydrolysis times. To protect tryptophan, 2% thioglycolic acid is added to the sample [41]. To remove hydrochloric acid, the hydrolysates are lyophilized or vacuum-dried in a centrifuge, then resuspended in small amounts of water, and dried again. The samples are taken up in special buffer solutions for subsequent amino acid analysis.

Because of the high sensitivity of modern amino acid analyzers, the purity of eluents becomes increasingly important. A possible source for contamination is the de-ionized water used for preparing the buffer solutions which may contain amino acid traces from the mixed-bed exchangers. This problem can be circumvented by applying reversed osmosis for water de-ionization. Microbial contamination in the finished buffer solutions is best avoided by cooling the eluents and by adding phenol. Reagents used to make eluents can also contaminate. According to one chemical manufacturer (Pierce, Rockford, IL, USA), it is citric acid which is primarily contaminated with traces of amino acids.

Further information about problems which may occur in amino acid analysis by ion-exchange chromatography can be found in the 1986 review paper by Williams [42].

5 Ion-Pair Chromatography (MPIC)

Ion-pair chromatography provides a useful alternative to ion-exchange chromatography. The selectivity of the separation in ion-pair chromatography is mainly determined by the make-up of the mobile phase, thus both anionic and cationic compounds can be separated. This universal applicability has helped ion-pair chromatography reach its present significance.

Haney et al. [1,2], Waters Associates [3], and Knox et al. [4,5] – to name just a few – found that by adding lipophilic ions such as alkanesulfonic acid or quaternary ammonium compounds to the mobile phase, solute ions of opposite charge can be separated on a chemically bonded reversed phase. The term "Reversed-Phase Ion Pair Chromatography" (RPIPC) has generally been adopted for this technique. The term "Mobile Phase Ion Chromatography" (MPIC) describes a method which combines the major elements of RPIPC with suppressed conductivity detection, previously described. Apart from the above-mentioned chemically bonded reversed phases, neutral divinylbenzene resins featuring a high surface area and a weakly polar character are used as stationary phases.

The physical-chemical phenomena, the basis for the retention mechanism, are still not fully understood. This mechanistic uncertainty is reflected by the many terms already proposed for this kind of separation method. Here, we discuss two hypothesis, although both of them lack an unequivocal experimental basis. Horvath et al. [6,7] take the view that solute ions form neutral ion pairs with the lipophilic ions in the aqueous mobile phase. These neutral ion pairs are retained at the non-polar stationary phase. In contrast, Huber, Hoffmann, and Kissinger [8-10] support the ion-exchange model, where the lipophilic reagent first adsorbs at the surface of the stationary phase, giving it an ion-exchange character. Both hypothesis represent limiting cases. It is not to be expected that the retention process is fully described by just one of the two limiting cases. One must also consider that the development of a retention model, that accounts for all experimental data in the same way, is impossible on the basis of chromatographic data only. A deeper insight into the very complex mechanistical events is severely obscured, first by the lack of physical-chemical data, and second by the difficulty in obtaining accurate values for the equilibrium constants of interest.

5.1 Survey of Existing Retention Models

According to the model of *ion pair formation* [6,11], the solute ion E interacts with the lipophilic ion H forming a complex EH. This complex can be reversibly bound at the non-polar surface of the stationary phase L (chemically bonded reversed phase or divinylbenzene resin), giving LEH. The equilibria thus established may be described as follows:

$$E + H \overset{K_1}{\rightleftharpoons} EH \tag{124}$$

$$EH + L \overset{K_2}{\rightleftharpoons} LEH \tag{125}$$

The respective equilibrium constants can be expressed by Eqs. (126) and (127). The concentration of individual species in the mobile and the stationary phases is indicated by the indices m and s, respectively.

$$K_1 = [EH]_m/[E]_m \cdot [H]_m \tag{126}$$

$$K_2 = [LEH]_s/[EH]_m \cdot [L]_s \tag{127}$$

$[L]_s$ is defined as the available free surface area of the stationary phase.

According to the ion-exchange model [8-10], the lipophilic ion is adsorbed at the surface of the stationary phase forming LH. This renders the non-polar resin a dynamic ion exchanger. The solute ion E may subsequently interact with LH:

$$H + L \overset{K_3}{\rightleftharpoons} LH \tag{128}$$

$$LH + E \overset{K_4}{\rightleftharpoons} LEH \tag{129}$$

The equlibria constants are defined in an analogous way:

$$K_3 = [LH]_s/[H]_m \cdot [L]_s \tag{130}$$

$$K_4 = [LEH]_s/[LH]_s \cdot [E]_m \tag{131}$$

Under the condition that solute ions and lipophilic ions can interact completely independently with the stationary phase, one obtains, according to Horvath [12], a very complicated equilibrium system (see Fig. 5-1).

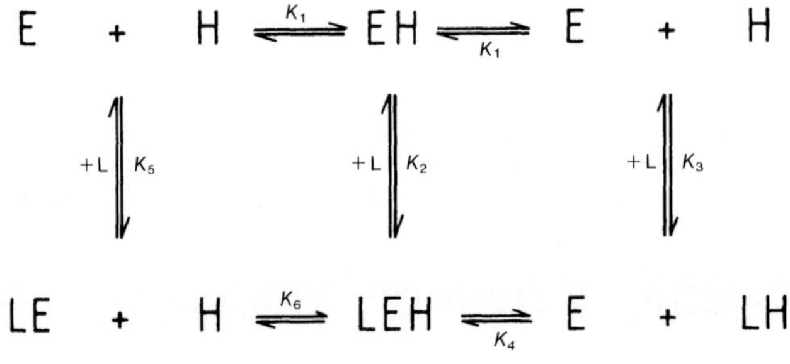

$$\begin{array}{ccccc} E + H & \overset{K_1}{\rightleftharpoons} & EH & \underset{K_1}{\rightleftharpoons} & E + H \\ \updownarrow +L \; K_5 & & \updownarrow +L \; K_2 & & \updownarrow +L \; K_3 \\ LE + H & \overset{K_6}{\rightleftharpoons} & LEH & \underset{K_4}{\rightleftharpoons} & E + LH \end{array}$$

Fig. 5-1. Schematic representation of the equilibrium system in ion-pair chromatography on non-polar stationary phases with lipophilic ions in the mobile phase.

Hence the equilibrium system must be extended as follows:

$$[E] + [L] \xrightleftharpoons{K_5} [LE] \tag{132}$$

$$[LE] + [H] \xrightleftharpoons{K_6} [LEH] \tag{133}$$

and

$$K_5 = [LE]_s/[E]_m \cdot [L]_s \tag{134}$$

$$K_6 = [LEH]_s/[LE]_s \cdot [H]_m \tag{135}$$

With the definition of the capacity factor k'

$$k' = \Phi \cdot K \tag{136}$$

Φ Phase volume ratio
K Distribution coefficient

the capacity factor for the solute ion E is expressed as

$$k' = \Phi \cdot \frac{[LEH]_s \cdot [LE]_s}{[E]_m \cdot [EH]_m} \tag{137}$$

When the stationary and mobile phases are in equilibrium, the concentration of lipophilic ions is constant. With

$$[E] \ll [H]$$

only a small fraction of lipophilic ions is complexed, so the concentration of these ions remains virtually unchanged:

$$[H]_m \equiv [H] \tag{138}$$

Assuming further that the fraction of solute ions adsorbed on the stationary phase is very small and that the total surface area L_T of the stationary phase is fixed, it holds:

$$[L]_T = [L]_s + [LH]_s \equiv [L] \tag{139}$$

If lipophilic ions and solute ions in the mobile phase form ion pairs, and if those ion pairs interact with the stationary phase, Eq. (137) can be expressed with the expressions in Eqs. (126), (127), (130), (131), (138), and (139) as a combination of the various equilibrium constants in dependence on [H] and [L]. For this, Eqs. (126), (127), (130), and (131) are transformed as follows:

$$[\mathrm{LE}]_s = K_1 \cdot [\mathrm{E}]_m \cdot [\mathrm{L}]_s \tag{140}$$

$$[\mathrm{E}]_m = [\mathrm{EH}]_m \,/\, K_2 \cdot [\mathrm{H}]_m \tag{141}$$

$$[\mathrm{L}]_s = [\mathrm{LH}]_s \,/\, K_3 \cdot [\mathrm{H}]_m \tag{142}$$

$$[\mathrm{LEH}]_s = K_4 \cdot [\mathrm{EH}]_m \cdot [\mathrm{L}]_s \tag{143}$$

Inserting Eqs. (140) to (143) into Eq. (137) and putting $[\mathrm{L}]_s/[\mathrm{E}]_m$ outside the brackets, one obtains:

$$k' = \Phi \cdot \frac{[\mathrm{L}]_s}{[\mathrm{E}]_m} \cdot \frac{K_4 [\mathrm{EH}]_m + K_1 \cdot [\mathrm{E}]_m}{1 + K_2 \cdot [\mathrm{H}]_m} \tag{144}$$

This expression is expanded with $(1 + K_3\,[\mathrm{H}]_m)$:

$$k' = \Phi \cdot \frac{[\mathrm{L}]_s}{[\mathrm{E}]_m} \cdot \frac{(1 + K_3 \cdot [\mathrm{H}]_m)(K_4 \cdot [\mathrm{EH}]_m + K_1 \cdot [\mathrm{E}]_m)}{(1 + K_2 \cdot [\mathrm{H}]_m)(1 + K_3 \cdot [\mathrm{H}]_m)} \tag{145}$$

With Eqs. (141) and (142) it follows:

$$k' = \Phi \cdot \frac{K_2 \cdot [\mathrm{LH}]_s}{K_3 \cdot [\mathrm{EH}]_m} \cdot \frac{(1 + K_3 \cdot [\mathrm{H}]_m)\,(K_4 \cdot [\mathrm{H}]_m + K_1 \cdot [\mathrm{E}]_m)}{(1 + K_2 \cdot [\mathrm{H}]_m)\,(1 + K_3 \cdot [\mathrm{H}]_m)} \tag{146}$$

$$k' = \Phi \cdot [\mathrm{LH}]_s \cdot \frac{\left(\frac{K_2}{K_3} + K_2 \cdot [\mathrm{H}]_m\right) \cdot \left(K_4 + \frac{K_1 \cdot [\mathrm{E}]_m}{[\mathrm{EH}]_m}\right)}{[1 + K_2 \cdot [\mathrm{H}]_m]\,(1 + K_3 \cdot [\mathrm{H}]_m)} \tag{147}$$

Inserting Eq. (141) and subsequent multiplication of the bracketed expression gives:

$$k' = \Phi \cdot [\mathrm{LH}]_s \cdot \frac{\frac{K_2 \cdot \mathrm{K}_4}{K_3} + \frac{K_1}{K_3 \cdot [\mathrm{H}]_m} + K_2 \cdot K_4 \cdot [\mathrm{H}]_m + K_1}{(1 + K_2 \cdot [\mathrm{H}]_m)\,(1 + K_3 \cdot [\mathrm{H}]_m)} \tag{148}$$

$$k' = \Phi \cdot \frac{[\mathrm{LH}]_s \cdot \left(\frac{K_2 \cdot \mathrm{K}_4}{K_3} + \frac{K_1}{K_3 \cdot [\mathrm{H}]_m}\right) + [\mathrm{LH}]_s \cdot (K_2\, K_4 \cdot [\mathrm{H}]_m + K_1)}{(1 + K_2 \cdot [\mathrm{H}]_m)\,(1 + K_3 \cdot [\mathrm{H}]_m)} \tag{149}$$

Inserting Eq. (142) it follows:

$$k' = \Phi \cdot \frac{K_2 K_4 [\mathrm{H}]_\mathrm{m} \cdot [\mathrm{L}]_\mathrm{s} + K_1 \cdot [\mathrm{L}]_\mathrm{s} + [\mathrm{LH}]_\mathrm{s} \cdot (K_2 K_4 \cdot [\mathrm{H}]_\mathrm{m} + K_1)}{(1 + K_2 \cdot [\mathrm{H}]_\mathrm{m})\,(1 + K_3 \cdot [\mathrm{H}]_\mathrm{m})} \tag{150}$$

After putting $[\mathrm{L}]_\mathrm{s}$ outside the brackets and considering Eq. (139) the intended dependence is derived:

$$k' = \Phi \cdot [\mathrm{L}] \frac{K_1 + K_2 K_4 \cdot [\mathrm{H}]}{(1 + K_2 \cdot [\mathrm{H}])\,(1 + K_3 \cdot [\mathrm{H}])} \tag{151}$$

According to the model of dynamic ion exchange, the following expression for the capacity factor k′ is obtained in much the same way using the expressions of Eqs. (126), (127), (130), (134), and (137) to (139):

$$k' = \Phi \cdot [\mathrm{L}] \frac{K_1 + K_3 K_5 \cdot [\mathrm{H}]}{(1 + K_2 \cdot [\mathrm{H}])\,(1 + K_3 \cdot [\mathrm{H}])} \tag{152}$$

Eqs. (151) and (152) show the dependence of the capacity factor k′ on the concentration of lipophilic ions. In a general form, this dependence can be described as

$$k' = (K_0 + B \cdot [\mathrm{H}])/(1 + K_2\,[\mathrm{H}]) \cdot (1 + K_3\,[\mathrm{H}]) \tag{153}$$

K_0 is the capacity factor of the solute ion if no lipophilic ions take part in the retention process. According to this model, K_2 is the formation constant for the respective ion pair, K_3 a constant characterizing the bonding between the lipophilic ions and the stationary phase. According to Eqs. (151) and (152), constant B represents the product of two equilibrium constants.

Plotting k' versus [H], a parabolic curve is obtained when, according to Eq. (153), $1/K_3 \cdot [\mathrm{H}] \ll 1/K_2$. Such dependences have been observed by Knox et al. [5,12].

Finally, in 1979, Bidlingmeyer et al. [13,14] introduced a third model which they termed the *ion interaction model*. It is based on conductivity measurements, the results of which rule out the formation of ion pairs in the mobile phase. This retention model, also used by Pohl [15] to interpret the retention mechanism on a MPIC phase, neither presupposes the formation of ion pairs nor is it based on classical ion-exchange chromatography.

According to the ion interaction model, a high surface tension is generated between the non-polar stationary phase and the polar mobile phase. From this, the stationary phase obtains a high affinity for those components of the mobile phase which are able to reduce the high surface tension. This comprises, for example, polar organic solvents, surfactants with their respective counter ions, and quaternary ammonium bases. Moreover, the model concept of ion interaction provides for an electrically charged double layer at the surface of the stationary phase. This phenomenon is schematically represented in Fig. 5-2 taking the analysis of surface-inactive anions as an example. The lipophilic ions (e.g., tetrabutylammonium cations) and acetonitrile as organic modifier are adsorbed in the inner region at the surface of the non-polar stationary phase. As all lipophilic cations are equally charged, the surface can be only partly covered with such ions because of the repulsive forces between these charges. The corresponding counter ions (typically OH^- ions when applying conductivity detection), as well as the analyte ions A^-, are found in the diffuse outer region. When increasing the lipophilic ion concentration in the mobile phase, the concentration of ions adsorbed to the surface will also increase because of the dynamic equilibrium between the mobile and the stationary phases. The transfer of a solute ion through the electrical double layer is, therefore, a function of electrostatic and van-der-Waals forces. If a solute ion with an opposite charge is attracted by the charged surface of the stationary phase, retention is a result of coulomb attractive forces and additional adsorptive interactions between the lipophilic part of the solute ion and the non-polar surface of the stationary phase. Adding a negative charge to the positively charged inner region of the double layer is tantamount to removing one charge from this region. To re-establish electrostatic equilibrium, another lipophilic ion can be adsorbed at the surface. Finally, two oppositely charged ions (not necessarily an ion pair) are adsorbed at the stationary phase.

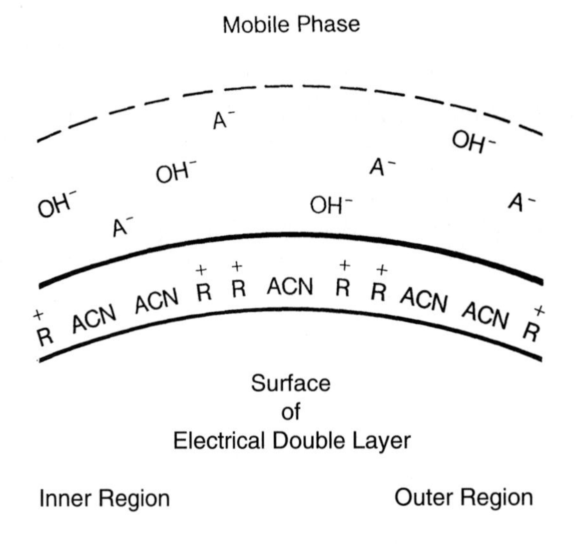

Fig. 5-2. Schematic representation of the electrically charged double layer when separating surface-inactive anions.

The separation of surface-inactive cations can be interpreted analogously. In this case, lipophilic anions are adsorbed at the resin surface, while the analyte cations are retained in the outer region of the double layer.

Unlike normal solute ions, surface-active ions may penetrate the inner region of the double layer, where they are adsorbed at the surface of the stationary phase. With this class of compounds, retention depends on the carbon chain length and, thus, on the degree of hydrophobicity. Retention increases with growing chain length. Acetonitrile as an organic modifier is also adsorbed at the resin surface, and is therefore involved in a competing equilibrium with the lipophilic ions. By blocking adsorption sites on the surface of the resin, the organic modifier serves to shorten retention when analyzing both surface-active and surface-inactive ions. In case of surface-active ions, this is achieved by direct competition; in case of surface-inactive ions via competition with the solvophobic counter ions $R\text{-}SO_3^-$ and R_4N^+, respectively.

5.2 Suppressor Systems in Ion-Pair Chromatography

In analogy to ion-exchange and ion-exclusion chromatography, the background conductance was chemically decreased in ion-pair chromatography to enable a sensitive conductivity detection. Conventional suppressor columns such as the ASC-1 (for anion determinations) and CSC-1 (for cation determinations) were initially used. After an average operation of about six hours, these columns had to be regenerated by the procedure described in Section 3.3.3.

The hollow fiber suppressors, developed for ion-exchange chromatography at the beginning of the 1980s, could not be employed in ion-pair chromatography because the occasionally high concentrations of organic solvents in the mobile phase were detrimental to the membrane material. Also, the membranes did not have adequate transport properties for the ion-pair reagents being used. To alleviate the problem of the conventional suppressor columns, discussed in Section 3.3.3, a sulfonated membrane was developed for the ion-pair chromatographic determination of anions. The membrane featured good transport properties for quaternary ammonium bases and was characterized by a high resistance to organic solvents. Such hollow fiber suppressors have been introduced under the trade name AFS-2. As in the AFS-1 for anion exchange chromatography, dilute sulfuric acid with c = 0.01 mol/L is employed as the regenerent. This means that in the suppressor reaction tetrabutylammonium cations are exchanged for regenerent protons. As in the process described in Section 3.3.3, water is the reaction product, whose formation provides the driving force for the proton diffusion through the membrane wall. The analyte anions are converted into their respective acids, thereby enabling a more sensitive and selective detection. Since hollow fiber suppressors are applied to ion-exclusion as well as ion-pair chromatography of anions, in case of alternate operation it is advisable to condition them overnight with the respective regenerent.

The CFS hollow fiber suppressor (see Section 3.4.3) that was developed for cation exchange chromatography can also be applied to cation analysis via ion-pair chromatography. It features good solvent stability and sufficient membrane transport properties for the anionic ion-pair reagent. This suppressor is regenerated with tetramethylammonium hydroxide using a concentration of c = 0.04 mol/L.

As described in Sections 3.3.3 and 3.4.3, hollow fiber suppressors no longer represent the state-of-the-art. Thus, a micromembrane suppressor was introduced under the trade name AMMS-MPIC for ion-pair chromatography of anions. Its structure corresponds to the systems developed for ion-exchange and ion-exclusion chromatography. Like the AFS-2, the AMMS-MPIC micromembrane suppressor contains a solvent-resistant membrane that is permeable to quaternary ammonium bases. Regarding the exchange capacity, no differences exist to the AFS-2, whereas the significantly smaller void volume contributes to the sensitivity increase. Regeneration of this micromembrane suppressor does not differ from that of the respective hollow fiber suppresor; it is also performed with dilute sulfuric acid of a concentration $c = 0.01$ mol/L.

Today, the CMMS micromembrane suppressor, introduced in Section 3.4.3, is employed for ion-pair chromatography of cations. Its solvent stability is sufficient, and it is regenerated with tetrabutylammonium hydroxide, also for ion-pair chromatographic applications. The regenerent concentration adheres to the concentration of the ion-pair reagent. A reagent concentration of $c = 0.002$ mol/L, for example, requires a TBAOH concentration of $c = 0.04$ mol/L.

5.3 Experimental Parameters that Affect Retention

The high flexibility in adjusting the chromatographic conditions to a given separation problem is the main advantage of ion-pair chromatography over ion-exchange chromatography. This flexibility results from the great variety of experimental parameters that affect retention. Thus, ion-pair chromatographic separations of ions on a MPIC phase are affected by the following parameters:

- Type of lipophilic counter ion in the mobile phase
- Concentration of lipophilic counter ion in the mobile phase
- Type of organic modifier
- Concentration of organic modifier in the mobile phase
- Type and concentration of inorganic additives
- Eluent pH
- Column temperature

The following discusses the effects of varying individual parameters on the retention process.

5.3.1 Type and Concentration of Lipophilic Counter Ions in the Mobile Phase

For the ion-pair chromatographic separation of *anions*, quaternary ammonium bases are preferably added as lipophilic ions to the mobile phase. According to the literature [16-18], one basically distinguishes between permanently coated and dynamically coated stationary phases. While dynamic coating is realized with reagents of low hydrophobicity such as tetraalkylammonium compounds, permanent coating is accomplished with strong-hydrophobic reagents such as cetyltrimethylammonium salts. After coating the stationary phase with these strong-hydrophobic salts, tetramethylammonium compounds are generally used to elute the solutes. As in liquid-liquid partition chromatography with pre-coated support material, these "permanently" coated phases are not stable for long periods, because even the strongly adsorbed cetyl compounds are slowly washed away.

The kind of counter ion of an ion-pair reagent is vitally important for selecting the appropriate detection method. If suppressed conductivity detection is applied, the ion-pair reagent is used in its hydroxide form. With direct conductivity detection, salicylate is perferred as the counter ion for tetraalkylammonium cations [19,20], since these salts exhibit a lower background conductance in aqueous solution. According to Wheals [21], eluents such as cetyltrimethylammonium bromide in combination with citric acid at pH 5.5 have proved suitable for UV-, RI-, and amperometric detection, as well as for direct conductivity detection. This example is an impressive illustration of the versatility of ion-pair chromatography.

The same applies to the ion-pair chromatographic separation of *cations*, which is performed either with long-chain alkanesulfonic acids or in the simplest manner with mineral acids. A survey of the most commonly used reagents is listed in Table 5-1 in the order of increasing hydrophobicity. A comparison by Cassidy and Elchuk [18] between chemically bonded reversed phases and organic divinylbenzene-based (Dionex IonPac NS1) or polystyrene/divinylbenzene-based polymers (Hamilton PRP-1) revealed that the latter have a lower chromatographic efficiency, but are characterized by a higher affinity for certain ion-pair reagents.

Table 5-1. Commonly used reagents for ion-pair chromatography on MPIC phases in the order of increasing hydrophobicity.

Anion analysis		**Cation analysis**
Ammonium hydroxide	Hydrophobicity ↓	Hydrochloric acid, perchloric acid
Tetramethylammonium hydroxide		Hexanesulfonic acid
Tetrapropylammonium hydroxide		Heptanesulfonic acid
Tetrabutylammonium hydroxide		Octanesulfonic acid

The choice of lipophilic ion depends solely on the degree of hydrophobicity of the analyte ion. For the separation of surface-inactive ions, a hydrophobic reagent is necessary; on the other hand, the separation of ions with long alkyl chains requires a strong-hydrophilic reagent. The hydrophobicity of an ion-pair reagent is generally determined by the number of its carbon atoms but, above all, by the length of its alkyl groups. This is illustrated in Fig. 5-3, which depicts the dependence of the k' value for diphenhydramine on the chain length of the ion-pair reagent upon application of a polymer phase. The sodium salts of various alkanesulfonic acids were used as ion-pair reagents. It appears from Fig. 5-3 that diphenhydramine retention significantly increases with growing chain length of the ion-pair reagent. According to the ion-pair model, such a dependence may be accounted for by the fact that the hydrophobic character of the formed ion pairs increases with growing chain length of the ion-pair reagent. This results in both a stronger interaction between the ion pair and the stationary phase and a reduced solubility in the eluent. A variation in the chain length of the ion-pair reagent affects the various solute ions in a different way. This parameter allows, therefore, to adjust very efficiently the selectivity of the chromatographic system. With alkaloids such as atropine and papaverine Jost et al. [22] even observed a change in the retention order

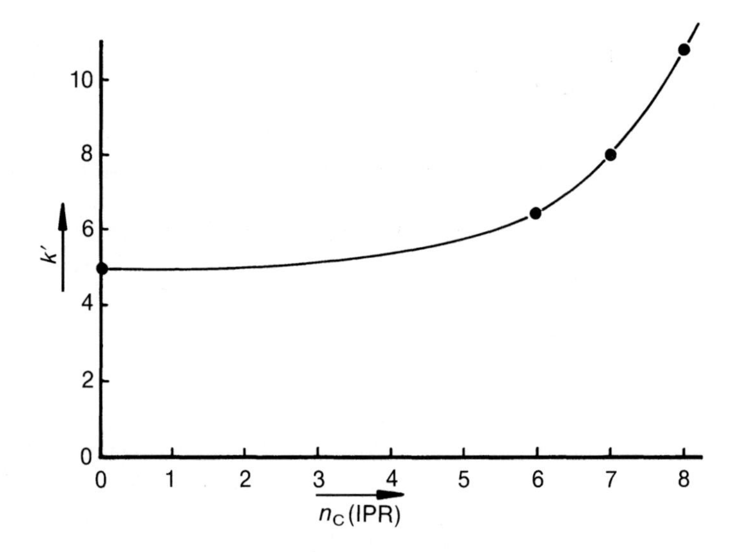

Fig. 5-3. Dependence of the diphenhydramine k' value on the chain length n_c of the ion-pair reagent. – Separator column: IonPac NS1 (10 μm); eluent: 0.05 mol/L KH_2PO_4 (pH 4.0) / acetonitrile (70:30 v/v) + 0.005 mol/L IPR; flow rate: 1 mL/min; detection: UV (220 nm); injection volume: 50 μL; solute concentration: 40 mg/L diphenhydramine hydrochloride.

with increasing chain length of the alkanesulfonates that were used as the ion-pair reagent.

Another method for controlling retention is by varying the concentration of the ion-pair reagent. Fig. 5-4 shows the dependence of the k' value on the concentration of the

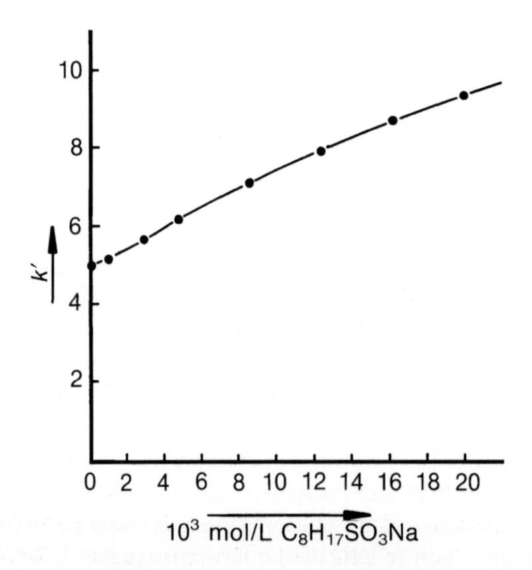

Fig. 5-4. Dependence of the diphenhydramine k' value on the ion-pair reagent concentration. – Separator column: IonPac NS1 (10 μm); eluent: 0.05 mol/L KH_2PO_4 (pH 4.0) / acetonitrile (70:30 v/v) + $C_8H_{17}SO_3Na$; flow rate: 1 mL/min; all other chromatographic conditions: see Fig. 5.3.

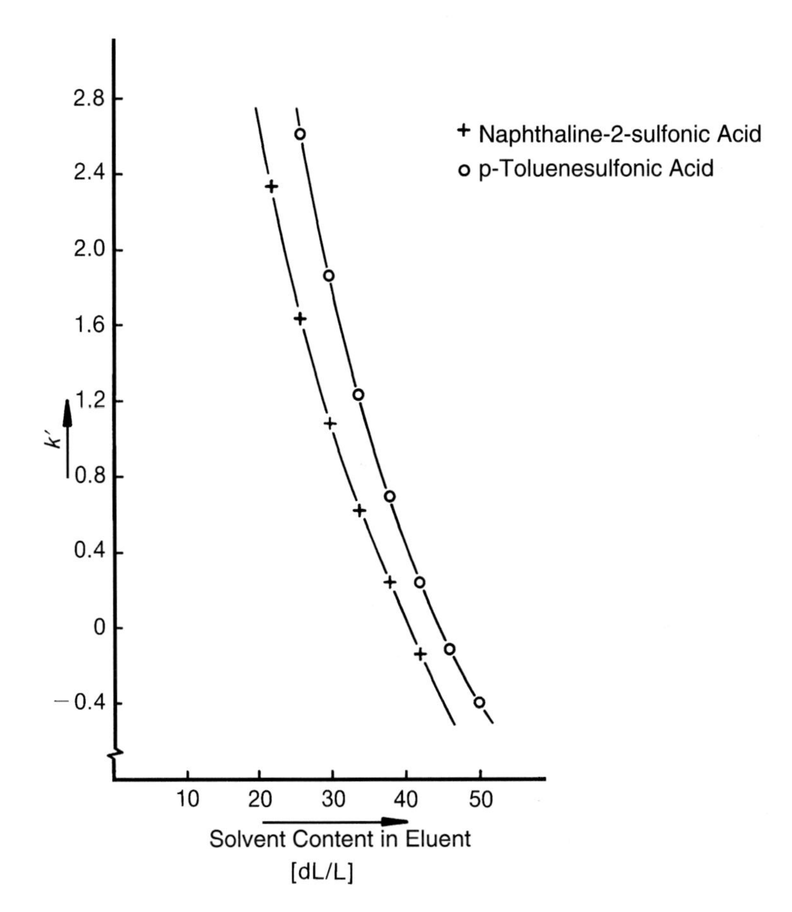

Fig. 5-5. Dependence of the retention of aromatic sulfonic acids on the organic solvent content in the mobile phase. – Separator column: IonPac NS1 (10 μm); eluent: 0.002 mol/L tetrabutylammonium hydroxide/acetonitrile; flow rate: 1 mL/min; detection: UV (254 nm); injection volume: 50 μL; solute concentrations: 40 mg/L each of toluene-*p*-sulfonic acid monohydrate and naphthaline-2-sulfonic acid.

ion-pair reagent, again using diphenhydramine as an example. A DVB resin was employed as the stationary phase, the eluent was comprised of a mixture of acetonitrile and a phosphate buffer with sodium octanesulfonate as the ion-pair reagent. As seen in Fig. 5-4, in the investigated concentration range between 0 and 0.02 mol/L the k' value depends almost linearly on the concentration. For some compounds such curves may pass a maximum [23]. The concentration of lipophilic ions to be used is limited by two factors:

- With increasing reagent concentration the surface of the stationary phase is increasingly blocked by undissociated molecules of the ion-pair reagent or by ion pairs from the ion-pair reagent and buffer ions.
- When applying conductivity detection, the reagent concentration is limited by suppression capacity.

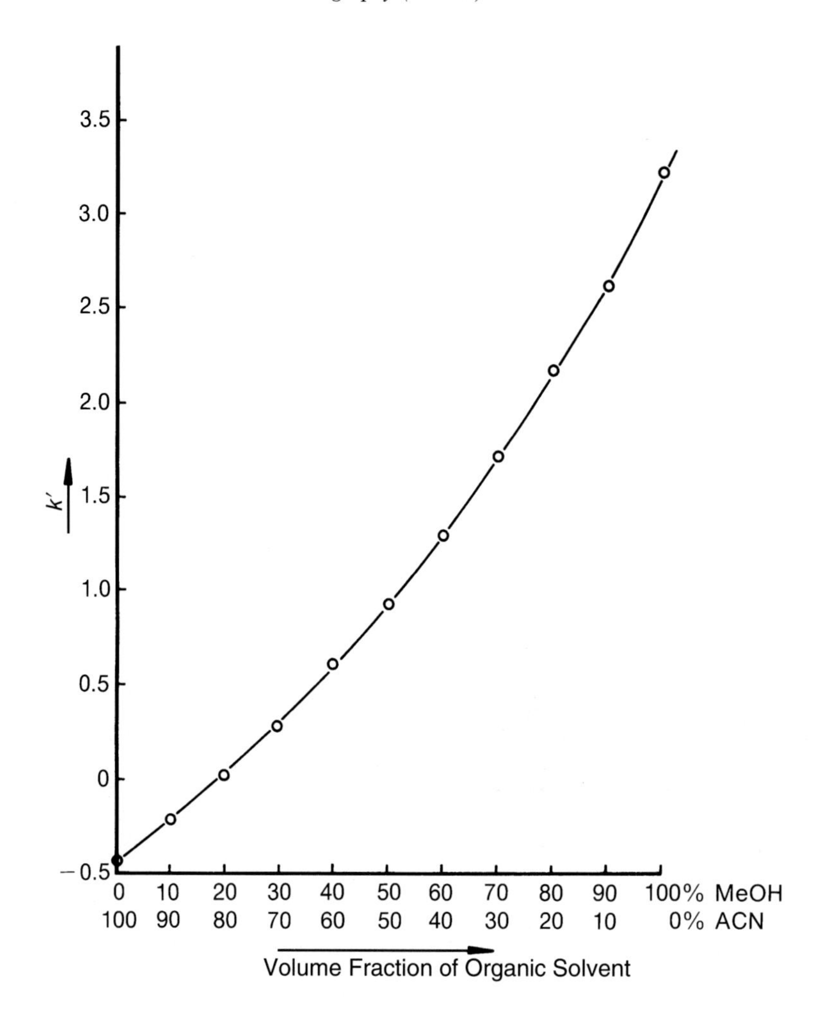

Fig. 5-6. Dependence of the naphthalene-2-sulfonic acid retention on the percentage distribution between methanol and acetonitrile in the volume fraction of the organic modifier. – Separator column: IonPac NS1 (10 μm); eluent: 0.002 mol/L TBAOH – acetonitrile/methanol (50:50 v/v); flow rate: 1 mL/min; detection: UV (254 nm).

Therefore, the compounds listed in Table 5-1 are employed in a concentration range between $5 \cdot 10^{-4}$ and 10^{-2} mol/L. In this range, large changes in the retention values with dependency on the ion-pair reagent concentration occur. In many cases, $c = 0.002$ mol/L has proved to be an appropriate reagent concentration.

5.3.2 Type and Concentration of the Organic Modifier

In analogy to reversed-phase chromatography, organic solvents such as acetonitrile or methanol are added to the aqueous mobile phase as organic modifiers. Because the solvent molecules are adsorbed at the surface of the stationary phase, they are in a

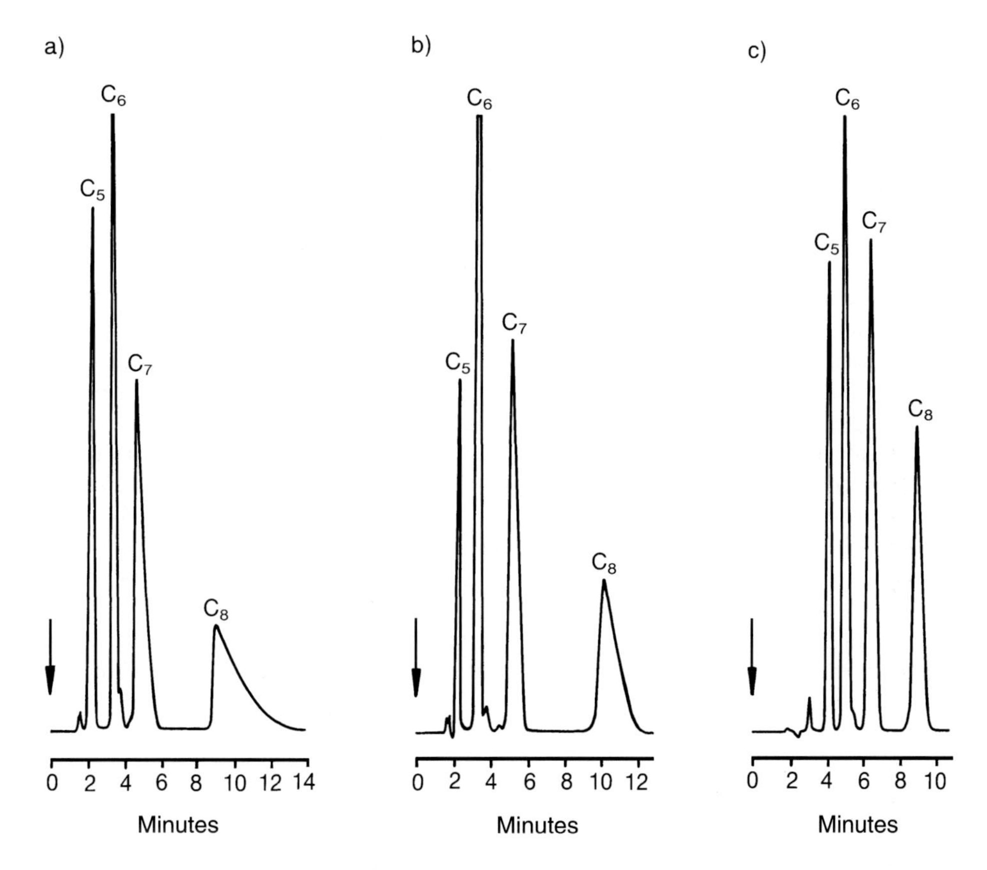

Fig. 5-7. Dependence of the separation of alkylsulfonates with medium chain lengths on the type of lipophilic ion and the concentration of organic modifier. – Separator column: IonPac NS1 (10 µm); eluent: a) 0.002 mol/L NH_4OH/acetonitrile (85:15 v/v), b) 0.002 mol/L TMAOH/acetonitrile (82:18 v/v), c) 0.002 mol/L TBAOH/acetonitrile (63:37 v/v).

competing equilibrium with lipophilic ions for active centers on the stationary phase available for adsorption. To illustrate this effect, the retention of two aromatic sulfonic acids – toluene-*p*-sulfonic acid and naphthalene-2-sulfonic acid – dependent on the water content in the mobile phase was investigated. In this experiment, an acetonitrile/water mixture was used to which tetrabutylammonium hydroxide (TBAOH) was added as the ion-pair reagent. Plotting ln k' versus the acetonitrile content in the mobile phase yields the parabolic dependence shown in Fig. 5-5.

When replacing acetonitrile by methanol, the solvent content in the mobile phase must be enhanced to obtain comparable retention times. This effect is illustrated by using the separation of naphthalene-2-sulfonic acid as an example. For its elution, aqueous TBAOH solution with a volume fraction of 500 mL/L acetonitrile as the organic modifier was selected. When this fraction is gradually replaced by methanol, a significant retention increase results. This is graphically depicted in Fig. 5-6 as a function of the percentage distribution between methanol and acetonitrile in the volume fraction of the organic solvent. The selectivity difference obtained with methanol as the organic modifier is due to the capability of methanol to form hydrogen bonds. Compared to acetonitrile, the higher viscosity of methanol is a disadvantage because it leads to a larger pressure drop along the column.

From the dependences shown in the Figs. 5-3, 5-4, and 5-5, it is obvious that the majority of analyte ions may be separated under different chromatographic conditions; i.e., with different ion-pair reagents and solvent contents. This is impressively demonstrated by the analysis of alkanesulfonates with medium chain lengths (C_5 to C_8). As these compounds have a hydrophobic character, the hydrophilic ammonium hydroxide suggests itself as the ion-pair reagent. The respective chromatogram is shown in Fig. 5-7a. It was obtained under isocratic conditions with an acetonitrile content of 150 mL/L in the mobile phase. All compounds are separated to baseline, but the heptane- and octanesulfonic acid peaks are characterized by a strong tailing. Changing to an ion-pair reagent of a higher hydrophobicity such as tetramethylammonium hydroxide leads to an enhanced resolution between the different signals. The retention time increase incident to it may be compensated, as shown in Fig. 5-7b, by raising the acetonitrile content in the eluent. The result is a chromatogram with comparable analysis times but distinctly reduced tailing of the two late-eluting components. This effect can be heightened by selecting tetrabutylammonium hydroxide as the ion-pair reagent. Fig. 5-7c shows the optimized separation of these compounds with regard to analysis time, resolution, and peak shape.

This example corroborates the above statement that analyte compounds can be separated by several methods. In many cases, the sample matrix will determine the choice of the chromatographic conditions.

5.3.3 Inorganic Additives

Inorganic additives such as sodium carbonate are added to the mobile phase to control the retention of di- and multivalent anions and to improve their peak shape. The sodium carbonate effect on retention is most simply illustrated with the analysis of inorganic anions as an example. Note the chromatogram in Fig. 5-8a which depicts the separation of fluoride, chloride, bromide, nitrate, and sulfate. It was obtained using tetrabutylammonium hydroxide as the ion-pair reagent and a small amount of acetonitrile as the organic modifier. Under these chromatographic conditions, the monovalent anions are separated to baseline and elute within 10 minutes. For the divalent sulfate, on the other hand, an extremely long retention time and a pronounced peak broadening is observed. Orthophosphate, which also exists as a divalent anion under these conditions, is even more strongly retained. After adding 0.001 mol/L sodium carbonate to this eluent, the chromatogram depicted in Fig. 5-8b is obtained. It reveals that the retention decrease is much higher for divalent anions than for monovalent ones, which avoids the unnecessarily high resolution between anions of different valency.

The addition of even minute amounts of sodium carbonate has a particularly strong effect on the retention behavior of multivalent anions. These comprise, for example, the two iron cyanide complexes $Fe(CN)_6^{3-}$ and $Fe(CN)_6^{4-}$, whose separation is obtained with an eluent containing only $3 \cdot 10^{-4}$ mol/L sodium carbonate (see Fig. 5-9), apart from tetrabutylammonium hydroxide and acetonitrile. Lowering the acetonitrile content in favor of sodium carbonate, the resolution between both signals will decrease drastically, although the peak shape of the iron(II) complex will be distinctly improved.

The effect of sodium carbonate as an inorganic additive is mechanistically not completely clear. According to the dynamic ion-exchange model, it is to be assumed that carbonate ions are found in a competing equilibrium with solute ions for the exchange groups that are adsorbed at the surface of the stationary phase. This is a plausible explanation for the strong effect of carbonate on the retention of divalent species.

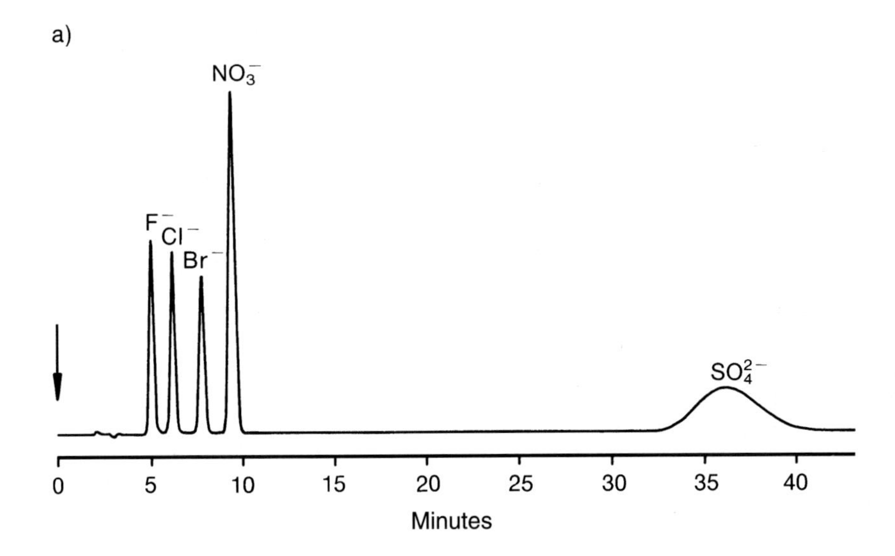

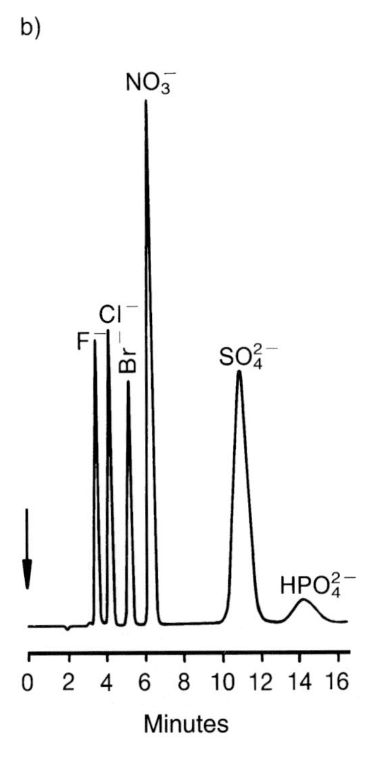

Fig. 5-8. Illustration of the sodium carbonate influence on retention exemplified with an inorganic anion separation. – Separator column: IonPac NS1 (10 μm); eluent: a) 0.002 mol/L TBAOH/acetonitrile (82:18 v/v), b) 0.002 mol/L TBAOH + 0.001 mol/L Na_2CO_3/acetonitrile (82:18 v/v); flow rate: 1 mL/min; detection: suppressed conductivity; injection volume: 50 μL; solute concentrations: 3 ppm fluoride, 4 ppm chloride, 10 ppm bromide, 20 ppm nitrate, 25 ppm sulfate, and 10 ppm orthophosphate.

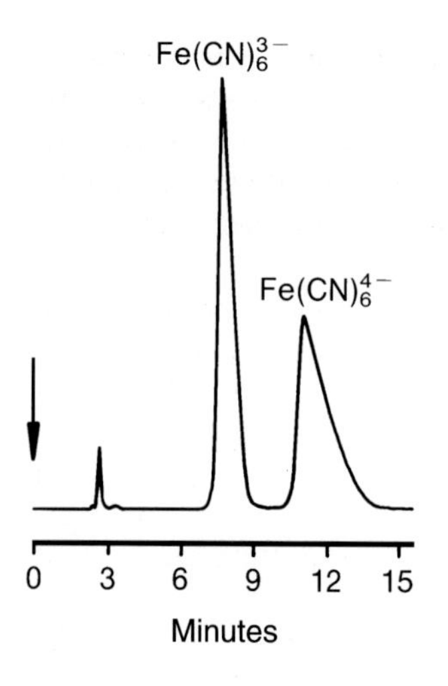

Fig. 5-9. Separation of the two iron cyanide complexes. – Separator column: IonPac NS1 (10 μm); eluent: 0.002 mol/L TBAOH + 0.0003 mol/L Na_2CO_3/acetonitrile (60:40 v/v); flow rate: 1 mL/min; detection: suppressed conductivity; injection volume: 50 μL; solute concentrations: 40 ppm each of $Fe(CN)_6^{3-}$ and $Fe(CN)_6^{4-}$.

5.3.4 pH Effects and Temperature Influence

For the analysis of multivalent ions it is often necessary to change the pH value of the mobile phase by adding appropriate acids or bases. As the retention of multivalent ions increases with the degree of dissociation, the pH value affects retention by determining the degree of dissociation. This effect is illustrated by the analysis of thiogly-

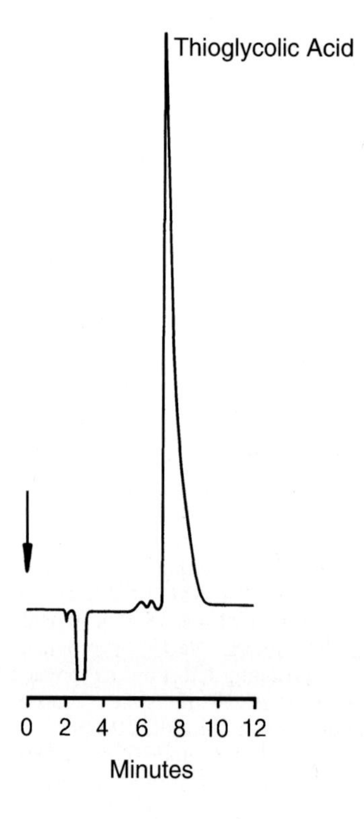

Fig. 5-10. Separation of thioglycolic acid. – Separator column: IonPac NS1 (10 μm); eluent: 0.002 mol/L TBAOH/acetonitrile (80:20 v/v), pH 7.25 with H_3BO_3; flow rate: 1 mL/min; detection: suppressed conductivity; injection volume: 50 μL; solute concentration: 50 ppm.

colic acid. This compound can be separated by adding tetrabutylammonium hydroxide as the ion-pair reagent and an appropriate amount of acetonitrile to the mobile phase. When detecting thioglycolic acid by suppressed conductivity detection, no chromatogram is obtained under the above-mentioned chromatographic conditions (pH 10.8). However, if the eluent pH is lowered to 7.25 by adding crystalline boric acid, the chromatogram in Fig. 5-10 results.

Boric acid is especially suitable for lowering the pH value. Because of its low degree of dissociation, it contributes only marginally to the increase in background conductance. In general, an increase in the eluent pH is accomplished with sodium hydroxide, because it suppresses to water and hardly affects the background conductance.

Apart from controlling the degree of dissociation of the analyte species, eluent pH changes are necessary to avoid unwanted side reactions in an acidic or alkaline medium. This applies, for example, to mercaptans which may react to form disulfides in alkaline medium.

In contrast to the eluent pH value, the column temperature is seldom relevant for optimizing the separation. Retention can be somewhat reduced by raising the column temperature. Generally speaking, the viscosity of the mobile phase will be reduced and the chromatographic efficiency will be increased when the column temperature is raised. For mechanistic investigations, however, a variation in column temperature offers the possibility to determine the temperature dependence of the retention and to derive important thermodynamic quantities such as the sorption enthalpies (see also Section 3.2).

5.4 Analysis of Surface-Inactive Ions

In the field of *anion analysis*, ion-pair chromatography is a significant alternative to ion-exchange chromatography. If a co-elution of two anions is suspected with one of the two methods, it is often possible to solve this separation problem with the other method, as two different compounds rarely show the same retention behavior under completely different chromatographic conditions.

A typical example is the analysis of nitrate and chlorate. With the exception of the IonPac AS9 (Dionex Corp.), these ions cannot be separated on conventional anion exchangers, making misinterpretation of the chromatogram possible. The only way to distinguish between nitrate and chlorate is to make use of their different absorption characteristics: chlorate is UV-transparent while nitrate can be detected at a wavelength of 215 nm. If both species are present in solution, determinations will only be possible via differential detection.

On the other hand, with ion-pair chromatography nitrate and chlorate are resolved using tetrabutylammonium hydroxide as the ion-pair reagent, with nitrate eluting prior to chlorate (Fig. 5-11). Pursuing the hypothesis that the stationary phase is transformed into a dynamic ion exchanger by adding lipophilic ions, this different selectivity compared to conventional ion exchangers seems to be related to the different type of exchange selectivity provided by the lipophilic ions.

Another example is the analysis of benzoic acid, which elutes on a conventional anion exchanger directly after nitrite. Interferences in the determination of this compound

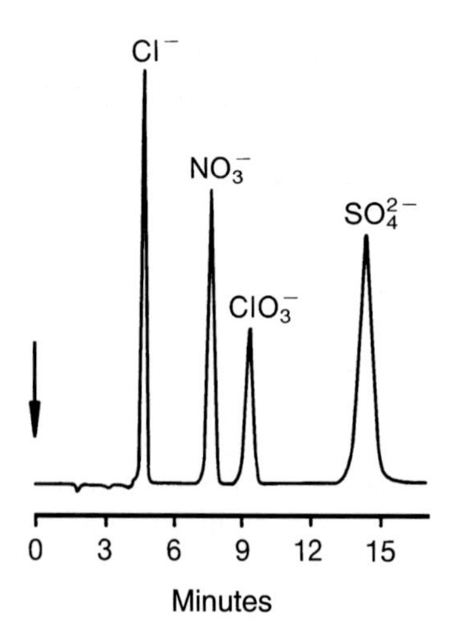

Fig. 5-11. Ion-pair chromatographic separation of nitrate and chlorate. – Separator column: IonPac NS1 (10 μm); eluent: 0.002 mol/L TBAOH + 0.001 mol/L Na_2CO_3 / acetonitrile (85:15 v/v); flow rate: 1 mL/min; detection: suppressed conductivity; injection volume: 50 μL; solute concentrations: 5 ppm chloride, 10 ppm nitrate, 10 ppm chlorate, and 15 ppm sulfate.

caused by a high electrolyte content in the sample are not to be ruled out unless the much more selective UV detection is applied. With ion-pair chromatography, benzoic acid exhibits a completely different retention behavior. This is due to π-π interactions between the aromatic benzoic acid ring and the aromatic skeleton of the stationary phase, so that benzoic acid is much more strongly retained. Under the chromatographic conditions given in Fig. 5-12, it elutes long after the mineral acids.

Although tetraalkylammonium salts, suitable as ion-pair reagents for ion-pair chromatography of simple inorganic anions, are predominantly employed in combination with polymer phases [24,25], they may equally be used when applying chemically bonded reversed phases [16,26]. The separation of inorganic anions on LiChrosorb RP18 shown in Fig. 5-13 serves as an example. In contrast to polymer phases, the fluoride determination is a problem because fluoride is poorly retained on chemically bonded reversed phases. If direct conductivity detection is applied, a quantitative evaluation of the fluoride signal is difficult.

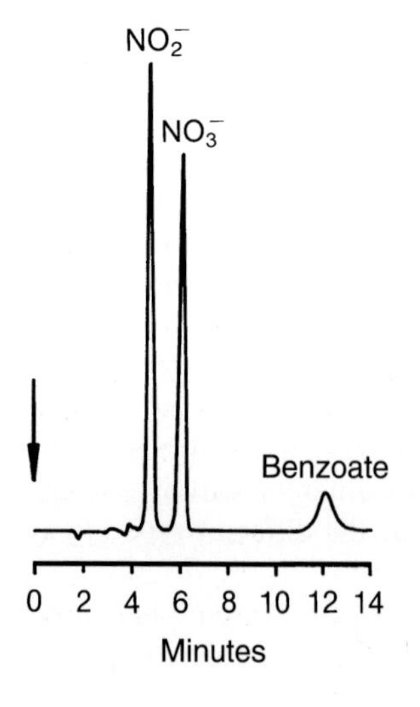

Fig. 5-12. Ion-pair chromatographic separation of nitrite, nitrate, and benzoate. – Separator column: IonPac NS1 (10 μm); eluent: 0.002 mol/L TBAOH + 0.001 mol/L Na_2CO_3 / acetonitrile (82:18 v/v); flow rate: 1 mL/min; detection: suppressed conductivity; injection volume: 50 μL; solute concentrations: 10 ppm nitrite, 10 ppm nitrate, and 20 ppm benzoate.

If the counter ion of the ion-pair reagent exhibits suitable absorption characteristics, indirect photometric detection is feasible. Based on a 1982 paper by Dreux et al. [27], Bidlingmeyer et al. [28] developed a procedure for separating inorganic anions with tetrabutylammounium salicylate as the eluent. This method is suitable for the simultaneous application of indirect photometric and direct conductivity detection. At a measuring wavelength of 288 nm, the authors obtained the chromatogram depicted in Fig. 5-14 using a radially compressed ODS cartridge as the stationary phase.

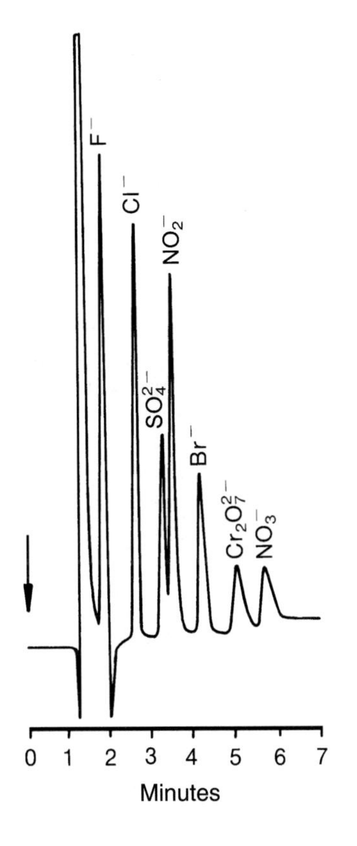

Fig. 5-13. Ion-pair chromatographic separation of inorganic anions on a chemically bonded reversed phase. – Separator column: LiChrosorb RP 18 (10 μm); eluent: 0.002 mol/L TBAOH + 0.05 mol/L phosphate buffer (pH 6.7); flow rate: 2 mL/min; detection: direct conductivity; injection volume: 20 μL; solute concentrations: 1000 ppm each of fluoride, chloride, sulfate, nitrite, bromide, dichromate, and nitrate; (taken from [26]).

Furthermore, ion-pair chromatographic separations of inorganic anions are performed on chemically bonded cyanopropyl phases [17,29]. Fig. 5-15 shows a typical result for a separation on such a stationary phase.

To analyze bromide and nitrate in foods, Leuenberger et al. [30] used a chemically bonded aminopropyl phase with a phosphate buffer eluent. Cortes [31] applied this method successfully to the separation of additional inorganic anions. The chromatogram depicted in Fig. 5-16 confirms the extraordinary selectivity of ion-pair chromatography for monovalent anions. A final assessment of the applicability of cyano- and aminopropyl phases for the separation of inorganic anions is not presently feasible, as only UV absorbing species have been investigated so far. Although such stationary phases exhibit the expected high chromatographic efficiency, they have a limited pH stability because of their silica structure.

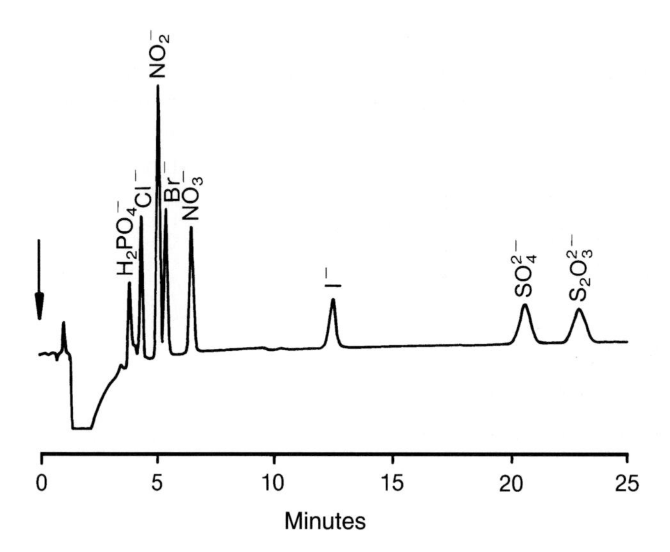

Fig. 5-14. Ion-pair chromatographic separation of inorganic anions with indirect photometric detection. – Separator column: Waters C_{18} Radial-PAK (5 µm); eluent: 0.0004 mol/L tetrabutylammonium salicylate (pH 4.62); flow rate: 2 mL/min; detection: UV (288 nm, indirect); injection volume: 50 µL; solute concentrations: 4 ppm orthophosphate, 2 ppm chloride, 4 ppm nitrite, bromide, nitrate and iodide, 6 ppm sulfate, and 6 ppm thiosulfate; (taken from [28]).

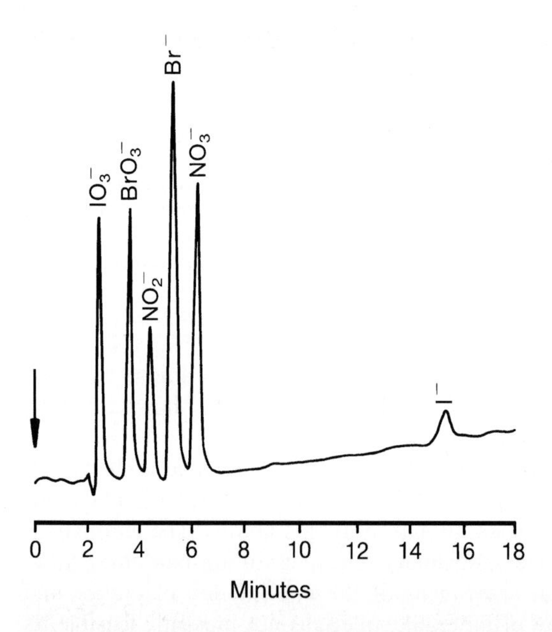

Fig. 5-15. Ion-pair chromatographic separation of inorganic anions on a chemically bonded cyanopropyl phase. – Separator column: Polygosil-60-D-10 CN; eluent: 0.1 mol/L Na_2HPO_4 + 0.1 mol/L KH_2PO_4 + 1 g/kg cetyltrimethylammonium chloride / acetonitrile (75:25 v/v); flow rate: 1.5 mL/min; detection: UV (205 nm); injection: not given; (taken from [29]).

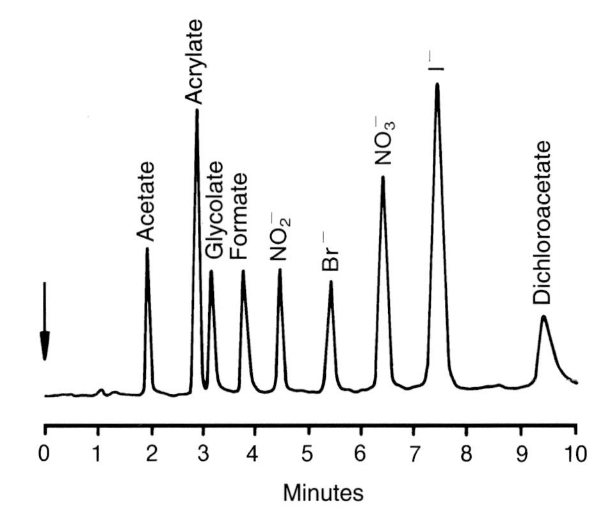

Fig. 5-16. Separation of inorganic anions on a chemically bonded aminopropyl phase. − Separator column: Zorbax NH_2; eluent: 0.03 mol/L H_3PO_4 (pH 3.2 with NaOH); flow rate: 2 mL/min; detection: UV (205 nm); injection volume: 20 µL; solute concentrations: 25 to 100 ppm; (taken from [31]).

All the aforementioned examples indicate the significance that ion-pair chromatography has gained in solving certain separation problems, even for the separation of simple inorganic anions.

Ion-pair chromatography assumes a special position in the analysis of strong polarizable anions. On conventional anion exchangers, they can only be eluted with strong eluents. The class of polarizable anions comprises perchlorate and citrate, oxidic sulfur anions, and metal complexes. For the ion-pair chromatographic analysis of these compounds, it generally suffices to increase the acetonitrile content in the mobile phase.

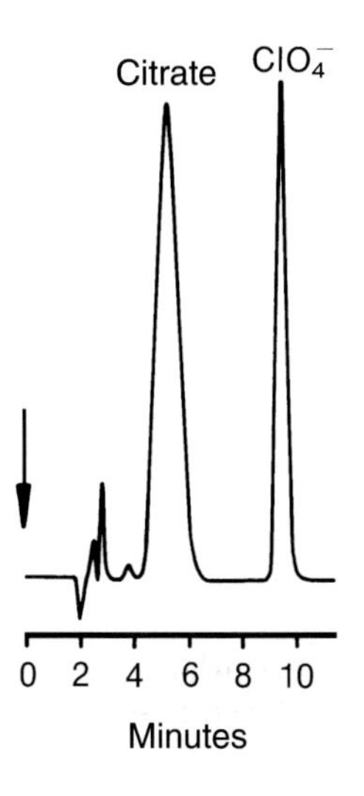

Fig. 5-17. Analysis of citrate and perchlorate. − Separator column: IonPac NS1 (10 µm); eluent: 0.002 mol/L TBAOH + 0.001 mol/L Na_2CO_3 / acetonitrile (66:34 v/v); flow rate: 1 mL/min; detection: suppressed conductivity; injection volume: 50 µL; solute concentrations: 50 ppm citrate and 10 ppm perchlorate.

This is indicated in Fig. 5-17, which shows the perchlorate separation using TBAOH as the ion-pair reagent and a comparably high solvent content of 340 mL/L. Remarkably, the trivalent citrate elutes prior to the monovalent perchlorate. The broader peak width of the citrate signal is caused by the higher valency of this compound.

In the field of inorganic sulfur compounds, ion-pair chromatography is applied to the analysis of dithionate [32], $S_2O_6{}^{2-}$, peroxodisulfate [33], $S_2O_8{}^{2-}$, and polythionates [34,35], $S_nO_6{}^{2-}$. In all cases, TBAOH is suited as the ion-pair reagent. Because the above-listed compounds are divalent, retention may be reduced with sodium carbonate. Another parameter affecting retention is the acetonitrile content in the mobile phase. The chromatogram in Fig. 5-18 shows the separation of sulfate, dithionate, and tetrathionate. Under these chromatographic conditions, peroxodisulfate elutes shortly before tetrathionate. In contrast, disulfite, $S_2O_5{}^{2-}$, and disulfate, $S_2O_7{}^{2-}$, decompose in aqueous solution according to

$$S_2O_5{}^{2-} + H_2O \rightleftharpoons 2\,HSO_3{}^- \tag{154}$$

$$2\,S_2O_7{}^{2-} \rightleftharpoons 2\,SO_4{}^{2-} + SO_3 \tag{155}$$

and thus cannot be analyzed via liquid chromatography. For analyzing polythionates, $S_nO_6{}^{2-}$ ($n > 2$), the acetonitrile content in the mobile phase must also be enhanced. A separation of higher polythionates ($n_S = 5$ to 11) is shown in Fig. 5-19, which was obtained under isocratic conditions by Steudel et al. [35]. Polythionates are sulfur chains bearing terminal sulfonate groups, thus measuring the light absorption at 254 nm may be applied as the detection method. Plotting the ln k' values of the investigated polythionates versus the sulfur atom number yields the parabolic dependence represented in Fig. 5-20. It is obvious that applying the gradient technique will allow higher polythionates to be detected with this method.

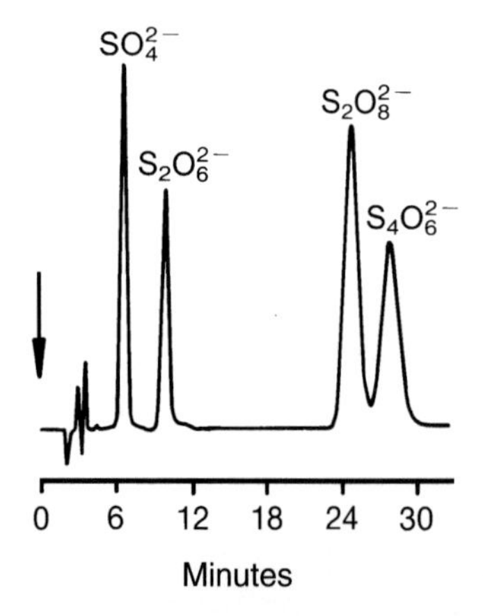

Fig. 5-18. Ion-pair chromatographic separation of sulfate, dithionate, peroxodisulfate, and tetrathionate. – Separator column: IonPac NS1 (10 μm); eluent: 0.002 mol/L TBAOH + 0.001 mol/L Na_2CO_3 / acetonitrile (77:23 v/v); flow rate: 1 mL/min; detection: suppressed conductivity; injection volume: 50 μL; solute concentrations: 5 ppm sulfate, 10 ppm dithionate, 20 ppm peroxodisulfate, and 20 ppm tetrathionate.

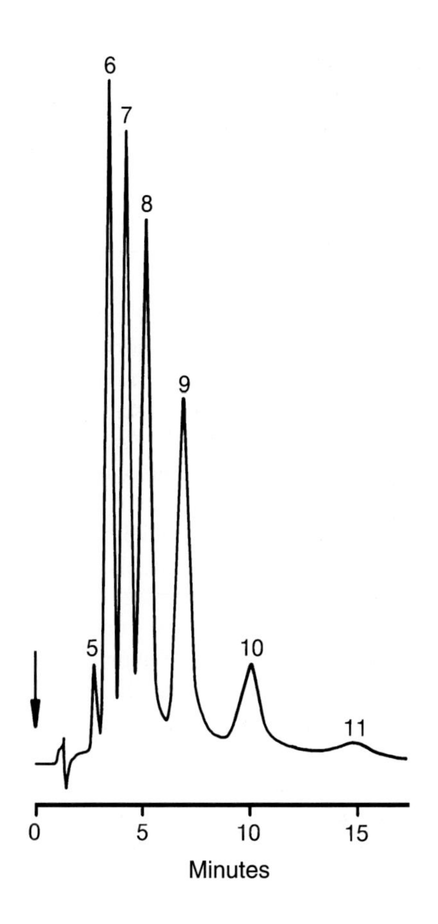

Fig. 5-19. Ion-pair chromatographic separation of higher polythionates, $S_nO_6^{2-}$ (n = 5 to 11). – Separator column: IonPac NS1 (10 µm); eluent: 0.002 mol/L TBAOH + 0.001 mol/L Na_2CO_3 / acetonitrile (60:40 v/v); flow rate: 1 mL/min; detection: UV (254 nm); (taken from [35]).

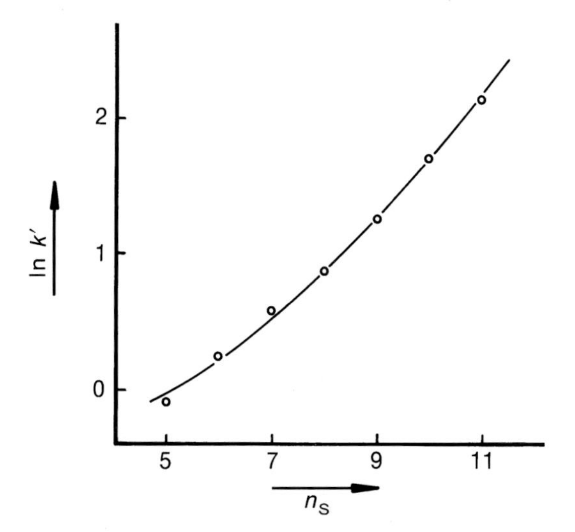

Fig. 5-20. Dependence of ln k' values for polythionates, $S_nO_6^{2-}$ (n = 5 to 11) on the sulfur atom number. – For chromatographic conditions see Fig. 5-19; (taken from [35]).

Ion-pair chromatography is also suited for the analysis of metal complexes. For their chromatographic separation, the complexes must be thermodynamically *and* kinetically stable. This means that complex formation must be thermodynamically possible and furthermore an irreversible process. Metal-ETDA and metal-DTPA complexes exhibit a corresponding high stability. To separate the Gd-DTPA complex (Fig. 5-21), which is of great relevance in the pharmaceutical industry, TBAOH was used as the ion-pair reagent [36]. Detection was carried out by measuring the electrical conductivity in combination with a suppressor system.

In the electroplating industry, applications have been specifically developed to analyze metal-cyanide complexes by ion-pair chromatography, which is interesting because the oxidation state of the metal can be determined via its complexation with cyanide. The two iron cyanide complexes, for example, are eluted in the order of increasing charge number. (See Fig. 5-9 in Section 5.3.3 for the corresponding chromatogram.) Apart from TBAOH as the ion-pair reagent, the addition of small amounts of sodium carbonate is of crucial importance. In contrast, for monovalent gold complexes, $Au(CN)_2^-$ and $Au(CN)_4^-$, the differing coordinaton number and, thus, the different spatial arrangement, is responsible for the separation of both complexes (Fig. 5-22). While the cyano-complexes of iron, cobalt, and gold exhibit a very high stability and therefore exist as distinct anions, the cyano-complexes of nickel, copper, and silver are not completely detected with the chromatographic analysis because of the low formation constants of these complexes, which lead to a slight dissociation into metal and ligand. By adding small amounts of potassium cyanide to the mobile phase, such an equilibrium is shifted to the complex side rendering it chromatographable. Fig. 5-23 illustrates a chromatogram of kinetically stable and unstable metal-cyanide complexes. Be advised, however, that the suppression product of an eluent containing potassium cyanide is HCN, and that the effluent from the chromatographic system should be collected in a waste container containing a strongly basic solution for safety.

Fig. 5-21. Separation of Gd-DTPA. – Separator column: IonPac NS1 (10 μm); eluent: 0.002 mol/L TBAOH + 0.001 mol/L Na_2CO_3 / acetonitrile (75:25 v/v); flow rate: 1 mL/min; detection: suppressed conductivity; injection: 50 μL sample (1:1000 diluted).

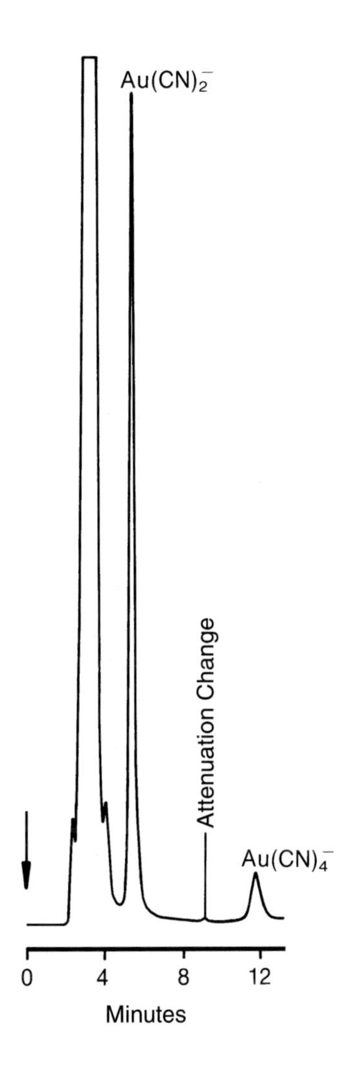

Fig. 5-22. Separation of gold(I) and gold(III) as cyanide complexes. – Separator column: IonPac NS1 (10 µm); eluent: 0.002 mol/L TBAOH + 0.002 mol/L Na_2CO_3 / acetonitrile (60:40 v/v); flow rate: 1 mL/min; detection: suppressed conductivity.

The analysis of thio- and selenometalates [37] provides an interesting application of ion-pair chromatography. These compounds play a significant role in some bioinorganic, nutritional physiological, and veterinary medical problems. Thio- and selenometalate ions are formed from electron-deficient transition metals such as vanadium, niobium, tantalum, molybdenum, tungsten, and rhenium in their highest oxidation states. Multimetal complexes having remarkable electronic properties can be obtained. Furthermore, it is noteworthy that poly(thiometalates) with mixed valencies are formed from thiometalates by novel redox processes with simultaneous condensation.

The starting compounds, tungstate and molybdate, can be separated by anion exchange chromatography (see Section 3.3.4.4). Substituting oxygen atoms with sulfur or selenium, however, causes such an increase in affinity toward the stationary phase, that even monothio- or monoseleno metalates are no longer eluted within an acceptable time using the usual carbonate/bicarbonate mixtures. As illustrated in Fig. 5-24 with the thiomolybdates $MoO_n\,S_{4-n}^{2-}$ (n = 0 to 4), all members of this series may be separated in a single run via ion-pair chromatography under isocratic conditions. The parabolic

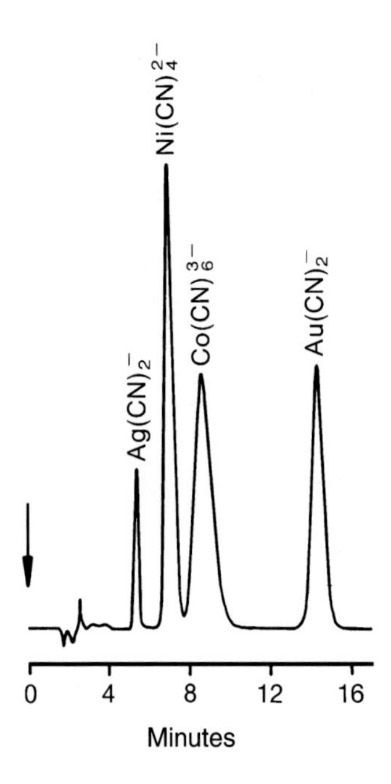

Fig. 5-23. Separation of kinetically stable and unstable metal-cyanide complexes. – Separator column: IonPac NS1 (10 µm); eluent: 0.002 mol/L TBAOH + 0.001 mol/L Na_2CO_3 + $2 \cdot 10^{-4}$ mol/L KCN / acetonitrile (70:30 v/v); flow rate: 1 mL/min; detection: suppressed conductivity; injection volume: 50 µL; solute concentrations: 80 ppm $KAg(CN)_2$, 40 ppm $K_2Ni(CN)_4$, 40 ppm $K_3Co(CN)_6$, and 80 ppm $KAu(CN)_2$; (taken from [36]).

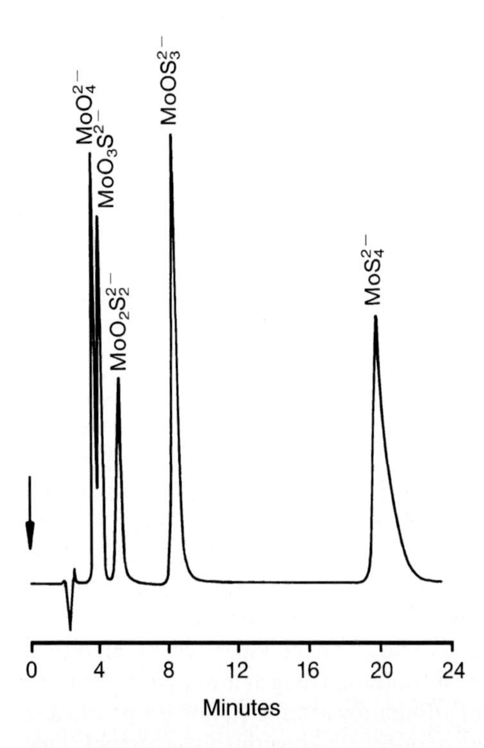

Fig. 5-24. Separation of molybdate and its thia-substituted derivatives. – Separator column: Ion Pac NS1 (10 µm); eluent: 0.002 mol/L TBAOH + 0.001 mol/L Na_2CO_3 / acetonitrile (75:25 v/v); flow rate: 1 mL/min; detection: suppressed conductivity; injection volume: 50 µL; solute concentrations: 50 mg/L $(NH_4)_2MoO_4$, 200 mg/L $(NH_4)_2MoOS_3$, and $(NH_4)_2MoS_4$; (taken from [37]).

retention increase observed with increasing substitution of oxygen atoms by sulfur or selenium was explained by Weiß et al. [37] on the basis of the retention model proposed by Bidlingmeyer et al. [13].

When the acetonitrile content in the mobile phase is further increased, the poly(thiometalate), $[Mo_2{}^VO_2S_2(S_2)_2]^{2-}$, is also accessible for analysis. This binuclear complex with η-S_2^{2-} ligands is formed from dithiomolybdate by intramolecular redox processes. Fig. 5-25 shows the chromatogram of this compound's tetramethylammonium salt. Obviously, molybdenum sulfur cluster can also be detected under the same chromatographic conditions. To these belong the binuclear cluster, $[Mo_2(S_2)_6]^{2-}$, in which molybdenum is only surrounded by disulfido ligands, as well as the later eluting trinuclear cluster, $[Mo_3S(S_2)_6]^{2-}$, that functions as a model compound for crystalline molybdenum(IV) sulfide. (MoS_2 catalysts are utilized in desulfurization of mineral oil). This is an impressive example of the potential of ion-pair chromatography to separate and detect inorganic anions with complex structures.

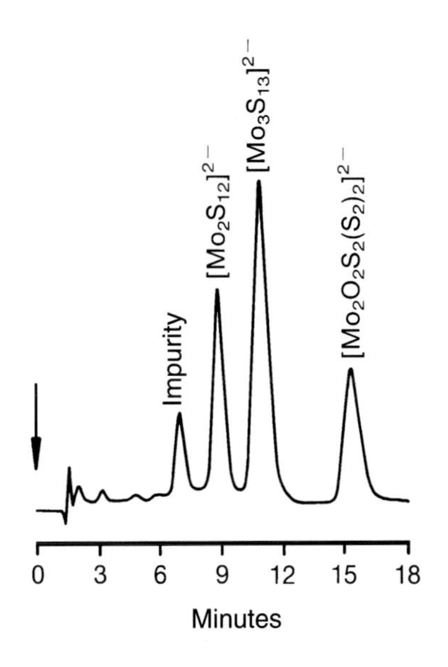

Fig. 5-25. Separation of various molybdenum disulfido complexes. – Separator column: IonPac NS1 (10 μm); eluent: 0.002 mol/L TBAOH + 0.001 mol/l Na_2CO_3 / acetonitrile (50:50 v/v); flow rate: 1 mL/min; detection: UV (254 nm); injection volume: 50 μL; solute concentrations: 50 mg/L $(NH_4)_2[Mo_2S_{12}]$, 50 mg/L $[(CH_3)_4N]_2[Mo_2O_2S_2(S_2)_2]$, and 100 mg/L $(NH_4)_2[Mo_3S_{13}]$; (taken from [37]).

In the field of *cation analysis*, ion-pair chromatography is the preferred method for the separation of amines of all types. While short-chain aliphatic amines (C_1 to C_3) and some smaller aromatic amines [39] can also be separated on surface-sulfonated cation exchangers, ion-pair chromatographic applications have been developed for the separation of structurally isomeric amines, alkanolamines, quaternary ammonium compounds, arylalkylamines, barbiturates, and alkaloids.

Fig. 5-26 shows the separation of mono-, di-, and triethylamine, accomplished by using octanesulfonic acid as the ion-pair reagent. The less-hydrophobic hexanesulfonic acid is used in combination with boric acid as the eluent for the separation of ethanolamines, as shown in Fig. 5-27. These compounds are detected by measuring the electrical conductance, thus the background conductance is generally lowered with a membrane suppressor. The addition of boric acid to both the eluent and the regenerent serves to enhance the sensitivity for di- and triethanolamine.

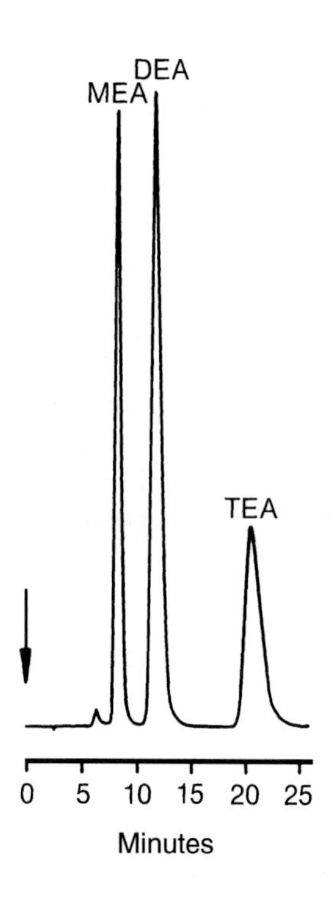

Fig. 5-26. Separation of mono-, di-, and triethylamine. – Separator column: IonPac NS1 (10 μm); eluent: 0.002 mol/L octanesulfonic acid / acetonitrile (92:8 v/v); flow rate: 1 mL/min; detection: suppressed conductivity; injection volume: 50 μL; solute concentrations: 50 ppm MEA, 100 ppm DEA and TEA.

Quaternary ammonium compounds can be analyzed under similar chromatographic conditions, but without adding boric acid. This includes choline (2-hydroxyethyl-trimethylammonium hydroxide), which is commonly found in fauna and flora as a basic constituent of lecithin-type phospholipids. The ion-pair chromatographic separation of choline requires an eluent that, in addition to hexanesulfonic acid as the ion-pair re-

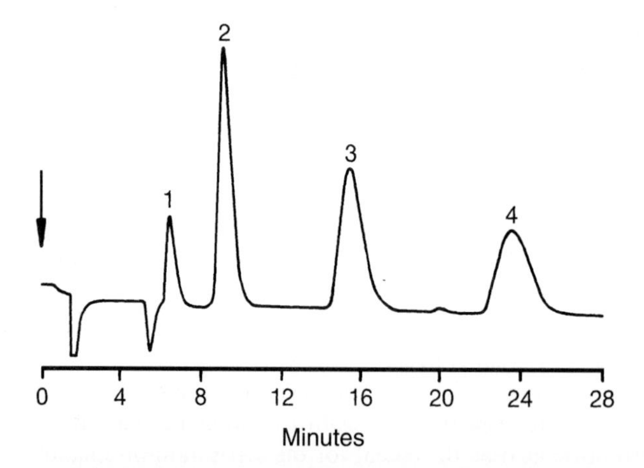

Fig. 5-27. Separation of ethanolamines. – Separator column: IonPac NS1 (10μm); eluent: 0.004 mol/L hexanesulfonic acid + 0.04 mol/L H_3BO_3; flow rate: 1 mL/min; detection: suppressed conductivity; injection volume: 100 μL; solute concentrations: ammonium (1), 2.5 ppm monoethanolamine (2), 5 ppm diethanolamine (3), and 10 ppm triethanolamine (4).

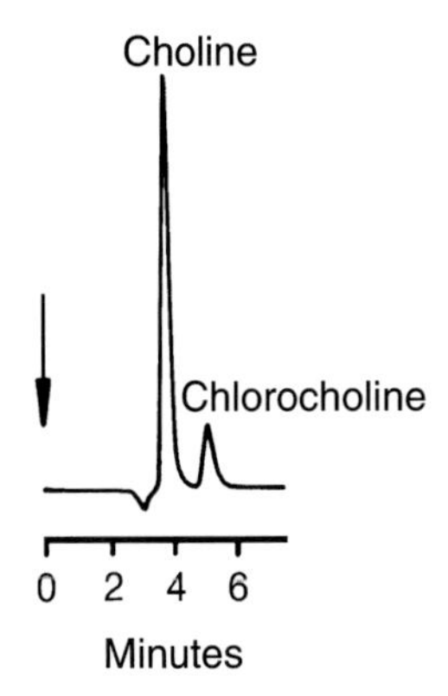

Fig. 5-28. Analysis of choline and chlorocholine. – Separator: IonPac NS1 (10 µm); eluent: 0.002 mol/L hexanesulfonic acid / acetonitrile (97:3 v/v); flow rate: 1 mL/min; detection: suppressed conductivity; injection volume: 50 µL; solute concentrations: 10 mg/L each of choline chloride and chlorocholine chloride.

agent, contains a small amount of acetonitrile as an organic modifier. Important choline derivatives such as acetylcholine and chlorocholine are more strongly retained under these chromatographic conditions and, thus, can be separated from choline. As an example, Fig. 5-28 shows the separation of choline and chlorocholine, which differ only in the substitution at the alkyl side-chain end:

$$[HO-CH_2-CH_2-\overset{+}{N}(CH_3)_3]OH^- \qquad [Cl-CH_2-CH_2-\overset{+}{N}(CH_3)_3]OH^-$$

Choline Chlorocholine

Much higher amounts of organic solvents are required for the elution of tetraalkylammonium compounds, which are are used as reagents in ion-pair chromatography of anions. While an amount of 60 mL/L acetonitrile suffices to elute tetramethylammonium hydroxide at a hexanesulfonic acid concentration of c = 0.001 mol/L (Fig. 5-29), 280 mL/L is needed for the elution of tetrapropylammonium ions and 480 mL/L for the even more hydrophobic tetrabutylammonium ions, when keeping the analysis time constant. Because of the drastic increase in retention time with a larger carbon skeleton, a simultaneous analysis of these compounds is only accomplished by applying a gradient technique.

Fig. 5-29. Separation of tetramethylammonium hydroxide. – Separator column: IonPac NS1 (10µm); eluent: 0.001 mol/L hexanesulfonic acid / acetonitrile (94:6 v/v); flow rate: 1 mL/min; detection: suppressed conductivity; injection volume: 50 µL; solute concentration: 20 ppm.

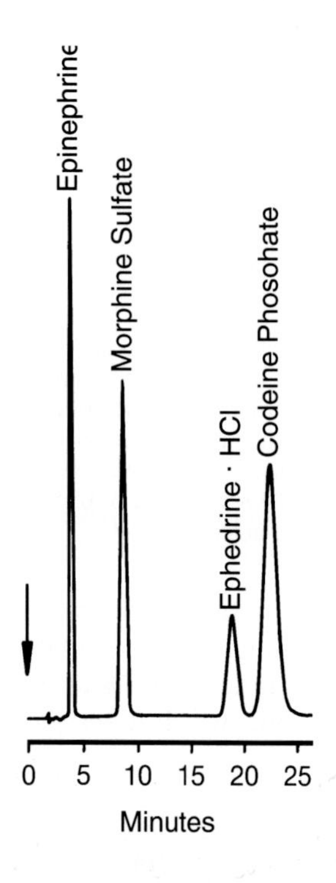

Fig. 5-30. Separation of epinephrine, ephedrine, and opium alkaloids. – Separator column: IonPac NS1 (10 µm); eluent: 0.005 mol/L sodium octanesulfonate + 0.05 mol/L KH_2PO_4 (pH 4.0) / acetonitrile (89:11 v/v); flow rate: 1 mL/min; detection: UV (220 nm); injection volume: 50 µL; solute concentrations: 10 mg/L epinephrine, 10 mg/L morphine sulfate, 20 mg/L ephedrine hydrochloride, and 20 mg/L codeine phosphate.

The various arylalkylamines and alkaloids can be separated with ion-pair chromatography by coating the stationary phase with sodium octanesulfonate. The effect of type and concentration of the ion-pair reagent on the retention of these compounds has been described in Section 5.3.1 at the diphenhydramine example. Fig. 5-30 reveals that under similar chromatographic conditions further arylalkylamines such as epinephrine and ephedrine can be separated in the presence of opium alkaloids such as morphine and codeine (3-monomethylester of morphine).

Morphine

To analyze atropine and cocaine, the acetonitrile content in the mobile phase has to be slightly increased (Fig. 5-31).

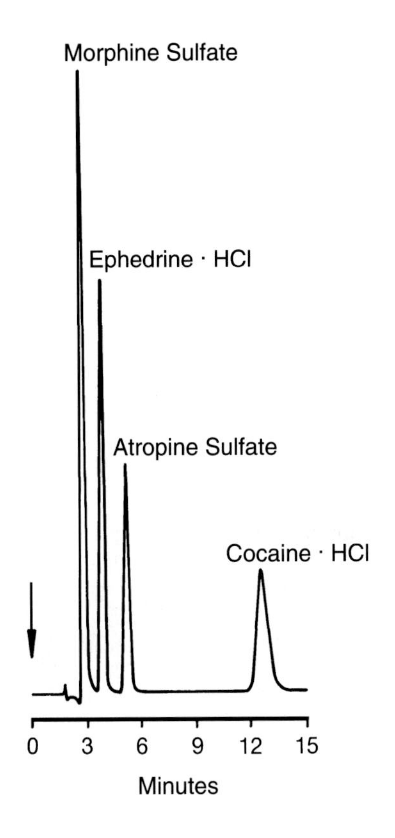

Fig. 5-31. Separation of atropine and cocaine. – Separator column: IonPac NS1 (10 μm); eluent: 0.005 mol/L sodium octanesulfonate + 0.05 mol/L KH_2PO_4 (pH 4.0) / acetonitrile (77:23 v/v); flow rate: 1 mL/min; detection: UV (220 nm); injection volume: 50 μL; solute concentrations: 10 mg/L morphine sulfate, 20 mg/L ephedrine hydrochloride, 20 mg/L atropine sulfate, and 20 mg/L cocaine hydrochloride.

Atropine

(2R,3S)-(−)-Cocaine

Much higher acetonitrile contents are required to elute the most important barbiturates: i.e., barbital, phenobarbital, and hexobarbital. Their separation is depicted in Fig. 5-32. Barbiturates are derived from barbituric acid (pK_a = 3.9 [40]) which is quite acidic because of its activated methylene group in 5-position.

Barbituric acid

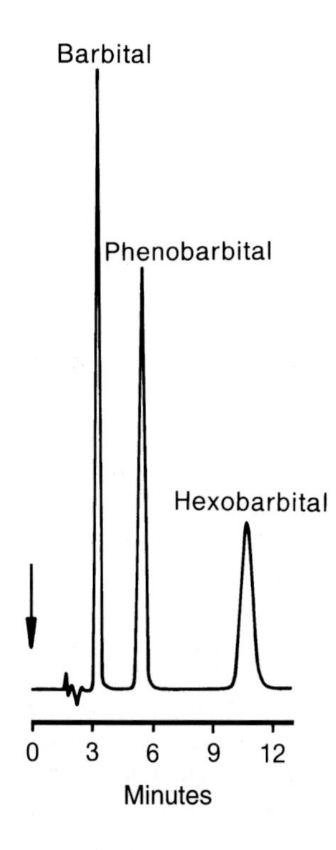

Fig. 5-32. Separation of various barbiturates. – Separator column: IonPac NS1 (10 μm); eluent: 0.005 mol/L sodium octanesulfonate + 0.05 mol/L KH_2PO_4 (pH 4.0) / acetonitrile (64:36 v/v); flow rate: 1 mL/min; detection: UV (220 nm); injection volume: 50 μL; solute concentrations: 10 ppm barbital, phenobarbital, and hexobarbital.

Substituting this group's H-atoms with various alkyl groups lowers the acidity of the resulting derivative. The acid strength of barbital (5,5-diethylbarbituric acid), for example, is given with $pK_a = 7.89$ [40]. The retention of barbiturates is completely unaffected by the type and concentration of lipophilc anions such as octanesulfonate present in the mobile phase. Nevertheless, it is logical to add sodium octanesulfonate to the eluent used for barbiturates, because barbiturates and alkaloids can then be analyzed in a single run. This applies, for example, to papaverine which has a retention time of about 6 minutes under the chromatographic conditions given in Fig. 5-32.

Papaverine

Because arylalkylamines, as well as alkaloides and barbiturates, have an aromatic skeleton, they can all be very sensitively detected by measuring the light absorption at a wavelength of 220 nm.

5.5 Analysis of Surface-Active Ions

The analysis of *surface-active anions* comprises the determination of simple aromatic sulfonic acids, hydrotropes (toluene, cumene, and xylene sulfonates), alkane- and alkene sulfonates, fatty alcohol ether sulfates, alkylbenzene sulfonates, and α-sulfofatty acid methyl esters. Many of these compounds are relevant predominantly in the detergent and cleansing industry.

Although surface-active anions with aromatic backbone have for sometime been separated by RPIPC and sensitively detected via their UV absorption [41], a chromatographic determination of the other compound classes mentioned above is only feasible by conductivity detection.

The retention behavior of surface-active anions depends on the alkyl side-chain length and, thus, on their hydrophobicity which increases with the chain length. In general, the hydrophilic ammonium hydroxide is employed as the ion-pair reagent, since the analyte compounds already have a strong hydrophobic character. The latter depends, apart from the alkyl chain length, on the kind and number of ionic and/or non-ionic substituents in the carbon skeleton. Fig. 5-33 shows the different retention behavior of alkyl sulfates and alkyl sulfonates of equal carbon chain length. It is obvious that alkyl sulfates are more strongly retained than the respective alkyl sulfonates. This effect is directly connected with a phenomenon observed during the investigation of the retention behavior of thioalkanes in reversed-phase chromatography. According to Fig. 5-34, each CH_2-S or CH-S entity in dialkyl sulfides and in alkane sulfonates represents a local polar center, which allows a stronger interaction with the polar eluent [42] via solvation. Thus, for each of these entities, a retention decrease is observed which, depending on the water content of the mobile phase, corresponds to the loss of 1 to 3 methylene groups. Moreover, the contribution to the retention of a methylene group is reduced by the presence of neighboring sulfur atoms. However, this effect does not arise in alkyl sulfates, which explaines the higher retention of these compounds.

Replacing a hydrogen atom in any of the methylene groups by a hydroxide function significantly reduces the retention time. The position of the hydroxide function exerts a noticeable influence on the resulting retention time. Fig. 5-35 shows the separation of two C_{16}-hydroxyalkane sulfonates, which are hydroxy-substituted in the 2- and 3-position of the alkyl chain. In comparison to non-substituted alkane sulfonates, the retention decrease corresponds to the loss of 2 to 3 methylene groups from the solvophobic alkyl groups.

The analysis of an olefin sulfonate, which is rarely a pure product, is much more difficult. The different double bond positions in the alkyl chain and the cis/trans isomerism generally lead to a product mixture of numerous compounds. Thus, satisfactory separations are only obtained if the spatial arrangement of both alkyl groups at the double bond and its position is defined. The chromatogram of a C_{16}-olefin sulfonate (trans-3-en) is shown in Fig. 5-36. It has a slightly shorter retention time than the alkane sulfonate of equal alkyl chain length, which is explained by the enhanced solubility of the unsaturated compound in the polar eluent.

The surface-active anions investigated so far include the class of *aryl sulfonates.* Fig. 5-37 shows a respective chromatogram with the separation of benzene, toluene, xylene, and cumene sulfonates. They are eluted in this order according to the number of carbon atoms of their substitutents. Tetrabutylammonium hydroxide was used as the ion-pair

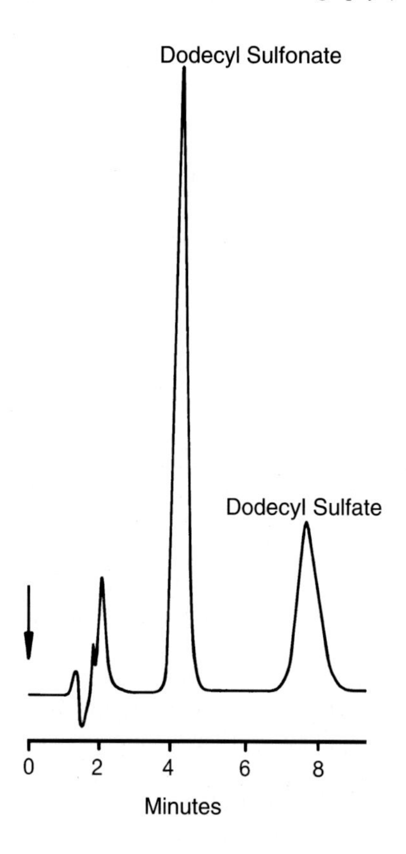

Fig. 5-33. Separation of dodecyl sulfonate and dodecyl sulfate. – Separator column: IonPac NS1 (10 µm); eluent: 0.01 mol/L NH_4OH / acetonitrile (65:35 v/v); flow rate: 1 mL/min; detection: suppressed conductivity; injection volume: 50 µL; solute concentration: 100 ppm each.

reagent to obtain high selectivity. The comparably small retention differences between the various aryl sulfonates allow an isocratic operation. Aryl sulfonates can be detected by applying the different modes of conductivity detection or UV detection at a wavelength of 254 nm, as shown in Fig. 5-37. The choice of the appropriate detection system usually depends on the nature of the matrix.

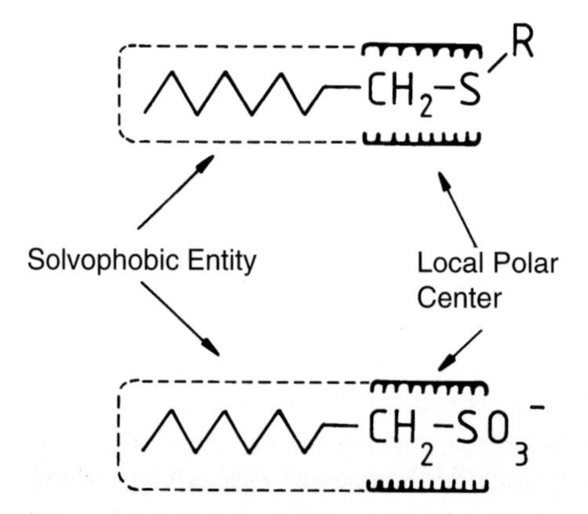

Fig. 5-34. Schematic representation of the loss of molecular surface area due to solvation of local polar centers comprising of a sulfur atom and neighboring CH-fragments.

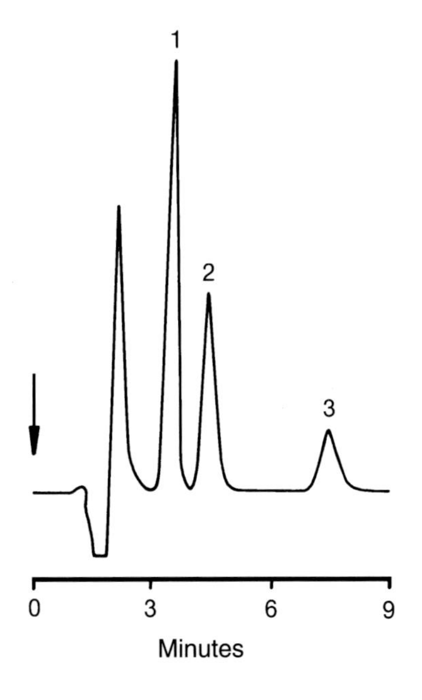

Fig. 5-35. Separation of two hydroxyalkane sulfonates with different hydroxide group position. – Separator column: IonPac NS1 (10 μm); eluent: 0.01 mol/L NH_4OH / acetonitrile (62:38 v/v); flow rate: 1 mL/min; detection: suppressed conductivity; injection volume: 50 μL; solute concentrations: 100 ppm each of C_{14}-alkane sulfonate (2-hydroxy) (1), C_{16}-alkane sulfonate (3-hydroxy) (2), and C_{16}-alkane sulfonate (2-hydroxy) (3).

The chromatographic separation of *fatty alcohol sulfates* requires the application of the gradient technique. The retention of these compounds increases exponentially with the alkyl chain length. Fatty alcohol sulfates such as lauryl sulfate are constituents of

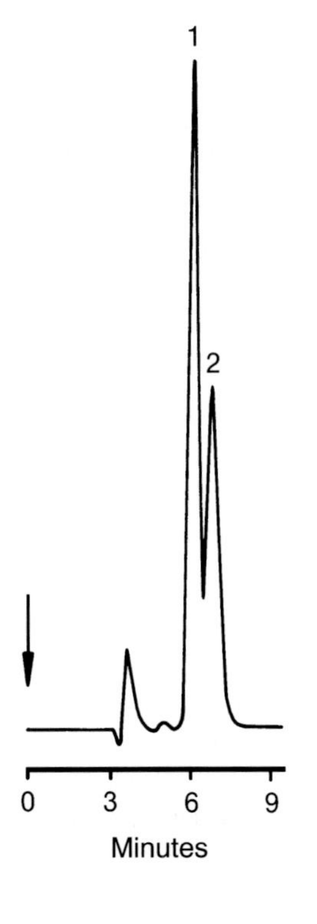

Fig. 5-36. Separation of alkane- and olefin sulfonates of the same chain length. – Separator column: IonPac NS1 (10 μm); eluent: 0.01 mol/L NH_4OH / acetonitrile (65:35 v/v); flow rate: 1 mL/min; detection: suppressed conductivity; injection volume: 50 μL; solute concentration: 100 ppm each of C_{16}-olefine sulfonate (trans-3-en) (1) and C_{16}-alkane sulfonate (2).

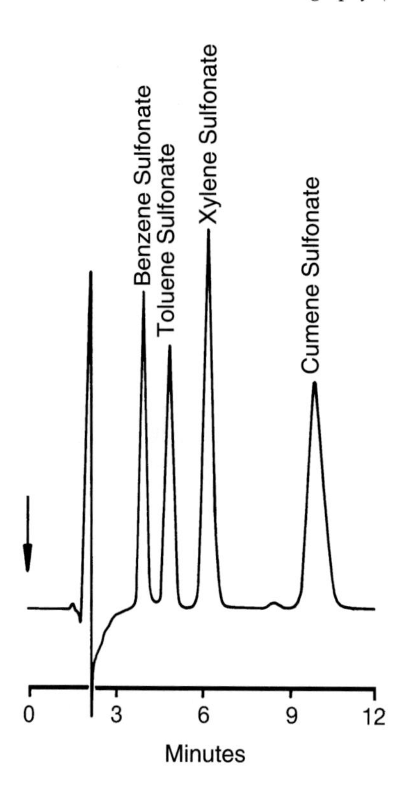

Fig. 5-37. Separation of various aryl sulfonates. – Separator column: IonPac NS1 (10 μm); eluent: 0.002 mol/L TBAOH / acetonitrile (70:30 v/v); flow rate: 1 mL/min; detection: UV (254 nm); injection volume: 50 μL; solute concentration: 50 ppm benzene sulfonate, 50 ppm toluene sulfonate, 100 ppm xylene sulfonate, and 100 ppm cumene sulfonate.

many cosmetics, thus their detection capability is of considerable interest. The separation of a standard mixture containing octyl, decyl, dodecyl, tetradecyl, and hexadecyl sulfate is shown in Fig. 5-38. The lower chromatogram represents an application of this gradient program, developed for this separation, to the analysis of a raw material sample (Rewopol NLS 28, Rewo, Steinau, Germany). In addition to lauryl sulfate, the major component, this product contains small amounts of tetradecyl and hexadecyl sulfate. Fatty alcohol sulfates have no chromophore, which makes sensitive detection of these compounds feasible only by measuring the electrical conductivity. The negative baseline drift during the chromatogram results from the rapidly increasing acetonitrile concentration in the mobile phase. It can be offset either by baseline subtraction or by increasing the concentration of the ion-pair reagent during the run.

Ethoxylation of fatty alcohols prior to their sulfonation leads to a class of compounds called *fatty alcohol ether sulfates,*

$$R\text{-}CH_2\text{-}O\text{-}(C_2H_4O)_n\text{-}SO_3Na \qquad R = C_{11} \text{ to } C_{13}, \quad n = 1 \text{ to } 5$$

Fatty alcohol ether sulfate

which are contained in a variety of detergents and cleansing agents. Depending on their degree of ethoxylation, fatty alcohol ether sulfates are extremely complex mixtures, for which the separation efficiency of a polymer phase is not sufficient. Good separations are obtained with silica-based, chemically bonded, reversed phases. The chromatographic conditions have to be adjusted accordingly. Thus, the free base ammonium

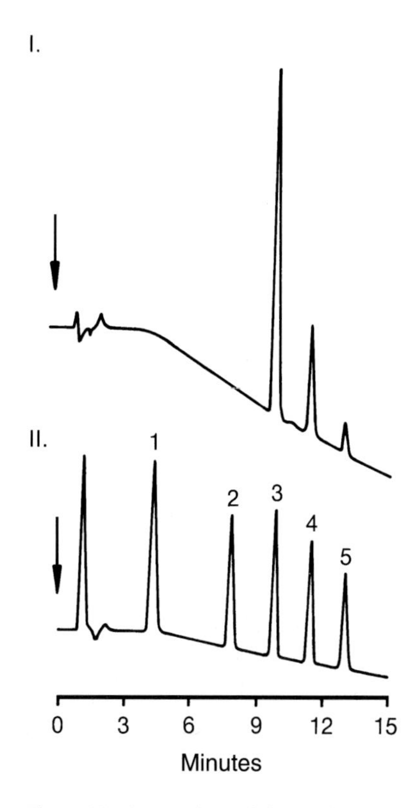

Fig. 5-38. Separation of fatty alcohol sulfates with various chain lengths. – Separator column: IonPac NS1 (10 μm); eluent: (A) 0.02 mol/L NH_4OH / acetonitrile (80:20 v/v), (B) 0.02 mol/L NH_4OH / acetonitrile (20:80 v/v); gradient: linear, 100% A to 100% B in 25 min; flow rate: 1 mL/min; detection: suppressed conductivity; injection volume: 50 μL; solute concentrations: I. 80 mg/L Rewopol NLS 28; II. 80 ppm each of octyl sulfate (1), decyl sulfate (2), dodecyl sulfate (3), tetradecyl sulfate (4), and hexadecyl sulfate (5).

hydroxide cannot be used as the ion-pair reagent because of the pH limitation of modified silica, ruling out the application of suppressor systems for the subsequent conductivity detection. For direct conductivity detection, sodium acetate has proved suited as the ion-pair reagent. It exhibits a sufficiently low background conductance at the required concentration of c = 0.01 mol/L. In combination with a solvent gradient, the separation of fatty alcohol ether sulfates according to their alkyl chain length and degree of ethoxylation is possible. Fig. 5-39 is an example of the analysis of Texapon N25 (Henkel KGaA, Düsseldorf, Germany), a raw material that contains, apart from dodecyl ether sulfate, the corresponding tetradecyl homologues. The resulting peak pattern thus consists of two overlapping peak series of ethoxylated compounds whose concentration decreases as the EO content increases. However, even the high separation efficiency of the stationary phase (Hypersil 5 MOS) is insufficient to resolve all compounds of this mixture. Co-elution is observed for tetradecyl sulfate, indicated as "C_{14}", and the higher ethoxylated dodecyl compound. Also, only a small retention difference exists between an alkyl ether sulfate with just one EO group and a pure alkyl sulfate with equal chain length. This means that the retention increase, which usually goes along with chain elongation by two methylene groups, is compensated by solvation due to the introduc-

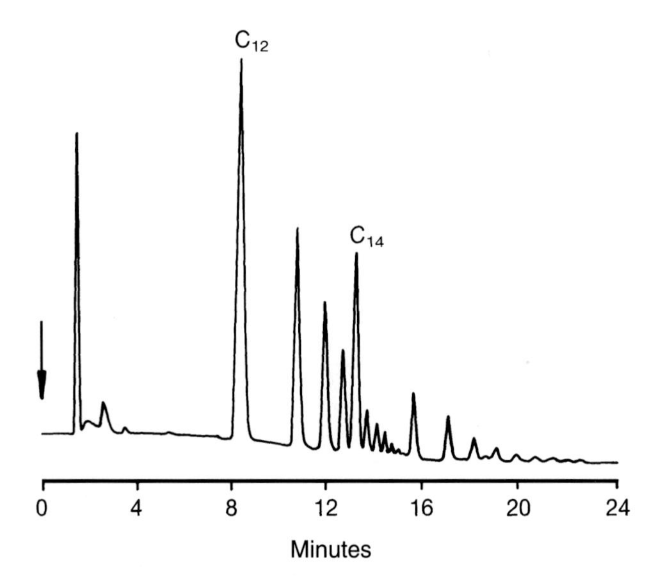

Fig. 5-39. Separation of a lauryl sulfate (Texapon N 25). – Separator column: Hypersil 5 MOS; eluent: (A) 0.001 mol/L NaOAc / acetonitrile (70:30 v/v), (B) 0.001 mol/L NaOAc / acetonitrile (60:40 v/v); gradient: linear, 100% A to 100% B in 10 min; flow rate: 1 mL/min; detection: direct conductivity; injection volume: 50 µL; solute concentration: 500 mg/L of the raw material.

tion of another polar center around the oxygen atom in the EO group. The retention-increasing effect of an EO group is noticed only at higher degrees of ethoxylation.

One of the most important anionic surfactants is the linear *alkylbenzene sulfonate* (LAS)

$H_3C-(CH_2)_n-CH(C_6H_4SO_3Na)-(CH_2)_m-CH_3$ $\quad m+n$ 7 till 10

Alkylbenzene sulfonate

that can be analyzed under similar chromatographic conditions. Alkylbenzene sulfonate is yet another complex mixture of compounds, as the aromatic ring may be attached to any non-terminal C-atom of the alkyl chain. Thus, the product distribution is statistical. Furthermore, the chain length may differ (C_{10} to C_{14}). Therefore, a complete separation of all components is not to be expected. Fig. 5-40 shows that the optimized separation of an LAS raw material sample, Marlon AFR (Chem. Werke Hüls, Marl, Germany), into homologues (C_{10} to C_{13}) is possible, although the respective major components of the different peak groups represent sum signals of the various isomers. Valuable information for the quality control of raw materials may be inferred from such chromatograms. UV detection in the wavelength region around 225 nm [41] provides an alternative to the direct conductivity detection utilized in Fig. 5-40.

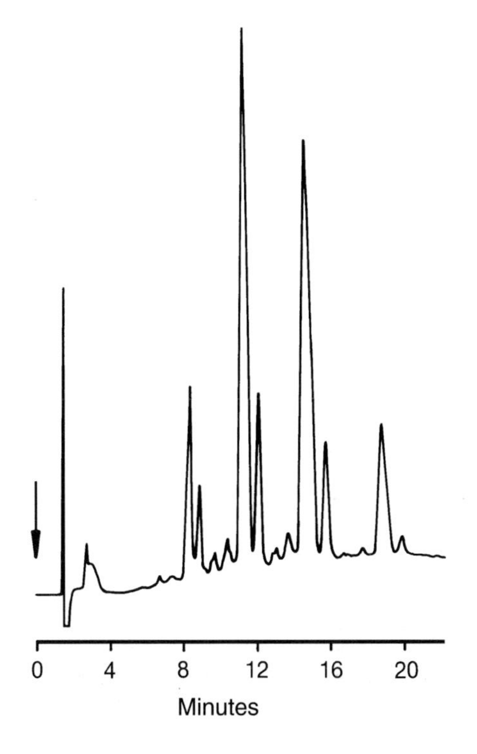

Fig. 5-40. Separation of a linear alkylbenzene sulfonate (Marlon AFR). – Separator column: Hypersil 5 MOS; eluent: (A) 0.001 mol/L NaOAc / acetonitrile (80:20 v/v), (B) 0.001 mol/L NaOAc / acetonitrile (60:40 v/v); gradient: linear, 60% B to 100% B in 15 min; flow rate: 1 mL/min; detection: direct conductivity; injection volume: 50 µL; solute concentrations: 1000 mg/L of the raw material.

Another constituent of many detergents and cleansing agents are *sulfosuccinic acid ester*.

$$\begin{array}{l} \quad SO_3Na \\ \quad\;\; | \\ \;\; HC-COOR \\ \quad\;\; | \\ H_2C-COOR \end{array} \qquad R = C_{10} \text{ to } C_{14}$$

Sulfosuccinic acid ester

Because of their good wetting properties, they are often a constituent of many quick-assay strips used in clinical chemistry. Their ion-pair chromatographic separation on Nucleosil 10 C_8 has recently been described by Steinbrech et al. [43], who used tetrabutylammonium hydrogensulfate as the ion-pair reagent and methanol as an organic modifier. Since sulfosuccinic acid ester do not have appreciable chromophores, this class of compounds was detected refractometrically. Fig. 5-41 shows the chromatogram of a raw material sample Rewopol SBF12 (Rewo, Steinau, Germany), a mixture of dialkyl sulfosuccinates of different chain length. As shown, the C_{12}-fraction represents the major component. The authors indicated a detection limit of 0.5 µg for this raw material.

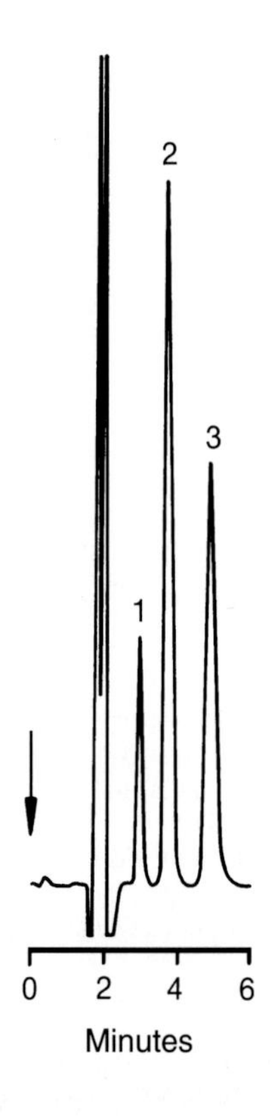

Fig. 5-41. Separation of fatty alcohol sulfosuccinates of different chain length. – Separator column: Nucleosil 10 C_8; eluent: 0.01 mol/L tetrabutylammonium hydrogensulfate (pH 3.0) / methanol (23:77 v/v); flow rate: 2 mL/min; detection: RI; injection: 50 μL of a Rewopol-SBF-12 solution with C_{10}-sulfosuccinate (1), C_{12}-sulfosuccinate (2), and C_{14}-sulfosuccinate (3); (taken from [43]).

While sulfosuccinic acid ester with chain lengths between C_{10} and C_{14} can be analyzed under isocratic conditions with the procedure described by Steinbrech et al., the separation of *fatty alcohol polyglycolether sulfosuccinates* requires the application of a gradient technique.

$$R-(OC_2H_4)_n-O-CO-CH_2-\underset{\displaystyle SO_3Na}{\underset{|}{CH}}-COONa \qquad R = C_8$$

Fatty alcohol polyglycolether sulfosuccinate

Fatty alcohol polyglycolether sulfosuccinates are used in many cosmetics because of their good compatibility with skin and mucous membrane. Due to the introduction of ethyleneoxide groups, the water solubility of these compounds is better than that of non-ethoxylated sulfosuccinates. Like fatty alcohol ether sulfates and alkylbenzene sulfonates, fatty alcohol polyglycolether sulfosuccinates are component mixtures of high complexity. Technical products of this kind are traded under the name Rewopol SBFA 30 (Rewo, Steinau, Germany). A remarkable chromatogram of this raw material was obtained by Janssen [44] on Hypersil 5 MOS with sodium acetate as the ion-pair reagent, applying an acetonitrile gradient. Although, as in the previous examples, not all compounds are baseline-resolved, the resolution obtained suffices to characterize this raw material unequivocally.

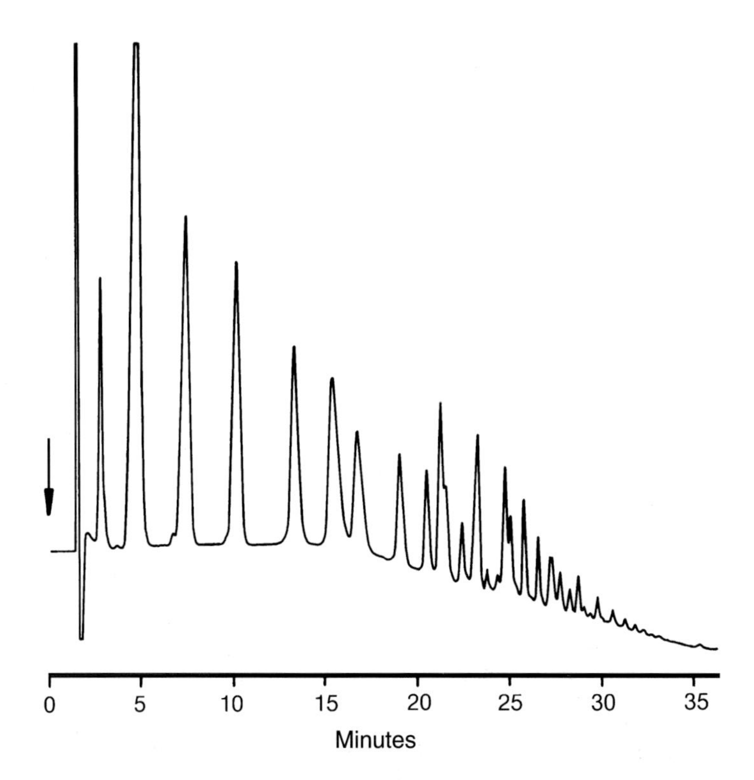

Fig. 5-42. Separation of a fatty alcohol polyglycolether sulfosuccinate (Rewopol SBFA 30). – Separator column: Hypersil 5 MOS; eluent: (A) 0.001 mol/L NaOAc / acetonitrile (78:22 v/v), (B) 0.001 mol/L NaOAc / acetonitrile (60:40 v/v); gradient: linear, 100% A isocratically for 10 min, then in 20 min to 100% B; flow rate: 1 mL/min; detection: direct conductivity; injection volume: 50 µL; solute concentration: 1000 mg/L of the raw material; (taken from [44]).

In the field of *surface-active cations*, ion-pair chromatography is predominantly applied to the analysis of quaternary ammonium compounds, pyridine, pyrrolidine, and piperidine quaternisates, and for sulfonium, phosphonium, ammonium, and hydrazinium salts.

The compound class of quaternary alkylammonium salts includes *alkyltrimethylammonium* and *dialkyldimethylammonium compounds*,

$$\left[R-\overset{+}{N}(CH_3)_3\right] Cl^- \qquad \left[(R)_2\overset{+}{N}(CH_3)_2\right] Cl^- \qquad R = C_{12} \text{ to } C_{18}$$

Alkyltrimethyl-ammonium chloride **Dialkyldimethyl-ammonium chloride**

which are found in low concentrations in many cosmetic products. In the past, cationic surfactants have played only a secondary industrial role and have recently been increasingly employed. Therefore, only a few liquid chromatographic methods for separating such compounds are described in the literature. As these quaternary ammonium compounds are comparably hydrophobic, hydrochloric acid is used as the ion-pair reagent for their separation. Because of the lack of suitable chromophores, quaternary ammonium compounds can only be sensitively detected via measuring the electrical conductivity. Fig. 5-43 shows the chromatogram of a tetradecyl-trimethylammonium chloride obtained under isocratic conditions with acetonitrile as the organic modifier. To elute a didodecyl-dimethylammonium bromide (Fig. 5-44) it is necessary to increase both the hydrochloric acid concentration and the organic solvent fraction.

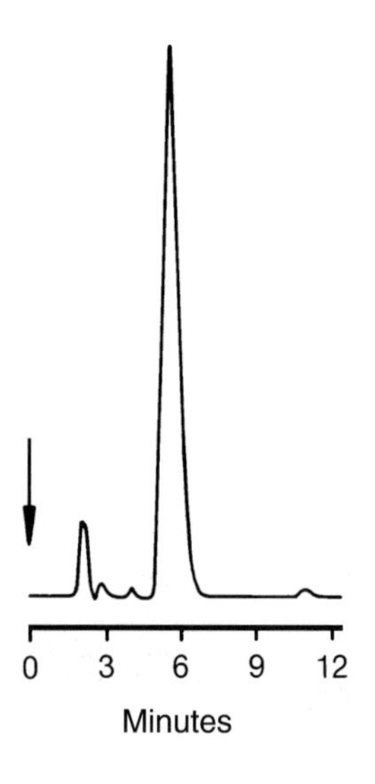

Fig. 5-43. Separation of a tetradecyl-trimethylammonium chloride. – Separator column: IonPac NS1 (10 μm); eluent: 0.005 mol/L HCl / acetonitrile (42:58 v/v); flow rate: 1 mL/min; detection: suppressed conductivity; injection volume: 50 μL; solute concentration: 200 mg/L of the raw material.

Alkyl-dimethyl-benzylammonium chlorides are constituents of many disinfectants because of their antibacterial activity.

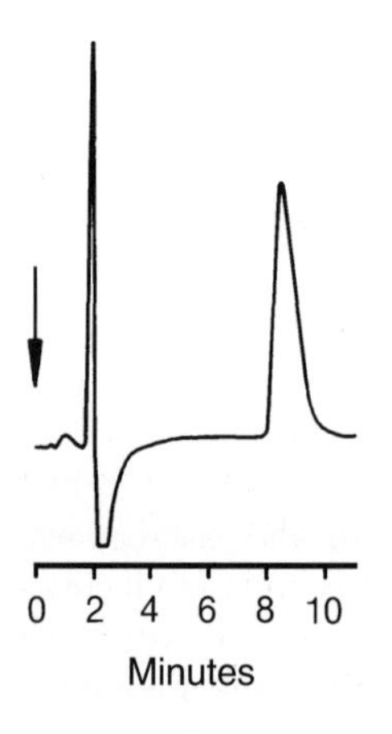

Fig. 5-44. Separation of a didodecyl-dimethylammonium bromide. – Separator column: IonPac NS1 (10 μm); eluent: (A) 0.02 mol/L HCl / acetonitrile (25:75 v/v); flow rate: 1 mL/min; detection: suppressed conductivity; injection volume: 50 μL; solute concentration: 300 ppm.

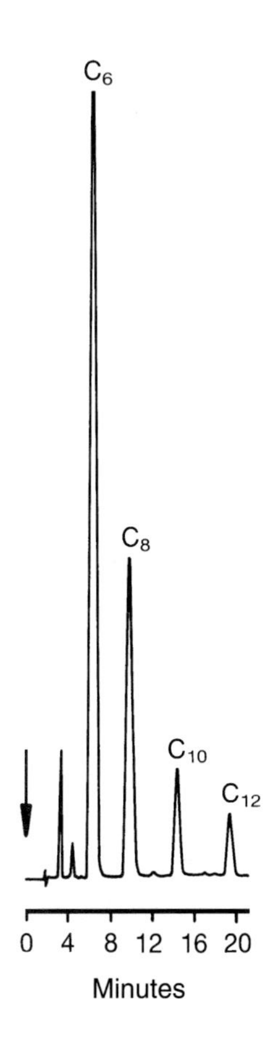

Fig. 5-45. Separation of an alkyl-dimethyl-benzylammonium chloride (Benzalkon A). – Separator column: IonPac NS1 (10 μm); eluent: (A) 0.02 mol/L HCl / acetonitrile (50:50 v/v), (B) 0.02 mol/L HCl / acetonitrile (20:80 v/v); gradient: linear, from 30% B to 100% B in 20 min; flow rate: 1 mL/min; detection: UV (215 nm); injection volume: 50 μL; solute concentration: 100 mg/L of the raw material.

$$\left[R-\overset{+}{N}(CH_3)_2-CH_2-C_6H_5 \right] Cl^- \qquad R = C_8 \text{ to } C_{18}$$

Alkyl-dimethyl-benzyl-ammonium chloride

Fig. 5-45 shows the chromatogram of a corresponding Benzalkon A raw material. The gradient technique was applied because of the carbon chain distribution between C_6 and C_{12} in this product. The two main components were identified as the C_6 and C_8 fractions by comparing with a standard consisting of decyl and dodecyl-dimethylbenzyl-ammonium chloride. Accordingly, the two minor components are the C_{10} and C_{12} fractions. It is remarkable that the peak symmetry of the individual signals can be significantly improved by raising the hydrochloric acid concentration to $c = 0.02$ mol/L. Since alkyl-dimethyl-benzylammonium chlorides contain an aromatic ring system, they can be detected with sufficient sensitivity by measuring the light absorption at 215 nm. Derivatives of these compounds, such as *dichlorobenzyl-alkyl-dimethylammonium chlorides*, may also be separated and detected under similar chromatographic conditions.

Fig. 5-46 provides an example showing the raw material Prevental RB50 (Henkel KGaA, Düsseldorf, Germany) which is an exceptionally pure product with a chain length C_{12}. A slightly shorter retention under these chromatographic conditions exhibits *diisobutyl-[2-(2-phenoxyethoxy)ethyl-] dimethylbenzylammonium chloride,*

$$H_3C-C(CH_3)_2-CH_2-C(CH_3)_2-C_6H_4-O-CH_2-CH_2-O-CH_2-CH_2-\overset{+}{N}(CH_3)_2-CH_2-C_6H_5$$

Hyamine 1622

which is utilized under the trade name Hyamine 1622 in analytical chemistry for the two-phase titration of anionic surfactants. It is occasionally also employed in the pharmaceutical and cosmetics industry. The chromatogram of this compound is illustrated in Fig. 5-47.

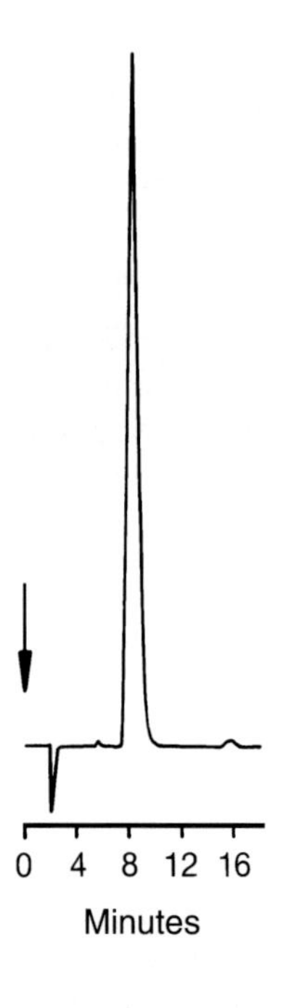

Fig. 5-46. Separation of a dichlorobenzyl-alkyl-dimethylammonium chloride (Prevental RB50). – Separator column: IonPac NS1 (10 μm); eluent: 0.005 mol/L HCl / acetonitrile (38:62 v/v); flow rate: 1 mL/min; detection: UV (215 nm); injection volume: 50 μL; solute concentration: 100 mg/L raw material.

Cyclic alkylammonium compounds resemble the linear products in their properties. They are widely found in antiseptic solutions, cremes, shampoos, and mouthwashes. As an example for this class of compounds, Fig. 5-48 displays the chromatogram of a dodecylpyridinium chloride. Its individual components, with the exception of the main component, exhibit a similar retention behavior to that of Benzalkon A. The minor components of this sample are presumably tetradecyl- and hexadecylpyridinium chloride.

In principle, modified silica may also be applied to all these separations. However, the high surface area of such materials require a much higher organic solvent fraction in the mobile phase to elute the compounds in a comparable time.

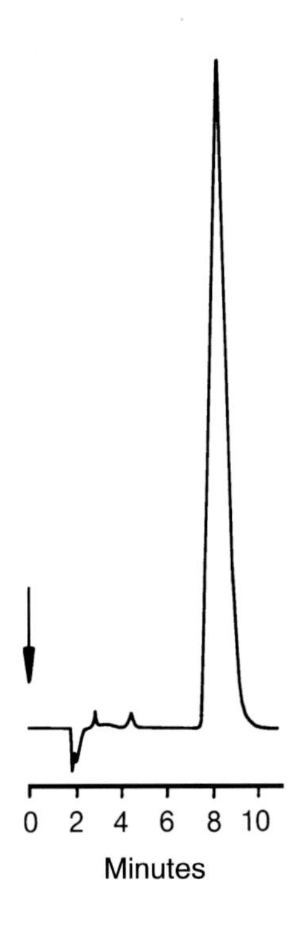

Fig. 5-47. Separation of Hyamine 1622. – Separator column: IonPac NS1 (10 μm); eluent: 0.005 mol/L HCl / acetonitrile (42:58 v/v); flow rate: 1 mL/min; detection: UV (215 nm); injection volume: 50 μL; solute concentration: 100 mg/L raw material.

As in cationic surfactants, very few liquid chromatographic methods are described in the literature for characterizing *sulfonium salts.* In a recent paper, Anklam et al. [45] showed that five- and six-membered cyclic sulfonium salts of the following structure are easily analyzed via ion-pair chromatography.

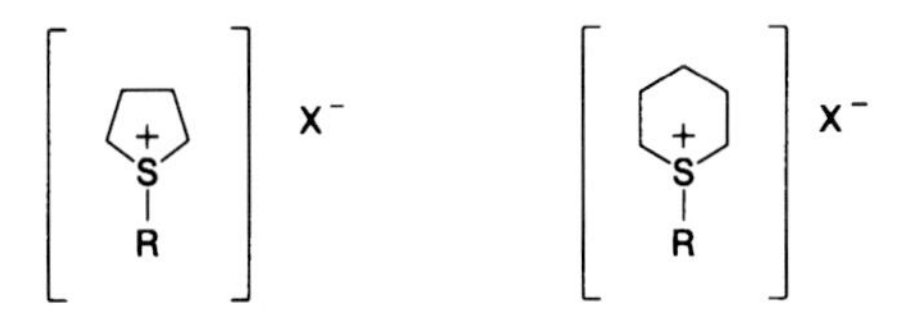

Structures of five- and six-membered sulfonium salts

The authors utilized a highly deactivated and stable ODS material, Inertsil ODS II (5 μm), as the stationary phase. With hexanesulfonic acid as the ion-pair reagent and acetonitrile as the organic modifier the chromatogram shown in Fig. 5-49 is obtained for six-membered sulfonium compounds bearing different alkyl rests R. Five-membered analogues show a correspondingly shorter retention. As expected, the retention increases as the chain length of the alkyl group increases. This is predicted by the theory of RPLC [47], since the molecular surface area and, thus, the solvophobic character of the solute increases with a larger R. Electrical conductance measurements utilizing a membrane-based suppressor system are a suitable detection method for non-chromophoric sulfonium salts.

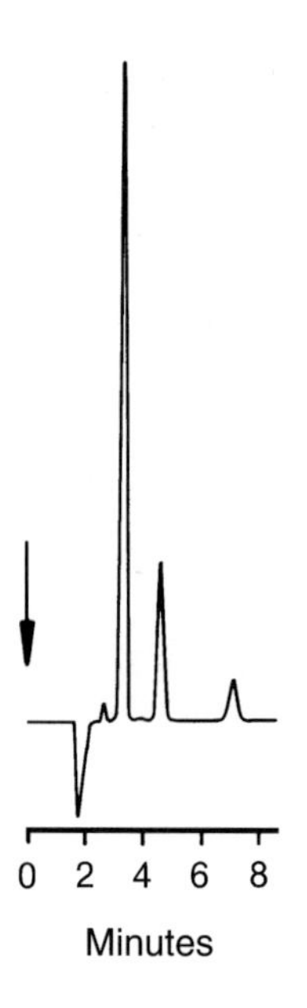

Fig. 5-48. Separation of dodecylpyridinium chloride. – Separator column: IonPac NS1 (10 µm); eluent: 0.02 mol/L HCl / acetonitrile (24:76 v/v); flow rate: 1 mL/min; detection: UV (215 nm); injection volume: 50 µL; solute concentration: 100 mg/L raw material.

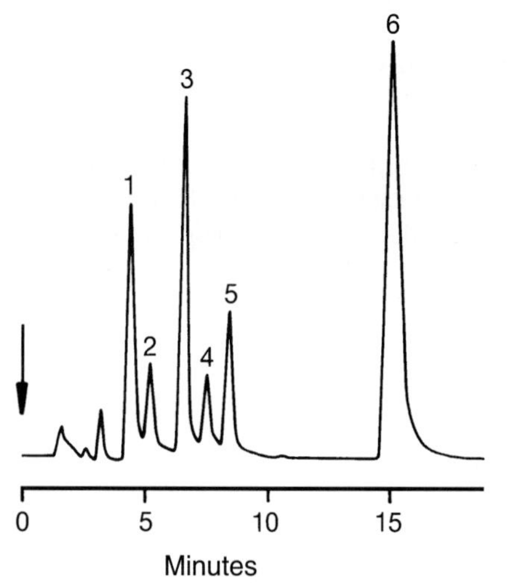

Fig. 5-49. Separation of various six-membered sulfonium salts on Inertsil ODS II (5 µm). – Eluent: 0.002 mol/L hexanesulfonic acid / acetonitrile (90:10 v/v); flow rate: 0.8 mL/min; detection: suppressed conductivity; compounds: R = Me (1), R = Et (2), R = *i*-Pr (3), R = *n*-Pr (4), R = *t*-Bu (5), and R = *n*-Bu (6); (taken from [46]).

$\overset{+}{P}(C_6H_5)_3$ Cl^-

Triphenyl-mono(β-jonyliden-ethylen)-phosphonium chloride

As an example for the ion-pair chromatographic analysis of *phosphonium compounds*, the separation of triphenyl-mono(β-jonyliden-ethylen)-phosphonium chloride is illustrated in Fig. 5-50.

This compound is a vitamin A precursor with a strong hydrophobic character. Therefore, sodium perchlorate was employed as the ion-pair reagent. This phosphonium compound contains three conjugated double bonds, thus UV detection is much more sensitive than suppressed conductivity detection. Fig. 5-50 reveals that the resulting signal is characterized by a strong tailing because of the π-π interactions with the aromatic backbone of the DVB resin that is used as the stationary phase. Aced [46] obtained much better peak shapes using the already-mentioned Inertsil ODS II column, with which she investigated the retention behavior of various triphenylphosphonium salts $[(Ph)_3P\text{-}R]^+X^-$. Fig. 5-51 illustrates this by showing the retention of some *n*-alkyl triphenylphosphonium salts of different chain length. Again, as with sulfonium salts, a retention increase is observed with growing chain length. In comparing triphenylphosphonium compounds with the corresponding *arsonium compounds* $[(Ph)_3As\text{-}R]^+X^-$, the latter exhibit a higher retention because of their larger surface area. It is noteworthy

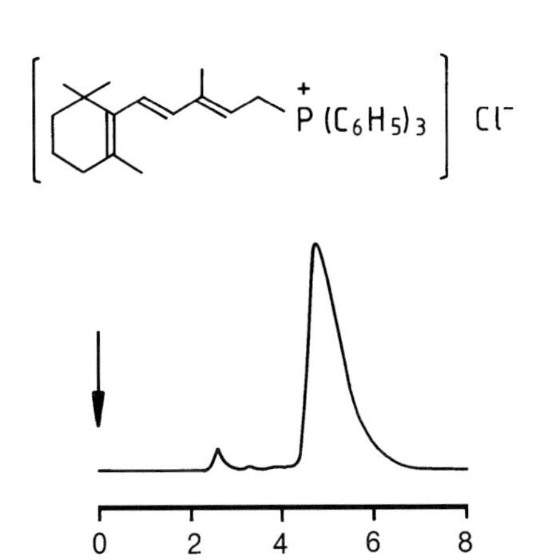

Fig. 5-50. Ion-pair chromatographic separation of a quaternary phosphonium salt. – Separator column: IonPac NS1 (10 µm); eluent: 0.005 mol/L $HClO_4$ / methanol (5:95 v/v); flow rate: 1 mL/min; detection: UV (280 nm).

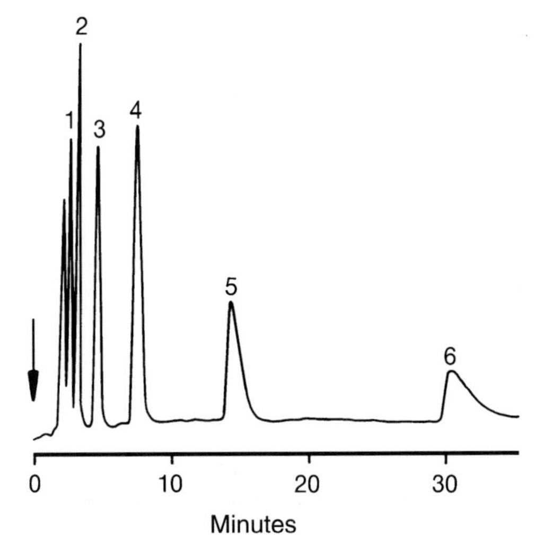

Fig. 5-51. Separation of various n-alkyl triphenylphosphonium compounds. – Separator column: Inertsil ODS II (5 µm); eluent: 0.002 mol/L HCl / acetonitrile (72:28 v/v); flow rate: 0.8 mL/min; detection: suppressed conductivity; compounds: R = Me (1), R = Et (2), R = *i*-Pr (3), R = *n*-Pr (4), R = *t*-Bu (5), and R = *n*-Bu (6); (taken from [46]).

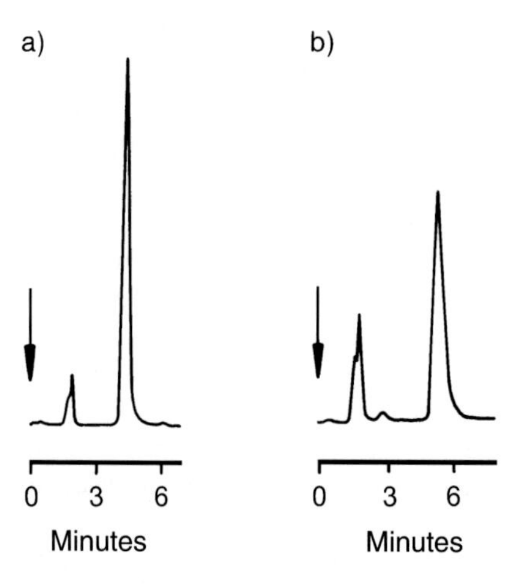

Fig. 5-52. Separation of two different *N,N'*-substituted hydrazinium compounds. – Separator column: IonPac NS1 (10 μm); eluent: (A) 0.002 mol/L HCl / acetonitrile (98:2 v/v), (B) 0.002 mol/L HCl / acetonitrile (93:7 v/v); flow rate: 1 mL/min; detection: suppressed conductivity; compounds: (A) $[C_9H_{19}N_2]^+BF_4^-$, (B) $[C_{13}H_{21}N_2]^+SO_3F^-$.

that the retention of triphenylphosphonium compounds that was observed by Aced [46] increases when hydrochloric acid is replaced by hydrogen bromide or hydrogen iodide as the ion-pair reagent.

Finally, note the class of *N,N'*-substituted *hydrazinium salts*, which may also be investigated by ion-pair chromatography. These compounds carry no chromophore, thus, a sensitive detection is only possible via conductometric detection with or without a suppressor technique. As an illustration for the great number of compounds that have already been investigated, Fig. 5-52 displays the chromatograms of two compounds having the following structures:

$C_9H_{19}N_2^+$ $C_{13}H_{21}N_2^+$

Although the carbon content of these compounds is fairly high, the bicyclic structure results in a comparatively low sorption area. In the present case, hydrochloric acid was used als the ion-pair reagent at a low organic solvent content.

5.6 Applications of the Ion Suppression Technique

Another application for neutral, non-polar silica-based stationary phases or organic polymers is the separation of weak acids or bases in their molecular form. Their dissociation is suppressed by choosing an appropriate pH value. The solute interactions with the stationary phase are determined solely by their adsorption and distribution

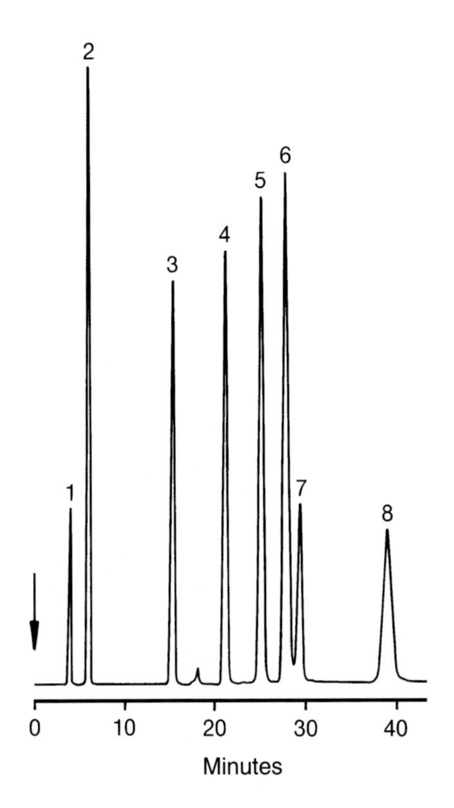

Fig. 5-53. Separation of various mono- and polyvalent phenols. – Separator column: Ion Pac NS1 (10 μm); eluent: (A) 0.01 mol/L KH_2PO_4 (pH 4.0) / acetonitrile (90:10 v/v), (B) 0.01 mol/L KH_2PO_4 (pH 4.0) / acetonitrile (20:80 v/v); gradient: linear, 15% B in 20 min to 55% B; flow rate: 1 mL/min; detection: UV (280 nm); injection volume: 50 μL; solute concentrations: 100 ppm each of pyrogallic acid (1), resorcinol (2), phenol (3), *o*-cresol (4), 2,4-dimethylphenol (5), *β*-naphthol (6), 2,4-dichloro-3-nitrophenol (7), and thymol (8).

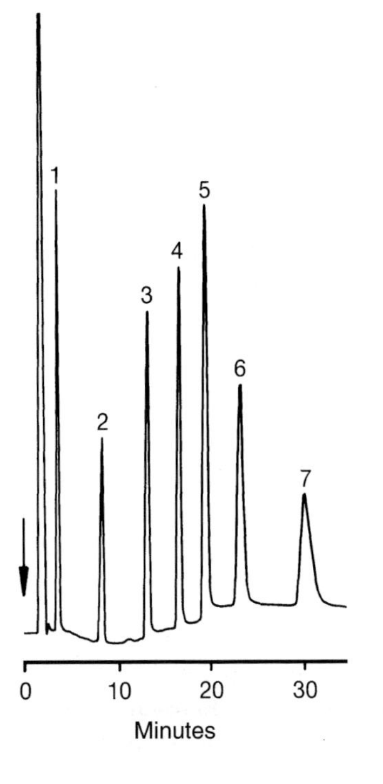

Fig. 5-54. Separation of some long-chain fatty acids by means of ion suppression technique. – Separator column: IonPac NS1 (10 μm); eluent: (A) $3 \cdot 10^{-5}$ mol/L HCl / acetonitrile / methanol (70:24:6 v/v/v), (B) $3 \cdot 10^{-5}$ mol/L HCl / acetonitrile / methanol (16:60:24 v/v/v); gradient: linear, 100% A in 15 min to 100% B; flow rate: 1 mL/min; detection: suppressed conductivity; injection volume: 50 μL; solute concentrations: 100 ppm butyric acid (1), 100 ppm caproic acid (2), 200 ppm caprylic acid (3), 200 ppm capric acid (4), 300 ppm lauric acid (5), 300 ppm myristic acid (6), and 400 ppm palmitic acid (7).

behavior. The term *ion suppression mode* has been coined for this technique. In some areas, it provides an alternative to ion-exchange chromatography.

One of the best known and most widely used applications of this technique is the separation of phenols, which are eluted on a non-polar phase using acetonitrile/water- or methanol/water-mixtures. A phosphate buffer is added to the solution to suppress the small dissociation of phenols into phenolate anions. Potassium dihydrogenphosphate at concentrations of about $c = 0.01$ mol/L are typically employed. Fig. 5-53 shows a separation of various mono-, di-, and trivalent phenols on an organic polymer.

Another application of the ion-suppression technique is the separation of long-chain fatty acids described by Slingsby [48]. While short-chain monocarboxylic acids up to valeric acid may be detected via ion-exclusion chromatography (see Section 4.6), long-chain fatty acids exhibit unacceptably long retention times under these conditions. In a protonated state, they can be easily separated on a non-polar stationary phase applying a solvent/water-mixture. Fig. 5-54 illustrates a respective separation of various fatty acids (butyric acid to palmitic acid) obtained on an organic polymer by applying a gradient technique. To ensure sufficient solubility of long-chain fatty acids in the mobile phase, a solvent mixture composed of acetonitrile *and* methanol was used. Small amounts of hydrochloric acid in the eluent served to suppress the dissociation of the compounds. If chemically modified silica (e.g. ODS) is used instead of an organic polymer, the enhanced solute interactions with such a material have to be compensated for. This is achieved both by raising the column temperature to 40 °C and by adding 2-propanol or THF to the mobile phase. The detection of long-chain fatty acids depends on the kind of carbon skeleton. If conjugated double bonds are present, as in case of sorbic acid, a sensitive UV-detection is feasible (Fig. 5-55).) On the other hand, if the carbon chain is purely paraffinic, as in case of the compounds represented in Fig. 5-54, conductometric detection must be utilized, since aliphatic fatty acids exhibit no appreciable absorption even at low wavelengths around 200 nm. A prerequisite for conductometric detection is the transformation of fatty acids that are separated as molecular compounds into their dissociated form. For this, the separator column effluent is passed through

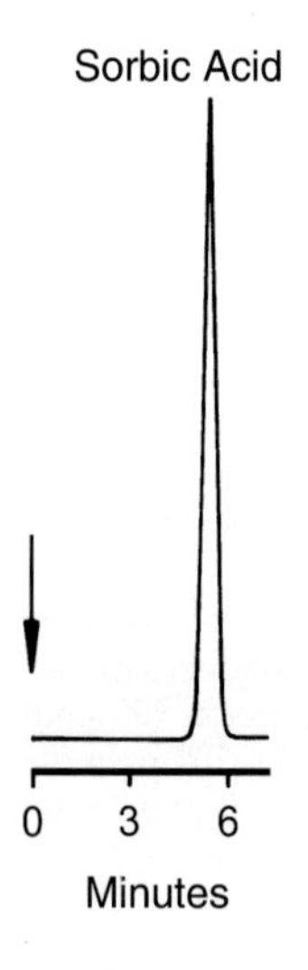

Fig. 5-55. Separation of sorbic acid. – Separator column: IonPac NS1 (10 μm); eluent: $5 \cdot 10^{-5}$ mol/L HCl / acetonitrile / methanol (57:34:9 v/v/v); flow rate: 1 mL/min; detection: UV (250 nm); injection volume: 50 μL; solute concentration: 5 ppm.

an AMMS-ICE micromembrane suppressor (see also Section 4.4) before entering the conductivity cell. The fatty acid oxonium ions are exchanged for potassium ions. The solutes enter the detector cell as fully dissociated potassium salts and may be detected conductometrically. In addition, the background conductance of the hydrochloric acid eluent is also reduced by converting it into the potassium salt. This is advantageous because it significantly improves the signal-to-noise ratio for the analyte compound. According to Slingsby [48], detection limits for this method are between 50 ppb for butyric acid and 50 ppm for stearic acid.

A remarkable application of ion suppression is the analysis of aminopolycarboxylic acids. In general, NTA and EDTA are separated on an anion exchanger and detected photometrically after derivatization with iron(III). Compounds such as iminodiacetic acid (IDA), ethylenediaminetriacetic acid (EDTriA), and ethylenediaminediacetic acid (EDDA), in contrast, cannot be detected under these chromatographic conditions, as they do not form the desired complexes with iron(III). This is not surprising, since EDTriA does not exist in acid pH range as an open-chain compound with complex-forming properties. Instead, it exists as a cyclic piperazine, which cannot be detected via post-column derivatization. The detection of all three complexing agents – IDA, EDTriA, and EDDA – is thus only possible via direct UV detection of the carboxylate group at low wavelengths around 215 nm. The first successful separations of these complexing agents were accomplished on a chemically bonded ODS phase by means of ion suppression.

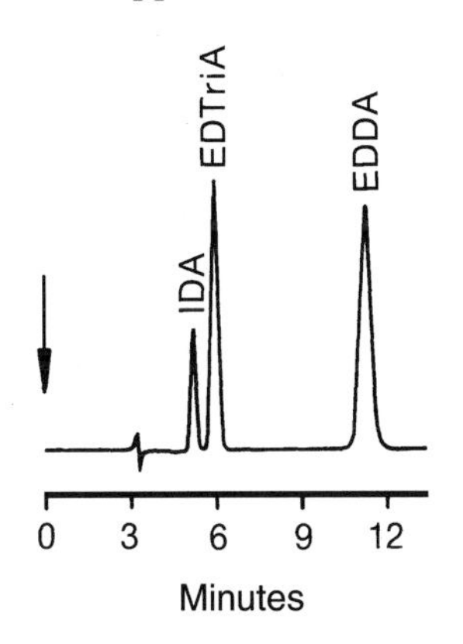

Fig. 5-56. Separation of iminodiacetic acid, ethylenediaminetriacetic acid, and ethylenediaminediacetic acid. – Separator column: Mikropak MCH-10 (Varian); eluent: 0.2 g/L sodium octylsulfate + 0.5 mL/L H_2SO_4 / methanol (92:8 v/v); flow rate: 1 mL/min; detection: UV (215 nm); injection volume: 50 μL; solute concentrations: 500 ppm IDA, 50 ppm EDTriA, and 50 ppm EDDA.

Due to the high polarity of these compounds, their retention is very small when the usual methanol/water-mixtures are used. The retention is markedly increased by coating the phase with an anionic surfactant. A good resolution of all three compounds within an acceptable total analysis time is obtained with sodium octylsulfate and methanol in sulfuric acid solution. Mikropak MCH-10 (Varian Co.) is suited as the stationary phase, for example, with a particle diameter of 10 μm. Fig. 5-56 shows the corresponding chromatogram of a standard with a baseline-resolved separation of all three components.

6 Detection Modes in Ion Chromatography

In the detection modes applied to ion chromatography one distinguishes between electrochemical and spectroscopic methods. Conductometric and amperometric detection are electrochemical methods, while the spectroscopic methods embrace UV/Vis, fluorescence, and refractive index detection. Added to this are the various applications of these detection methods, described in detail below.

In most cases, the choice of a suitable detection mode depends on the separation method and the correspopnding eluents. If detection is to be carried out by directly measuring a physical property of the solute ion (e.g. UV absorption), it must differ substantially in this property from eluent ions which are present in much higher concentration. However, eluent and solute ions often exhibit similar properties, so that direct detection is only feasible where selective detection of a limited number of solute ions is desirable.

A much broader range of applications have detection methods, with which one measures the change in a certain physical property of the eluent (e.g. conductance) that is due to the solute ion elution. A sufficient difference between eluent and solute ions is a prerequisite in the measuring values of this property. Most of the detection methods applied to ion chromatography are based on this technique. For the ensuing discussion a further subdivision into *direct* and *indirect* methods is made. Direct detection methods are those, in which eluent ions exhibit a much *smaller* value than solute ions for the property to be measured. On the other hand, detection methods are called indirect, if eluent ions exhibit a much *higher* value for the property to be measured than do solute ions.

6.1 Electrochemical Detection Methods

6.1.1 Conductometric Detection

As a universal method for the detection of ionic species, conductometric detection assumes a central position in ion chromatography. Thus, the fundamental theoretical principles of this detection method are summarized in the following.

6.1.1.1 Theoretical Principles

Electrical Conductance of Electrolyte Solutions [1]

The electrical resistance of an electrolyte solution is described by Ohm's law

$$R = \frac{U}{I} \tag{156}$$

R Resistance
U Voltage
I Current strength

As the resistance depends on the type of conductor, the resistivity ρ is defined as a material-specific quantity as follows

$$\rho = \frac{A \cdot R}{l} \tag{157}$$

A Cross section of conductor
l Length of conductor

The electrical conductance $\varkappa$ with a unity S cm^{-1} represents the reciprocal of the resistivity.

$$\varkappa = \frac{1}{\rho} \tag{158}$$

The electrical conductance of electrolytes is strongly dependent on the concentration. To compare the conducting power of different electrolyte solutions, the electrical conductance is divided by the equivalent concentration c_{ev} which yields the conductivity (S cm^2 val^{-1}):

$$\Lambda = \frac{\varkappa}{c_{ev}} \tag{159}$$

When two electrodes immersed in an electrolyte solution are connected to a power source, an electrical field of strength E is created between them. In this field a directed mass transport occurs. Anions drift to the positive pole while cations drift to the negative pole. The quantity for the mobility of ions that is independent on the field strength is obtained by dividing the ion velocitity v by the field strength. It is called ion mobility u (cm^2 V^{-1} s^{-1}):

$$u = \frac{v}{E} \tag{160}$$

In a strong electrolyte, cations and anions are present in solution in the concentration c_+ and c_-, respectively. The following applies:

$$c = c_+ = c_- \tag{161}$$

When the electrolyte is in an electrical field, the number of cations and anions passing the given cross section A of the electrolyte within the time t is given by $N_A \cdot c_+ \cdot v_+ \cdot A \cdot t$ and $N_A \cdot c_- \cdot v_- \cdot A \cdot t$, respectively. Each ion carries an electrical charge, so this directed motion is associated with a charge transport. The ratio of the sum of charges and time represents the electrical current strength I:

$$I = I_+ + I_- \tag{162}$$

$$I = e \cdot N \cdot c_+ \cdot v_+ \cdot A + e \cdot N \cdot c_- \cdot v_- \cdot A \tag{163}$$

N Loschmidt number
e Elemental charge

With the Faraday constant $F = N \cdot e$ and taking into account the ion mobility according to Eq. (160), one obtains:

$$I = F \cdot c \cdot A \cdot E\,(u_+ + u_-) \tag{164}$$

Since:

$$\varkappa = \frac{I}{A \cdot E} \tag{165}$$

one obtains with Eq. (164) the electrical conductance for the solution of a strong electrolyte:

$$\varkappa = c \cdot F \cdot (u_+ + u_-) \tag{166}$$

Correspondingly, for a polyvalent electrolyte

$$\varkappa = c \cdot z \cdot F \cdot (u_+ + u_-) \tag{167}$$

where z is the number of charges. In case of incomplete dissociation, the degree of dissociation also has to be taken into account:

$$\varkappa = \alpha \cdot c \cdot z \cdot F \cdot (u_+ + u_-) \tag{168}$$

Thus, the electrical conductance increases

- with increasing ion concentration,
- with increasing charge numbers of the ions, and
- with increasing mobility of the ion.

With Eqs. (159) and (167) one obtains:

$$\Lambda = \frac{\varkappa}{z \cdot c} = F \cdot (u_+ + u_-) \tag{169}$$

According to Kohlrausch, the quantities $F \cdot u_+$ and $F \cdot u_-$ are the equivalent ionic conductances Λ_+ and Λ_-, respectively. The conductivity is a summation:

$$\Lambda = \Lambda_+ + \Lambda_- \tag{170}$$

The conductivity of strong electrolytes decreases continuously with increasing concentration; with decreasing electrolyte concentration it approaches a material-specific limiting value, the conductivity at infinite dilution Λ_∞. This behavior may be attributed to the decrease in ion mobility with increasing concentration. At increasing electrolyte concentration the ions move so close to each other that they affect each other electrostatically. On time average, the Coulomb interaction increases with a reduction in the ionic distance. This interionic interaction is enhanced at high electrolyte concentrations because the dielectric constant of the solvent is decreased by the electrolyte. This electrostatic interaction may even cause solvated ions to form *ion pairs* which do not contribute to the conductance [2].

Kohlrausch observed the following empirical relation describing the concentration dependence of strong electrolytes:

$$\Lambda = \Lambda_\infty - k \cdot \sqrt{c} \tag{171}$$

(Kohlrausch square root law)

Here k depends on the ionic charge numbers. Plotting Λ versus $\sqrt{c}$ yields a straight line, the slope of which depends on the electrovalence of the electrolyte. Eq. (171) only applies to low concentrations ($c < 10^{-2}$ mol/L).

An increasing electrical conductance is usually observed with increasing temperature, as the viscosity of the solution decreases exponentially with rising temperature.

Conductivity measurements only define the sum of the equivalent ionic conductances; no information about their individual values can be derived. However, it is known that the equivalent conductances of ions may differ significantly. According to Kohlrausch's law of independent ion drift, all ions move independent of each other in an infinitely diluted solution. Since the equivalent conductances of ions differ, they contribute differently to the current transport. The contribution of an ionic species i to the total current is called the transport number t_i:

$$t_\mathrm{i} = \frac{I_i}{I} \tag{172}$$

With Eqs. (162) and (163) applies:

$$I_+ = F \cdot c \cdot A \cdot E \cdot u_+ \tag{173}$$

With Eq. (164) follows the contribution to the current that is apportioned to the cations, denoted as cation transport number t_+:

$$t_+ = \frac{I_+}{I} = \frac{\Lambda_+}{\Lambda_+ + \Lambda_-} = \frac{\Lambda_+}{\Lambda} \tag{174}$$

Similarly, it applies for the anion transport number t_-:

$$t_- = \frac{I_-}{I} = \frac{\Lambda_-}{\Lambda_+ + \Lambda_-} = \frac{\Lambda_-}{\Lambda} \tag{175}$$

The equivalent ionic conductances of anions and cations are typically between 35 S cm^2 val^{-1} and 80 S cm^2 val^{-1} (see Table 6-1). The oxonium ion with 350 S cm^2 val^{-1} and the hydroxide ion with 198 S cm^2 val^{-1} (25 °C) provide the only exceptions. Since these values are only that high in aqueous solution, a special transport mechanism has to be assumed that is correlated with the water structure. Hydrogen bonds between associated water molecules allow the exchange of protons and hydroxide ions over a long chain of water molecules without hydrated ions actually migrating. The higher equivalent ionic conductances of H^+ and OH^- are used, especially in the suppressor technique to convert the investigated salts into higher conductive species while the eluent is converted into a less conductive form (see Sections 3.3.3 and 3.4.3).

Table 6-1. Equivalent conductances of some selected anions and cations.

Anions	Λ_- [S cm^2 val^{-1}]	Cations	Λ_+ [S cm^2 val^{-1}]
OH^-	198	H^+	350
F^-	54	Li^+	39
Cl^-	76	Na^+	50
Br^-	78	NH_4^+	73
I^-	77	K^+	74
NO_3^-	71	Mg^{2+}	53
SO_4^{2-}	80	Ca^{2+}	60
Benzoate	32	Sr^{2+}	59
Phthalate	38	Ba^{2+}	64

Interionic Interactions

For dilute electrolyte solutions, Lewis and Randall observed that the mean activity coefficient of a strong electrolyte does not depend on the kind of ion, but only on the concentration and charge numbers of *all* ions present in solution. So, the individual properties of the ions are not decisive for interionic interactions in dilute electrolyte solutions. These observations paved the way for the introduction of the concept of ionic strength I:

$$I = \frac{1}{2} \sum_i c_i \cdot z_i^2 \tag{176}$$

This is a very functional quantity, as the mean activity coefficient of an electrolyte may easily be represented. Lewis discovered the following empirical relation:

$$\log f_{\pm} = -A \cdot z_+ \cdot |z_-| \cdot \sqrt{I} \tag{177}$$

$f_{\pm}$ Mean activity coefficient

When $\log f_{\pm}$ is plotted versus $\sqrt{I}$, a straight line is obtained with a slope that only depends on the charge numbers of ions present in the electrolyte.

According to the Debye-Hückel theory, it is assumed that the mutual influence of ions present in the crystal lattice of the solid is not fully cancelled in electrolyte solution. Accordingly, owing to electrostatic attractive forces each ion in solution is surrounded preferably by ions of opposite charge. Although these ions are subjected to thermal motion, in time average a positive ion, for example, will form the central ion of an oppositely charged ion cloud. Each ion in this ion cloud interacts with the central ion. The total charge of the ion cloud equals that of the central ion because of the electroneutrality condition.

Following Debye-Hückel, the distribution of ions can be calculated via the Boltzmann energy distribution. The application of this law is based on the concept that the ion cloud represents a space charge which is most dense in the vicinity of the central ion and decreases with growing distance to the central ion. A number of simplifying assumptions concerning the state of ions is made:

- The electrolyte is fully dissociated.
- Only electrostatic forces acting between ions are responsible for the interionic interactions.
- The electrostatic interaction energy is small compared to the thermal energy.
- The ions are regarded as point charges with an electrical field of spherical symmetry and, thus, as non-polarizable.
- The dielectric constant of the electrolyte solution is equal to that of the pure solvent.

With increasing electrolyte concentration, these conditions become increasingly inadequate:

- Strong electrolytes form ion pairs at high concentration.
- Apart from electrostatic interactions, ion molecule interactions also occur at high electrolyte concentrations which affect the solvation state of the ions and the solvent structure.
- The electrostatic interaction energy becomes so high at high electrolyte concentration, that ions no longer execute an unhindered thermal motion.
- With increasing electrolyte concentration, ions may polarize each other in case of a close encounter.
- Interactions of ions with the solvent change the dielectric constant of the solvent.

According to Lewis, the constant A in Eq. (177) had to be determined empirically. On the other hand, with the aid of the Debye-Hückel theory, it can be based on physical quantities and, thus, may be calculated. The theoretically derived dependence

$$\log f = -A \cdot z^2 \cdot \sqrt{I} \tag{178}$$

allows the following conclusions:

- At finite ionic strength $\log f$ always has a negative value; that is, the activity coefficient is smaller than one in the validity range of the equation.
- The logarithm of the activity coefficient decreases from its limiting value $\log f = 0$ (for $I = 0$) with the square root of the ionic strength.
- Ions with high charge numbers deviate more strongly from the ideal behavior than do ions with small charge numbers.
- The activity coefficient is strongly affected by temperature and dielectric constant.

Eq. (178) only represents a limiting law owing to the reduced validity of the prerequisites on which the Debye-Hückel theory is based. Thus, Eq. (178) is applicable to monovalent 1-1 electrolytes in aqueous solution to, at most, an ionic strength of 10^{-2} mol/L.

Also, the Debye-Hückel theory enables one to calculate quantitatively the effect of the ion cloud on ionic mobility.

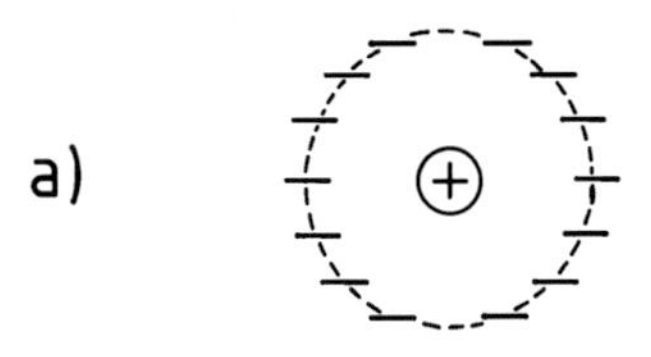

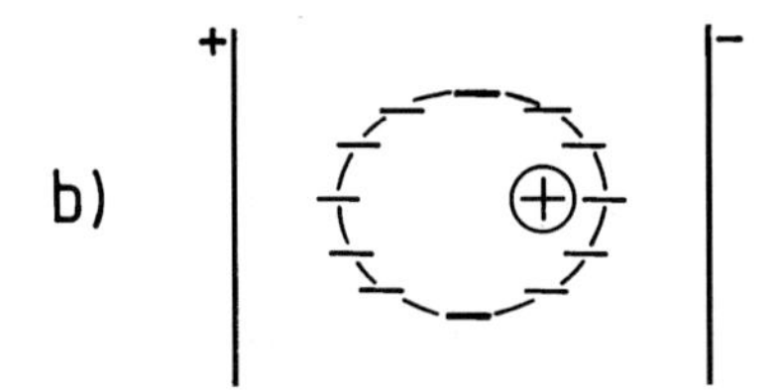

Fig. 6-1. Schematic representation of a central ion with its ionic cloud: a) without an external electrical field, b) within an external electrical field.

When a central ion moves in an electric field the ionic cloud surrounding the ion is permanently formed. This requires a certain time called the relaxation time. Therefore, as illustrated in Fig. 6-1, the charge density around the central ion is no longer symmetrical, but is lower in front of the central ion than behind it. This dissymmetry in charge distribution leads to an electrostatic deceleration of the central ion which reduces the ion mobility.

The ion cloud with the solvation shells of its ions moves in a direction opposite to the central ion. Therefore, the central ion does not move relative to a resting medium but rather against a solvent flow. The resulting reduction in mobility is called the electrophoretic effect. Both effects become more important with increasing electrolyte concentration and result in a decrease in conductivity.

The quantitative analysis according to Onsager yields the following equation for the conductivity of a 1-1 electrolyte:

$$\Lambda = \Lambda_\infty - \underbrace{\frac{8.2 \cdot 10^5}{\left(\varepsilon \cdot \frac{T}{K}\right)^{\frac{3}{2}}} \cdot \Lambda_\infty \cdot \sqrt{\frac{c}{\text{mol L}^{-1}}}}_{\text{relaxation effect}}$$

$$- \underbrace{\frac{82.48\ cm^2\ S\ mol^{-1}}{\frac{\eta}{\text{g cm}^{-1}\ \text{s}^{-1}} \cdot \left(\varepsilon \cdot \frac{T}{K}\right)^{\frac{1}{2}}} \cdot \sqrt{\frac{c}{\text{mol L}^{-1}}}}_{\text{electrophoretic effect}} \tag{179}$$

$$\Lambda = \Lambda_\infty - (\text{relaxation effect}) - (\text{electrophoretic effect})$$

The lower the dielectric constant ε of the solvent, the stronger the interionic interactions and, thus, the relaxation and electrophoretic effects. For the latter, solvent viscosity also plays a decisive role. When both the relaxation and the electrophoretic constants are known, the conductivity coefficient is obtained:

$$f_\Lambda = \frac{\Lambda}{\Lambda_\infty} \tag{180}$$

$$= 1 - \left(A' + \frac{B}{\Lambda_\infty}\right) \cdot \sqrt{c}$$

A' Relaxation constant
B Electrophoretic constant

6.1.1.2 Application Modes of Conductometric Detection

According to Fritz et al. [3], when the separator column effluent is passed into the conductivity cell *without* applying a suppressor system, the electrical conductance $\varkappa$ of a solution is given by:

$$\varkappa = \frac{(\Lambda_+ + \Lambda_-) \cdot c \cdot \alpha}{10^{-3} K} \tag{181}$$

Λ_+, Λ_- Equivalent ionic conductances of cations and anions
c Concentration
α Dissociation constant of the eluent
K Cell constant

Under the prerequisite that the cell constant and the equivalent ionic conductances of eluent anions and eluent cations are known, Eq. (181) makes it possible to calculate the conductivity of typical eluents for this kind of detection method. To determine the cell constant, the conductance of a potassium chloride solution with defined concentration is usually measured, as the equivalent ionic conductances of potassium and chloride ions are known with 74 S cm^2 val^{-1} and 76 S cm^2 val^{-1}, respectively (see Table 6-1).

When the eluent of the concentration c_E contains cations E^+ and anions E^-, the background conductances $\varkappa_E$ may be calculated via Eq. (181) as follows:

$$\varkappa_E = \frac{(\Lambda_{E^+} + \Lambda_{E^-}) \cdot c_E \cdot \alpha_E}{10^{-3} K} \tag{182}$$

When the concentration of solute ions passing the detector is denoted as c_S and their degree of dissociation as α_S, the eluent concentration in the measuring cell during the elution of solute ions is given by $(c_E - c_S \alpha_S)$. Hence, in this instance the measured conductivity is caused by eluent ions *and* solute ions, as well as by the eluent cations that are required to maintain electroneutrality. In anion exchange, solute cations do not have to be taken into account because they are not retained at the anion exchanger. The conductivity resulting from the elution of a solute ions is given by:

$$\varkappa_S = \frac{(\Lambda_{E^+} + \Lambda_{E^-})(c_E - c_S \cdot \alpha_S)\,\alpha_E}{10^{-3} K} + \frac{(\Lambda_{E^+} + \Lambda_{E^-})\, c_S\, \alpha_S}{10^{-3} K} \tag{183}$$

The change in conductance associated with the elution of a solute ion is

$$\begin{aligned} \Delta\varkappa &= \varkappa_S - \varkappa_E \\ &= \left[\frac{(\Lambda_{E^+} + \Lambda_{E^-})\,\alpha_S - (\Lambda_{E^+} + \Lambda_{E^-})\,\alpha_E - \alpha_S}{10^{-3} K}\right] c_S \end{aligned} \tag{184}$$

In principle, Eq. (184) applies to all ion chromatographic methods. It reveals that the detector signal not only depends on the solute ion concentration, but also on the equivalent ionic conductances of eluent cations and eluent and solute anions, as well as the degree of dissociation of eluent and solute ions. The latter two parameters are determined by the pH value of the mobile phase.

Interestingly, the degree of eluent dissociation significantly affects the detector signal. A sensitivity increase with decreasing degree of dissociation of the eluent is derived from Eq. (184). This can be confirmed by the work of Fritz and Gjerde [4], who obtained a much higher sensitivity with pure boric acid as the eluent than with sodium benzoate of comparable elution strength.

According to the above definition, direct detection is feasible when using carefully selected eluents such as phthalate [5] or benzoate [6], which exhibit a low equivalent ionic conductance (see Table 6-1). This results in a conductivity increase when a solute ion passes the conductivity cell.

Alternatively, a strongly conducting eluent may be used. In this case, elution of the solute ions is associated with a negative conductivity change. This indirect detection method is applied to the separation of anions with potassium hydroxide as the eluent [7]. A corresponding chromatogram is displayed in Fig. 6-2. This indirect detection method is also utilized in the analysis of mono- and divalent cations, which are eluted by dilute nitric acid or nitric acid/ethylenediamine-mixtures.

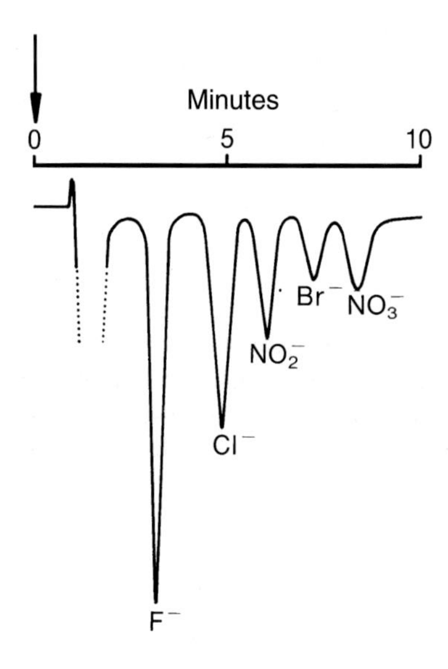

Fig. 6-2. Separation of various inorganic anions with indirect conductometric detection. – Separator column: TSK Gel 620 SA; eluent: 0.002 mol/L KOH; flow rate: 1 mL/min; injection volume: 100 μL; solute concentrations: 5 ppm each of the various anions; (taken from [7]).

Upon application of a suppressor system, the observed sensitivity enhancement is caused by two processes. On the one hand, the eluent is converted into a lower conductive form in the suppressor system which, according to Eq. (184), results in a sensitivity increase. A further enhancement is obtained by converting the solute ions into their corresponding acids or bases. In anion analysis, the associated conductivity change, according to Eq. (185), is to be attributed mainly to the presence of strongly conductive oxonium ions:

$$\Delta\varkappa = \left[\frac{(\Lambda_{H^+} + \Lambda_{S^-}) - (\Lambda_{H^+} + \Lambda_{E^-})}{10^{-3}\,K}\right] c_S \tag{185}$$

Finally, it should be pointed out that chemical suppression in form of protonation reactions is also applicable to zwitterionic eluents [8,9].

6.1.2 Amperometric Detection

Amperometric detection ist generally used for the analysis of ions with pK values above 7 which, owing to their low dissociation, can hardly be detected or not detected at all via conductivity.

In conventional amperometric detection, a three-electrode detector cell consisting of a working electrode, a reference electrode, and a counter electrode is used. The electrochemical reaction at the working electrode is either an oxidation or a reduction. The necessary potential is applied between the working electrode and the Ag/AgCl reference electrode. This Ag/AgCl reference electrode represents a second kind of electrode. In this electrode design, apart from the element and the electrolyte solution, a second solid phase in the form of a sparingly soluble salt contributes to the electrode reaction. Thus, the activity of the potential-determining cation depends via the solubility product on the activity of the anion involved in the formation of the sparingly soluble salt. The Ag/AgCl electrode is utilized as a reference electrode since it is characterized by a good potential constancy at current flow. The purpose of the counter electrode, which is usually made of glassy carbon, is to maintain the potential. Furthermore, it inhibits a current flow at the reference electrode, which could destroy it. When an electroactive species passes the flow cell it is partly oxidized or reduced. This reaction results in an anodic or cathodic current which is proportional to the species concentration over a certain range, and which may be represented as a chromatographic signal.

Such detectors are employed for analyzing a wealth of inorganic and organic ions in the ppb range. This includes environmetally relevant anions such as sulfide and cyanide [10,11], but also arsenic(III) [12], halide ions, hydrazine, and phenols. A survey of electrochemically active compounds and the working electrodes and potentials is given in Table 6-2.

Table 6-2. Electroactive compounds and the required working electrodes and potentials.

Compound	Working electrode	Working potential [V]
HS^-, CN^-	Ag	0
Br^-, I^-, SCN^-, $S_2O_3^{2-}$	Ag	0.2
SO_3^{2-}	Pt	0.7
OCl^-	Pt	−0.2
N_2H_4	Pt	0.5
Phenols	GC	1.2

6.1.2.1 Fundamental Principles of Voltammetry

Information about suitable working potentials for the amperometric detection of electroactive species are obtained in voltammetric experiments. The term "voltammetry" refers to the investigation of current-voltage curves in dependence of the electrode reactions, the concentrations and its exploitation for analytical chemistry. Of the different types of voltammetry, information from the hydrodynamic and pulsed voltammetry can best be applied to amperometry. In both cases, the analyte ions are dissolved in a supporting electrolyte which has several functions:

- Lowering the resistance of the solution ensures that the voltage drop $i \cdot R_L$ is kept low. This is the case for electrolyte concentrations of 0.1 mol/L.
- It has to prevent that *depolarisators* (anions or cations) reach the electrode during their reduction or oxidation. Therefore, the concentration of the supporting electrolyte should exceed that of the depolarisators by a factor of 50 to 100.

The chlorides, chlorates, and perchlorates of alkali and alkaline-earth metals, alkali hydroxides, and carbonates, as well as quaternary ammonium compounds are utilized as supporting electrolytes.

In general, the following reaction occurs in an amperometric detection:

$$\mathrm{A} \rightleftharpoons \mathrm{B} + n \cdot e \tag{186}$$

A transfers n electrons to the working electrode and is oxidized to B. The relation between the concentration of oxidized and reduced species and the applied potential is described by the Nernst equation:

$$E = E_0 + \frac{0.059}{n} \cdot \log \frac{[\mathrm{B}]}{[\mathrm{A}]} \tag{187}$$

E Working potential
[A], [B] Equilibrium concentrations of both species at the electrode surface
E_0 Standard potential, at which the concentrations in A and B are equal. (Since chromatography is carried out at low concentrations, concentrations may be used instead of activities in Eq. (187) in good approximation.)

The working potential selected must have A fully oxidized to B to obtain as high a current yield as possible. For this, the working potential E must always be higher than the standard potential E_0.

A voltammogram of the oxidation of A is recorded with the aid of hydrodynamic voltammetry. This is done by pumping the substance dissolved in the supporting electrolyte through a flow cell, which contains the respective working electrodes. The potential applied to the electrodes is continuously raised while the current flow is registered. Fig. 6-3 shows the sigmoid curve, which is characteristic for substance A. The curve may be understood in terms of the concentration of species A at the electrode surface (Fig. 6-4). Only a small current flows, denoted as residual current, until potential E_A is reached, which is due to the charging of the double layer at the electrode surface and to the reaction of solution impurities. The concentration of A at the electrode surface is equal to that in the bulk of the solution. At potential E_A, the concentration ratio B to A changes until at the standard potential E_0 the concentrations of A and B at the electrode surface are the same. The concentration of A remains constant a few micrometers away from the electrode surface. The distance between the electrode and the point at which the concentration is equal to that in the bulk solution is known as the thickness δ of the diffusion layer. Because of the concentration gradient, the transport of A to the electrode surface occurs only by diffusion within this distance. When the potential is

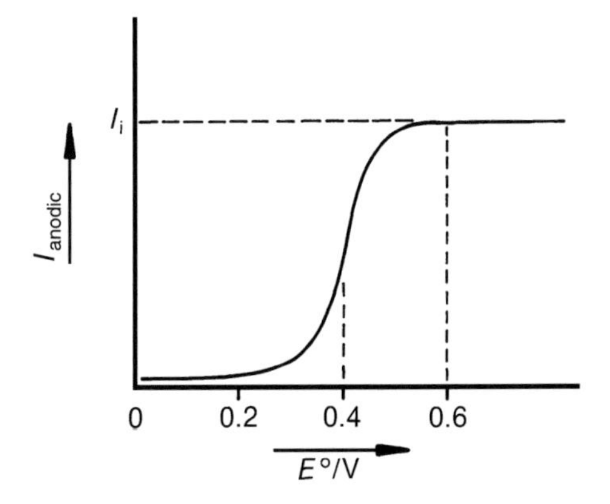

Fig. 6-3. Voltammogram of the oxidation of a species A.

increased in the direction of E_B, the current strength reaches a maximum at E_B and remains nearly constant as A is immediately oxidized to B after it has diffused to the electrode surface. Thus, the current strength is limited by the speed of diffusion as the transport process being the rate-determining step. The resulting current is called the diffusion current. According to Eq. (188), this diffusion current is proportional to the concentration of A in the bulk solution.

$$I = \frac{n \cdot F \cdot A \cdot D \cdot c}{\delta} \tag{188}$$

F Faraday constant
A Electrode surface
D Diffusion coefficient of A

Pulsed amperometry in a three-electrode detector cell is carried out in a non-flowing solution, in which solute ions are dissolved in the supporting electrolyte. A pulsed potential is applied and is increased stepwise after each pulse. The resulting current

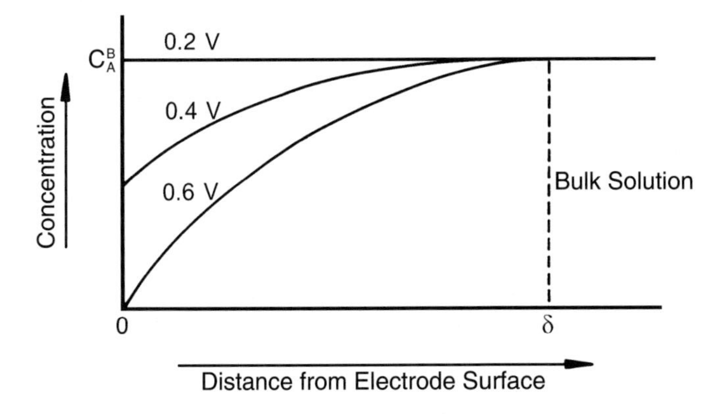

Fig. 6-4. Dependence of solute concentration c from the distance to the electrode surface at various working potentials.

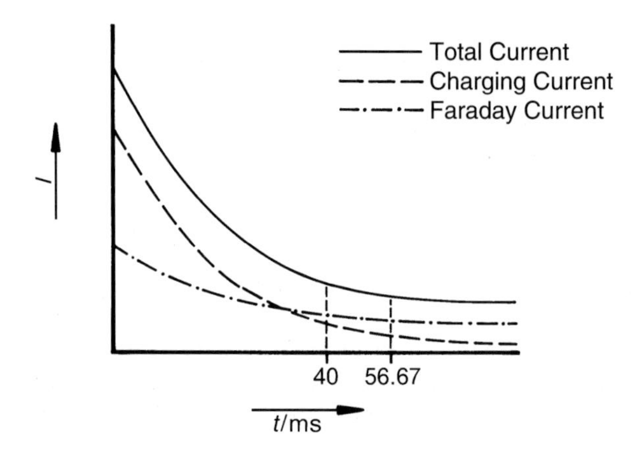

Fig. 6-5. Drop of the total current, the Faraday current, and the charging current after applying a pulse in the diffusion-controlled plateau region.

strength is measured at each step. The voltammogram does not differ from that shown in Fig. 6-3. The measured current strength is very high immediately after a pulse in the diffusion-controlled plateau region, as the molecules A near the electrode are oxidized. Molecules further from the electrode reach the electrode surface only by diffusion and are oxidized there, resulting in a respective drop of the current strength. This drop is illustrated in Fig. 6-4.

Two kinds of current are generated during the oxidation of A to B: the charging current and the Faraday current. While the charging current results from the charging of the interface between the electrode and the bulk solution that acts as a condensator, the Faraday current represents the electron transfer in the oxidation of A to B. As illustrated in Fig. 6-5, the drop of the charging current follows an exponential time law. The drop of the Faraday current with time is described by the relation of Cotrell:

$$I(t) = \frac{n \cdot F \cdot A \cdot D^{1/2} \cdot c}{\pi^{1/2} \cdot t^{1/2}} \tag{189}$$

t Time

Although the current strength is a function of $1/t^{1/2}$, a direct proportionality to the solute ion concentration exists. The sum of both currents is measured as the total current. Since the charging current is much greater than the Faraday current, measurement of the total current is delayed by more than 40 ms after application of the pulse. During this time, the ratio of Faraday current to charging current is enhanced as the latter drops much faster with time.

6.1.2.2 Amperometry

While the working potential required for the desired electrochemical reaction may be determined with voltammetric experiments, amperometry is used as the detection method in ion chromatography. A distinction is made between amperometry with *constant* working potential and pulsed amperometry.

Amperometry with Constant Working Potential

This kind of amperometry is the most widely used electrochemical detection method in liquid chromatography. A constant DC potential is continuously applied to the electrodes of the detector cell. The theory of amperometry with constant working potential does not differ from the theory of hydrodynamic voltammetry, even though the applied potential remains constant.

The working potential is selected so that it is in the diffusion-controlled plateau region for the analyte ion. When several ions with different standard potentials are to be detected in the same run, the working potential must be high enough to cover the plateau regions of *all* ions to be analyzed. The amperometric detection with constant working potential is routinely applied to the species listed in Table 6-2. Fig. 6-6 shows the application of this detection method to the analysis of iodide in a highly concentrated sodium chloride solution. To eliminate interfering effects by the chloride matrix, the iodide oxidation is carried out at a Pt working electrode and not, as usually done, at a Ag working electrode. The necessary sensitivity for this application is obtained by conditioning the electrode for one hour with a saturated potassium iodide solution.

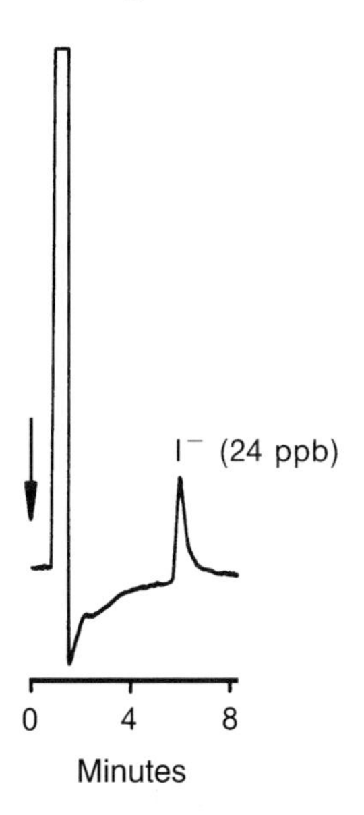

Fig. 6-6. Analysis of iodide in a concentrated NaCl solution. – Separator column: IonPac AS7; eluent: 0.2 mol/L HNO_3; flow rate: 1.5 mL/min; detection: amperometry at a Pt working electrode; oxidation potential: +0.8 V; injection: 50 µL of a 10-fold diluted saturated NaCl solution.

Pulsed Amperometry

Amperometric detection of electroactive species requires that reaction products from the oxidation or reduction of solutes do not precipitate at the electrode surface. Contaminated electrodes change their surface characteristics, thus leading to an enhanced baseline drift, increased background noise, and a constantly changing response. This behavior is particularly pronounced in the amperometric detection of carbohydrates.

Pulsed amperometric detection, on the other hand, utilizes a rapidly repeating sequence of three different working potentials, $E1$, $E2$, and $E3$, which are applied for the times t_1, t_2, and t_3 in increments of 60 ms. In contrast to conventional amperometry, the resulting current is only registered in short time intervals. By applying additional,

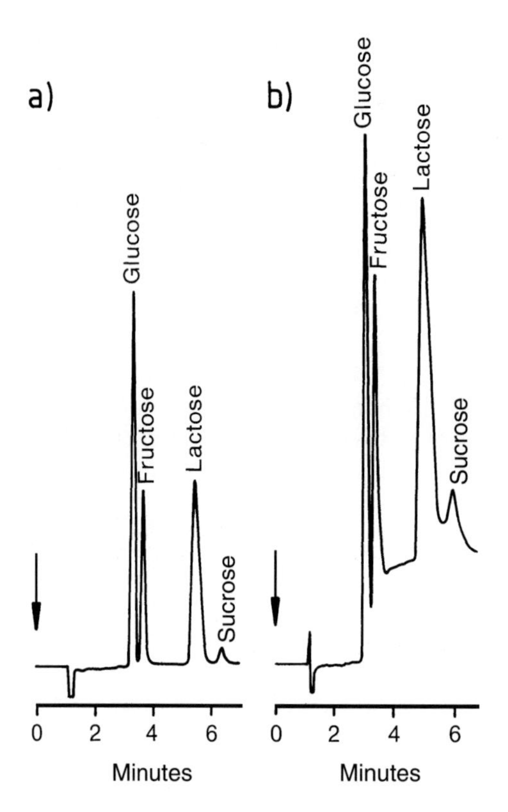

Fig. 6-7. Separation of monosaccharides in a chocolate milk on a latex-based anion exchanger.
a) with pulsed amperometric detection at a Au working electrode,
b) with conventional amperometric detection applying a constant working potential.

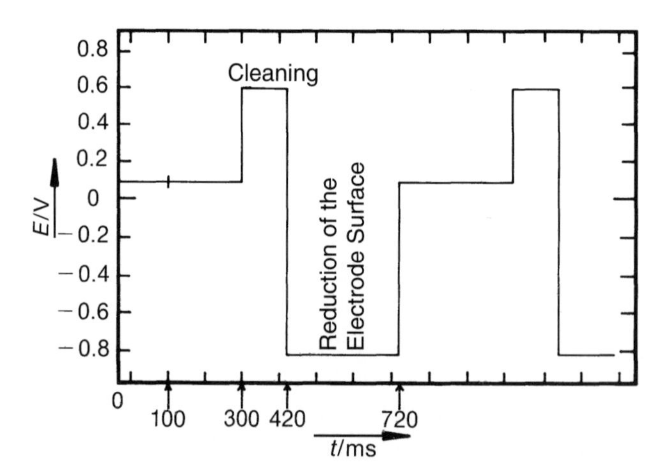

Fig. 6-8. Potential sequence at a Au working electrode for carbohydrate detection.

more positive and more negative potentials, oxidizable and reducible species may be removed from the electrode surface. The advantage of this technique is illustrated in Fig. 6-7, which is a comparison of both amperometric detection methods taking as example the chromatographic separation of sugars in chocolate milk. The working potentials for the carbohydrate analyses are determined with the aid of cyclic voltammetry. The respective pulse sequence is depicted in Fig. 6-8.

This potential sequence not only allows the analysis of carbohydrates and amino acids (see Section 4.9.2) but it enables the detection of organic sulfur species, provided they carry a free electron pair at the sulfur. As an illustration, the chromatogram in Fig. 6-9 displays the separation of lipoic acid which carries a disulfide bridge as a structural element.

COOH
S—S

Lipoic acid

With the aid of pulsed amperometry, this compound can be directly oxidized without prior conversion into the dihydrolipoic acid. The sensitivity of this method is in the lower pmol range for such compounds.

With the development of the pulsed amperometric detector (PAD) [13] a new detector cell was also designed. It is schematically shown in Fig. 6-10. To facilitate replacing the

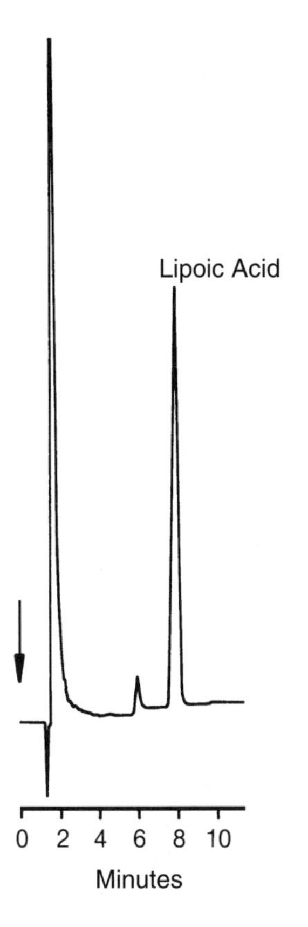

Fig. 6-9. Analysis of lipoic acid utilizing pulsed amperometric detection. – Separator column: CarboPac PA-1; eluent: 0.1 mol/L NaOH + 0.5 mol/L NaOAc; flow rate: 1 mL/min; detection: pulsed amperometry at a Au working electrode; injection volume: 50 μL; solute concentrations: 40 mg/L.

working electrode, the detector cell consists of two halves separated by a spacer. While one part houses the reference electrode and the capillary connections, the other houses the working electrode. This second cell block may be replaced as a whole. Two stainless steel connectors at the inlet and outlet boreholes serve as the counter electrode.

The choice of the working electrode suitable for a given application is determined by the following factors:

- the potential limits for the working electrode in the eluent,
- the possibility of oxidation of the working electrode by formation of complexes with solute species,
- the kinetics of the electrochemical reaction, and
- the background noise.

Potential Limits

The applied potential is limited in the negative range by the potential at which the supporting electrolyte or the solvent is reduced. The limiting factor in the positive potential range is the potential at which the solvent, the working electrode itself, or the supporting electrolyte is oxidized. The potential limits for Au-, Ag-, Pt-, and GC working electrodes in acidic and alkaline solution are listed in Table 6-3.

The potential limits are particularly dependent on the solution pH. In principle, more negative potentials may be applied in alkaline solution than in acidic solution and, conversely, in acidic solution more positive potentials may be selected than in alkaline solution. For oxidation reactions at high positive potentials, GC- and Pt electrodes are suited; reduction reactions at very negative potentials can be performed at GC-, Ag-, and Au electrodes.

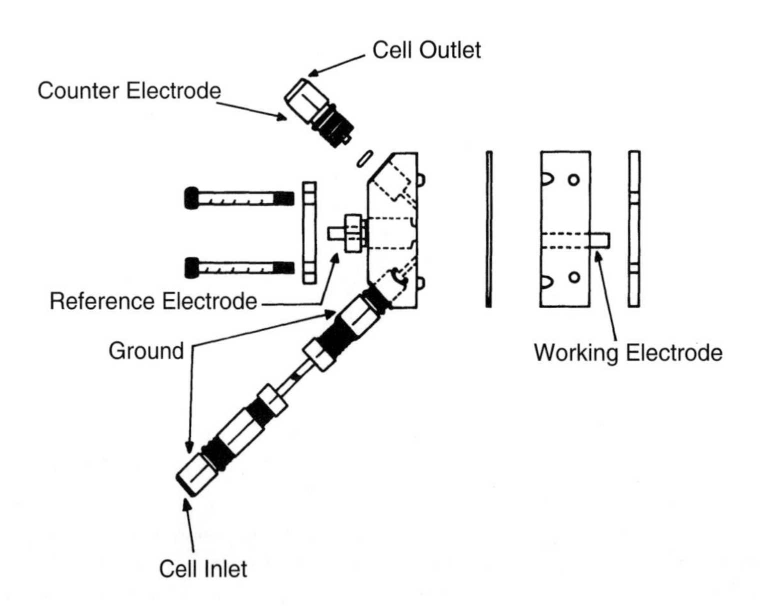

Fig. 6-10. Schematic representation of the detector cell for pulsed amperometry.

Possibility of Complex Formation

At positive potentials, metals such as silver and gold may form complexes with suitable anions in the analyte solution. This may be a disadvantage because it lowers the positive potential limit. For the analysis of complex-forming anions such as sulfide and cyanide, on the other hand, this possibility is advantageous.

Table 6-3. Potential limits in acidic and alkaline solution (reference electrode: Ag/AgCl).

Working electrode	Solution [0.1 mol/L]	Potential range [V]
Gold (Au)	KOH	−1.25...0.75
	$HClO_4$	−0.35...1.10
Silver (Ag)	KOH	−1.20...0.10
	$HClO_4$	−0.55...0.40
Platinum (Pt)	KOH	−0.90...0.65
	$HClO_4$	−0.20...1.30
Glassy carbon (GC)	KOH	−1.50...0.60
	$HClO_4$	−0.80...1.30

Kinetics

If the reaction from A to B is reversible according to Eq. (186), the concentration ratio of A to B at the electrode surface is in the equilibrium state, as described by the Nernst equation (187). As a prerequisite for this, the electron transfer between electrode and solute ions must be kinetically favored. Therefore, a certain species may be oxidized or reduced with different speed at various electrode materials. If an electrode material is chosen at which the electrochemical reaction is very slow, an acceleration of the reaction is only possible by raising the working potential.

The deviations from the equilibrium Galvani potentials[1)] and the respective electrode potentials are denoted as overvoltage. If the transport of reaction partners to the electrode and away from it is slower than the passage reaction, this is called concentration overvoltage. The cause for the passage voltage is a slow passage reaction (i.e., the transfer of species from one side to the other side requires a certain amount of activation energy). If the overvoltage exceeds the permissible potential limit for this electrode material, another material must be selected.

Background Noise

The background noise is typically caused by small fluctuations in the eluent flow rate, vortexing in the detector cell or small temperature variations, respectively. Electrode materials with slow kinetics produce less background noise by its very nature than do

1) The term galvani potential denotes the potential difference between two electrically conducting phases. The equilibrium galvani potential is the galvani potential in the electrochemical equilibrium.

fast-kinetic materials, as their response times with regard to interferences are also slower.

6.2 Spectroscopic Detection Methods

6.2.1 UV/Vis Detection

Direct UV/Vis Detection

In contrast to RPLC, UV detection is of little importance in ion chromatography, but may be considered a wellcomed supplement to conductometric detection. Direct UV detection is at a disadvantage because most inorganic anions do not possess an appropriate chromophore. Thus, they absorb generally at wavelengths below 220 nm [14]. This was corroborated by the work of Reeve [15] and Leuenberger et al. [16], who separated inorganic anions on a chemically bonded cyano- or aminopropyl phase, respectively, and detected them at a wavelength of 210 nm. Recent work by Williams [17] demonstrated the advantages of a simultaneous UV and conductometric detection in combination with a suppressor system. The absorption wavelength of 195 nm chosen by Williams, however, is only applicable to samples with comparatively simple matrices. Therefore, this method is only of academic interest, since in the area of surface water and waste water analysis several organic species absorbing at 195 nm often represent a potential interference.

Table 6-4. Optimal UV measuring wavelengths for some selected inorganic anions.

Anion	Measuring wavelength [nm]
Bromide	200
Chromate	365
Iodide	236
Metal-chloro complexes	215
Metal-cyano complexes	215
Nitrate	215
Nitrite	207
Thiocyanate	215
Thiosulfate	215

Direct UV detection gained great significance in the determination of nitrite and nitrate [18,19] as well as bromide and iodide in the presence of high chloride concentrations. The optimal measuring wavelength for the determination of those anions is listed in Table 6-4. Fig. 6-11 illustrates the superiority of direct UV detection over conductometric detection with a nitrite determination in the presence of a 100-fold chloride excess. Determinations of this kind may be performed in all saline samples such as body fluids, sea water, meat products, sausages, etc. It is worth mentioning that also metal-cyano and metal-chloro complexes [20] can also be detected at a wavelength of 215 nm.

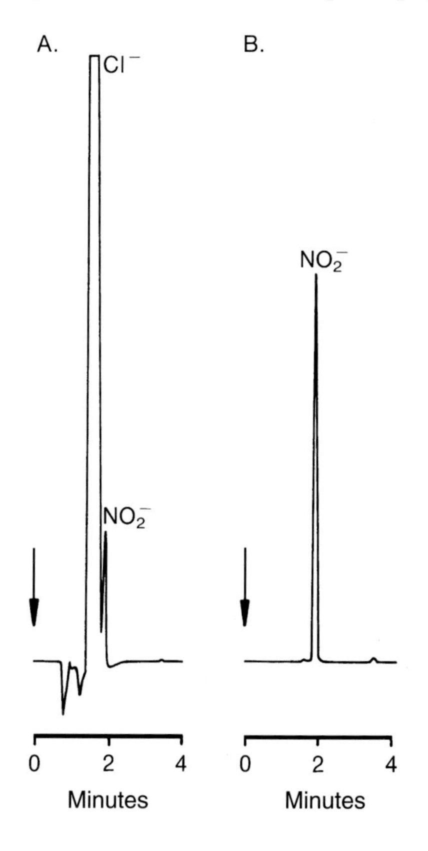

Fig. 6-11. Comparison of direct UV detection and suppressed conductometric detection in the nitrite determination at 100-fold chloride excess. – Separator column: IonPac AS4A; eluent: 0.0017 mol/L $NaHCO_3$ + 0.0018 mol/L Na_2CO_3; flow rate: 2 mL/min; detection: (A) suppressed conductivity (sensitivity: 10 μS/cm), (B) UV detection at 215 nm (sensitivity: 0.05 aufs); injection volume: 50 μL; solute concentrations: 100 ppm chloride and 1 ppm nitrite.

UV/Vis Detection in Combination with Derivatization Techniques

One of the most important applications of UV/Vis detection is the photometric determination after derivatization of the column effluent. First of all, this concerns the determination of heavy and transition metals after reaction with 4-(2-pyridylazo)-resorcinol (PAR).

N=N
OH
OH
PAR

The metal ions are separated on an ion exchanger as oxalate or PDCA complexes (PDCA: pyridine-2,6-dicarboxylic acid). They are then mixed with the PAR reagent and form chelate complexes which absorb in the wavelength range between 490 and 530 nm. A uniform response factor for the different metal ions is obtained by adding Zn-EDTA to the PAR reagent [21]. The metal ions in the column effluent dispel an equivalent zinc ion concentration from the Zn-EDTA complex, which then form the respective chelate

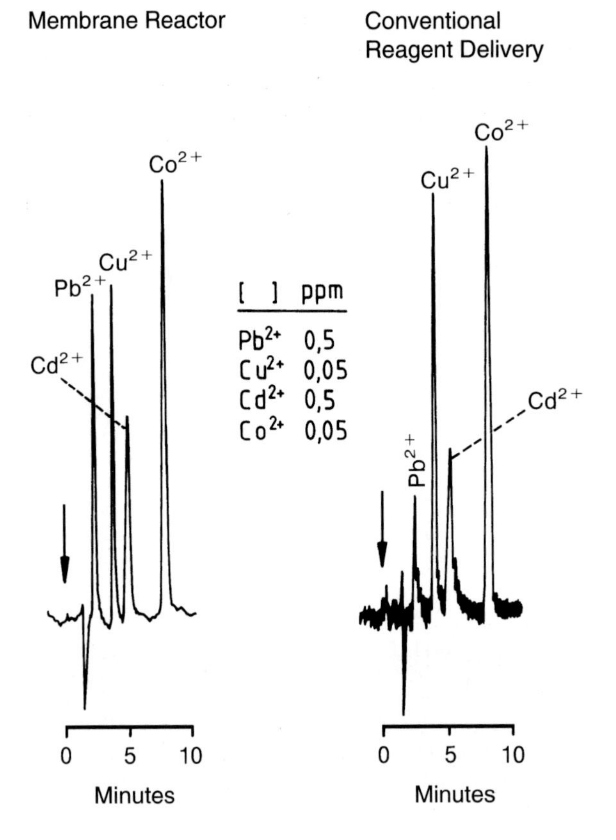

Fig. 6-12. Comparison of the signal-to-noise ratios between a conventional reagent delivery and a delivery via a semipermeable membrane.

complex with PAR. It can be detected at its absorption maximum at 490 nm. Using this method it should be noted that alkaline-earth metals, such as magnesium and calcium, that are present in real samples in much higher concentrations also release Zn ions, thus interfering with the heavy and transition metal analysis.

The reagent may be delivered in two ways. A truly pulsation-free delivery is achieved pneumatically by supplying the pressurized reagent to the column effluent through a capillary. Alternatively, a pump with a puls dampener may be employed. When the reagent is delivered via a tee connector, the reaction with the reagent occurs in an appropriately dimensioned reaction coil. This is typically filled with chemically inert plastic beads to reduce band spreading and to optimize mixing.

The reagent delivery is improved significantly upon application of a membrane reactor. This contains a semipermeable membrane which is permeable for certain reagents. Furthermore, for the operation of a membrane reactor a pressurized container is required, from which the reagent solution reaches the membrane reactor pulsation-free. There the reagent diffuses into the interior of the membrane where mixing with the column effluent takes place. This kind of reagent delivery significantly reduces the short-term noise in the subsequent photometric detection. Fig. 6-12 illustrates this effect in comparison to a conventional reagent delivery. It is particularly noticeable when the reagent being used exhibits a high background absorption at the measuring wavelength. A much smaller reagent volume is delivered when using a membrane reactor, so a significant sensitivity increase is observed because of the smaller dilution factor. Example chromatograms on the analysis of heavy and transition metals upon application of derivatization with PAR are depicted in Section 3.4.5.3.

An extremely selective reagent for chromium(VI) is 1,5-diphenylcarbazide which, in acidic solution, forms an analytically useful inner complex with dichromate. Like the metal-PAR complexes, it is red-purple colored and, thus, may be photometrically detected at 520 nm. The chromate separation is performed – as already mentioned – by means of anion exchange.

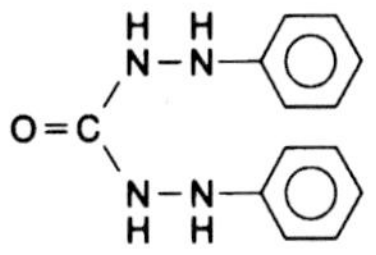

1,5- Diphenylcarbazide

A post-column derivatization with subsequent photometric detection has also been developed for the determination of aluminium. Using a mixture of ammonium sulfate and sulfuric acid, aluminium is separated as $AlSO_4^+$ ion on a cation exchanger. It forms a stable complex with the disodium salt of 4,5-dihydroxy-1,3-benzenedisulfonic acid (Tiron) at pH 6.2 which can be detected at 313 nm.

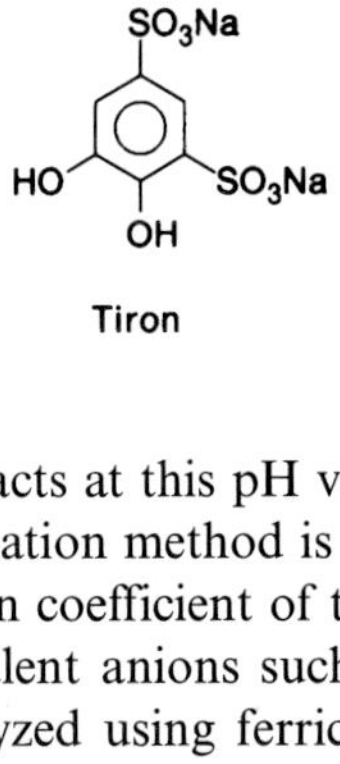

Tiron

Tiron reacts at this pH value only with aluminium and iron(III), hence this aluminium determination method is very selective. The only disadvantage is the comparatively low extinction coefficient of the resulting complex at 310 nm being $\varepsilon = 6000$ L/(mol cm).

Polyvalent anions such as polyphosphates, polyphosphonates, and complexing ions are analyzed using ferric nitrate in acidic solution as the reagent [22] after the compounds have been separated by means of anion exchange chromatography. The reaction of these compounds with iron(III) causes a bathochromic shift of the background absorbance up to the wavelength range between 310 and 330 nm. Since other inorganic anions such as chloride, sulfate, etc. also cause such a shift under these conditions, the detection of complexing agents via derivatization with iron(III) is only possible in matrices with relatively low electrolyte content. Occasionally, the selectivity of the method is not sufficient for the analysis of complexing agents in detergents and cleansing agents as well as in the waste waters resulting from their production and use, as these matrices partly have very high concentration differences between the complexing agents and the electrolytes.

Phosphorus-containing complexing agents, in contrast, can be determined very selectively. The method is based on the selective determination of polyphosphonates developed by Baba et al. [23-25]. It was modified appropriately and was successfully applied to the analysis of polyphosphates *and* polyphosphonates in detergents and cleansing agents by Vaeth et al. [26]. Using nitric acid or peroxodisulfate at 105 °C polyphosphates and polyphosphonates are hydrolyzed to orthophosphate, which is then reacted in a second step with a molybdate/vanadate reagent to the known phosphorvanadatomolybdic acid. Its absorption is measured at 410 nm. The polyphosphate hydrolysis is quantitative under these conditions; the degradation of phosphonates reaches about 90% to 100% and, therefore, must be checked for each particular phosphonate under the given instrumental parameters. In relation to the derivatization with iron(III), simple chromatograms are also obtained with this phosphorus-specific detection for such complex matrices as detergents, since only compounds containing phosphorus are detected. Quantitative analysis is also quite simple. Calibration may be effected with orthophosphate since the peak areas are directly proportional to the phosphorus content of the compound. Additionally, the phosphorus-specific detection allows the use of KCl/EDTA-mixtures as the eluent instead of nitric acid. This results in shorter conditioning times for the separator column and rules out a potential hydrolysis of polyphosphates during the separation.

The derivatization with sodium molybdate can also be utilized for the analysis of water soluble silicate. Its separation can be achieved with both ion-exchange and ion-exclusion chromatography. Silicate with sodium molybdate also forms the yellow heteropolyacid $H_4[Si(Mo_3O_{10})_4]$ with an absorption maximum at 410 nm.

Finally, attention is drawn to the extremely selective method of post-column derivatization of alkaline-earth metals with Arsenazo I [*o*-(1,8-dihydroxy-3,6-disulfo-2-naphthylazo)-benzenearsonic acid], which is particularly suited for the determination of these elements

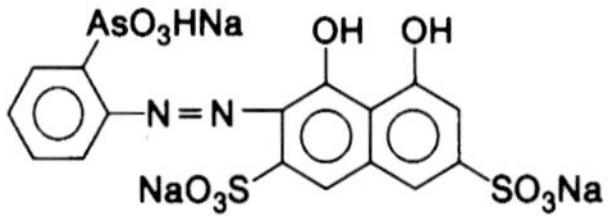

Arsenazo I

at high alkali metal excess [27]. Characteristic applications include the determination of magnesium and calcium in very pure sodium chloride solutions used for the electrolysis of alkali metal chlorides, and the determination of strontium in sea water. With alkaline-earth metals, Arsenazo I forms red-purple complexes absorbing at a wavelength of 570 nm. Fig. 6-13 reveals that the reagent exhibits a high background absorption even at this wavelength, which suggests the use of a membrane reactor. Arsenazo I does not form complexes with alkali metals, thus ensuring the selectivity for alkaline-earth metals that is required for the above-mentioned matrices. Arsenazo I may also be utilized as a derivatization reagent for the analysis of lanthanides. Fig. 6-14 shows an example chromatogram, which was obtained by Wang et al. [28] using a silica-based anion exchanger with 2-methyllactic acid as the eluent.

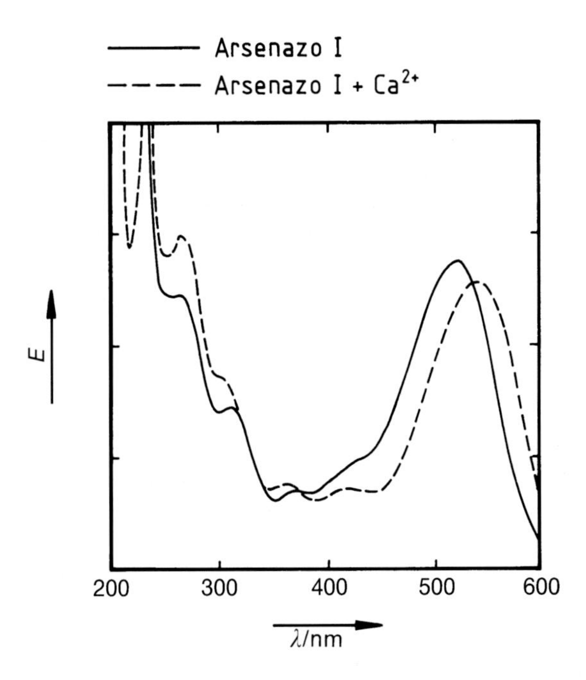

Fig. 6-13. Absorption spectra of Arsenazo I and its corresponding calcium complex.

Indirect UV Detection

UV detection may also be performed indirectly. This method is called indirect photometric chromatography (IPC). Introduced independently by Small et al. [29] and by Cochrane and Hillman [30] in 1982, a UV absorbing eluent is utilized for the determination of UV transparent ions. In the case of anion analysis, the anion exchanger being used is equilibrated with a UV active eluent Na^+E^-. According to Fig. 6-15a, at a constant flow rate and an appropriate absorption wavelength a constant signal at the detector outlet is observed. If a sample Na^+S^- is injected, the solute anion S^- is retained under suitable chromatographic conditions and is eluted with a certain retention time. The nearly Gaussian-shaped elution profile is seen in Fig. 6-15b. Owing to the principle of electroneutrality and the exchange equilibrium, a change in total absorption must accompany the appearance of S^-, as both the total anion concentration (E^- and S^-) and the concentration of counter ions (Na^+) remain constant. In this way, the concentration of S^- is determined *indirectly* by measuring the decreasing UV absorption of E^-.

For the analysis of inorganic anions via indirect UV detection, one normally utilizes aromatic carboxylic acids as in direct conductometric detection, since they possess a correspondingly high light absorptivity. Small et al. illustrated this with the separation of various inorganic anions (shown in Fig. 6-16), which were separated on a latexed anion exchanger using sodium phthalate as the eluent. The anions shown in Fig. 6-16 exhibit no intrinsic absorption at the measuring wavelength of 285 nm.

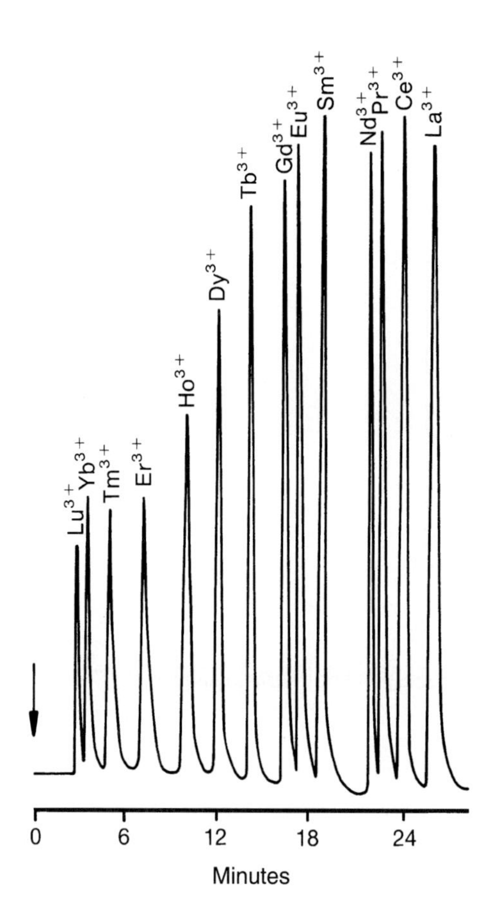

Fig. 6-14. Detection of lanthanides by post-column derivatization with Arsenazo I. – Separator column: Nucleosil 10 SA; eluent: 2-methyllactic acid; gradient: linear, 0.01 mol/L in 30 min to 0.04 mol/L; detection: photometry at 600 nm after reaction with Arsenazo I; solute concentrations: 10 ppm each of the various lanthanides; (taken from [28]).

When selecting suitable eluents for indirect photometric detection the chemical and photometric properties of the eluent ions have to be taken into consideration. Knowing the absorption properties of the organic acids and the dependence of its degree of dissociation on the pH of the mobile phase is essential for a successful application of this method. These criteria were investigated by Small et al. [29] for a series of eluent ions. Phthalate, *o*-sulfobenzoate, iodide, and trimesate proved to be particularly suited for the inorganic anion analysis (see also Section 3.3.4.3). Fig. 6-17 reveals that indirect photometric detection also allows the detection of carbonate. Owing to its high p*K* value this ion cannot be determined together with strong mineral acids via suppressed conductometric detection.

The precise measurement of eluent ion absorption is also extremely important in indirect photometric detection. It is known from classical spectrophotometry [31] that the photometric error is only small in an absorbance range between 0.2 and 0.8. There-

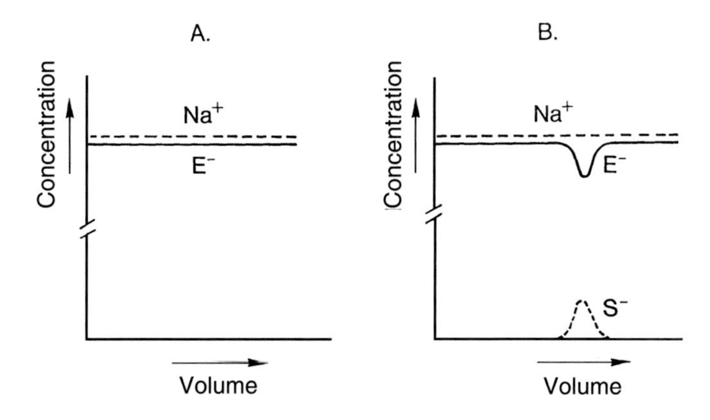

Fig. 6-15. Principle of the indirect photometric detection.

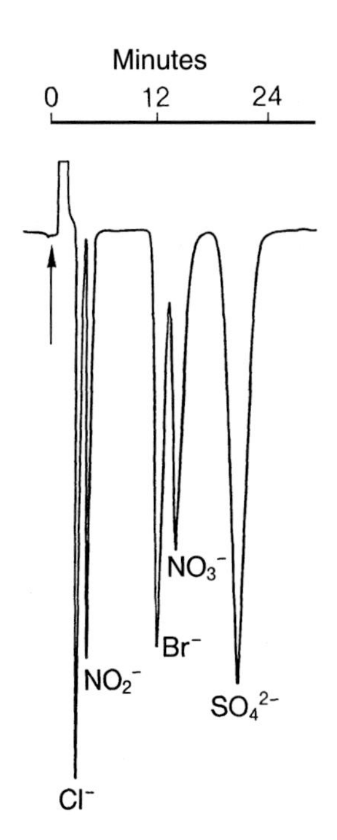

Fig. 6-16. Indirect photometric detection of various inorganic anions. – Separator column: 250 mm x 4 mm I.D. SAR-40-0.6; eluent: 0.001 mol/L sodium phthalate (pH 7 to 8); flow rate: 2 mL/min; detection: UV (285 nm, indirect); injection volume: 20 μL; solute concentrations: 106 ppm chloride, 138 ppm nitrite, 400 ppm bromide, 310 ppm nitrate, and 480 ppm sulfate; (taken from [29]).

fore, the eluent ion concentration chosen should have a background absorption that lies within this range. However, because of the exchange capacity of the separator column and the required elution power, the eluent ion concentration may not be freely selected. The background absorption must be adjusted accordingly by choosing a suitable measuring wavelength. Indirect photometric detection is compatible with separator columns of differing exchange capacities and with eluents of high and low concentrations. This detection method may be adapted to the respective chromatographic conditions.

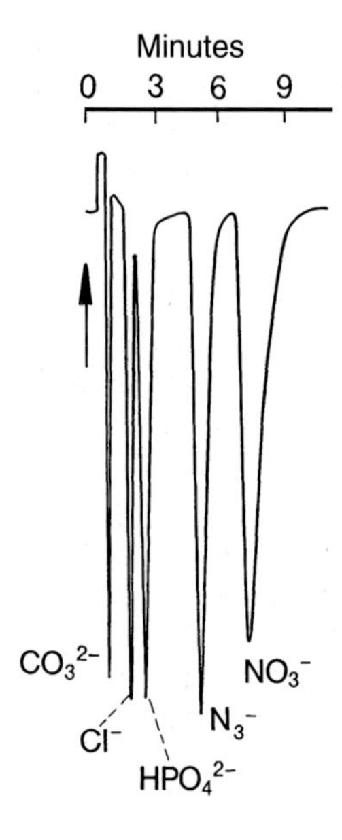

Fig. 6-17. Indirect photometric detection of weak and strong inorganic acids. – Separator column: see Fig. 6-16; eluent: 0.001 mol/L sodium phthalate + 0.001 mol/L H_3BO_3 (pH 10); flow rate: 5 mL/min; detection: UV (285 nm, indirect); injection volume: 20 µL; solute concentrations: 90 ppm carbonate, 70 ppm chloride, 190 ppm orthophosphate, 250 ppm azide, and 500 ppm nitrate; (taken from [29]).

The measuring wavelength directly affects the detection sensitivity. If the molar extinction coefficient of the eluent ion increases due to changes in the measuring wavelength, the detection sensitivity is enhanced, as the absorbance difference becomes larger at constant eluent ion concentration. The highest sensitivity is obtained in the wavelength region of the absorption maximum of the eluent ion.

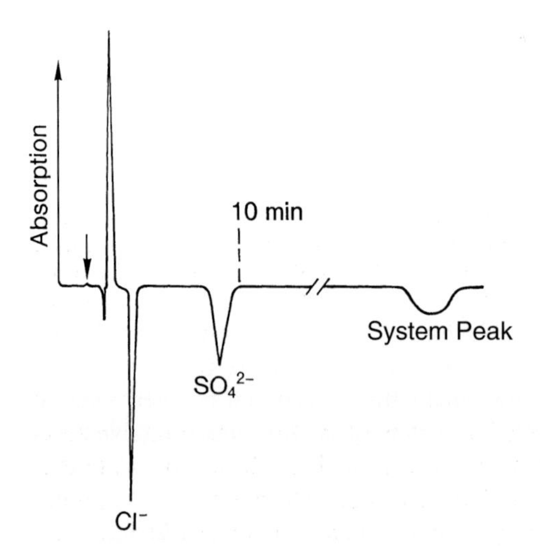

Fig. 6-18. Ion-pair chromatographic separation of chloride and sulfate with indirect photometric detection. – Separator column: 200 mm x 4.6 mm I.D. Zorbax C8; eluent: 0.001 mol/L *N*-methyloctylammonium-*p*-toluenesulfonate; flow rate: 1.6 mL/min; injection volume: 20 µL; solute concentrations: 35 ppm chloride and 96 ppm sulfate; (taken from [32]).

The principle of indirect photometric detection can also be applied to ion-pair chromatography provided the counter ion of the ion-pair reagent exhibits the respective absorption properties. For example, *N*-methyloctylammonium-*p*-toluenesulfonate was utilized as the ion-pair reagent to analyze chloride and sulfate [32] upon application of indirect photometric detection. The corresponding chromatogram is shown in Fig. 6-18.

In connection with indirect photometric detection, a signal is observed in anion separations both in ion-exchange and in ion-pair chromatography which may not be attributed to any solute ion [32-34]. This signal, usually referred to as the system peak, changes its size and position depending on the chosen chromatographic conditions. It represents a potential interference for solute ions eluting in this range. A detailed explanation of this phenomenon is found in Section 3.3.4.3.

Indirect photometric detection was also utilized for cation determination. Fig. 6-19 illustrates the separation of sodium, ammonium, and potassium with copper sulfate as the eluent obtained by Small et al. [29]. Aromatic bases as cationic analogues to aromatic acids used in anion analysis are inappropriate as eluents for this detection method. As monovalent cations, they are already eluted by the oxonium ions. Thus, a significant absorption change is only observed when the protonated base cation contributes to the ion-exchange process.

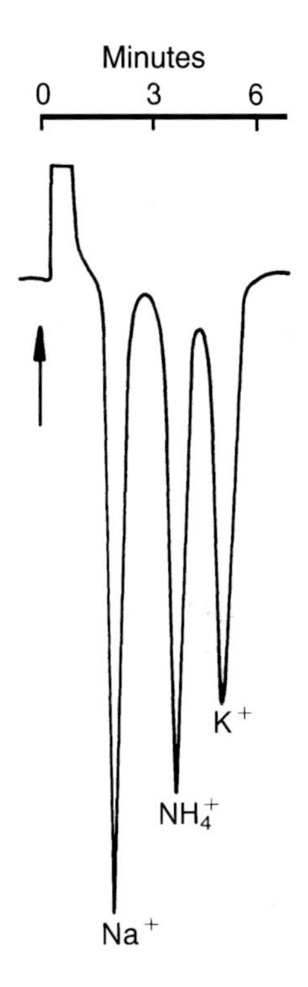

Fig. 6-19. Indirect photometric detection of sodium, ammonium, and potassium. – Separator column: Dowex 50; eluent: 0.005 mol/L copper sulfate; flow rate: 0.7 mL/min; detection: UV (252 nm, indirect); injection volume: 20 µL; solute concentrations: 230 ppm sodium, 180 ppm ammonium, and 391 ppm potassium; (taken from [29]).

In summary, it must be pointed out that indirect photometric detection is characterized by a higher sensitivity compared to direct conductivity detection, while it is markedly less sensitive than supressed conductivity detection. The use of indirect photometric detection is preferred for the determination of main components to prevent a strong dilution of the sample. With direct sample injection, typical detection limits for inorganic anions are in the medium to higher ppb range.

6.2.2 Fluorescence Detection

Fluorescence detection is utilized in ion chromatography mainly in combination with post-column derivatization, since inorganic anions and cations, with the exception of the uranyl cation UO_2^{2+}, do not fluoresce. Fluorescence results from the excitation of molecules via absorption of electromagnetic radiation. It is the emission of fluorescence radiation when the excited system returns to the energetic ground level. The emitted wavelength is characteristic for the kind of molecule while the intensity is proportional to the concentration.

The best known and most widely used fluorescence method was developed by Roth and Hampai [35] for the detection of primary amino acids, and was described in Section 4.9.2. It is based mainly on the reaction of α-amino acids with *o*-phthaldialdehyde (OPA) and 2-mercaptoethanol to yield an intensively blue-fluorescing complex. Even at room temperature, the reaction occurs within seconds. At an excitation wavelength of 340 nm and an emission wavelength of 455 nm this method has a detection limit in the lower pmol range. The derivatization with *o*-phthaldialdehyde applies to all compounds carrying a primary amino group. This includes the ammonium ion, primary amines, polyamines, and peptides.

The only drawback of the fluorescence method developed by Roth and Hampai is the relatively small stability of the *N*-substituted 1-alkylthioisoindoles that are formed from α-amino acids after reaction with OPA. This severely limits the applicability of OPA for the *pre-column derivatization* of primary amines.

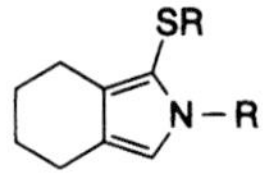

SR

N−R

N-Substituted alkylthioisoindole

Much higher stabilities and partly higher fluorescence yields were obtained by Stobough et al. [36] with naphthaline-2,3-dialdehyde (NDA) as the reagent, which reacts in the presence of cyanide ions with primary amines to *N*-substituted 1-cyanobenz[f]isoindoles (CBI):

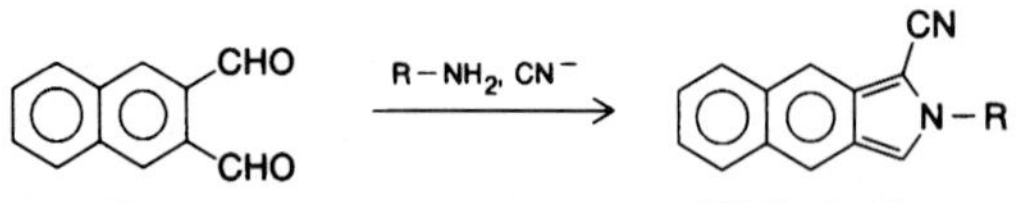

CHO CHO $\xrightarrow{R-NH_2,\ CN^-}$ CN N−R

NDA CBI-Derivative

The excitation spectrum of these derivatives shows maxima at 246 nm and 420 nm; the emission is measured at a wavelength of 490 nm. At an excitation wavelength of 246 nm detection limits are obtained in the medium to lower fmol range. However, secondary amines can only be subjected to both derivatization methods described above after oxidation.

Section 6.2.1 illustrated that with the phosphorus-specific detection of polyphosphates and polyphosphonates, two-step post-column derivatizations are now possible. The analysis of a secondary amine via a two-step derivatization and fluorescence detection is exemplified with a herbicide – the glyphosate [*N*-(methylphosphono)-glycine].

$$\begin{array}{l} HN-CH_2-PO_3H_2 \\ | \\ CH_2-COOH \end{array}$$

Glyphosate

After its separation on an anion exchanger it is oxidized with hypochlorite and subsequently reacted with OPA/2-mercaptoethanol. As seen in the corresponding chromatogram in Fig. 6-20, even the main glyphosate metabolite – aminomethanesulfonic acid (AMPA) – may be detected in the same run.

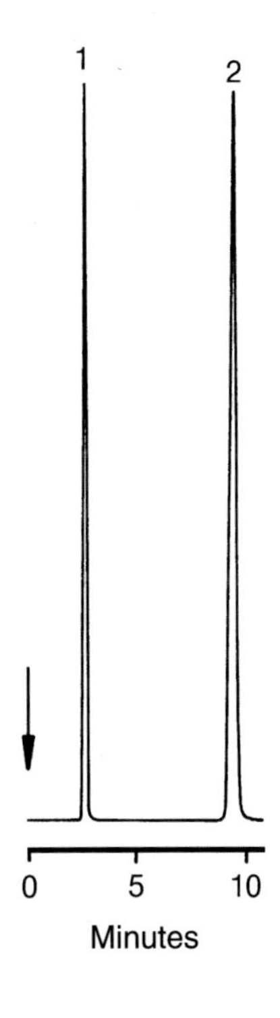

Fig. 6-20. Two-step derivatization of glyphosate with subsequent fluorescence detection. – Separator column: IonPac AS6 (CarboPac); eluent: 0.01 mol/L H_3PO_4; flow rate: 1 mL/min; detection: fluorescence after oxidation with NaOCl and reaction with OPA; injection volume: 100 μL; solute concentrations: 2.5 ppm each of aminomethanephosphonic acid (1) and glyphosate (2).

A further derivatization technique for forming fluorophors has been described by Lee and Fields [37]. They reacted oxidizable inorganic anions such as nitrite, thiosulfate, and iodide with cerium(IV), thereby forming fluorescing cerium(III) according to Eqs. (190), (191), and (192)

$$HNO_2 + 2\,Ce^{4+} + H_2O \rightarrow NO_3^- + 2\,Ce^{3+} + 3\,H^+ \tag{190}$$

$$2\,S_2O_3^{2-} + 2\,Ce^{4+} \rightarrow S_4O_6^{2-} + 2\,Ce^{3+} \tag{191}$$

$$2\,I^- + 2\,Ce^{4+} \rightarrow I_2 + 2\,Ce^{3+} \tag{192}$$

Cerium(III) may be detected at an excitation wavelength of 247 nm and an emission wavelength of 350 nm. To stabilize the cerium(IV) reagent, it is prepared in sulfuric acid of the concentration c = 0.5 mol/L. The addition of sodium bismutate serves to oxidize possible cerium(III) traces in the reagent to keep the residual fluorescence as low as possible.

For the reaction of the column effluent with the cerium(IV) reagent, instead of a simple injection loop, Fields et al. used a solid-bed reactor with a volume of 2.8 mL. This relatively large volume is necessary to allow the required reaction time of at least two minutes for the oxidation of nitrite ions with cerium(IV). While the reaction of nitrite ions with cerium(IV) is comparatively slow, the maximum fluorescence yield with iodide is obtained in less than ten seconds. On the other hand, the reaction kinetics with thiosulfate appears to be completely different. As seen in the respective diagram in Fig. 6-21, this reaction is characterized by a fast rise of the fluorescence yield within a short time, which increases as the reaction product from Eq. (191), tetrathionate, also reacts slowly with cerium(IV).

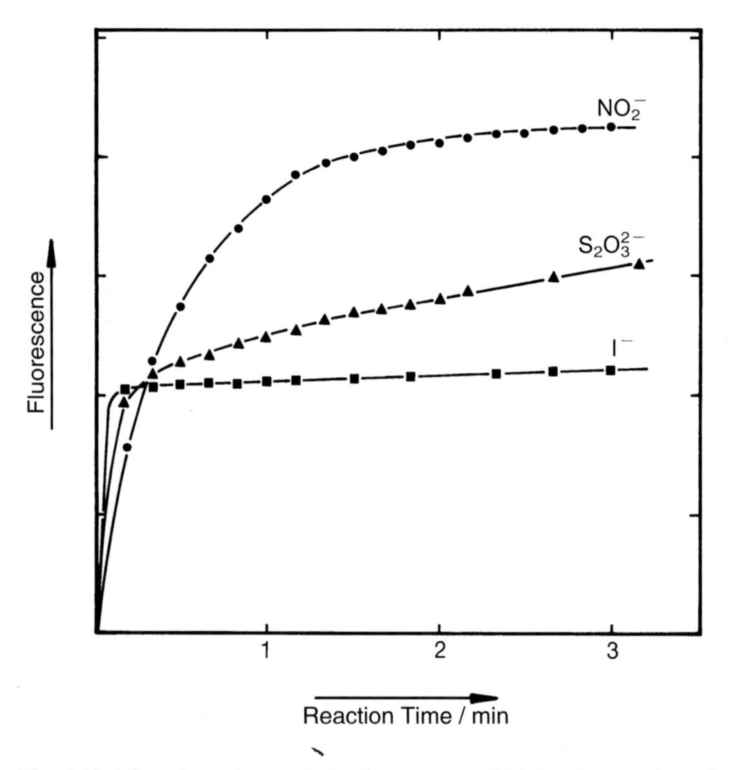

Fig. 6-21. Time dependence of the fluorescence yield for the reaction of nitrite, thiosulfate, and iodide with cerium(IV); (taken from [37]).

The choice of the eluent is decisively important for achieving maximum sensitivity. According to Fields et al., for the exchange chromatographic separation of the investigated anions in a single run potassium hydrogenphthalate and succinic acid may be used as the eluent, with succinic acid exhibiting the lower residual fluorescence. A 1.2-nmol standard of these three anions is shown in the following chromatogram (Fig.

6-22), that was obtained on Vydac 302 IC with potassium hydrogenphthalate as the eluent. The detection limit obtained with this method is in the lower ppb range for all three investigated anions. Even though it does not differ considerably from that of direct UV detection the specificity of this method is significantly better especially for samples having a high sodium chloride content.

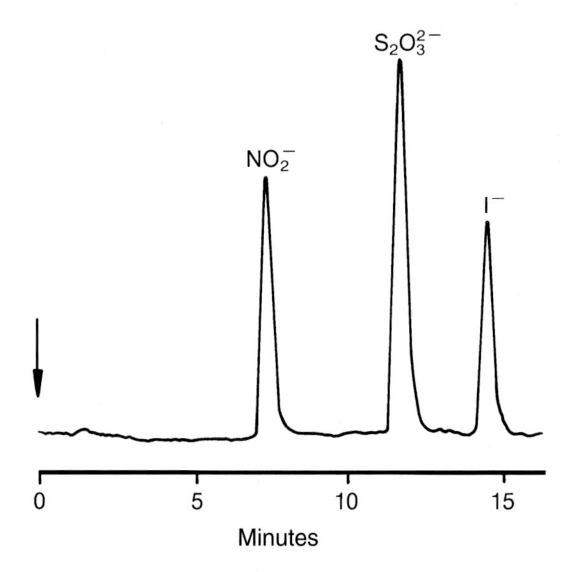

Fig. 6-22. Analysis of nitrite, thiosulfate, and iodide upon application of fluorescence detection after derivatization with cerium(IV). – Separator column: Vydac 302 IC; eluent: 0.001 mol/L KHP + 0.01 mol/L Na_2SO_4 pH 5.5 with $Na_2B_4O_7$; flow rate: 1 mL/min; detection: fluorescence after reaction with cerium(IV); injection volume: 100 µL; solute concentrations: 0.5 ppm nitrite, 1.1 ppm thiosulfate, and 1.5 ppm iodide; (taken from [37]).

Another important application of the redox pair cerium(III)/cerium(IV) is the indirect fluorescence chromatography (IFC) of alkali and alkaline-earth metals. In analogy to indirect photometric detection (IPC), an eluent is utilized with a high residual fluorescence which is lowered during the solute ion elution. Hence, negative signals are registered for analyte species. While it has been known for some time that an aqueous cerium(III) solution fluoresces [38,39] and, as described above, is used analytically [37], only the recently published work of Danielson et al. [40,41] confirmed that strongly diluted cerium(III) solutions can also be employed to elute alkali metals and ammonium. The cerium(III) concentration required for an almost baseline-resolved separation of monovalent cations is only 10^{-5} mol/L. Fig. 6-23 shows such a chromatogram with a total analysis time of less than eight minutes. Lithium cannot be determined under these chromatographic conditions because it elutes within the void volume. The detection limits obtained for monovalent cations (3 ppb for sodium and 100 ppb for cesium) are comparable to those of other detection methods such as indirect photometric detection and conductometric detection. Again, in samples with complex matrices, the higher specificity of IFC is advantageous. This also applies to the indirect fluorescence detection of alkaline-earth metals that was recently described by Bächmann et al. [42]. Since a higher cerium(III) concentration is necessary to elute divalent cations, a significantly higher residual fluorescence results that adversely affects the baseline quality.

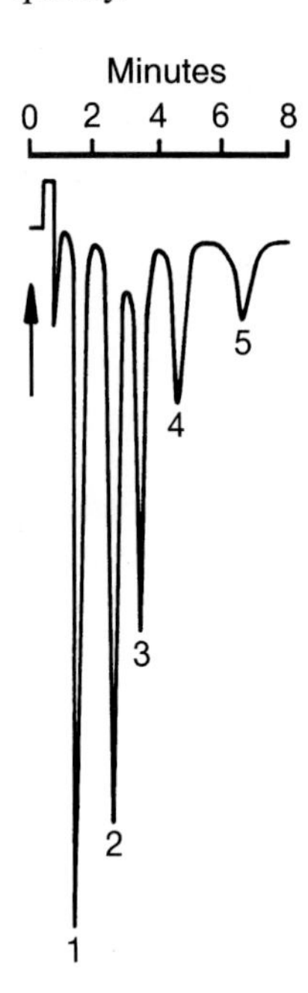

Fig. 6-23. Analysis of monovalent cations with indirect fluorescence detection. – Separator column: 100 mm x 3.2 mm I.D. ION-210; eluent: 10^{-5} mol/L cerium(III) sulfate; flow rate: 1 mL/min; detection: indirect fluorescence; injection volume: 20 µL; solute concentrations: 0.16 ppm sodium (1), 0.15 ppm ammonium (2), 0.21 ppm potassium (3), 0.71 ppm rubidium (4), and 1.2 ppm cesium (5); (taken from [41]).

The indirect fluorescence detection also applies to anion analysis [43,44]. Mho and Yeung [43] utilized salicylate as the fluorescing eluent ions. Owing to their structural similarities to phthalate and benzoate, these ions exhibit good elution properties. Salicylate shows a strong absorption at 325 nm and a fairly efficient emission at 420 nm. A HeCd laser with double-beam optics was used as the excitation source that emits polarized UV light with a wavelength of 325 nm at a power of about 7 mW. However, with regard to sensitivity, the method does not differ from the above-described cation analysis technique using cerium(III); i.e., the detection limits obtained for inorganic anions such as iodate and chloride correspond to those of indirect photometric detection and conductometric detection, because the eluent ion concentration of about $2.2 \cdot 10^{-4}$ mol/L sodium salicylate that is needed for eluting solute ions is normally too high. The full potential of this analytical method will only be exploited when it is possible to manufacture separator columns with extremely low ion-exchange capacities that allow the use of very dilute eluents.

6.3 Other Detection Methods

Apart from the detection methods described thus far, refractive index detection is also mentioned in the literature, although this detection method is very insensitive and, above all, unspecific. In contrast to other detection methods, refractive index detection is subject to much fewer restrictions. The refractive index of the solvent has no influence because the measurement is usually carried out by comparing the refractive index of the column effluent with that of the pure solvent present in the reference cell. Thus, the sensitivity of this detection method depends solely on the detector performance. Commercially available detectors are usually designed for the detection of higher concentrations. For the separation of inorganic anions and their detection via changes in the refractive index, organic acids such as phthalic acid [33,34,45], salicylic acid [45], *p*-hydroxybenzoic acid [45], and *o*-sulfobenzoic acid [45] are utilized as the eluent. Fig. 6-24 illustrates such a chromatogram. The only advantage of refractive index detection is the feasibility to use comparatively concentrated eluents, which allow the use of high-capacity ion exchangers.

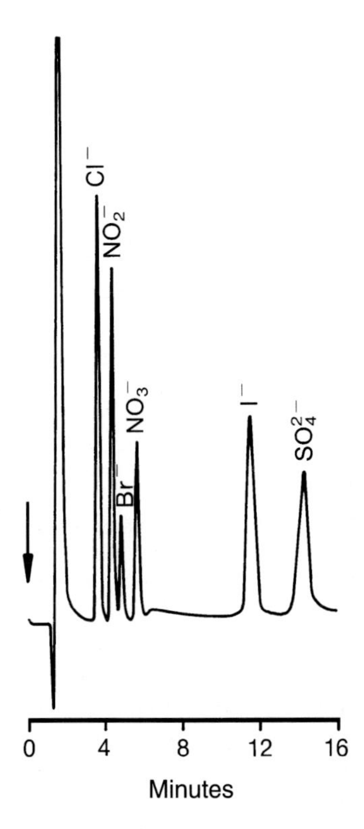

Fig. 6-24. Refractive index detection of various inorganic anions. – Separator column: 250 mm x 4.6 mm I.D. Vydac 302 IC; eluent: 0.006 mol/L sodium hydrogenphthalate, pH 4.0; flow rate: 2 mL/min; injection volume: 100 µL; solute concentrations: 2.5 to 7.5 ppm; (taken from [46]).

The combination of an ion chromatographic separation with a radioactivity monitor (type LB505, Berthold, Germany) for the analysis of radiostrontium was described by Stadlbauer et al. [47]. Their objective was to develop a method for the simple separation of the fission products Sr-90 and Sr-89 from other radionuclides. One of them is barium (Ba-133) which, like Cs-137, is a companion of Sr-90/Sr-89 and is formed in a similar yield (about 6%) in a nuclear reactor. Stadlbauer et al. separated these ions on a surface-sulfonated cation exchanger which was connected to a scintillation detector cell. Fig. 6-25 shows the chromatogram of a Sr-90 standard that was obtained with this setup. The peak volume in this chromatogram corresponds to 76.4 Bq Sr-90. Although on-line detection of radiostrontium is only feasible at medium to high activity concentrations, this methods provides advantages for process control and self-monitoring of nuclear-medical and nuclear power plants. The fractionated collection of the column effluent makes it possible to obtain Sr-90 in pure form from samples having a high calcium excess or a complex radioactive contamination by various nuclides. It may be determined off-line via a methane flow counter.

Finally, reference is made to the element-specific detection via atom spectroscopic techniques, which include atomic absorption and atomic emission spectroscopy.

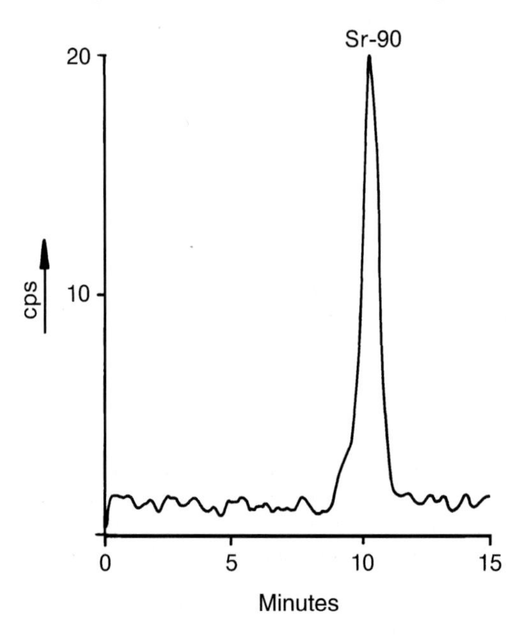

Fig. 6-25. Analysis of Sr-90 with the aid of a radioactivity monitor. – Separator column: IonPac CS2; eluent: 0.03 mol/L HCl + 0.002 mol/L histidine hydrochloride; flow rate: 1.5 mL/min; detection: scintillation measurement for Sr-90; injection volume: 50 μL; solute concentration: 76.4 Bq Sr-90 with inactive $SrCl_2$ as the carrier.

The coupling of an atomic absorption spectrometer with an ion chromatograph is relatively straightforward. It requires a capillary which connects the separator column end with the nebulizer of the AAS instrument [48]. When selecting the eluent to be used to separate the analyte species, it is important to prevent a high background signal and a quick sooting of the burner by the mobile phase. The advantage of this coupling is that it allows the separation and detection of metals in different oxidation states. Woolson and Aharonson [49], for example, utilized this method for the determination of arsenic(III) and arsenic(V).

The same compounds have also been examined by Urasa et al. [50], who coupled an ion chromatograph to a DCP atom emission spectrometer[2)] by connecting the column end capillary with the nebulizer of the spectrometer, which was slightly modified [51]. A principle problem of DC plasma detection is the relatively low sensitivity, which may be attributed to the inadequate efficiency of the nebulization and to the sample dilution during the chromatographic process. The dilution effect alone accounts for a reduction

2) DCP: direct current plasma

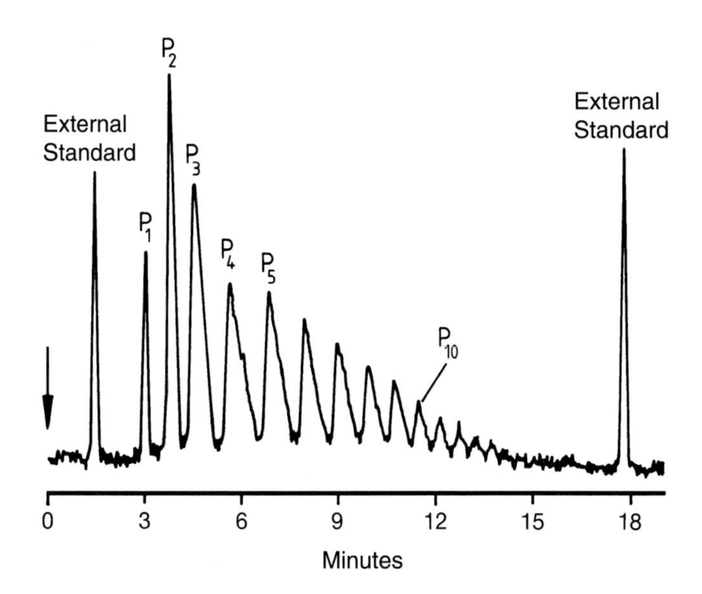

Fig. 6-26. Analysis of polyphosphoric acid upon application of DC plasma detection. – Separator column: 150 mm x 4.1 mm I.D. PRP-1; eluent: (A) 0.01 mol/L tetraethylammonium nitrate (pH 9), (B) 0.01 mol/L tetraethylammonium nitrate + 0.1 mol/L KNO_3 (pH 9.0); gradient: linear, 1% B to 40% B in 18 min; flow rate: 1.2 mL/min; detection: DC plasma; emission wavelength: 214.9 nm; injection: 20 μL of a solution with ca. 60 μg P; external standard: 8.24 μg P as orthophosphate; (taken from [52]).

in signal intensity by 50% to 75% compared to direct DCP-AES. Urasa et al. compensated this by injecting large sample volumes (up to 1000 μL). Alternatively, they examined the applicability of concentrator columns for improving the detection sensitivity. It turned out that 10 ppb arsenic(V) may be correctly determined by concentrating a volume of 100 mL. This represents a significant advance with regard to direct DCP-AES which is less sensitive by two orders of magnitude.

An interesting application of direct current plasma detection is the method developed by Biggs et al. [52] for the determination of polyphosphates, which were separated on a neutral PS/DVB-based polymer phase with the aid of ion-pair chromatography. The separation of the phosphate oligomers P_1 to P_{12} in neutralized polyphosphoric acid was carried out with tetraethylammonium nitrate as the ion-pair reagent and by applying a potassium nitrate gradient. Contrary to expectations, the rise of the salt concentration during the gradient run did not cause a baseline drift nor did it affect the response factor. A representative chromatogram of a polyphosphoric acid sample neutralized with tetraethylammonium hydroxide is shown in Fig. 6-26. The detection limit of this method is 0.2 μg P.

7 Quantitative Analysis

7.1 General

The following discussion deals with the various steps necessary for developing a procedure for the quantative analysis [1] of a compound. It is assumed that the chromatographic conditions for the separation of this compound have been established and optimized.

A quantitative analysis may be subdivided into the following individual steps:

- sampling,
- sample preparation,
- separation,
- detection, and
- signal processing.

Chemical analysis starts with sampling which is aimed at obtaining a sample that represents the bulk composition. Errors are encountered, above all in the selection of a representative sample as well as in the fields of sample storage and sample preparation. The sampling of inhomogeneous solids such as soil samples, detergents, etc., poses a problem. Frequently, physiological liquids such as urine are sampled for a prolonged time to obtain a representative sample. Errors in the sample storage are often encountered with volatile, reactive, or oxygen-sensitive samples, respectively. Likewise, adsorption at and desorption from container walls must be considered as possible error sources.

Sample preparation embraces all processes that convert the analyte into a form suitable for analysis. This includes a possible comminution, homogenization, digestion, dissolution, and filtration of the sample. Ideally, the sample should be dissolved in the eluent being used since the negative peaks that are observed during the dead time in conductivity detection mode do not occur. Before injecting the sample solution, it should be micro-filtrated (0.22 or 0.45 µm) to prevent particulate matter from entering the column.

Sample injection is typically performed via a sample loop. This allows the injection volume to be reproducibly introduced. Also, the injection process may easily be automated.

The subsequent chromatographic separation also involves a series of error sources. The major causes are the co-elution of ions and their possible decomposition in the mobile or at the stationary phase, respectively.

The chromatographic separation is followed by a measurement in which a physical property (electrical conductance, UV absorption, etc.) of the sample is determined. This property must correlate in a clear and known way with the chemical composition and structure of the sample. The measured signals are then processed and listed in an analysis report.

7.2 Analytical Chemical Information Parameters

In the evaluation of the measured signals, quantities must be determined to provide information about the chemical composition of the sample. These properties are called information parameters because they carry analytical chemical information.

In chromatography, a series of signals from the detector output is registered as the chromatogram. The qualitative information is derived from the retention time, t_{ms}, which is determined by the chromatographic process and which depends on the thermodynamic properties of both the stationary phase and the solutes. The quantitative information stems from the area under the signal, and is determined by the measurement process and depends on the detector properties. In so doing, it is assumed that the sample component elutes quantitatively from the analytical column.

According to Kucera [2] or Grushka [3], respectively, the area of peak i corresponds to the zero moment m_{0i} of the distribution function describing the peak.

$$m_{0i} = \int_{-\infty}^{+\infty} y_i(t)\, \mathrm{d}t \tag{193}$$

The first moment normalized to m_0 corresponds to the retention time:

$$m_{1i} = \int_{-\infty}^{+\infty} t \cdot y_i(t)\, \mathrm{d}t / m_{0i} \tag{194}$$

m_{0i} Peak area A_i
m_{1i} Retention time of the component i
y_i Intensity of the component signal i as a function of time
t Time

The second moment m_{2i} of a peak i with the distribution function $y_i(t)$ is related to the retention time m_{1i} and as the central moment $\overline{m_{2i}}$ represents the peak time variance σ_{ti}^2:

$$\overline{m_{2i}} = \int_{-\infty}^{+\infty} (t - t_{ms})^2 \cdot \mathrm{y}(t)\, \mathrm{d}t / m_{0i} \tag{195}$$

The interpretation of the retention time as the first moment is more comprehensive than the use of the peak time (t_{max}). m_1 indicates the position of the peak center on the time scale that, in the case of unsymmetrical peaks, may deviate significantly from the position of the peak maximum. The thermodynamically correct retention time is actually derived from the position of the Gaussian component in a deconvoluted total peak profile. According to a more recent opinion of Jönsson [4], the correct retention time equals the median value of the peak.

Several approaches exist (Dorsey et al. [5]; Yau [6]; Grushka [3]) for the determination of moments from the exponential peak function that are retrievable from laboratory data systems with slightly different results.

7.3 Determination of Peak Areas

The first step in the quantitative evaluation of a chromatogram is the peak area determination or the measurement of quantities that are proportional to the peak area (peak height, peak area by multiplication of peak height and half width, triangulation). The various methods for determining peak areas or quantities proportional to the peak area differ with respect to

- accuracy,
- precision,
- applicability for area determination, and
- time expenditure / costs.

7.3.1 Manual Determination of Peak Areas and Peak Heights

Peak Height

As seen in Fig. 7-1 the peak height is measured as the distance between baseline and the peak maximum. A baseline drift is compensated for by interpolation of the baseline between peak start and peak end. The determination of the peak height yields a quantity that is proportional to the peak area only in case of constant peak shapes, and therefore has only the character of an approximation method. In overlapping peaks, the measurement of peak height is superior to the peak area measurement, since the same accuracy is obtained at lower resolution. The precision of a determination via peak height measurement is – depending on the quality of the calibration – between 1 and 2% rel.

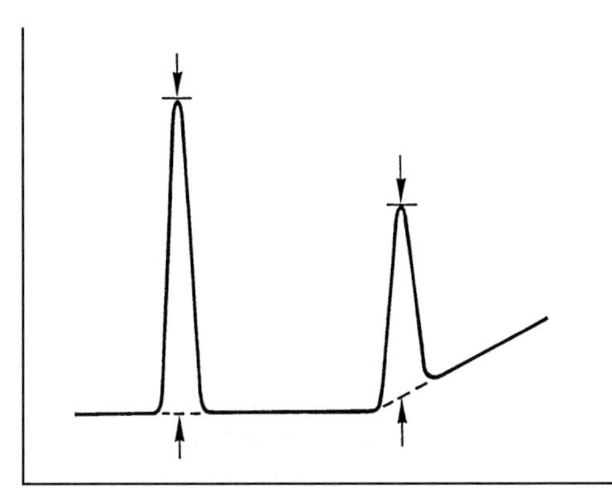

Fig. 7-1. Peak height measurement.

Peak Area by Multiplication of Peak Height and Half Width

Often, a chromatographic signal may be approximated by a triangle. The calculation of the area is performed as shown in Fig. 7-2:

$$A = H \cdot w_{1/2} \tag{196}$$

H Peak height
$w_{1/2}$ Half width of the peak

Since the peak width on the baseline may be affected by adsorption and tailing effects, the peak's half width is used in the area calculation. This procedure for peak area calculation should only be applied to symmetrical signals. The area thus calculated is proportional to the concentration but slightly smaller than the true value. For an accurate determination of the half widths, a high chart speed must be chosen.

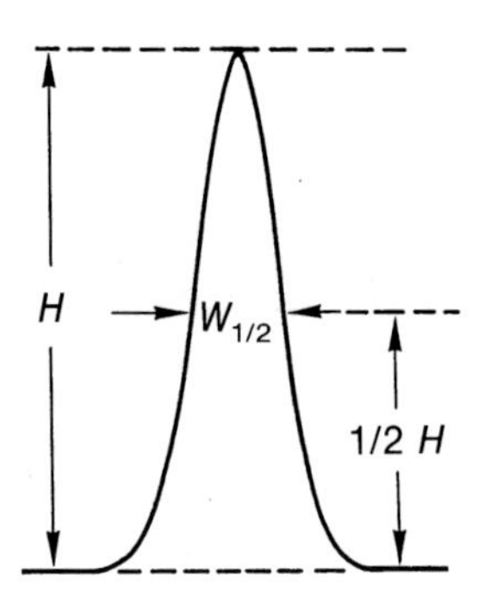

Fig. 7-2. Determination of the peak area by multiplication of peak height and half width.

Triangulation

This procedure is also based on the approximation of a chromatographic signal by a triangle. For the area calculation, tangents in the two inflection points of the peak are positioned. As seen in Fig. 7-3, the peak height is measured as the distance between the baseline and the point of intersection of both tangents, the peak width is given by the baseline section as determined by the inflectional tangents. The peak area is then calculated as follows:

$$A = 1/2\, B \cdot H \tag{197}$$

B Peak width on the baseline as determined by the inflectional tangents
H Peak height

The procedure of triangulation is subject to the same limitations as those already described. It has an additional drawback in that inflectional tangents have to be drawn. Small errors in the positioning of these inflectional tangents significantly affect the peak height measurement; therefore, this procedure is not recommended.

7.3.2 Electronic Peak Area Determination

For the electronic evaluation of chromatographic data, one uses either dedicated systems, integrators, or personal computers equipped with the respective software.

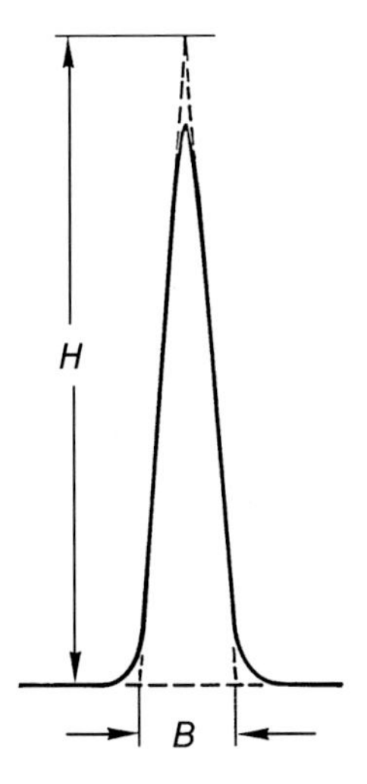

Fig. 7-3. Triangulation.

Digital Integrators

A digital electronic integrator is an instrument that allows one to measure peak areas and retention times automatically and to compensate for baseline drifts. With the aid of a voltage-to-frequency converter, an output signal is generated with a given frequency that is proportional to the input signal. During integration all pulses from the V/F-converter are counted. The total number of pulses during a peak represents the peak area. By means of an integrator peaks of several millivolts up to about one volt may be detected; thus, the linear range covers more than five orders of magnitude. In addition, the recorder signal can be attenuated independent of the integrator signal. Digital integrators offer the advantage of high precision and the conversion of the chromatographic signal into a numeric result. If the control parameters for the integrator are erraneously set, there is a possibility that trends like baseline drifts and jumps will not be recognized and, consequently, the baseline is not adjusted. This could lead to errors in the area calculation of non-resolved or tangential peaks.

Apart from normal integration and the printing of chromatograms, modern compact integrators provide the possibility to customize the calculations and analysis reports. This includes the multi-level calibration, linear or non-linear, using peak areas or peak heights; the possibility to calibrate the peak area via an internal standard, the peak area normalization; and the calculation of simple statistical quantities such as the relative standard deviation. Nearly all commercial instruments can be freely programmed in BASIC and contain a power-failure back-up system to prevent the loss of data in case of short-term power failures. Options are 256 K RAM for storing chromatograms and for their blockwise reintegration and for storing the raw data of a second channel that may be connected. Many digital integrators can be equipped with a timed-function module that, with the aid of a BASIC program, can be used to control external instrument modules such as an autosampler. Data may be transferred to other computer systems via a RS-232C-interface.

A survey over the methods for determining peak areas and quantities that are proportional to peak area is given in Tab. 7-1.

Table 7-1. Survey of various methods for calculating peak areas.

Method	Required time	Max. precision in routine operation σ [%]	Accuracy	Applicability to unsymmetrical peaks
Peak height measurement	small	0.5 (depending on peak symmetry)	good (with symmetrical peak shape)	not suited
Peak height × half width	small	2	good (with constant and almost symmetrical peak shape)	fairly suited
Triangulation	small	3	good (with constant and almost symmetrical peak shape)	fairly suited
Electronic integration	zero	0.1	very good (depending on the algorithm for curve smoothing, peak detection, and baseline drift)	suited

Computer

While digital integrators serve to measure peak areas and retention times, they are of limited use for instrument control purposes and advanced data processing. Therefore, interest in computer-assisted chromatography systems has risen rapidly. The personal computer has to fulfill the following tasks:

- control and monitoring of the chromatographic system,
- calculation of peak areas and retention times,
- calculate sample composition on the basis of the data obtained, taking into account the various calibration methods.

This requires a software package that normally features graphic control. The simultaneous use of other software programs (e.g., word processors and spreadsheets) and data transfer to these programs is easily realized owing to the underlying user interface (e.g., MS Windows). With a RAM-buffered interface, it is possible to control several single-channel and multi-channel chromatographs. This interface is programmed via the personal computer. So, on each channel analyses can be performed independent of and without reverting to the computer. If the instrument modules are equipped with a parallel interface, they may be completely controlled via a 50-pin cable. Other instruments may be controlled by means of short-circuit relays, TTL ports, or switching supplies. It is interesting to connect digital detectors since their data may be directly processed at automatic range switching, thus allowing to make use of the full dynamic range of such a detector.

7.4 Determination of Sample Composition

At the quantitative evaluation of chromatograms the peak areas A_i or quantities proportional to the peak area are related to the amount Q_i of the individual sample components:

$$\begin{aligned} A_i &= f(Q_i) \\ &= \overline{S_i} \cdot Q_i \end{aligned} \tag{198}$$

$S_i = (\mathrm{d}S/\mathrm{d}Q)_{iQ}$ Sensitivity of the detection or registration system for the sample component i at a given amount Q

$\overline{S_i}$ Mean value of the sensitivity as defined by the slope of the regression straight line $A = A_0 + \overline{S_i} \cdot Q$

The value for $\overline{S_i}$ must be determined by calibration. In general, the mean sensitivity is not constant. As it depends from the amount Q_i, a single calibration with any amount Q_{i0} does not suffice. Either the amount Q_{i0} of the unknown sample is approximated to the unknown sample amount Q_{ix} or a calibration function with n amounts $Q_{i1}, \ldots, Q_{in}$ is used which may be applied to samples with different amounts Q_i within the calibrated range.

The calibration function describes the relation between peak area and the amount of sample component. In general, it is not exactly linear [7]. It may often be approximated by the linear relation given in Eq. (198). When the spanned calibration range increases, deviations from the linear function become noticeable.

Calibration may be performed by one of the following four different methods with the aid of modern electronic integrators:

- area normalization,
- internal standard,
- external standard, and
- standard addition.

7.4.1 Area Normalization

The method of area normalization requires that all sample components eluted from the analytical separator column have been detected. If response factors are not taken into account this method can only be applied to the calibration of sample components having the same response.

Area normalization is mainly used, therefore, in gas chromatographic analyses of carbohydrates. Its application to ion chromatography is limited, as sample components only rarely exhibit the same response at the detection methods employed.

The calculation of the area for the unknown peak *X* in Fig. 7-4 is performed according to Eq. (199):

$$\% X = \frac{A_x \cdot 100}{A_x \cdot A_y \cdot A_z} \tag{199}$$

$$= \frac{A_x \cdot 100}{\sum_{i=1}^{n} A_i}$$

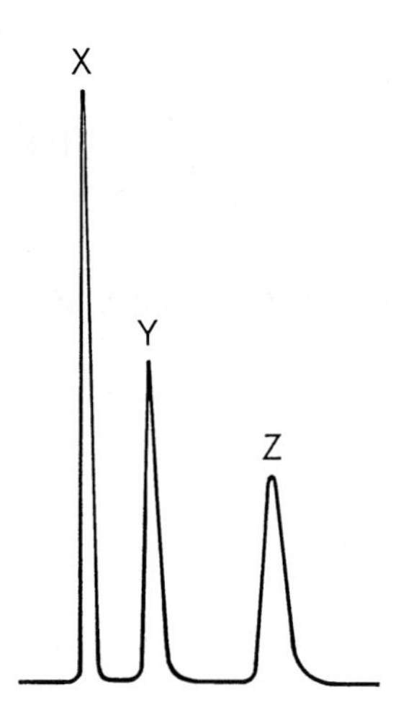

Fig. 7-4. Area normalization.

7.4.2 Internal Standard

In this method a standard substance of known concentration is added to the unknown sample. The sample composition is determined by comparing the peak areas of standard and sample component. This method does not require all signals to be eluted and detected. Therefore, the method of internal standard is also suited for samples in which not all of the components must be determined.

The internal standard should meet the following requirements:

- chemical stability of the standard under the given chromatographic conditions,
- elution near the substance that is to be determined,

- baseline-resolved separation from both neighboring signals,
- the substance-specific correction term is known or can be established,
- similarity in the concentration and response to the component to be analyzed,
- high purity.

To calibrate according to the internal standard method, the analyte components must exist as reference compounds in sufficiently high purity. To determine the correction terms, a solution is prepared that contains known amounts of the substances of interest and the internal standard. This solution is chromatographed and the correction terms are calculated based on the peak areas with Eq. (200):

$$f_x = \frac{A_x \cdot W_{is}}{A_{is} \cdot W_x} \tag{200}$$

f_x Correction term for component x
A_x Peak area of component x
W_x Concentration of component x
A_{is} Peak area of internal standard
W_{is} Concentration of internal standard

To plot the calibration curve, solutions with different concentrations of the compounds to be determined and a known concentration of the internal standard are chromatographed. The ratio of the peak area of the analyte substance to that of the internal standard, A_x/A_{is}, is plotted as a function of the concentration W_x.

After the calibration a known amount of an internal standard is added to a known sample amount and the sample is chromatographed again. The calculation of the composition is carried out with Eq. (201):

$$\% X = \frac{A \cdot f_x \cdot W_{is} \cdot 100}{A_{is} \cdot W_s} \tag{201}$$

Based on peak areas, with careful calibration a relative precision of at most 0.1% can be obtained.

7.4.3 External Standard

The calibration method most often used in ion chromatography is direct comparison of the peak area in an unknown sample with that of a solution with a known content of the same substance. This method requires the injection of constant volumes under constant chromatographic conditions. Errors in the sample delivery, however, are almost excluded upon application of a sample loop valve. A prerequisite is the existence of reference compounds for all sample components to be analyzed. In practice, several different standard solutions in the investigated concentration range are prepared and chromatographed [8]. When the resulting peak area is plotted versus the concentration of the standards, one obtains a substance-specific calibration function.

The peak area of the analyte substance should lie between that of the calibration standards so that the concentration of the compound to be determined can be calculated via interpolation. Extrapolation is not recommended because the curve of the function outside the investigated concentration range is not known. The slope of the calibration straight line in Fig. 7-5 equals the substance-specific correction term, which depends on the properties of both the substance itself and the detector. If possible errors in the sample delivery are excluded, a precision of 0.3% relative can be obtained in routine operation.

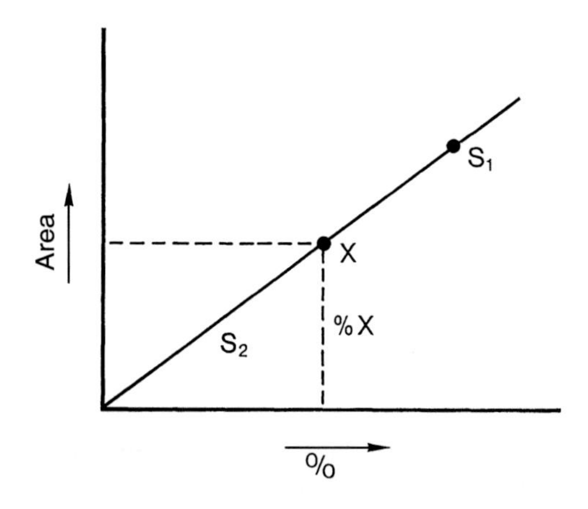

Fig. 7-5. Calibration with the aid of an external standard.

7.4.4 Standard Addition

The standard addition method represents a combination of the calibration with the aid of both an external *and* an internal standard. It is used in ion chromatography predominantly in matrix problems. After analysis of the analyte sample under suitable chromatographic conditions, a known amount of the compoment of interest is added and the sample is again chromatographed. In the calculation of the concentration, the volume change has to be taken into account. A peak that is not of interest serves as internal standard to compensate for the dosing error. Calculation is performed according to Eq. (202):

$$Q_i = \frac{\Delta Q_i}{\frac{(A + \Delta A)_{i2}}{A_{i1}} \cdot \frac{A_{s1}}{A_{s2}} - 1} \tag{202}$$

Q_i	Concentration of the analyte ion i before addition
ΔQ_i	Concentration of the ion i added to the sample volume Q
A_{i1}, $(A + \Delta A)_{i2}$	Peak areas of the analyte ion i in the chromatograms 1 and 2
A_{s1}, A_{s2}	Peak areas of the reference compound in the chromatograms 1 and 2

The concentration of the analyte ion may also be determined graphically. As seen in Fig. 7-6, the peak area of the compound of interest is plotted versus the added concentration. A linear correlation exists if the measuring quantity (electrical conductance, UV absorption, etc.) is proportional to the concentration of the analyte ion. The concentration of the ion is obtained by extrapolating the straight line to the abscissa.

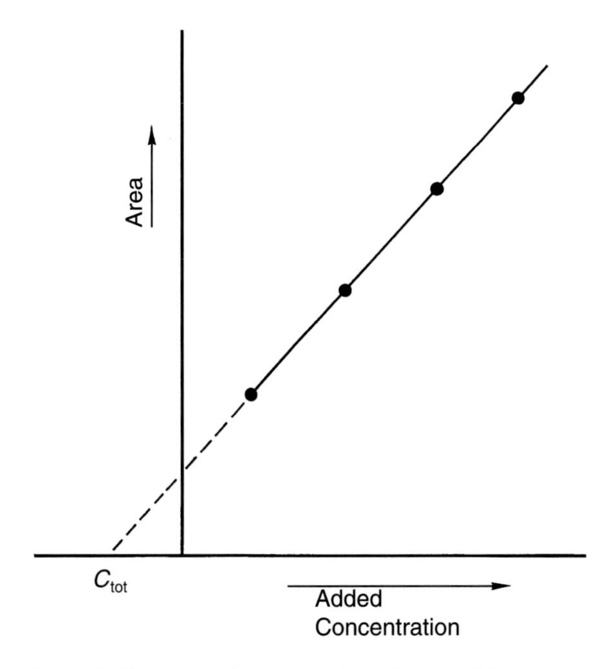

Fig. 7-6. Calibration by means of standard addition.

7.5 Statistical Data Assessment

When a sample is repeatedly analyzed in the laboratory using the same measuring method, results are collected that deviate from each other to some extent. The deviations, representing a scatter of individual values around a mean value, are denoted as statistical or random errors, a measure of which is the *precision*. Deviations from the true content of a sample are caused by *systematic* errors. An analytical method only provides true values if it is free of systematic errors. Random errors make an analytical result less precise while systematic errors give incorrect values. Hence, precision of a measuring method has to be considered separately. Statements relative to the accuracy are only feasible if the true value is known.

Standard Deviation

The repeatability is expressed typically as the standard deviation of the mean value. At an infinite number of measurements, the frequency of individual results is represented by a Gaussian curve. As a measure for the scatter, therefore, the distance of the inflection points from the true value is used. This is denoted as theoretical standard deviation σ. However, in practice, neither an infinite number of measuring values exists nor is the true value μ of a sample known. The frequency distribution of a limited number of individual values no longer corresponds to a normal distribution but rather to a t-distribution. At n repetitions of a determination, one obtains $\overline{x}$ and s as estimators for the mean value and the standard deviation, respectively [9].

$$\overline{x} = \frac{x_1 + x_2 + x_3 + \cdots + x_n}{n} \tag{203}$$

$$= \frac{1}{n} \cdot \sum_{i=1}^{n} x_i$$

$$s = \sqrt{\frac{1}{n-1} \cdot \sum_{i=1}^{n} (x_i - \overline{x})^2} \tag{204}$$

As the number of measurements increases, the assessment values $\overline{x}$ and s approach the true values μ and σ, respectively.

The standard deviation is a measure for the magnitude of the random error of an analysis, which depends on the method and sample composition. The relative random error ε describes the precision of an analysis, which increases with decreasing relative random error:

$$\varepsilon = \frac{\sigma}{\mu} \tag{205}$$

Using the relative random error, an analyst may quickly establish, independent of the magnitude of the measuring results obtained, if they are within the precision of the method.

Scatter and Confidence Range

The statistical significance of a result is expressed by the scatter range T and the confidence range. The confidence range means that P% of all individual measurements, the arithmetic mean $\overline{x}$ of which is given by the result, can be expected to be within the range $\overline{x} + T$ and $\overline{x} - T$. With details on T, a statement is obtained about the quality of the raw data material, without knowing the uninteresting individual data. The scatter range T is calculated by:

$$T = s \cdot t \tag{206}$$

t Student factor

The Student t-factor depends on the statistical probability P and the number of degree of freedoms $f = n-1$. As a certain statistical relevance P must be chosen for the calculation of T, and has to be taken into account in the interpretation of the result, the result is given in %P:

$$\text{result} = \overline{x} + T\,(\text{unit});\ (\pm\, s;\ P\%;\ n)$$

In constrast to the scatter range T of individual values, the scatter range of the arithmetic mean value is denoted as confidence range $T/\sqrt{n}$.

If a series of m arithmetic mean values μ_i is determined each from n individual measurements, the following relation results for the estimator $s_{\mu i}$ of the standard deviation of the arithmetic mean values:

$$s_{\mu i} = s/\sqrt{m} \tag{207}$$

This means, that the estimator of the standard deviation of the arithmetic mean is lowered with m in comparison to the standard deviation of the individual measurements.

7.5.1 Tests for Outliers

From time to time, in practice, measuring values are obtained that deviate significantly from the others, without being able to clarify the reason for this. Outlier tests aid in deciding if the value is part of a homogeneous data material.

Nalimov Test

Given a series of measuring values with the characteristic standard data n, $\overline{x}$, and s, the test quantity r^* is calculated by putting the value x^*, a suspected outlier, into Eq. (208):

$$r^* = \frac{|x^* - \overline{x}|}{s} \cdot \sqrt{\frac{n}{n-1}} \tag{208}$$

The distinction is made by comparing r^* with r_i. The value r_i which depends on the number n of individual measurements and on the statistical relevance P, may be taken from the r-table [9].

Example:

The following data with arithmetic mean and standard deviation are obtained as the measuring result of a chloride determination:

$x_1 = 30.4$ ppm Arithmetic mean: $\overline{x} = 30.04$ ppm
$x_2 = 30.0$ ppm Standard deviation: $s = \pm\, 0.313$
$x_3 = 30.5$ ppm
$x_4 = 30.2$ ppm
$x^* = 29.1$ ppm

For the Nalimov test, the degree of freedom is calculated by:

$$f = n - 2 \tag{209}$$

With $n = 5$ individual data, the degree of freedom is $f = 3$. Putting the values for $\overline{x}$ and s into Eq. (208), one obtains:

$$r^* = \frac{|29.1 - 30.04|}{0.313} \cdot \sqrt{\frac{5}{5-1}} \tag{210}$$
$$= 3.358$$

The statistical factors r_i are taken from the r-table to decide if the suspected value is an outlier. For $f = 3$ one finds:

$r(95\%) = 1.757$
$r(99\%) = 1.918$
$r(99.9\%) = 1.982$

To determine if x^* represents an outlier, compare r^* with r_i.

$r(95\%) < r^* < r(99\%)$	probable outlier
$r(99\%) < r^* < r(99.9\%)$	significant outlier
$r^* > r(99.9\%)$	highly significant outlier

From the comparison of the value calculated for r^* with these statistical factors it is concluded that the suspected value is a highly significant outlier. Thus, arithmetic mean and standard deviation must be calculated again without the outlier:

Arithmetic mean	$\overline{x} = 30.28$ ppm
Standard deviation	$s = \pm 0.049$

The corrected values differ markedly from those that have been distorted by the outlier.

Test according to Grubbs

The test according to Grubbs differs from that of Nalimov by the test quantity r_m, which is calculated by Eq. (211) as follows:

$$r_m = \frac{|x^* - \overline{x}|}{s} \tag{211}$$

From the table "Comparison values $r_m(P)$ for the outlier test according to Grubbs" one selects the values $r_m(90\%)$, $r_m(95\%)$, and $r_m(99\%)$ for $f = n-2$. It is then tested for the following criteria:

$r_m(90\%) < r_m < r_m(95\%)$	probable outlier
$r_m(95\%) < r_m < r_m(99\%)$	significant outlier
$r^* > r_m(99\%)$	highly significant outlier

If outliers that are possibly present have been eliminated from the measuring series, $\overline{x}$ and s are again calculated.

7.5.2 Regression Calculation

Section 7.4 mentions that the calibration function describes the relation between the peak area and the amount of sample component. As systematic deviations from a linear function may occur, a measure for the linearity of the calibration function is needed. This is done by using the maximum relative deviation of the linearized calibration function from the true calibration function in a given range of signals. The linear range is defined by the maximum signal value, up to which the relative deviation does not exceed a predefined value. The straight line equation $y = a + b \cdot x$ is obtained by linear regression of the individual data. The regression coefficients are calculated by the Gaussian least-squares method. With n pair of data (x_i, y_i), it is demanded that

$$\sum_{i=1}^{n} (y_i - Y_i)^2 = \sum_{i=1}^{n} \left[y_i - (a + b \cdot x_i)^2 \right] \tag{212}$$

be minimized. From this follows:

$$b = \frac{\sum_{i=1}^{n} (x_i - \bar{x})(y_i - \bar{y})}{\sum_{i=1}^{n} (x_i - \bar{x})^2} = \frac{\sum_{i=1}^{n} x_i y_i - \frac{1}{n}\left[\sum_{i=1}^{n} x_i \cdot \sum_{i=1}^{n} y_i\right]}{\sum_{i=1}^{n} x_i^2 - \frac{1}{n}\left[\sum_{i=1}^{n} x_i\right]^2} \tag{213}$$

$$a = \bar{y} - b \cdot \bar{x} \tag{214}$$

The quantities $\bar{x}$ and $\bar{y}$ represent the mean values of the data values x_i and y_i. The measure for the relation between x_i and y_i is given by the correlation coefficient $r_{x,y}$:

$$r_{x,y} = \frac{\sum_{i=1}^{n} (x_i - \bar{x})(y_i - \bar{y})}{\sqrt{\sum_{i=1}^{n} (x_i - \bar{x})^2 \cdot \sum_{i=1}^{n} (y_i - \bar{y})^2}} = \frac{\sum_{i=1}^{n} x_i y_i - \frac{1}{n}\left[\sum_{i=1}^{n} x_i \cdot \sum_{i=1}^{n} y_i\right]}{\sqrt{\sum_{i=1}^{n} x_i^2 - \frac{1}{n}\left(\sum_{i=1}^{n} x_i\right)^2} \cdot \sqrt{\sum_{i=1}^{n} y_i^2 - \frac{1}{n}\left(\sum_{i=1}^{n} y_i\right)^2}} \tag{215}$$

This coefficient is independent of the unit of the characteristics and may assume any value between -1 and $+1$. If $r_{x,y} = 1$, the relation is directly linear.

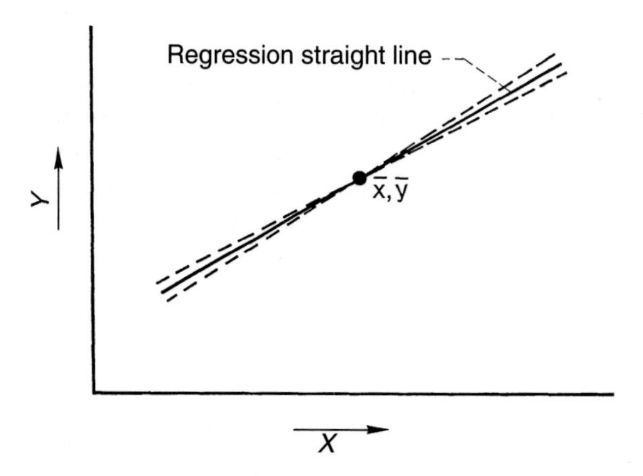

Fig. 7-7. Representation of the confidence interval.

The process characteristic values [10–13] can be determined with the aid of regression coefficients. This includes the parameters

- *sensitivity*, which is defined as the slope b of the regression line
- *residual standard deviation* S_y

$$S_y = \sqrt{\frac{\sum_{i=1}^{n} (y_i - \bar{y})^2}{N-2}} \tag{216}$$

N number of calibration points

- *process standard deviation* S_{x0}, and

$$S_{x0} = S_y / b \tag{217}$$

- *relative process standard deviation* V_{x0}

$$V_{x0} = \frac{S_{x0} \cdot 100}{\bar{x}} [\%] \tag{218}$$

Statements are thus obtained regarding the quality of the analytical method used at a single concentration level and for the whole working range from x_{min} to x_{max}.

7.5.3 Calculation of the Confidence Interval

In order to calculate the confidence interval for the regression coefficient b, a confidence number γ (95%, 99%) is chosen. Then the solution c of the equation

$$F(c) = \frac{1}{2}(1 + \gamma) \tag{219}$$

is determined from the table with Student *t*-factors for $n - 2$ degrees of freedom. Calculate

$$1 = c \cdot \frac{\sqrt{a}}{s} \cdot \sqrt{(n-1)(n-2)} \tag{220}$$

The confidence interval is then

$$\text{conf}\,\{b - 1 \leq b \leq b + 1\}$$

and is depicted in Fig. 7-7. The relative error L of the slope is given by:

$$L = \frac{1}{b} \cdot 100\ [\%] \tag{221}$$

8 Applications

The analytical method known as ion chromatography is suited for the analysis of a variety of inorganic and organic anions and cations. It is characterized by a high selectivity and sensitivity. The great number of applications portrayed below are examples of this method.

Ion chromatography has become an indispensable tool for the analytical chemist in the area of anion analysis. In many cases this method has superseded conventional wet chemical methods such as titration, photometry, gravimetry, turbidimetry, and colorimetry, all of which are labor-intensive, time-consuming, and occasionally susceptible to interferences. Publications by Darimont [1] and Schwedt [2] have shown, that ion chromatographic methods yield results comparable to conventional analytical methods, thus dissolving the scepticism with which this analytical method was initially met. In the field of cation analysis, ion chromatography is attractive because of its simultaneous detection and sensitivity. It provides a welcome complement to atomic spectroscopic methods such as AAS and ICP.

The emphasis of ion chromatographic applications is in the following areas:

- environmental analysis,
- power plant chemistry,
- semiconductor industry,
- household products and detergents,
- electroplating,
- pharmaceuticals,
- basic and luxury foods,
- biotechnology,
- agriculture,
- pulp and paper industry,
- mining, and
- metal processing.

The separation and detection methods employed in ion chromatography, as well as a selection of ionic species that may be analyzed with this method, are summarized in Tables 8-1 and 8-2.

8.1 Ion Chromatography in Environmental Analysis

Environmental analysis is one of the most important fields of application of ion chromatography; it is divided into water hygiene, soil hygiene, and air hygiene.

Table 8-1. Overview of the ion chromatographic separation methods.

Separation methods	Mechanism	Functionality of resin	Recommended eluents	Analyzable species
HPIC Anions	Ion-exchange	$-\overset{+}{N}R_3$	Na_2CO_3/ $NaHCO_3$	F^-, Cl^-, Br^-, I^-, SCN^-, CN^-, $H_2PO_2^-$, HPO_3^{2-}, HPO_4^{2-}, $P_2O_7^{4-}$, $P_3O_{10}^{5-}$, NO_2^-, NO_3^-, S^{2-}, SO_3^{2-}, SO_4^{2-}, $S_2O_3^{2-}$, $S_2O_6^{2-}$, $S_2O_8^{2-}$, OCl^-, ClO_2^-, ClO_3^-, ClO_4^-, SeO_3^{2-}, SeO_4^{2-}, $HAsO_3^{2-}$, WO_4^{2-}, MoO_4^{2-}, CrO_4^{2-}, carbohydrates, peptides, proteins, etc.
HPIC Cations	Ion-exchange	$-SO_3^-$	HCl/ HCl/DAP [a]	Li^+, Na^+, NH_4^+, K^+, Rb^+, Cs^+, Mg^{2+}, Ca^{2+}, Sr^{2+}, Ba^{2+}, small aliphatic amines
		$-SO_3^-$/ $-\overset{+}{N}R_3$	Oxalic acid PDCA [b]	Fe^{2+}, Fe^{3+}, Cu^{2+}, Ni^{2+}, Zn^{2+}, Co^{2+}, Pb^{2+}, Mn^{2+}, Cd^{2+}, Al^{3+}, Ga^{3+}, V^{5+}, UO_2^{2+}, lanthanides
HPICE	Ion-exclusion	$-SO_3^-$	HCl Octane-sulfonic acid	aliphatic carboxylic acids, borate, silicate, carbonate, alcohols, aldehydes
MPIC Anions	Ion-pair formation	neutral	NH_4OH TMAOH [c] TPAOH [d] TBAOH [e]	In addition to the anions listed under HPIC: anionic surfactants, metal-cyano complexes, aromatic carboxylic acids
MPIC Cations	Ion-pair formation	neutral	HCl Hexane-sulfonic acid Octane-sulfonic acid	In addition to the cations listed under HPIC: alkylamines, alkanolamines, quaternary ammonium compounds, cationic surfactants, sulfonium compounds, phosphonium compounds

[a] DAP 2,3-diaminopropionic acid
[b] PDCA pyridine-2,6-dicarboxylic acid
[c] TMAOH tetramethylammonium hydroxide
[d] TPAOH tetrapropylammonium hydroxide
[e] TBAOH tetrabutylammonium hydroxide

The main focus of applications in environmental analytical chemistry is the qualitative and quantitative analysis of anions and cations in all kinds of water [3−8]. For example, the anions chloride, nitrite, bromide, nitrate, orthophosphate, and sulfate, from the concentration of which the water quality depends, may be separated and determined in less than ten minutes. In a simple drinking water analysis of the main components (chloride, nitrate, and sulfate), it is possible, as illustrated in Fig. 8-1, to carry out a determination every three minutes. The high sensitivity of this method (detection limit with a *direct injection* of 50 μL sample: ca. 10 ppb) and the possibility for automation contributed much to the rapid spreading of ion chromatography as an analytical tool.

Table 8-2. Overview of the ion chromatographic detection methods.

Detection method	Principle	Applications
Conductometric	Electrical conductance	Anions and cations with pK_a or $pK_b < 7$
Amperometry	Oxidation and reduction at Ag-/Pt-/Au- and glassy carbon electrodes	Anions and cations with pK_a or $pK_b > 7$
UV/Vis detection with or without post-column derivatization	UV/Vis light absorption	UV active anions and cations, heavy metals after reaction with PAR and Tiron, polyvalent anions after reaction with iron(III), silicate and phosphate after reaction with molybdate, alkaline-earth metals after reaction with Arsenazo I
Fluorescence in combination with post-column derivatization	Excitation and emission	Ammonium, amino acids, and polyamines after reaction with *o*-phthaldialdehyde

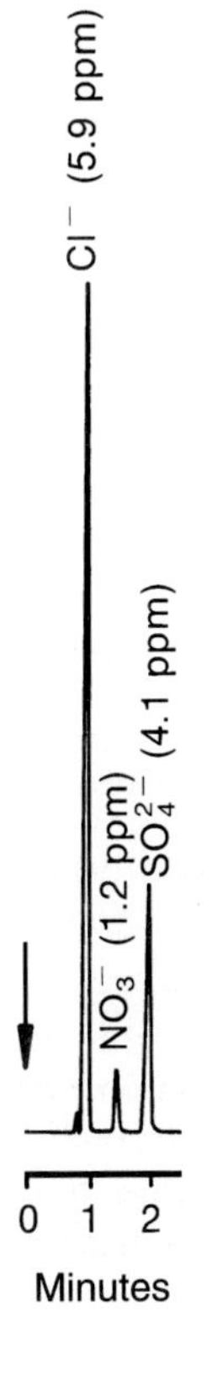

Fig. 8-1. Anion analysis of a drinking water. – Separator column: Fast-Sep Anion; eluent: 0.00015 mol/L $NaHCO_3$ + 0.002 mol/L Na_2CO_3; flow rate: 2 mL/min; detection: suppressed conductivity; injection: 25 µL potable water from Idstein (undiluted).

In addition to salts from mineral acids, it is also possible to determine salts from weak inorganic acids via ion chromatography. This includes, for example, orthosilicate [9, 10]. Anion exchange and ion-exclusion chromatography are both suited for the separation. In both cases, photometric detection after post-column reaction with sodium molybdate is employed as an extremely selective identification. Both methods also allow the separation of orthophosphate, which may also be determined with this detection method. When the effluent of an anion exchanger is derivatized after passing the suppressor system and conductivity cell, a simultaneous analysis of strong and weak inorganic acids becomes possible (see Fig. 8-2). Alternatively, Fig. 8-3 displays the chromatogram of a drinking water sample obtained by ion-exclusion chromatography. A combination of conductometric detection and post-column derivatization is also feasible in this case if, for example, inorganic acids such as orthophosphate, orthosilicate, and carbonate are to be determined together with aliphatic monocarboxylic acids in a single run.

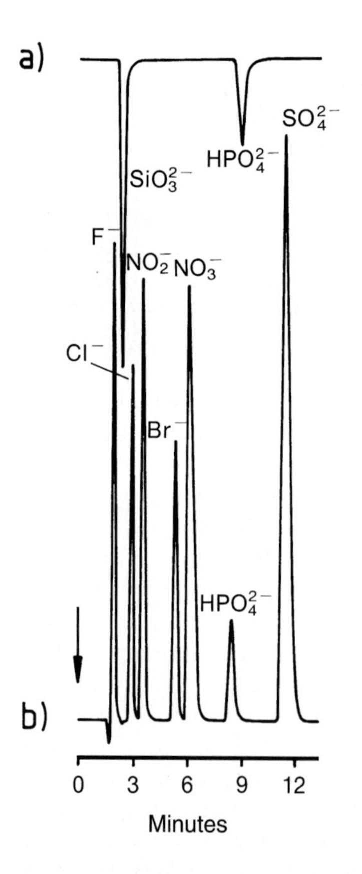

Fig. 8-2. Simultaneous analysis of weak and strong inorganic acids. – Separator column: IonPac AS4A; eluent: 0.0017 mol/L $NaHCO_3$ + 0.0018 mol/L Na_2CO_3; flow rate: 1 mL/min; detection: (A) suppressed conductivity, (b) photometry at 410 nm after post-column reaction with sodium molybdate; injection volume: 50 μL; solute concentrations: 3 ppm fluoride, 4 ppm chloride, 10 ppm nitrite and bromide, 20 ppm nitrate, 10 ppm orthophosphate, 25 ppm sulfate, and 27 ppm orthosilicate.

The simultaneous analysis of alkali and alkaline-earth metals is another important ion chromatographic application in the field of drinking and surface water analysis. The corresponding chromatogram in Fig. 8-4 shows the separation of sodium, ammonium, potassium, magnesium, and calcium in less than 20 minutes. It was obtained with a separator column IonPac CS10 which was described in Section 3.4.4.

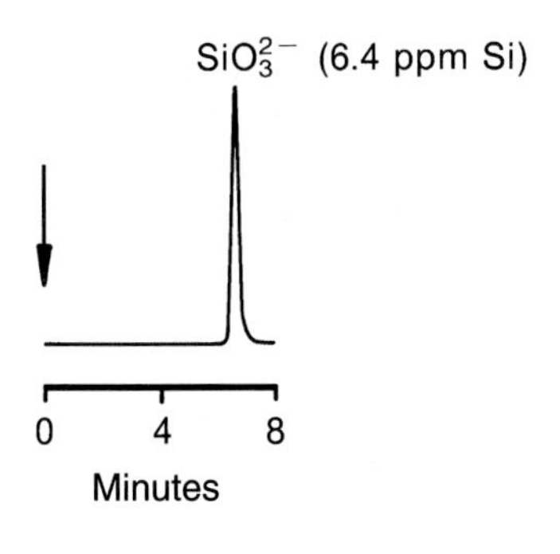

Fig. 8-3. Analysis of orthosiliate in drinking water. – Separator column: IonPac ICE-AS1; eluent: 0.0005 mol/L HCl; flow rate: 1 mL/min; detection: photometry at 410 nm after post-column reaction with sodium molybdate; injection volume: 50 μL potable water from Idstein (undiluted)

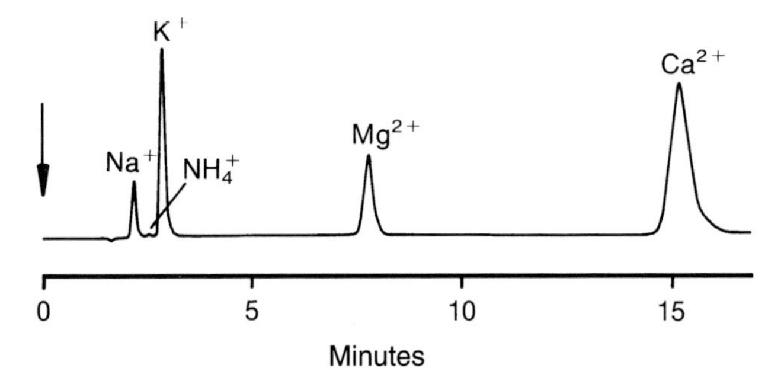

Fig. 8-4. Simultaneous analysis of alkali and alkaline-earth metals in drinking water. – Separator column: IonPac CS10; eluent: 0.04 mol/L HCl + 0.005 mol/L 2,3-diaminopropionic acid; flow rate: 1 mL/min; detection: suppressed conductivity; injection volume: 50 μL potable water from Vienna (undiluted).

The analysis of anions and cations in snow and ice samples [11], as well as in rain, ground, and bathing water, is similarly simple. The partly very low electrolyte content in snow and ice samples sometimes requires the application of the concentrator technique (see also Section 8.2) or the injection of very high sample volumes (5 mL to 10 mL). Apart from the obligatory membrane filtration (0.45 μm), the only sample preparation in the analysis of rain, ground, and bathing water is often a dilution with deionized water. Fig. 8-5 shows an anion chromatogram of a rain water, which was injected without any dilution because of its low electrolyte content. The increased presence of bromide in such samples is remarkable. Ground water – especially from wood rich areas – should be injected through an extraction cartridge OnGuard-P (Dionex) to remove humic acids. This cartridge contains a polyvinylpyrrolidone resin (PVP) which selectively retains humic acids. As an example, Fig. 8-6 shows the chromatogram of a correspondingly diluted sample from a sampling station at the river Main. Apart from the analysis of the main components chloride, nitrate, and sulfate, a peculiarity of bathing water analysis is the determination of the two chlorine species – chlorite, ClO_2^-, and chlorate, ClO_3^-. These are disproportionation products of chlorine dioxide, which some countries use to fumigate drinking and bathing water for disinfection purposes [12,13]. Chlorine dioxide easily dissolves in water, but decomposes slowly with formation of chlorous acid and chloric acid:

$$2ClO_2 + H_2O \rightarrow HClO_2 + HClO_3 \qquad (222)$$

Chlorous acid rapidly peaks down yielding hydrochloric acid and chloric acid:

$$6ClO_2 + 3H_2O \rightarrow HCl + 5HClO_3 \quad (223)$$

Chlorous acid is retained only in alkaline solution since the corresponding chlorites are more stable than the corresponding acid:

$$2ClO_2 + 2OH^- \rightarrow ClO_2^- + ClO_3^- + H_2O \quad (224)$$

For water treatment purposes, only small amounts of chlorine dioxide are necessary; therefore, the reaction products chlorite and chlorate are only present at low concentrations. Their separation from the main components is performed with a separator column IonPac AS9. Fig. 8-7 shows the chromatogram of a test mixture developed by the American **E**nvironmental **P**rotection **A**gency (EPA). This test mixture is adjusted to the matrix of such water, thus allowing to test the applied chromatographic method with regard to separation efficiency and sensitivity.

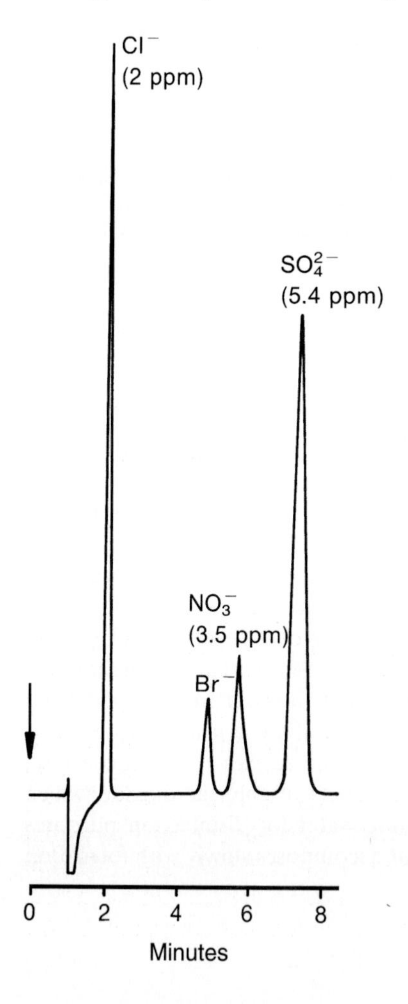

Fig. 8-5. Anion analysis of rain water. – Separator column: IonPac AS4; eluent: 0.0028 mol/L $NaHCO_3 + 0.0022$ mol/L Na_2CO_3; flow rate: 2 mL/min; detection: suppressed conductivity; injection: 50 µL ground sample.

Surface waters are mostly analyzed using ion exchangers with slightly higher exchange capacity to take into account the occasionally high electrolyte concentration in such samples. High differences in the concentrations between major and minor components require the capability of injecting such samples undiluted without overloading the separator column. This is illustrated in Fig. 8-8 with a chromatogram of a Rhine water sample in which small quantities of orthophosphate and bromide could be detected in addition to the main components chloride, nitrate, and sulfate.

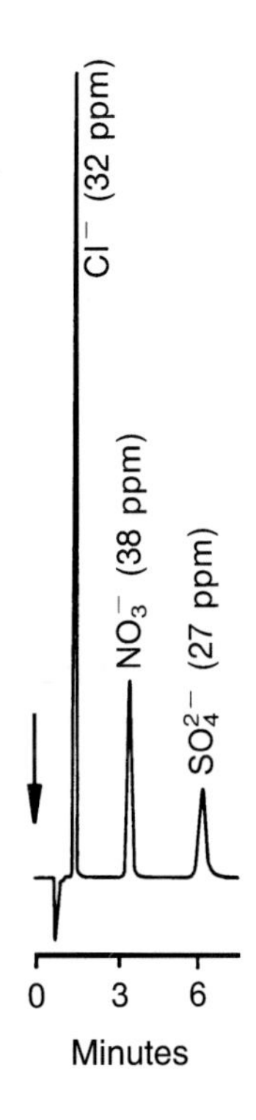

Fig. 8-6. Anion analysis of ground water. – Separator column: IonPac AS4A; eluent: 0.0017 mol/L $NaHCO_3$ + 0.0018 mol/L Na_2CO_3; flow rate: 2 mL/min; detection: suppressed conductivity; injection: 50 µL ground water (1:10 diluted).

The determination of the three nitrogen parameters – nitrite, nitrate, and ammonium – is of special interest in the field of waste water analysis. Waste water samples normally show a high chloride content. Nitrite, which elutes immediately afterwards, can therefore only be determined unequivocally by means of UV detection. Fig. 8-9 shows the chromatogram of a municipal sewage sample, which had to be injected undiluted owing to its low nitrite content. Despite of the resulting overloading of the anion exchanger with the main components chloride and sulfate, the selective detection renders possible a quantitative evaluation of the nitrite signal. The nitrate ion elutes later and is present in much higher concentrations, but may also be detected at a wavelength of 215 nm owing to its optical absorption properties. The negative peak occuring after the nitrate signal is caused by the UV transparent sulfate which reduces the absorption of the column effluent for a short time due to its high concentration.

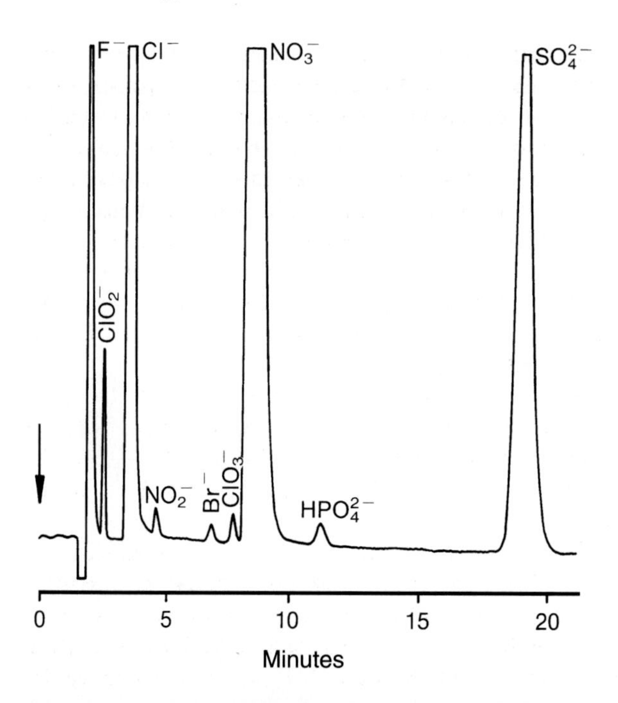

Fig. 8-7. Separation of chlorite and chlorate from other mineral acids. – Separator column: IonPac AS9; eluent: 0.00075 mol/L $NaHCO_3$ + 0.002 mol/L Na_2CO_3; flow rate: 1 mL/min; detection: suppressed conductivity; injection: 50 µL; solute concentrations: 5.6 ppm fluoride, 0.5 ppm chlorite, 45.4 ppm chloride, 0.06 ppm nitrite, 0.08 ppm bromide, 0.14 ppm chlorate, 42.1 ppm nitrate, 0.17 ppm orthophosphate, and 5 ppm sulfate.

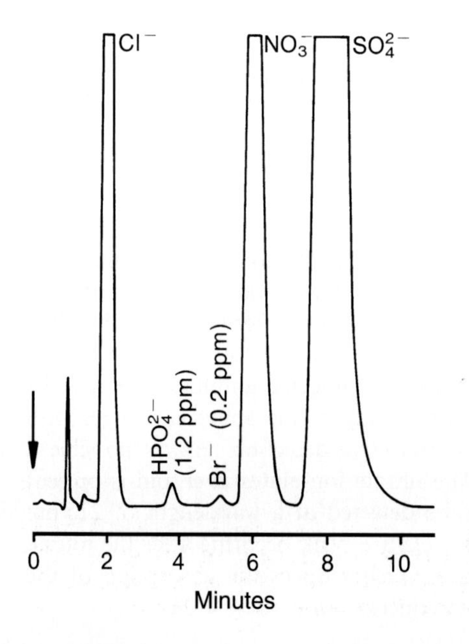

Fig. 8-8. Anion analysis of a Rhine water sample. – Separator column: IonPac AS3; eluent: 0.0028 mol/L $NaHCO_3$ + 0.0022 mol/L Na_2CO_3; flow rate: 2.3 mL/min; detection: suppressed conductivity; injection: 50 µL Rhine water (Düsseldorf).

Unlike nitrite, it is possible to determine conductometrically ammonium even in high sodium excess, since a high resolution between sodium, ammonium, and potassium is obtained with modern cation exchangers. Fig. 8-10 illustrates this with another municipal sewage sample, in which ammonium was detected after appropriate dilution.

Apart from municipal sewages, industrial waste waters may be also be analyzed via ion chromatography. However, the way of looking at the problem may differ strongly to some extent. An unusual example from the area of anion analysis is the separation of monoisopropyl sulfate and sulfate in the waste water from an isopropanol synthesis that is shown in Fig. 8-11. Analogous to the retention behavior of inorganic and organic phosphates (see Fig. 3-91 in Section 3.3.5.1), a much shorter retention is observed for the organic component in the case of sulfates.

A first glance, the separation of heavy and transition metals shown in Fig. 8-12 merely illustrates the applicability of ion chromatography to a class of compounds which are normally analyzed using atom spectroscopic methods. However, in view of matrix problems, these methods are also subject to limitations when the analyte sample, as in the present case, is an aqueous eluate from a technical process in which small amounts of nickel are to be determined in the presence of iron as the main component. Ion chromatographically, this poses no difficulty since the nickel determination is not interferred with by high iron concentrations when oxalic acid is applied as the complexing agent. It appears from Fig. 8-12, that under the chromatographic conditions used, other heavy and transition metals such as copper, cadmium, cobalt, and zinc could be detected.

In principle, care must be taken in the analysis of waste water samples so that the ion exchanger being used does not be contaminated by organic material such as fats, oils, surfactants, etc. They can be removed from the sample via solid phase extraction on suitable materials (e.g. OnGuard cartridges). Further details regarding the subject of sample preparation may be found in Section 8.9

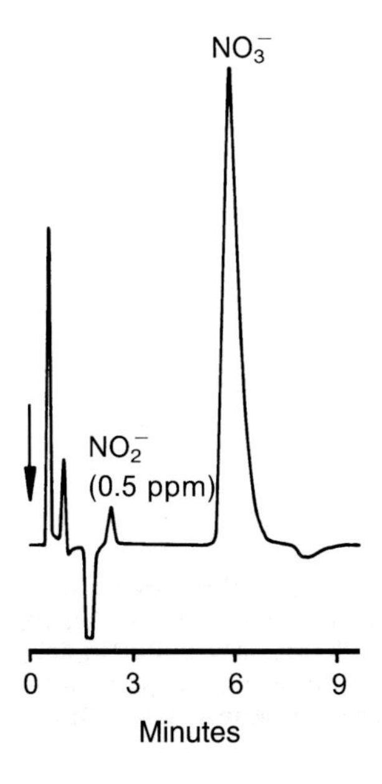

Fig. 8-9. Analysis of nitrite and nitrate in a municipal waste water via UV detection. – Separator column: IonPac AS3; eluent: see Fig. 8-8; flow rate: 2.3 mL/min; detection: UV (215 nm); injection: 50 µL waste water (undiluted).

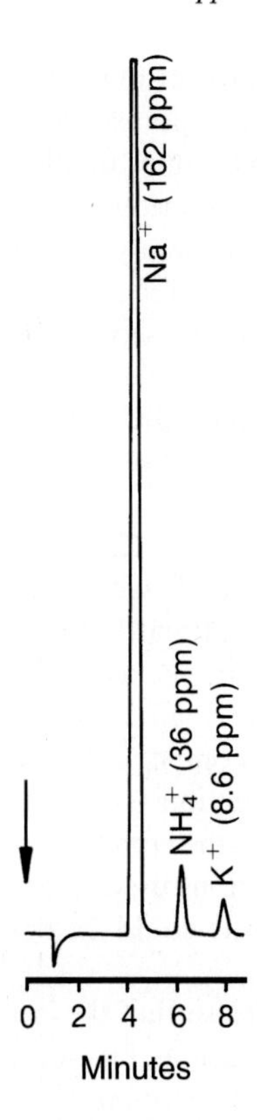

Fig. 8-10. Analysis of ammonium in munipal sewage. – Separator column: IonPac CS3; eluent: 0.03 mol/L HCl; flow rate: 1.5 mL/min; detection: suppressed conductivity; injection: 50 µL waste water (1:10 diluted).

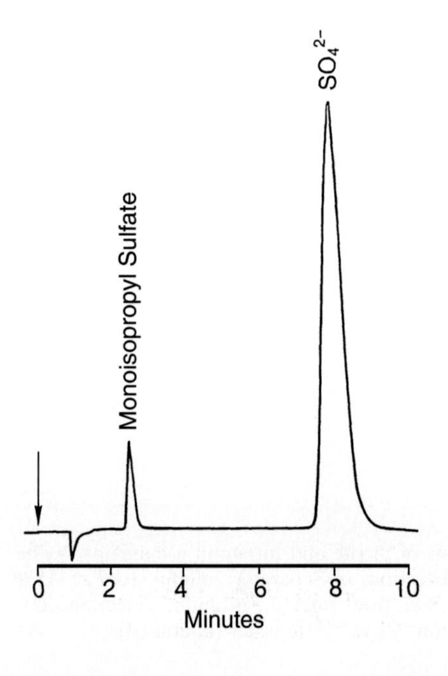

Fig. 8-11. Anion analysis of a waste water from the isopropanol synthesis. – Separator column: IonPac AS3; eluent and flow rate: see Fig. 8-8; detection: suppressed conductivity; injection: 50 µL waste water (1:5000 diluted).

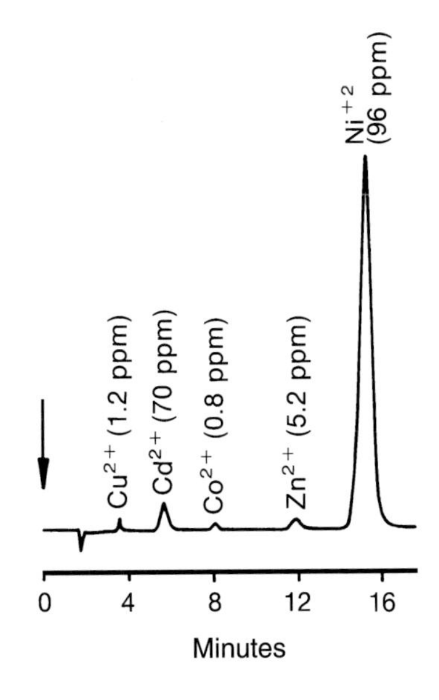

Fig. 8-12. Analysis of heavy and transition metals in industrial waste water. – Separator column: IonPac CS5; eluent: 0.05 mol/L oxalic acid + 0.095 mol/L LiOH; flow rate: 1 mL/min; detection: photometry at 520 nm after post-column reaction with PAR; injection: 50 µL waste water (1:50 diluted).

Sample preparation is even more important in the field of soil analysis. When the sample is a neutral sludge, it may be diluted due to the high electrolyte content. The quality of the resulting chromatogram in Fig. 8-13, therefore, hardly differs from that of a drinking water chromatogram. In contrast, soil samples, are often extracted with a 10% potassium chloride solution. Due to the chloride matrix it is not possible to analyze the nitrogen parameters of interest, nitrite and nitrate, using a separation system with an eluent based on a carbonate/bicarbonate-mixture. This problem is easily solved, however, by employing potassium chloride as the eluent and determining the analyte ions

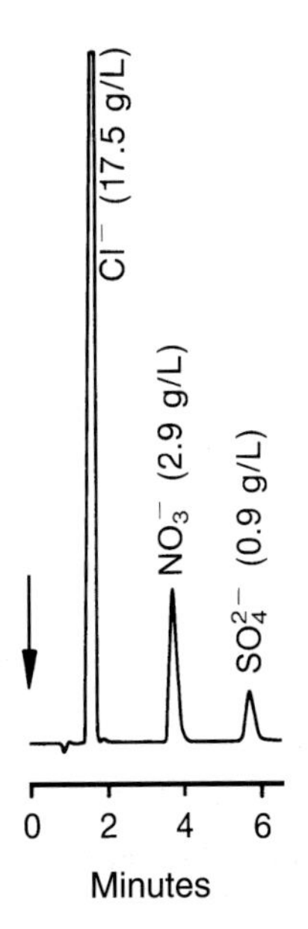

Fig. 8-13. Separation of inorganic anions in the filtrate of a neutral sludge. – Separator column: IonPac AS4A; eluent: 0.0017 mol/L $NaHCO_3$ + 0.0018 mol/L Na_2CO_3; flow rate: 2 mL/min; detection: suppressed conductivity; injection: 50 µL sample (1:250 diluted).

nitrite and nitrate photometrically. When a stationary phase is used that exhibits both anion and cation exchange capacities (e.g., IonPac CS5), the ammonium ion is also retained, which may be detected very sensitively by derivatization with *o*-phthaldialdehyde and subsequent fluorescence detection. By combining both detection methods all three nitrogen parameters can be determined in one run. Fig. 8-14 illustrates this with chromatograms from a real sample and a standard, which were obtained applying simultaneous detection.

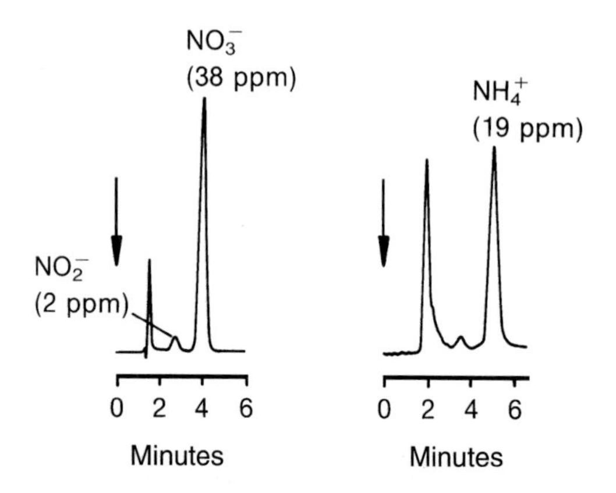

Fig. 8-14. Simultaneous detection of nitrite, nitrate, and ammonium in a KCl soil extract. – Separator column: IonPac CS5; eluent: 0.035 mol/L KCl; flow rate: 1 mL/min; detection: (A) UV (215 nm), (B) fluorescence after reaction with OPA; injection: 50 µL sample of a 1:10 diluted KCl soil extract (10%).

Interesting ion chromatographic applications in the area of air hygiene comprise the analysis of inorganic anions and cations in fly ashes [14] and atmospheric aerosols [15, 16], the analysis of nitrate and nitric acid in the atmosphere utilizing the Denuder technique [17], and the analysis of formaldehyde and acetaldehyde after appropriate sampling [18]. Gases such as cyanic acid [19], sulfur dioxide, and nitroxides [20] can also be determined ion chromatographically by passing the gases over a filter prepared with carbonate, which is extracted afterwards with de-ionized water. The separation of arsenic(V) in an aqueous extract from fly ash is provided in Fig. 8-15 as an example for this kind of application. Under the chosen chromatographic conditions arsenate elutes shortly after sulfate, which together with chloride and orthophosphate is one of the main sample components.

Ion chromatography is also employed as detection method in the sum detection of organically bonded halogen and sulfur compounds (AOX and AOS) [21–23]. In these methods the water sample is treated with nitrate solution and nitric acid to displace interfering chloride and sulfate ions. An activated carbon that is low in chlorine, bromine, and sulfur and that is almost ash-free is added, at which time the organically bonded halogen and sulfur compounds are adsorbed. After this concentration step the loaded activated carbon is filtered off over a polycarbonate filter and incinerated together with the filter at 1000°C in an oxygen stream. Until now the resulting reaction gases were introduced into the titration cell of a micro-coulorimeter, where chloride, bromide, and iodide are precipitated with silver ions present in the electrolyte. The silver ion concentration is determined potentiometrically. The silver ion concentration is readjusted to its initial value via anodic oxidation of a silver electrode. The amount of charge measured during the electrolysis is equivalent to the amount of chloride, bromide, and iodide which entered the cell. In this type of analysis chlorine, bromine, and iodine are detected as molar sum; it is impossible to differentiate the species. In the AOS method the reaction gases are passed into an aqueous solution and are analyzed using appropriate detection method.

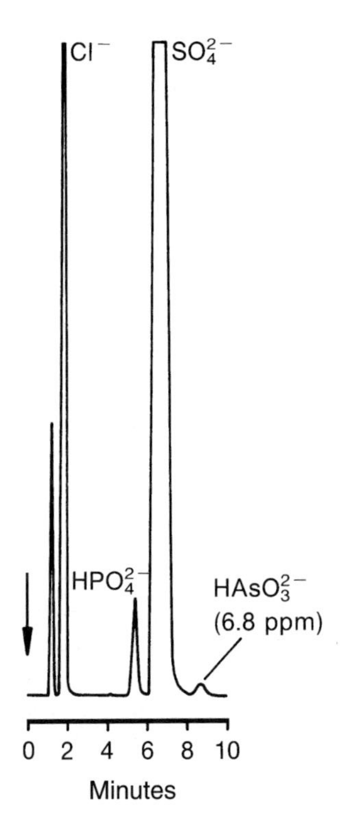

Fig. 8-15. Separation of arsenate in an aqueous extract from fly ash. – Chromatographic conditions: see Fig. 8-13; injection: 50 μL sample (undiluted).

However, if ion chromatography is used as the detection method the AOX and AOS parameters may be determined in one working step. There is an added possibility to differentiate between organically bonded chlorine, bromine, and iodine, respectively. As absorption solution for the acidic reaction gases, one applies the basic buffer solution made of sodium bicarbonate and sodium carbonate that is used for the ion chromatographic analysis. To achieve complete oxidation of sulfite to sulfate a small amount of hydrogen peroxide is added. A typical chromatogram obtained in the simultaneous determination of AOX and AOS is shown in Fig. 8-16. In addition to chloride and sulfate, signals for nitrite and nitrate are recognized. The latter are due to the application of nitrate, which was added to prevent chloride and sulfate adsorption.

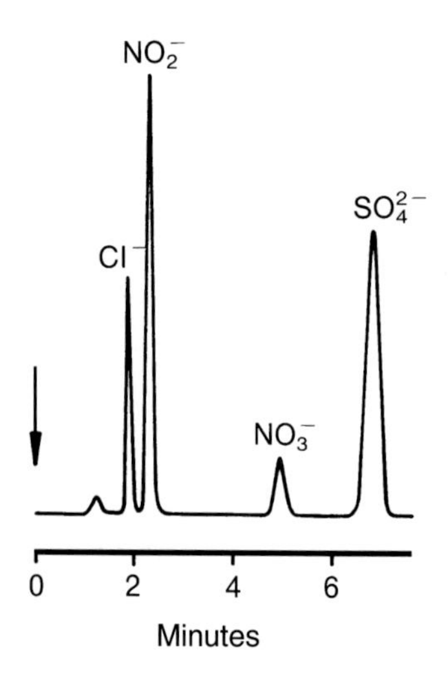

Fig. 8-16. Typical chromatogram of a simultaneous AOX/AOS determination. – Separator column: IonPac AS4; eluent: 0.0028 mol/L $NaHCO_3$ + 0.0022 mol/L Na_2CO_3; flow rate: 2 mL/min; detection: suppressed conductivity; injection: 100 µL absorption solution; (taken from [23]).

8.2 Ion Chromatography in Power Plant Chemistry

The evaluation of the water, steam, and condensate quality is one of the most important applications of ion chromatography in power plant technology. In the past, water quality was monitored via registration of the electrical conductance. Routine plant control by this method allows leaks in condensers to be identified and the funtioning of water purification devices to be checked. However, conductometric detection is unspecific, since a rise of the signal provides no information about the kind of contaminant. In addition, sodium chloride or sulfate impurities resulting from inflowing cooling water or insufficiently regenerated condensate purification filters cannot be reliably detected in the sub-ppb range.

By means of ion chromatography, high purity water can be assayed both for their chloride and sulfate content and their sodium content after pre-concentration on suitable concentrator columns. To do this, the instrument should be connected to the sampling line to pass, with the aid of a delivery pump, a defined amount of high purity water free of contamination through the concentrator column. The concentrator column is connected to the injection valve in place of the sample loop (Fig. 8-17). By switching the valve, all ions that have accumulated on the concentrator column are flushed onto the analytical separator column where they are separated. A tandem switching with two concentrator columns, as indicated in Fig. 8-17, allows a simultaneous analysis and loading. If the instrument cannot be directly connected to a sampling line, the samples must be transported in FEP-Teflon (Nalgene) containers, as only this material has proven to exhibit no wall adsorption effects.

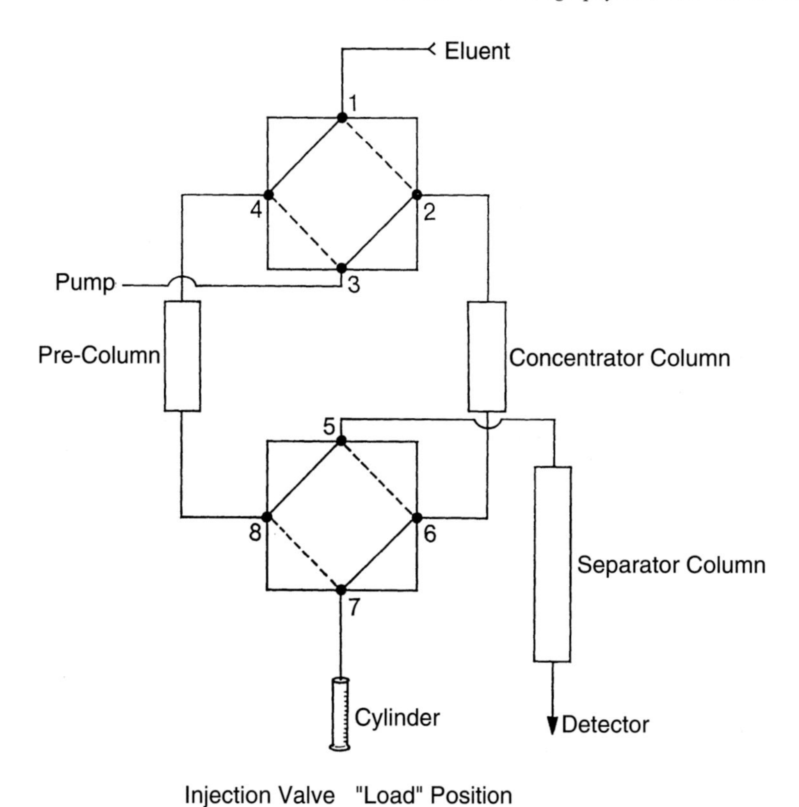

Fig. 8-17. Valve switching for applications utilizing a concentrator technique.

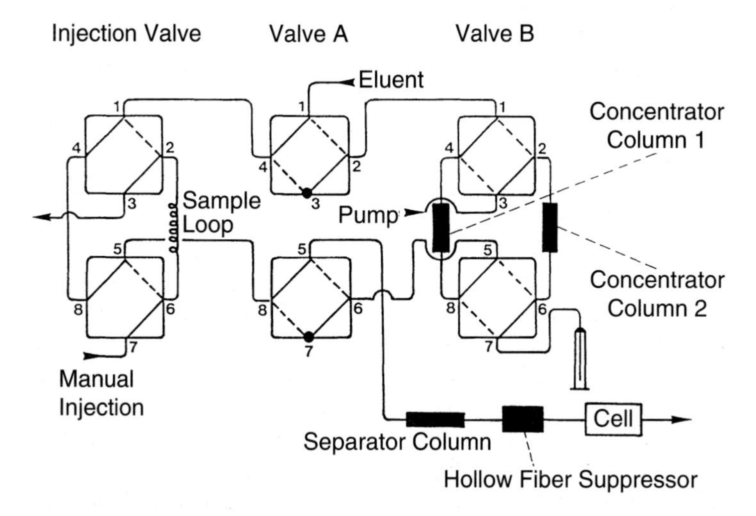

Fig. 8-18. Valve switching for alternating application of a sample loop and a concentrator technique.

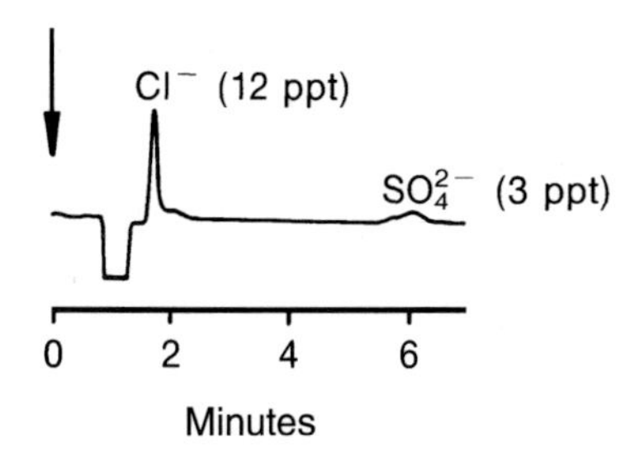

Fig. 8-19. Anion analysis of high purity water after pre-concentration. – Separator column: IonPac AS4A; chromatographic conditions: see Fig. 8-13; concentrated volume: 50 mL high purity water (SERAL).

The valve switching shown in Fig. 8-18 allows both the injection using a sample loop and the application of a concentrator technique simply by switching an auxiliary valve. Such a valve switching is only useful if the ion chromatograph is used for high purity water analyses and also for application in higher concentration ranges. An example is the alternating analysis of high purity water and cooling water.

Fig. 8-19 shows the chromatogram of such a high purity water with chloride and sulfate concentrations in the sub-ppb range. This chromatogram was obtained after

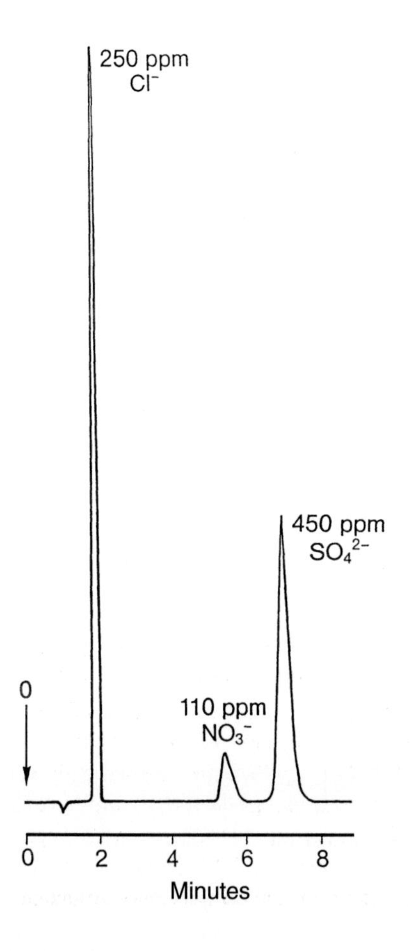

Fig. 8-20. Anion analysis of a cooling water. – Separator column: IonPac AS4; chromatographic conditions: see Fig. 8-16; injection: 10 µL sample (1:10 diluted).

pre-concentration of 50 mL high purity water [24]. This method allows the lower detection limits for inorganic anions and cations to be reduced from about 10 ppb at direct injection of 50 μL sample volume to about 10 ppt after pre-concentration [25].

The chromatogram of a cooling water is displayed in Fig. 8-20. The sample was diluted 1:10 with de-ionized water to avoid overloading the separator column. A direct sample injection is not feasible due to its high electrolyte content, since the conductance is not proportional to the solute concentration in this concentration range.

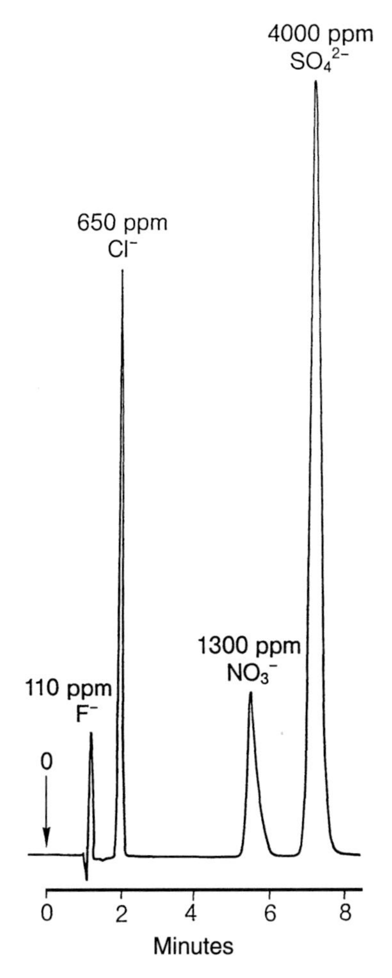

Fig. 8-21. Analysis of inorganic anions in the scrubber solution of a flue gas desulfurization plant. – Chromatographic conditions: see Fig. 8-16; injection: 50 μL sample (1:100 diluted).

Another important application is the monitoring of the scrubber solutions in flue gas desulfurization plants that operate according to the wet absorption principle. The respective scrubber solutions contain a variety of ionic components that can be analyzed via ion chromatography. In flue gas desulfurization using gypsum as the absorber, for example, sulfite is oxidized to sulfate. As is seen in Fig. 8-21, sulfate may be determined in one run together with other mineral acids. Also, anion exchange chromatography can be employed to test such scrubber solutions for toxic chromium(VI), although the high sulfate loading forces one to apply a selective detection method. This is based on the derivatization of the chromate with 1,5-diphenylcarbazide and subsequent photometric detection at 520 nm of the complex being formed. The chromatogram shown in Fig. 8-22 impressively demonstrates the high selectivity of this method, as only chromium(VI) is indicated in spite of the high electrolyte content of the investigated sample.

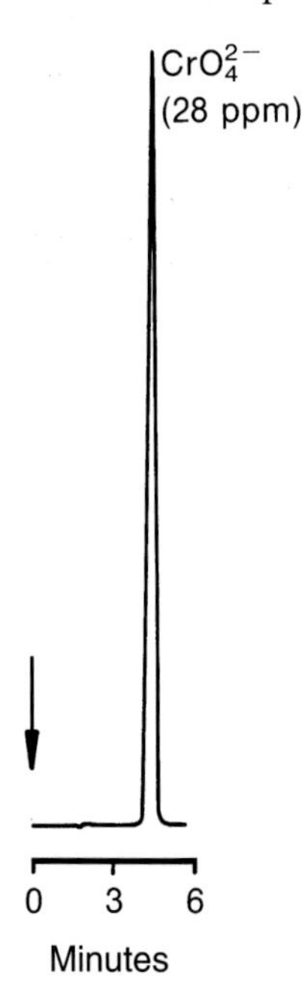

Fig. 8-22. Analysis of chromium(VI) in a scrubber solution from the flue gas desulfurization. – Separator column: IonPac CS5; eluent: 0.002 mol/L pyridine-2,6-dicarboxylic acid + 0.002 mol/L Na_2HPO_4 + 0.01 mol/L NaI + 0.05 mol/L NH_4OAc + 0.0028 mol/L LiOH; flow rate: 1 mL/min; detection: photometry at 520 nm after reaction with 1,5-diphenylcarbazide; injection: 50 μL sample (1:10 diluted).

A derivatization technique must be used if ammonium is to be determined in this sample. The high concentration of alkali and alkaline-earth metals does not allow a conductometric detection of this ion. To detect ammonium selectively, the derivatization technique using *o*-phthaldialdehyde with subsequent fluorescence detection is suited that has been described repeatedly. This is illustrated in Fig. 8-23 with the chromatogram of a sample that was diluted 1:5 in which a primary aliphatic amine is detected as well as ammonium using this detection method.

Additionally, sulfur compounds may be detected in scrubber solutions from flue gas desulfurization. Dithionate, $S_2O_6^{2-}$, is formed, for example, via recombination of two hydrogensulfite radicals. It may also be analyzed ion chromatographically [26]. Dithionate exhibits a strong affinity toward the stationary phase of an anion exchanger, hence, either electrolytes with high elution power or high electrolyte concentrations must be used. In both cases, a sensitive detection by measuring the electrical conductance is extremely difficult. Since there is no alternative to this detection method due to the UV transparency of dithionate, another separation method must be used. Ion-pair chromatography proved to be particularly suited, because this method allows the retention behavior of the analyte species to be influenced by a variety of experimental retention-determining parameters. As dithionate is a surface-inactive substance, tetrabutylammonium hydroxide is used as the ion-pair reagent. Organic solvents such as acetonitrile or methanol serve to adjust the polarity of the mobile phase. Fig. 8-24 shows an example chromatogram for the determination of dithionate in a sample from the flue gas scrubbing. For the analysis of polythionates, $S_nO_6^{2-}$ ($n > 2$), only the acetonirile content in the mobile phase must be enhanced.

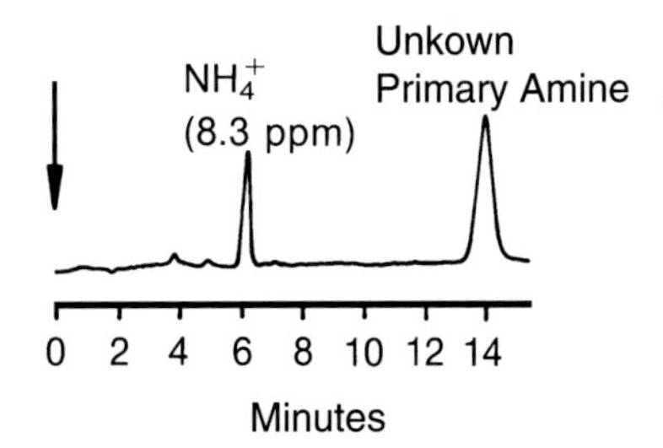

Fig. 8-23. Analysis of ammonium in a scrubber solution from flue gas desulfurization. – Separator column: IonPac CS2; eluent: 0.03 mol/L HCl; flow rate: 1 mL/min; detection: fluorescence after reaction with OPA; injection: 50 μL sample (1:5 diluted).

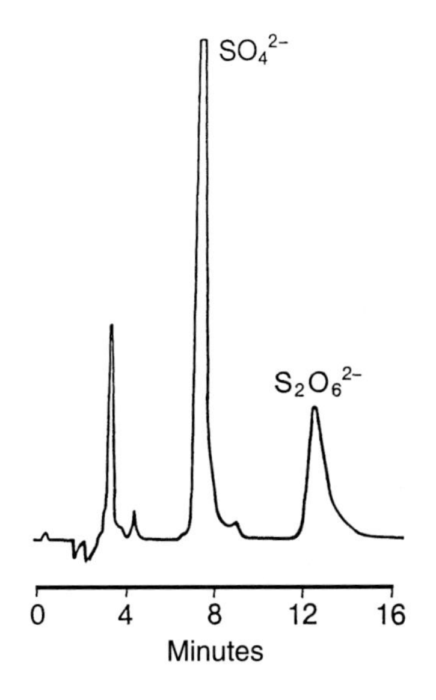

Fig. 8-24. Analysis of dithionate in a scrubber solution from flue gas desulfurization. – Separator column: Ion Pac NS1 (10 μm); eluent: 0.002 mol/L TBAOH + 0.001 mol/L Na_2CO_3 – acetonitrile (85:15 v/v); flow rate: 1 mL/min; detection: suppressed conductivity; injection: 50 μL sample (1:100 diluted).

Under similar chromatographic conditions, it is possible to detect the class of sulfur-nitrogen-compounds, which includes compounds such as amidosulfonic acid, hydroxylaminedisulfonic acid, and nitrilotrisulfonic acid. Although the presence of all three compounds in scrubber solutions from flue gas desulfurization has been postulated, only amidosulfonic acid and nitrilotrisulfonic acid have clearly been identified in these samples. While the former elutes in the retention range of chloride, nitrilotrisulfonic acid as a trivalent species exhibits a much higher retention than dithionate.

Formic acid, added to some scrubber solutions for pH control, can be separated from inorganic anions and thus be determined with the aid of ion-exclusion chromatography. A corresponding chromatogram is depicted in Fig. 8-25.

Ion chromatography may also contribute analytically to the monitoring of the flue gas denitrification process. The most important parameter – the concentration ratio between nitrite and nitrate – can be established within five minutes as revealed in Fig. 8-26.

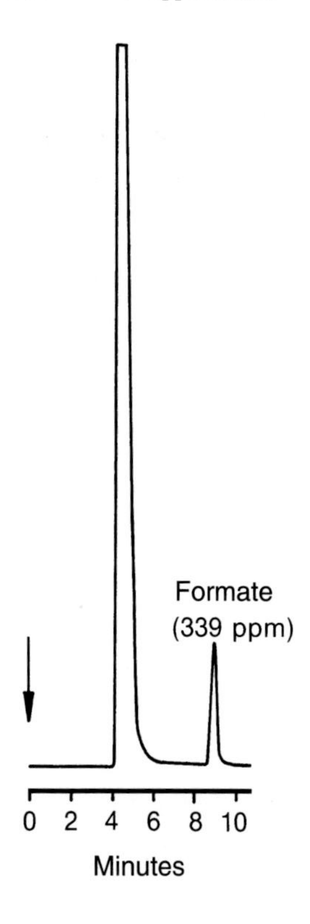

Fig. 8-25. Analysis of formic acid in the scrubber solutions from flue gas desulfurization. – Separator column: IonPac ICE-AS1; eluent: 0.0005 mol/L octanesulfonic acid; flow rate: 1 mL/min; detection: suppressed conductivity; injection: 50 µL sample (1:25 diluted).

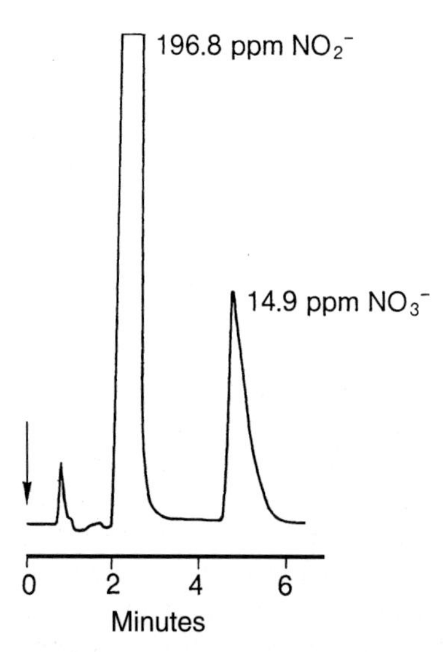

Fig. 8-26. Analysis of nitrite in a scrubber solution from flue gas denitrification. – Separator column: IonPac AS3; eluent: 0.0028 mol/L $NaHCO_3$ + 0.0022 mol/L Na_2CO_3; flow rate: 2.3 mL/min; detection: suppressed conductivity; injection: 50 µL sample.

Another interesting application is the analysis of trace amounts of cooling water conditioners. Typically, these products are mixtures of polycarboxylic acids and polyphosphonic acids that are added to cooling waters as corrosion inhibitors. The application of these conditioners stabilizes calcium bicarbonate at thermally stressed positions. Scaling at the metal surfaces is prevented.

Polyphosphonic acids can be separated on a special latex-based anion exchanger. They are photometrically determined after complexation with ferric nitrate [27] (see Section 3.3.5.2). This method enables a qualitative analysis of inorganic and organic phosphates that are contained in conditioners. Fig. 8-27 illustrates this with the chromatogram of a commercial product. Clearly, it is a multicomponent mixture with two main components and several minor components.

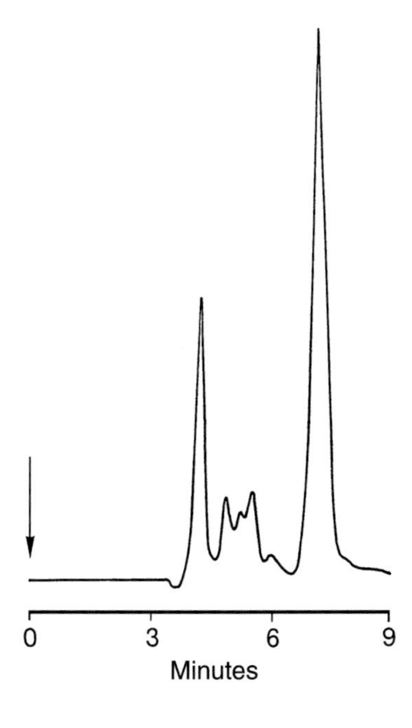

Fig. 8-27. Analysis of a cooling water conditioner. – Separator column: IonPac AS7; eluent: 0.03 mol/L HNO_3; flow rate: 0.5 mL/min; detection: photometry at 300 nm after reaction with ferric nitrate; injection: 50 µL sample (1:1000 diluted).

Since the concentration of these products in cooling water is in the upper ppb range, the sensitivity of this method is insufficient upon direct injection of a 50-µL sample. Therefore, the concentrator technique described above must be applied for determining the concentrations of these components. At a concentrator volume of 2 mL, the chromatogram of a cooling tower circulation water represented in Fig. 8-28 is obtained. The polyphosphonic acid is clearly detectable. The other two signals visible in the chromatogram are chloride and sulfate, which are present in much higher concentrations and are also enriched. Both anions are complexed by ferric nitrate, so they appear as large signals in the chromatogram. The possibility of polyphosphonates co-eluting with inorganic anions may be eliminated by using a phosphorus-specific detection [28]. This also facilitates the quantitative analysis of the polyphosphonates. Because of their hydrolysis to orthophosphate, the concentration of the relevant components does not have to be determined via external calibration with appropriately pure reference substances.

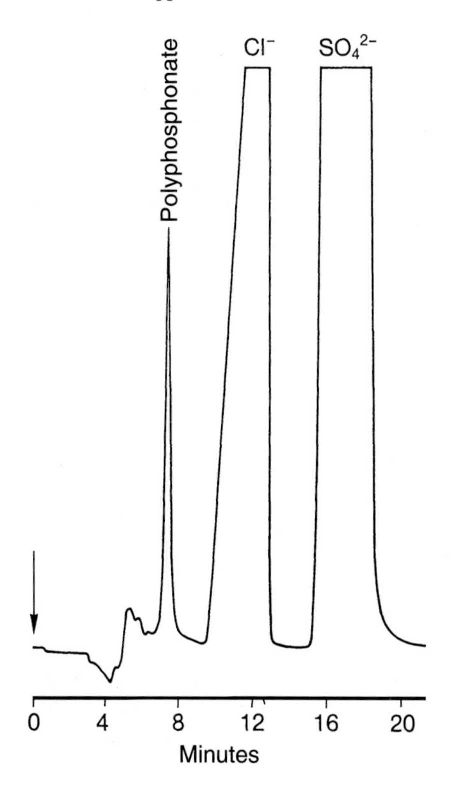

Fig. 8.28. Analysis of conditioners in a cooling tower circulating water. – Chromatographic conditions: see Fig. 8-27; concentrated volume: 2 mL.

8.3 Ion Chromatography in the Electroplating Industry

Within a short time, the electroplating industry has become one the most important fields of application for ion chromatography [29,30]. Attention is being focused on the routine monitoring of plating baths and the improvement of the plating quality in dependence of the chemical composition. With rising production costs for high quality products, it became necessary to use novel analytical methods that contribute to a better understanding of the complex chemical processes occuring in a plating bath.

Ion chromatographic analysis methods ensure speed and high precision in the analysis of main components as well as of reaction and decomposition products in electroplating baths. The advantage of ion chromatography relative to the partly unspecific wet chemical methods utilized so far lies in the selectivity of the stationary phases and the detection systems being used. Thus, in most cases sample preparation is limited to a simple dilution with de-ionized water and subsequent filtration. A variety of applications on the basis of electrodeposition and electroless plating is summarized in Table 8-3.

Table 8-3. Fields of application and examples of analyses in the electroplating industry.

Type of plating bath	Examples of analyses
Copper (electroless)	Determination of formic acid, tartaric acid, triethanolamine, EDTA, and $Cu(EDTA)^{2-}$
Copper sulfate (electrolytical)	Determination of the Cl^-/SO_4^{2-} ratio, of metallic impurities such as Ni^{2+} and Zn^{2+}, organic additives
Copper pyrophosphate (electrolytical)	Determination of ammonium, orthophosphate, and nitrate
Copper cyanide (electrolytical)	Determination of Cu^{2+} after preparation with strong acids, hexacyano ferrates
Nickel sulfamate (electrolytical)	Determination of sulfamate, chloride, sulfate, and ammonium
Nickel/iron (electrolytical)	Determination of the Ni^{2+}/Fe^{2+} ratio, Na^+, Cu^{2+}, boric acid, saccharine, and lauryl sulfate
Nickel nickel/copper (electroless)	Determination of hypophosphite, phosphite, citric acid, succinic acid, Ni^{2+}/Co^{2+} ratio
Gold cyanide (electrolytical)	Determination of $Au(CN)_2^-$, $Au(CN)_4^-$, $Co(CN)_6^{3-}$, CN^-, chloride, orthophosphate, carbonate, and hexacyano ferrates

Analysis of Inorganic Anions

Applications for the determination of mineral acids in the field of plating bath analyses are manifold. In copper sulfate plating baths for electrodeposition, it is possible to simultaneously analyze the small amount of chloride contained in these baths apart from the main component sulfate (Fig. 8-29).

An important constituent in copper pyrophosphate baths is nitrate, which enhances the maximum permissible current density [31]. Fig. 8-30 shows the respective chromatogram with the separation of nitrate and orthophosphate. The latter is the hydrolysis product of pyrophosphate that is formed during the plating process. The main component pyrophosphate may also be separated on a latexed anion exchanger. It is detected after complexation with ferric nitrate in a post-column reaction by measuring the light absorption (see Section 3.3.5.2).

Nickel plating baths based on electroless deposition contain a reducing agent and a catalyst. The reducing agent acts as electron donor for the reduction of metal ions to the metal. Hypophosphite, formaldehyde, hydrazine, and boron hydride are suited for this. A colloidal Pd/Sn^{2+} solution usually serves as the catalyst. This solution is adsorbed to the material surface and catalyzes the deposition of a monolayer of the respective metal. Further plating occurs autocatalytically with the formation of hydride ions.

The following reaction scheme has been proposed for the electroless deposition of nickel with hypophosphite as reducing agent:

$$
\begin{aligned}
&H_2PO_2^- + H_2O \xrightarrow{Pd/Sn^{2+}} HPO_3^{2-} + 2\,H^+ + H^- \\
&2\,H^- + Ni^{2+} \longrightarrow Ni + H_2 \uparrow \\
\hline
&2\,H_2PO_2^- + 2\,H_2O + Ni^{2+} \longrightarrow Ni + H_2 \uparrow + 4\,H^+ + 2\,HPO_3^{2-}
\end{aligned}
\tag{225}
$$

It is obvious from this reaction scheme that the deposition rate depends on the concentration of the reducing agent. Since both nickel and hypophosphite are exhausted in the plating process, occasionally both compounds must be added to the bath as corresponding salts. The electroless plating process is accompanied by an increase in the concentration of phosphite as the hypophosphite oxidation product, which limits the effectiveness of such a plating bath. Both phosphorus species can be determined simultaneously with the aid of ion chromatography, thus allowing the plating bath control to be optimized. Fig. 8-31 shows the difference in the amounts of both compounds between a new and an exhaused plating bath after one and five metal turn overs, respectively.

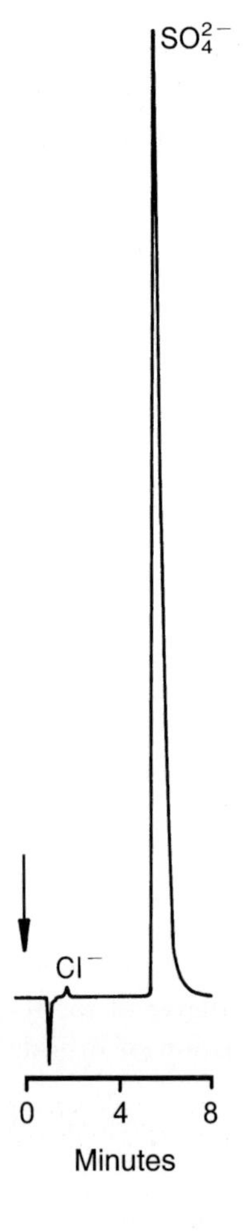

Fig. 8-29. Trace analysis of chloride in an acid copper sulfate plating bath. – Separator column: IonPac AS4; eluent: 0.0028 mol/L $NaHCO_3$ + 0.0022 mol/L Na_2CO_3; flow rate: 2 mL/min; detection: suppressed conductivity; injection: 10 µL sample (1:1000 diluted).

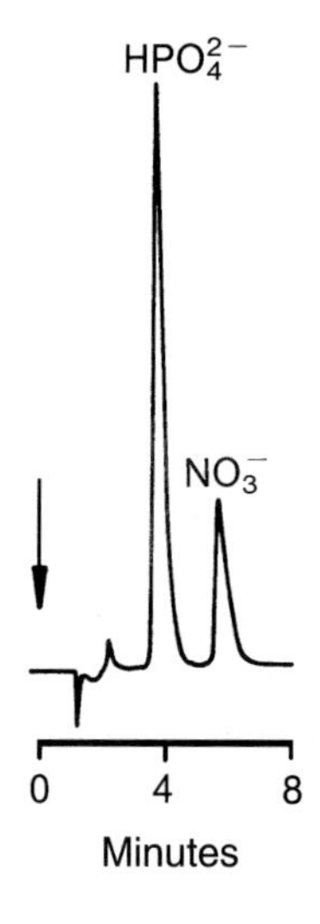

Fig. 8-30. Analysis of mineral acids in a copper pyrophosphate bath. – Separator column: IonPac AS3; eluent: 0.003 mol/L $NaHCO_3$ + 0.0028 mol/L Na_2CO_3; flow rate: 2.3 mL/min; detection: suppressed conductivity; injection: 10 µL sample (1:1000 diluted).

If the nickel bath contains organic acids such as lactic acid, it is necessary to apply a gradient technique, as lactic acid and hypophosphite have the same retention time under the isocratic conditions chosen in Fig. 8-31.

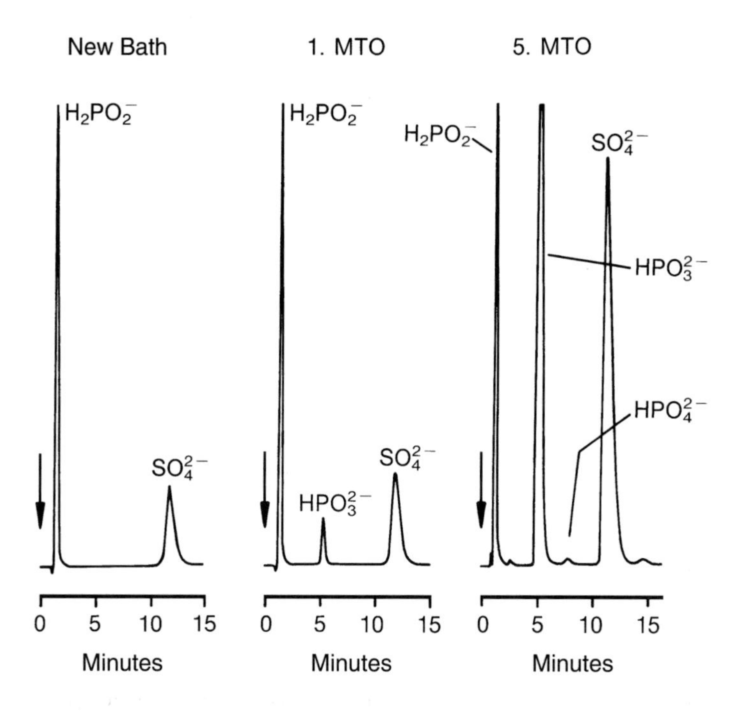

Fig. 8-31. Analysis of hypophosphite and phosphite in an electroless nickel bath. – Separator column: IonPac AS3; eluent: 0.003 mol/L Na_2CO_3; flow rate: 2.3 mL/min; detection: suppressed conductivity; injection: 50 µL sample (1:200 diluted).

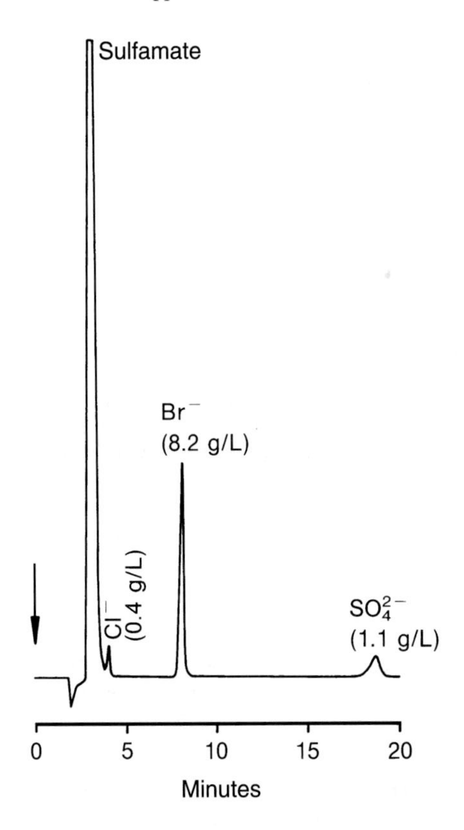

Fig. 8-32. Analysis of chloride, bromide, and sulfate in a nickel sulfamate bath. – Separator column: 2 IonPac AS4A; eluent: 0.0017 mol/L $NaHCO_3$ + 0.0018 mol/L Na_2CO_3; flow rate: 1.5 mL/min; detection: suppressed conductivity; injection: 50 µL sample (1:1000 diluted).

The main component in nickel sulfamate baths – sulfamate – can be determined in a single run, in addition to the decomposition product sulfate and other bath constituents such as chloride and bromide (see Fig. 8-32). To achieve a sufficient separation between the sulfamate ions and chloride, both of which elute near the void volume, two identical anion exchangers were used in series, even though this increases the total analysis time to about 20 minutes.

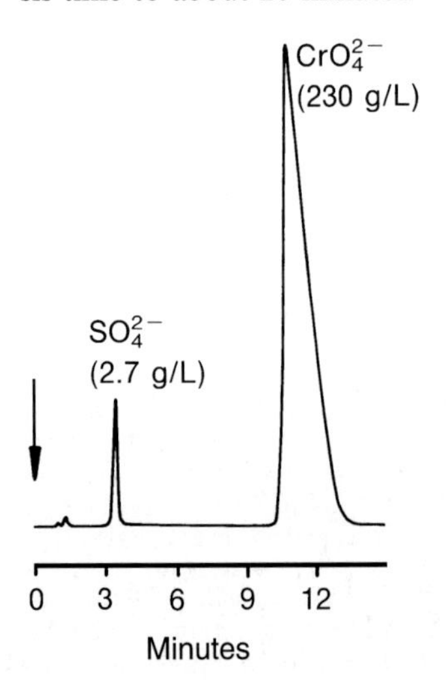

Fig. 8-33. Analysis of sulfate and chromate in a chromic acid bath. – Separator column: IonPac CS5; eluent: 0.001 mol/L $NaHCO_3$ + 0.005 mol/L Na_2CO_3; flow rate: 1.5 mL/min; detection: suppressed conductivity; injection: 50 µL sample (1:2000 diluted).

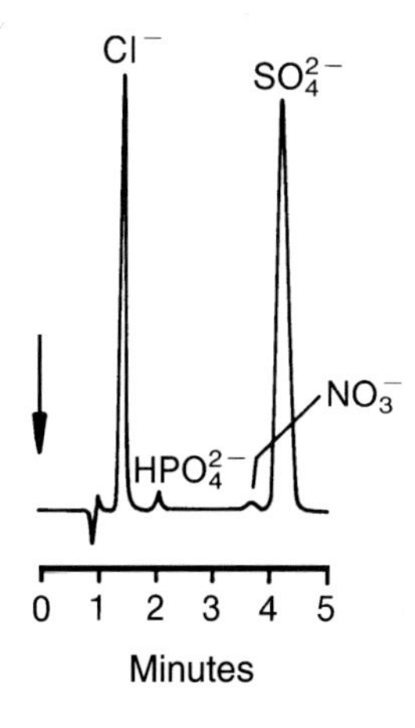

Fig. 8-34. Analysis of inorganic anions in an electroplating waste water. – Separator column: IonPac AS4; chromatographic conditions: see Fig. 8-29; injection: 50 μL sample (1:2500 diluted).

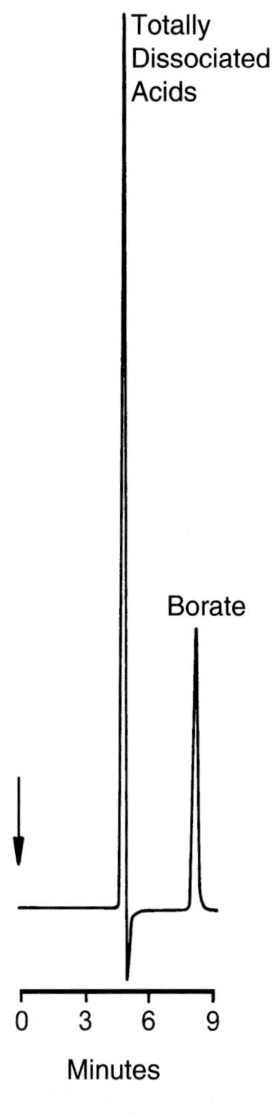

Fig. 8-35. Boric acid determination in a nickel/iron plating bath. – Separator column: IonPac ICE-AS1; eluent: 0.001 mol/L octanesulfonic acid; flow rate: 1 mL/min; detection: suppressed conductivity; injection: 50 μL sample (1:1000 diluted).

In sulfuric acid chromium baths, ion chromatography allows the simultaneous determination of the two main components sulfate and chromate, even if they are usually present in very different concentrations. This is illustrated in Fig. 8-33. In the analysis of such samples, it should be noted that after dilution the solutions are alkaline. Chromium(VI) is present as dichromate in the acidic pH range, thus a repeated injection of this strongly oxidizing substance would damage the separator column irreversibly.

Fig. 8-34 shows the chromatogram of an electroplating waste water that can be directly injected after appropriate dilution and filtration (0.45 µm). In addition to the main components chloride and sulfate, orthophosphate and nitrate can also be detected.

Weak inorganic acids such as borate and carbonate can be determined by means of ion-exclusion chromatography. In addition to chlorate and sulfate, pure and alloyed nickel plating baths contain high concentrations of boric acid with its weak buffer effect. Conventional analysis methods for boric acid such as titrimetric determination are time-consuming and suffer from poor precision, because the titration end point is often difficult to recognize. The ion chromatographic analysis of boric acid in a nickel/iron plating bath is illustrated in Fig. 8-35. The sole sample preparation here was, again, dilution with de-ionized water. Carbonate that is an ingredient for example in gold plating baths as a buffer agent can be determined with a gross retention time of t_{ms} = 13.5 min under the same chromatographic conditions.

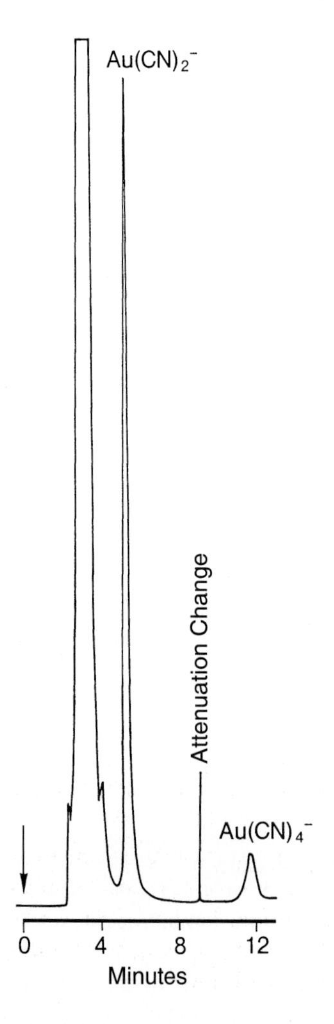

Fig. 8-36. Separation of gold(I) and gold(III) as cyano complexes in a gold plating bath. – Separator column: IonPac NS1; eluent: 0.002 mol/L TBAOH + 0.002 mol/L Na_2CO_3/acetonitrile (60:40 v/v); flow rate: 1 mL/min; detection: suppressed conductivity; injection: 50 µL sample (1:200 diluted).

Analysis of Metal Complexes

Ion-pair chromatography is capable of determining the concentration ratio between gold(I) and gold(III). In the presence of an excess of cyanide, both gold species exist as anionic cyano complexes, $Au(CN)_2^-$ and $Au(CN)_4^-$. To determine the gold content, conventional analysis methods such as atomic absorption spectroscopy or precipitation and titration techniques require time-consuming sample preparation to break the complex bond. Moreover, such methods do not allow individual oxidation states to be specified. Owing to their stability, anionic gold complexes may be separated with the aid of ion-pair chromatography and may be detected conductometrically. Fig. 8-36 illustrates the chromatogram of a gold plating solution buffered with orthophosphate and citrate that was diluted and filtrated prior to the measurement. Cobalt sulfate is added as a hardening agent to some plating baths. The cyano complex $Co(CN)_6^{3-}$ that is formed during the plating process and that is ineffective as a hardening agent can be separated from the gold-cyano complexes (Fig. 8-37).

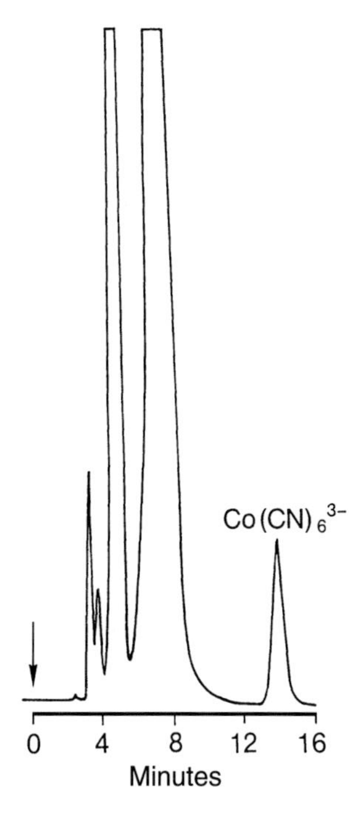

Fig. 8-37. Determination of the cobalt-cyano complex in a gold plating bath. – Separator column: IonPac NS1; eluent: 0.002 mol/L TPAOH/acetonitrile (90:10 v/v); flow rate: 1 mL/min; detection: suppressed conductivity; injection: 50 μL sample (1:400 diluted).

Copper- and silver-cyanide complexes may also be analyzed using ion-pair chromatography. Owing to their low stability, however, some sodium cyanide is added to the mobile phase to shift the complexing equilibrium to the complex side.

Metal-EDTA complexes may be separated with the aid of anion exchange chromatography (see Fig. 3-78 in Section 3.3.4.4). They are detected again via electrical conductivity measurements. In electroless copper baths, for example, it is thus possible to distinguish between free and complex-bound EDTA. Other bath constituents do not interfere with the determination of the copper-EDTA complex, $Cu(EDTA)^{2-}$.

Analysis of Organic Acids

Organic acids are usually separated by means of ion-exclusion chromatography and detected via their electrical conductivity. Applications are mainly in the field of copper and electroless nickel plating baths.

Normally, copper baths contain formaldehyde and a complexing agent such as tartaric acid. During the plating process, formaldehyde is oxidized to formic acid. Ion chromatography provides simultaneous detection of both acids. Therefore, aging processes of the plating bath, which are indicated by an increased formate content, can be immediately recognized. Fig. 8-38 shows a chromatogram of organic mono-, di-, and tricarboxylic acids, mild reducing agents commonly used in an electroless nickel bath. Again, in this case, the simultaneous detection capability of ion chromatography is impressive.

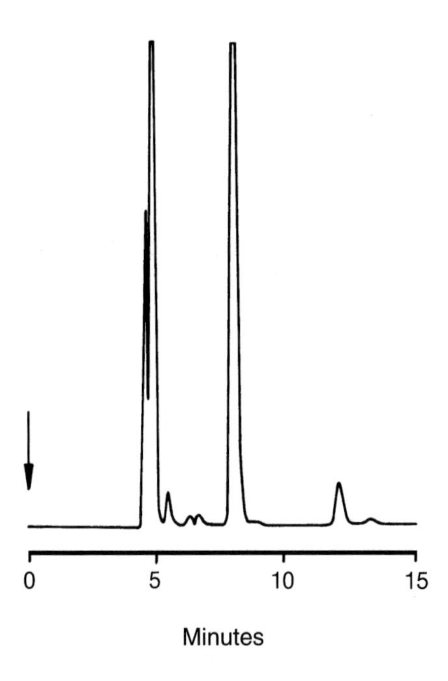

Fig. 8-38. Analysis of organic acids in an electroless nickel bath. – Separator column: IonPac ICE-AS1; chromatographic conditions: see Fig. 8-35; injection: 50 μL sample (1:200 diluted).

Ion-pair chromatography is capable of separating short-chain sulfonic acids such as allyl- and vinylsulfonic acid that are sometimes contained in nickel baths. Conventional anion exchangers and ion-exclusion phases cannot be used, since sulfonic acids as strong organic acids elute near the void volume, making it impossible to separate them from the electrolytic bath constituents. As short-chain sulfonic acids exhibit only a low hydrophobicity, the comparatively solvophobic tetrabutylammonium hydroxide is utilized as the ion-pair reagent. Detection is accomplished by measuring the electrical conductance since olefinic sulfonic acids also do not possess suitable chromophores that enable a sensitive UV detection. The chromatogram of a vinyl- and allylsulfonic acid standard in Fig. 8-39 clearly illustrates that the retention behavior of these compounds resembles that of inorganic anions such as chloride and sulfate.

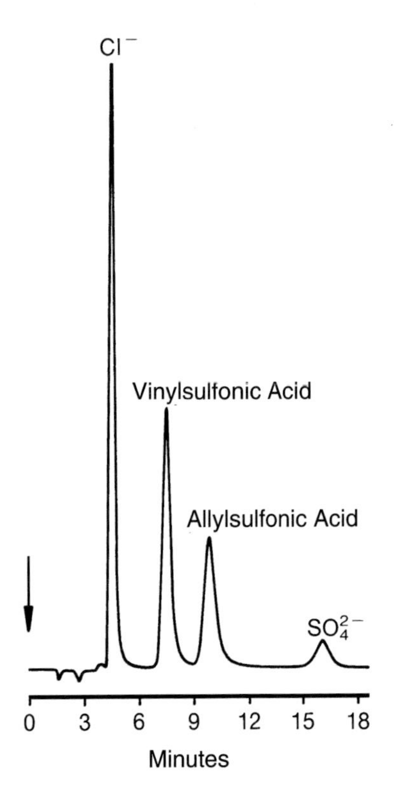

Fig. 8-39. Separation of short-chain olefinic sulfonic acids. – Separator column: IonPac NS1 (10 µm); eluent: 0.002 mol/L TBAOH + 0.001 mol/L Na_2CO_3/acetonitrile (92:8 v/v); flow rate: 1 mL/min; detection: suppressed conductivity; injection volume: 50 µL; solute concentrations: 30 ppm each of vinyl- and allylsulfonic acid.

Analysis of Inorganic Cations

While alkali and alkaline-earth metals can also be rapidly and very sensitively detected by other instrumental analysis methods, the advantage of ion chromatography lies in the simultaneous detection of the ammonium ion. In copper pyrophosphate baths, for example, the addition of ammonia improves the plating evenness. However, as the ammonia concentration continuously decreases at higher bath temperatures, it must be added to maintain optimal bath conditions. As seen in Fig. 8-40, after separation on an anion exchanger the ammonium ion can be detected quickly and reliably separated from sodium and potassium.

Ion chromatography provides an alternative to atom spectroscopic methods for the determination of heavy and transition metals. Its main advantage is the simultaneousness of the procedure and the possibility to distinguish between different oxidation states. For example, the determination of iron(III), copper, and zinc in a chromic acid bath (Fig. 8-41) may be performed free of interferences despite the high chromium(VI) load. Fig. 8-42 illustrates the determination of heavy and transition metals in a nickel/iron plating bath. Both oxidation states of iron can be clearly distinguished via ion chromatography.

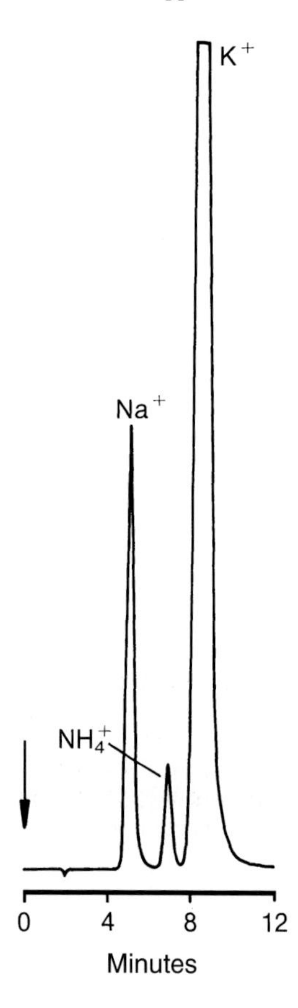

Fig. 8-40. Analysis of ammonium in a copper pyrophosphate bath. – Separator column: IonPac CS1; eluent: 0.005 mol/L HCl; flow rate: 2.3 mL/min; detection: suppressed conductivity; injection: 50 μL sample (1:2500 diluted).

Analysis of Organic Additives

Organic additives play an important role for maintaining the plating quality of plating baths. If these compounds exhibit an ionic character, there is a possibility to detect them by ion chromatography. If conductometric detection is impossible because of a high electrolyte content in the sample, direct UV detection is used.

Nickel/iron plating baths, for example, contain saccharin and surfactants such as sodium laurylsulfate. A study by Becker and Bolch [32] showed that the appropriate concentration range must be observed to ensure a constant plating quality. For the determination of these two compounds ion chromatographic methods were developed. The corresponding chromatograms in Fig. 8-43 and 8-44 reveal a baseline-resolved separation of saccharin and sodium laurylsulfate in a nickel/iron plating bath. In both cases, detection was accomplished by measuring the electrical conductance.

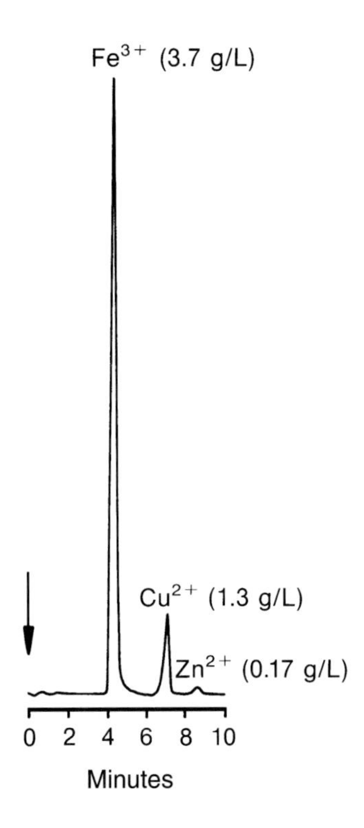

Fig. 8-41. Determination of iron(III), copper, and zinc in a chromium plating bath. – Separator column: IonPac CS5; eluent: 0.006 mol/L pyridine-2,6-dicarboxylic acid + 0.0086 mol/L LiOH; flow rate: 1 mL/min; detection: photometry at 520 nm after reaction with PAR; injection: 50 µL sample (1:1000 diluted).

Apart from these two substances, a variety of further organic compounds are added to plating baths as brightening agents, levelling agents, and stabilizers. Their chemical structure is known only in a few cases. Of great importance are thiourea derivatives such as 2-imidazolidinthione (ethylenethiourea) and *N,N'*-diphenylthiourea, which are employed for example in electroless nickel and nickelborohydride plating baths as brightening agents.

2-Imidazolidinthione **_N,N'_-Diphenylthiourea**

To separate these basic substances, a neutral polymer is again suitable as the stationary phase; detection is performed by measuring the light absorption because of existing chromophores. The chromatographic conditions that have been developed for the elution of both compounds differ only in their acetic acid and organic solvent content in the mobile phase. The corresponding chromatograms are shown in Fig. 8-45 and 8-46, respectively.

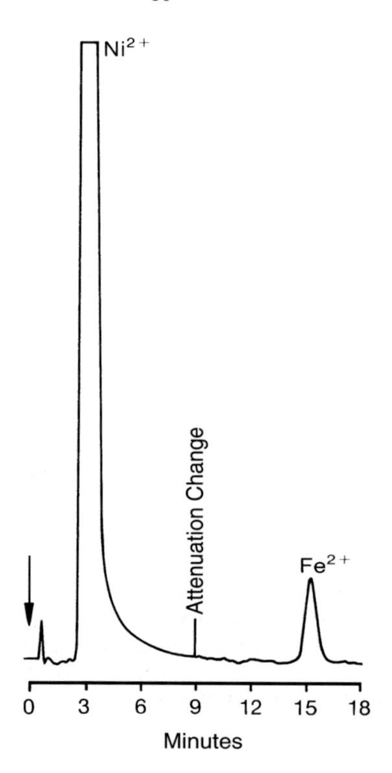

Fig. 8-42. Transition metals in a nickel/iron plating bath. – Separator column: IonPac CS2; eluent: 0.01 mol/L citric acid + 0.0075 mol/L oxalic acid (pH 4.3 with LiOH); flow rate: 1 mL/min; detection: see Fig. 8-41; injection: 50 μL sample (1:500 diluted).

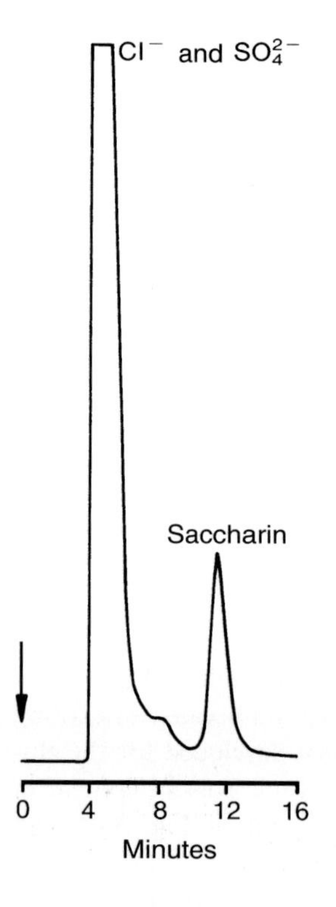

Fig. 8-43. Determination of saccharin in a nickel/iron plating bath. – Separator column: IonPac NS1 (10 μm); eluent: 0.002 mol/L TMAOH (pH 12 with NaOH)/acetonitrile (95:5 v/v); flow rate: 1 mL/min; detection: suppressed conductivity; injection: 50 μL sample (1:50 diluted).

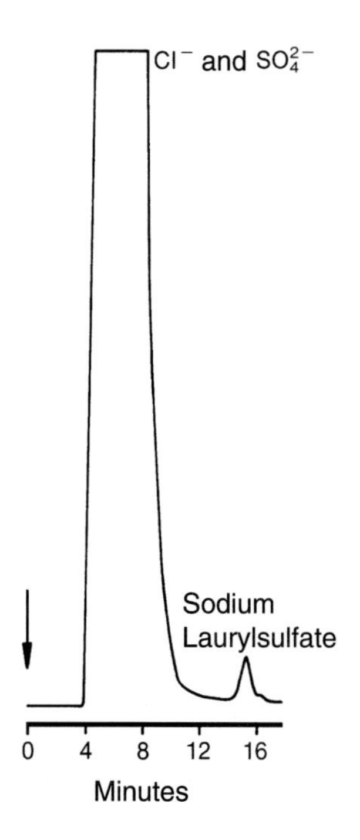

Fig. 8-44. Determination of sodium laurylsulfate in a nickel/iron plating bath. – Separator column: IonPac NS1 (10 µm); eluent: 0.01 mol/L NH_4OH/acetonitrile (72:28 v/v); flow rate: 1 mL/min; detection: suppressed conductivity; injection: 50 µL sample (1:10 diluted).

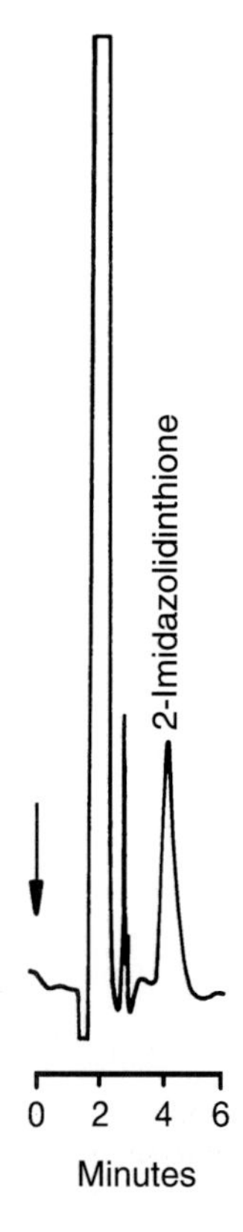

Fig. 8-45. Determination of 2-imidazolidinthione in an electroless nickel plating bath. – Separator column: IonPac NS1; eluent: acetic acid/methanol/water (1:5:94 v/v/v); flow rate: 0.8 mL/min; detection: UV (254 nm); injection: 50 µL sample with 3 ppm of the relevant substance.

Fig. 8-46. Determination of *N,N'*-diphenylthiourea in an electroless nickel-borohydride plating bath. – Separator column: IonPac NS1; eluent: acetic acid/methanol/acetonitrile/water (1:39:31:29 v/v/v/v); flow rate: 0.8 mL/min; detection: UV (254 nm); injection: 50 μL sample with 2 ppm of the relevant substance.

8.4 Ion Chromatography in the Semiconductor Industry

A key problem in semiconductor industry is the lifetime of the semiconductor components being produced which is decisively affected by ionic contaminations caused by

- insufficient quality of the water being used,
- incomplete rinsing processes,
- use of impure chemicals,
- contamination during passivation, and
- poor quality of the employed polymers.

The complexity, speed of development, and the diversification of this industry have demanded the highest quality; in the employed water as well as in the chemicals and solvents being used.

Ion chromatography offers the most efficient method for analyzing ionic species for the production of semiconductor components and printed circuit boards. Conventional methods, based on the SEMI (**S**emiconductor **E**quipment and **M**aterials **I**nstitute, Inc.) regulations for identifying and quantifying ionic species responsible for corrosion, are labor-intensive and time-consuming, whereas it is possible to obtain complete anion and cation profiles within ten minutes using ion chromatographic techniques. The various areas of application and typical analytical examples are summarized in Table 8-4. The many possible applications for ion chromatography in the field of microelectronics are illustrated by characteristic examples.

Table 8-4. Application areas and typical analytical examples in the semiconductor industry.

Application area	Analytical example
Water analysis	Determination of mineral acids in de-ionized water, process liquors, rinsing and waste waters
Etching solutions	Determination of the main components and impurities in HF/HNO_3- and $HF/HNO_3/HOAc$-mixtures
Solvents	Determination of chloride (directly or after extraction)
Acids	Determination of anions such as chloride, nitrate, and sulfate in phosphoric acid (reagent grade)
Hydroxides	Determination of sodium in 46% KOH
Polymers	Determination of anionic and cationic impurities in encapsulation plastics

The quality of the pure water required in high quantities in the manufacture of printed circuit boards is of great importance [35]. As in the field of power plant chemistry, in semiconductor industry ion chromatographic methods for determining mineral acids [34], silicate, alkali metals, and heavy and transition metals [35] have been used for some years to ensure this quality. Today, these methods are fully automated. The equipment necessary for that may be configured for both laboratory and on-line operation. The latter allows the continuous monitoring of the pure water quality at several sampling locations connected to the chromatograph via appropriately dimensioned tubings. A personal computer serves as the control unit of such an on-line chromatograph, and aids in the selection and subsequent analysis of the individual sample streams in any sequence and frequency and may be analyzed subsequently. The calibration of the system can also be fully automated. The necessary standards are prepared in this case by freely programmable dilution of concentrated standard solutions in a specially designed module. The standards are delivered to the chromatograph as separate streams. The analysis report is usually not obtained in the form of actual chromatograms but rather as numeric printouts of the determined solute concentrations or graphically as trending diagrams.

In this connection, the on-line procedure that has been developed by Johnson et al. [36] for the simultaneous analysis of mineral acids and silicate is very interesting. As depicted in Fig. 8-47, the sample is passed through two different concentrator columns. The first one, the TAC-1, retains the mineral acids, which are separated on a conventional anion exchanger by using a carbonate/bicarbonate-mixture after switching the respective injection valve; detection is carried out conductometrically. Silicate is not retained at this concentrator column, since the capacity of a TAC-1 in the carbonate/bicarbonate-form is too low. Therefore, after passing the TAC-1, the analyte sample is passed through a concentrator column IonPac AG5, which is suited for concentrating silicate because of its high capacity. Elution is then accomplished with a mixture of boric acid and sodium hydroxide; the sensitive detection in the lowest ppb range is performed by measuring the light absorption at 410 nm after reaction with sodium molybdate. Some glycerol is added to the reagent to prevent the precipitation of sodium molybdate in the pump head of the delivery pump. Any precipitates that might arise can be rendered soluble with sodium hydroxide solution of the concentration $c = 0.1$ mol/L. Fig. 8-48 shows the two chromatograms of a standard containing fluoride, chloride, bromide, nitrate, orthophosphate, sulfate, and silicate with mass concentrations in the lower µg/L range that have been obtained using this method. If the silicate standard is prepared from ultrapure de-ionized water that has been additionally passed through an IonPac AG5 column to remove silicate traces, a detection limit in the sub-ppb range may also be realized for this ion.

Upon application of the concentrator technique, also heavy and transition metals may be determined in the sub-ppb range [35]. Based on the photometric detection of metal ions after derivatization with PAR, detection limits of about 10 ppt are obtained for metals such as iron, copper, nickel, zinc, cobalt, and manganese after corresponding pre-concentration. Hence, this method is more sensitive than graphite furnace AAS, which is utilized as the traditional method for analyzing trace amounts of metals in this area. Ion chromatography has the additional advantage of being able to detect several

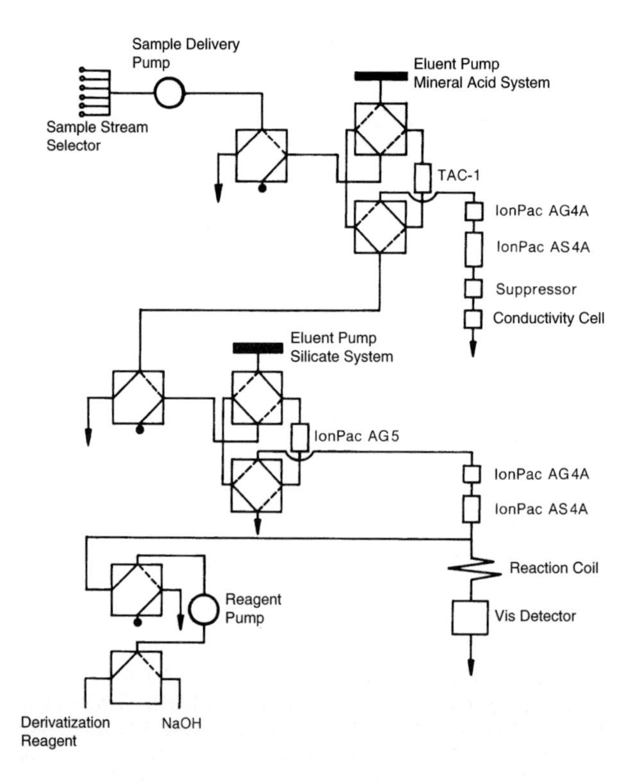

Fig. 8-47. Valve switching for the simultaneous analysis of mineral acids and silicate in trace amounts.

metals *on-line* in one run. The chromatogram shown in Fig. 8-49 with metal concentrations between 2 ng/L and 1600 ng/L was obtained after concentrating 180 mL of deionized water.

Ion chromatography is not only used to monitor the water quality, but also to analyze a variety of process liquors that are employed in the manufacture of printed circuit boards. This includes cleansers, palladium-based activators, and various electroplating baths such as acidic and electroless copper baths, tin/lead baths, electrolytical nickel baths, and gold baths. The analytical chemistry of the key substances contained in these baths is described in detail in the preceding chapter.

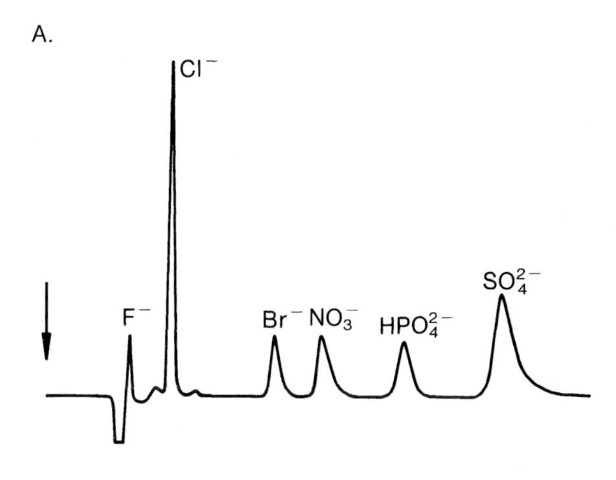

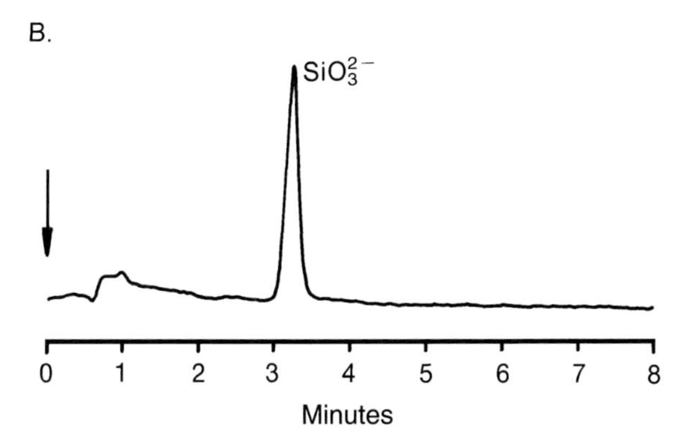

Fig. 8-48. On-line determination of mineral acids and silicate. – Separator columns: 2 IonPac AS4A; concentrator columns: (A) TAC-1, (B) IonPac AG5; eluent: (A) 0.0017 mol/L $NaHCO_3$ + 0.0018 mol/L Na_2CO_3, (B) 0.015 mol/L H_3BO_3 + 0.015 mol/L NaOH; flow rate: (A) 2 mL/min, (B) 1 mL/min; detection: (A) suppressed conductivity, (B) photometry at 410 nm after reaction with sodium molybdate; concentrated volume: 23 mL; solute concentrations: 10 ppb fluoride, 10 ppb chloride, 10 ppb bromide, 10 ppb nitrate, 20 ppb orthophosphate, 20 ppb sulfate, and 23 ppb silicate; (taken from [36]).

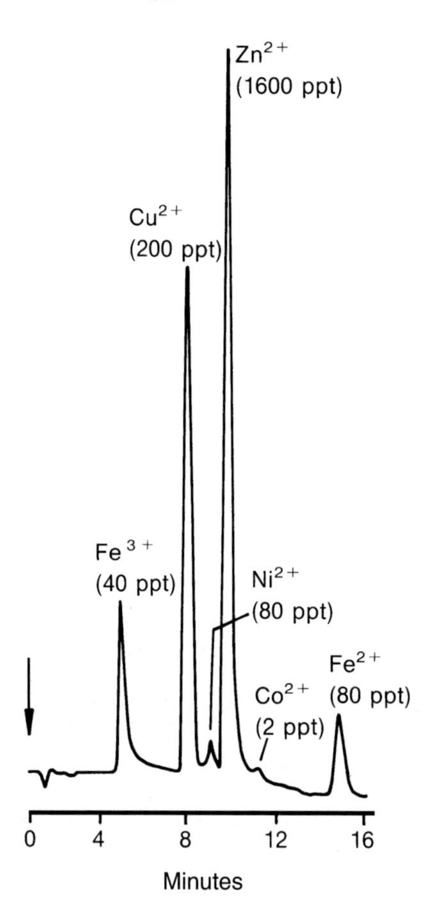

Fig. 8-49. Trace analysis of metals in de-ionized water. – Separator column: IonPac CS5; concentrator column: IonPac CG2; eluent: 0.006 mol/L pyridine-2.6-dicarboxylic acid + 0.0086 mol/L LiOH; flow rate: 1 mL/min; detection: photometry at 520 nm after reaction with PAR; concentrated volume: 180 mL; (taken from [35]).

Another area of application of ion chromatography is the analysis of etching solutions. These are mixtures of hydrofluoric and nitric acid or of hydrofluoric, nitric, and acetic acid used to remove metal oxides and other impurities from the metal surface. The effectiveness of the process depends both on the content of the respective acids and on the impurities in the etching mixtures. As an example, Fig. 8-50 shows the chromatogram of a HF/HNO_3/HOAc-solution diluted in the ratio 1:1,000 with de-ionized water that was obtained by applying conductometric detection with the aid of ion-exclusion chromatography. It should be noted that the total fluoride concentration, not only the free hydrofluoric acid concentration, is analyzed by ion chromatography, since the metals dissolved during the etching process are present in the form of fluoride complexes (e.g., FeF_2^+). These complexes decompose in alkaline solution to form the corresponding metal hydroxides. Indirect analytical methods, such as that developed by Dulski [37], show good agreement with conventional wet-chemical methods [38].

Even after production of printed circuit boards and semiconductor components, ion chromatography may be taken advantage of. Malfunctions are often caused by corrosive anions on the surface of printed circuits or inside, where the microchips are imbedded. Identification and quantification of these compounds with the aid of ion chromatographic techniques often represents a significant contribution for localizing the sources

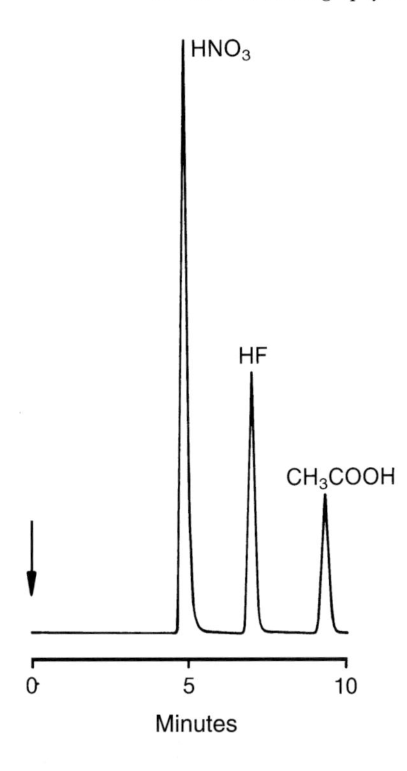

Fig. 8-50. Chromatogram of a $HF/HNO_3/HOAc$ etching solution. – Separator column: IonPac ICE-AS1; eluent: 0.001 mol/L octanesulfonic acid; flow rate: 1 mL/min; detection: suppressed conductivity; injection: 50 μL sample (1:1000 diluted).

of contamination. One of the key sources of contamination is the soldering process, as the soldering agents used may contain corrosive anions. After soldering they lead to corrosion phenomena inside the encapsulation casing. Fig. 8-51 demonstrates this effect taking the analysis of mineral acids in semiconductor components as an example. For anion determation the packages of components showing malfunctions are cracked and extracted with 2 mL of de-ionized water. The chromatogram shown in Fig. 8-51 is obtained after direct injection of such an extract. Apart from relatively high amounts of chloride and nitrate, a number of other mineral acids and oxalic acid can be detected in low concentrations. Thus, by employing ion chromatography, the soldering process can be optimized by selecting suitable soldering agents.

8.5 Ion Chromatography in the Detergent and Household Product Industry

8.5.1 Detergents

Detergents for household and industrial use consist of a large number of very different individual components. As a result of the technical development as well as ecological and economical restrictions, these products are continuously changing. Thus, it is not surprising that in this field analytical methods are constantly under further development and that product-related analytical methods come of age fairly rapidly.

Ionic detergent components such as

- surfactants,
- builders,
- bleaching agents, and
- fillers and finishing materials

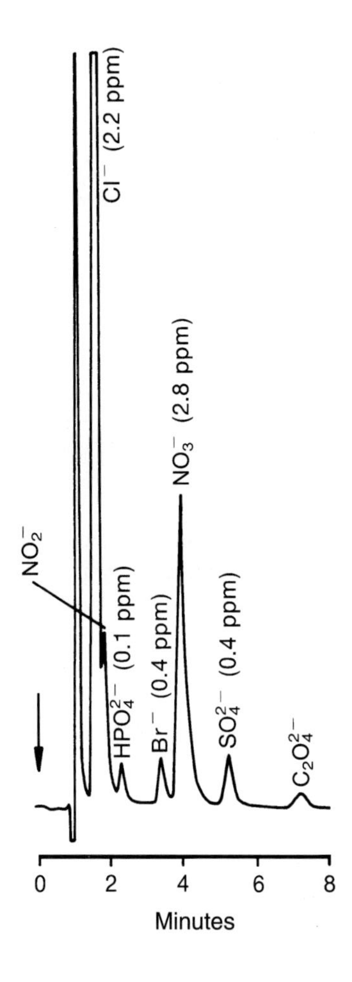

Fig. 8-51. Analysis of mineral acids in dual inline enclosures. – Separator column: IonPac AS4; eluent: 0.0028 mol/L $NaHCO_3$ + 0.0022 mol/L Na_2CO_3; flow rate: 2 mL/min; detection: suppressed conductivity; injection: 50 μL sample (undiluted).

play a key role in the washing process. The determination of these compounds represents a classical analytical problem in the detergent industry. The conventional wet-chemical and instrumental methods for analyzing ionic detergent components embrace gravimetry for sulfate determination [39]; potentiometry for determining chloride, phosphate, and borate [40,41]; complexometry for the quantification of NTA and EDTA [42]; and photometry again for sulfate determination [43]. Such methods are labor-intensive and time-consuming. Therefore, they are replaced by much faster and more sensitive ion chromatographic techniques [44].

Table 8-5. Summary of surfactants that can be analyzed by ion chromatography.

Formula		Chemical type
$R-CH_2-COONa$	$R = C_{10}$ to C_{16}	Soap
$R-C_6H_4-SO_3Na$	$R = C_{11}$ to C_{13}	Alkylbenzene sulfonate
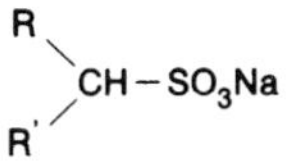 $(R)(R')CH-SO_3Na$	$R, R' = C_{10}$ to C_{17}	Alkane sulfonate
$R-CH_2-CH=CH-(CH_2)_n-CH_2-SO_3Na$	$R = C_{10}$ to C_{14} $n = 1$ to 3	α-Olefin sulfonate
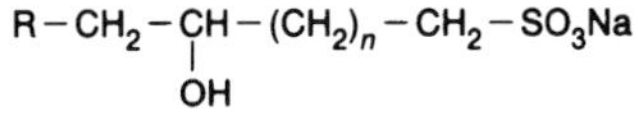 $R-CH_2-CH(OH)-(CH_2)_n-CH_2-SO_3Na$	$R = C_9$ to C_{13} $n = 1$ to 2	Hydroxyalkane sulfonate
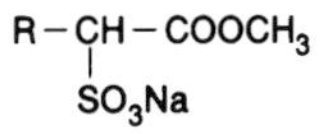 $R-CH(SO_3Na)-COOCH_3$	$R = C_{14}$ to C_{16}	α-Sulfo-fatty acid methylester
$R-CH_2-O-SO_3Na$	$R = C_{11}$ to C_{17}	Alkyl sulfate
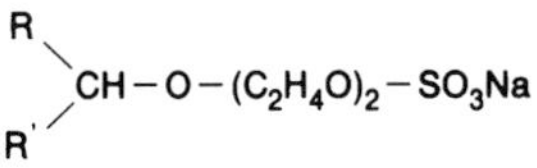 $(R)(R')CH-O-(C_2H_4O)_2-SO_3Na$	a) $R' = H$ $R = C_{11}$ to C_{13} b) $R, R' = C_{10}$ to C_{14}	Alkyl ether sulfate a) Fatty alcohol ether sulfate b) *sec.* Alkyl ether sulfate
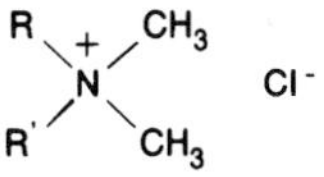 $(R)(R')N^+(CH_3)_2\ Cl^-$	$R = C_{16}$ to C_{18}	Dialkyl-dimethylammonium chloride
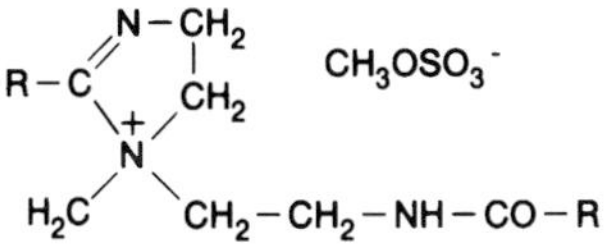 $R-C$ (ring: $=N-CH_2-CH_2-N^+$) with N^+ bearing H_2C and $CH_2-CH_2-NH-CO-R$; $CH_3OSO_3^-$	$R = C_{16}$ to C_{18}	Imidazolinium salts
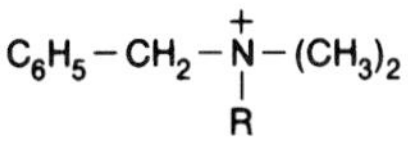 $C_6H_5-CH_2-N^+(R)-(CH_3)_2$	$R = C_8$ to C_{18}	Alkyl-dimethyl-benzylammonium chloride

Surfactants

Surfactants are the most important group of detergent components; they are constituents in all detergents. Generally, they are water-soluble, surface-active compounds that carry both a hydrophilic functional group and a long alkyl chain as a hydrophobic rest. Distinction is made between the following surfactant groups:

- Anionic surfactants
- Cationic surfactants
- Non-ionic surfactants
- Amphoteric surfactants.

Surfactants that can presently be analyzed by means of ion chromatography are summarized in Table 8-5. Up until now, surfactants were analyzed by a two-phase titration [45,46]. This technique is based on an equilibrium reaction between surfactant dyes and anionic/cationic surfactant salts and their distribution between chloroform and water in a two-phase mixture. Hyamine 1622 (bis-*i*-butyl-phenoxyethyl-dimethyl-benzylammonium chloride) has proved to be a suitable titrant for anionic surfactants, whereas lauryl sulfate is used for cationic surfactants.

A chromatographic determination of the classes of compounds listed in Table 8-5 can be achieved using ion-pair chromatography in combination with conductometric or UV detection. The retention behavior of surface-active compounds has already been discussed in detail in Section 5.5. This section discusses a series of example chromatograms of the analyses of various surfactant raw materials. For an unequivocal identification of surfactants in detergent products, however, a wet-chemical pre-separation into the individual surfactant types is necessary, since detergent formulations often contain a combination of surfactants such as anionic and non-ionic surfactants. The carbon-chain distribution can be determined by ion-pair chromatography. If a crude material consists of several compound classes, a comparison of the chromatographic "fingerprints" can provide valuable information for the raw material control.

Builders

The builders include complexing agents such as sodium triphosphate, ion exchangers such as Zeolite A, and washing alkalies such as sodium carbonate and sodium silicate.

Inorganic and organic complexing agents are of central importance in the course of the washing process, since magnesium and calcium ions stemming from the water, dirt, or textiles, respectively, are complexed with their help.

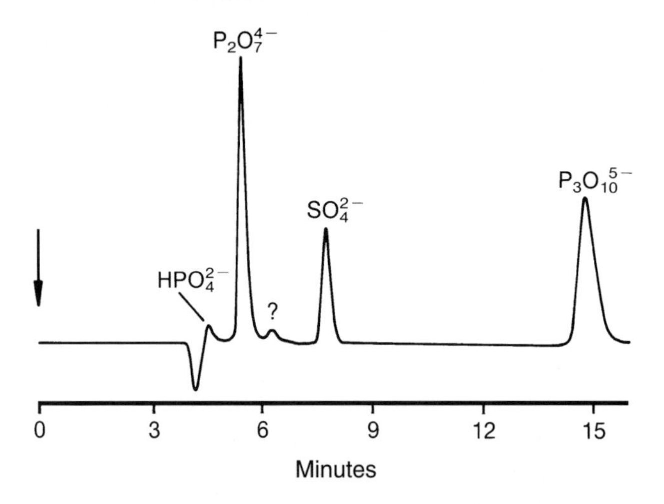

Fig. 8-52. Analysis of condensed phosphates in detergents. – Separator column: IonPac AS7; eluent: 0.07 mol/L HNO_3; flow rate: 0.5 mL/min; detection: photometry at 330 nm after reaction with ferric nitrate; injection: 50 μL of a 0.1% washing powder solution.

Until now, both sodium triphosphate and polyphosphonic acids were determined with time-consuming ion-exchange chromatographic techniques [47]. Owing to their polyvalent character, they were not amenable to ion chromatography. However, the analysis of these compounds has become possible in combination with a recently developed latex-based anion exchanger and a specific post-column derivatization [48]. Eluting condensed phosphates in finished products requires the use of nitric acid with a concentration $c = 0.07$ mol/L. As revealed by Fig. 8-52, sulfate, which is added to the washing powder as formulating agent, can also be detected under these chromatographic conditions. Calibration for the quantification of triphosphate is difficult, since a commercial triphosphate also contains noticeable amounts of diphosphate. In the absence of a high-purity triphosphate, one can use the phosphorus-specific detection method developed by Vaeth et al. [50], in which di- and triphosphate are hydrolyzed to orthophosphate. The latter is detected photometrically after reaction with sodium molybdate (see Section 3.3.5.2). This method is extremely well suited for determining the degree of phosphate conservation. Fig. 8-53 shows the chromatogram of a NTA-containing detergent that was obtained with nitric acid of concentration $c = 0.03$ mol/L. Under these chromatographic conditions, polyphosphonic acids, which are often present in detergents at low concentrations, can also be separated. Here, the polyphosphonic acid contained in the detergent sample elutes in the sulfate retention range. Triphosphate has a very high retention time under these conditions, thus the separator column must occasionally be

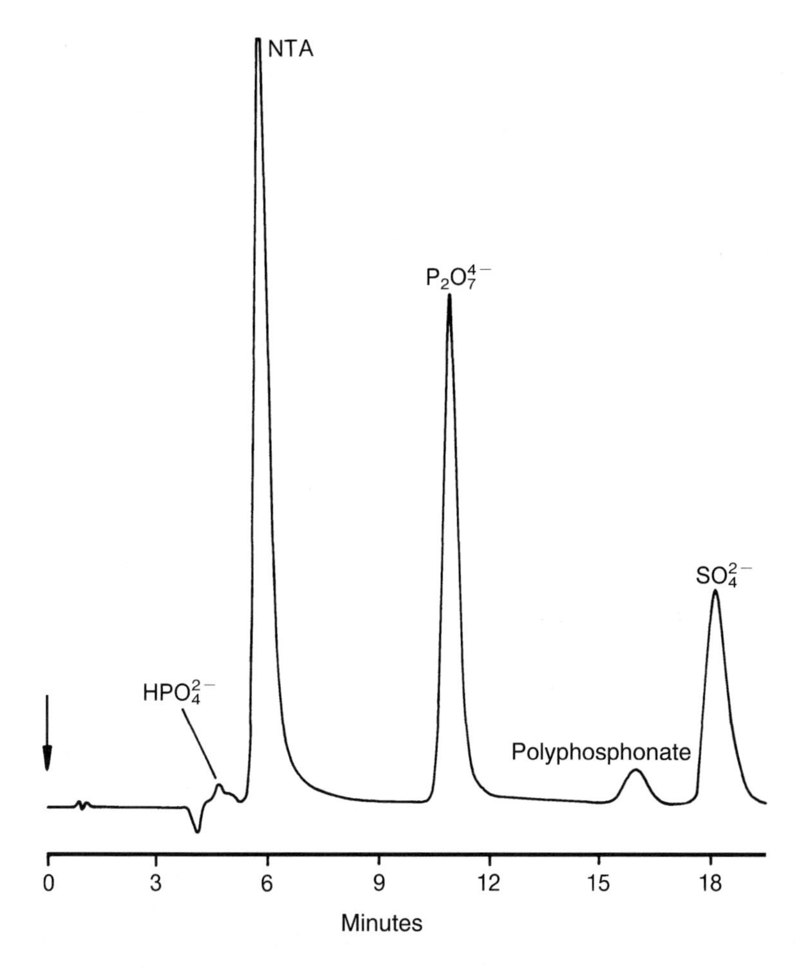

Fig. 8-53. Analysis of inorganic and organic complexing agents in detergents. – Separator column: IonPac AS7; eluent: 0.03 mol/L HNO_3; flow rate and detection: see Fig. 8-52; injection: 50 µL of a 0.2% washing powder solution.

flushed with concentrated nitric acid. It should be noted that the analysis of NTA in the finished product with conventional methods is impossible in the presence of other complexing agents such as citric acid.

Carbonate and silicate can be analyzed ion chromatographically via ion-exclusion chromatography. While carbonate is detected by electrical conductivity (see Fig. 8-54), silicate is detected photometrically – as described above – after derivatization with sodium molybdate.

Bleaching Agents

In an alkaline medium, the perhydroxide anion of the peroxide bleach [51] that predominates in Europe is formed as an active intermediate from hydrogen peroxide:

$$H_2O_2 + OH^- \rightleftharpoons H_2O + HO_2^- \qquad (226)$$

The hydrogen peroxide is provided by sodium perborate, which exists as peroxoborate in crystalline state.

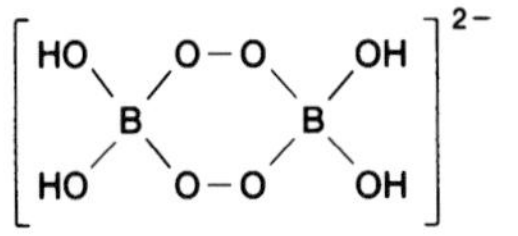

$$\left[(HO)_2B(O{-}O)_2B(OH)_2 \right]^{2-}$$

Peroxoborate anion

In aqueous solution, the peroxoborate anion hydrolyzes to form hydrogen peroxide. "Bleaching activators" are used to obtain a good bleaching action at temperatures below 60 °C, and to form with hydrogen peroxide organic peracid intermediates at pH values between 9 and 12, that exhibit a good bleaching action in the low temperature region. Tetraacetylethylene diamine (TAED) is preferred in this regard, and reacts as shown in the following scheme:

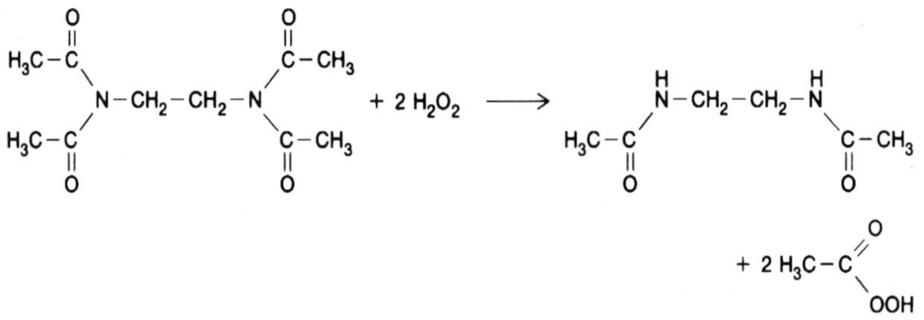

$$(H_3C{-}CO)_2N{-}CH_2{-}CH_2{-}N(CO{-}CH_3)_2 + 2\,H_2O_2 \longrightarrow H_3C{-}CO{-}NH{-}CH_2{-}CH_2{-}NH{-}CO{-}CH_3 + 2\,H_3C{-}C(=O){-}OOH \qquad (227)$$

When stored for some time, TAED decomposes with the loss of acetic acid. As illustrated in Fig. 8-54, perborate and acetic acid can be easily and rapidly determined via ion chromatography in the finished detergent product. Carbonate can also be determined in the same run.

The effort required to analyze perborate, NTA, carbonate, and acetate, including the necessary calibrations, takes about 0.8 hours per sample [52]. In comparison, the effective working time for the conventional borate analysis via the extraction method alone is about 2.5 hours; to determine all the above components requires about 5.8 hours [52]. This example illustrates the significant cost saving brought about by the introduction of ion chromatographic techniques.

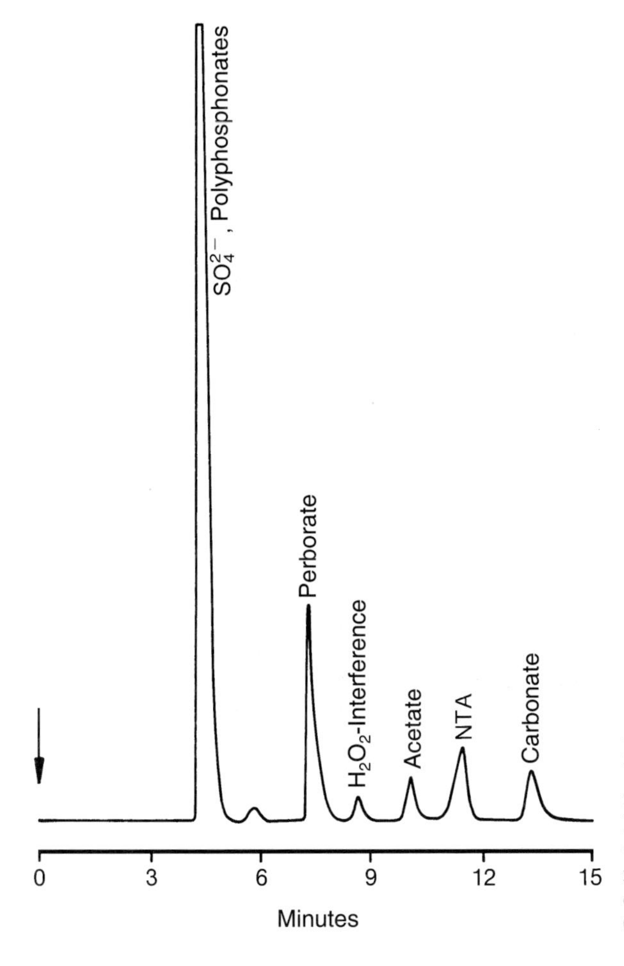

Fig. 8.54. Analysis of sodium perborate by ion-exclusion chromatography. – Separator column: IonPac ICE-AS1; eluent: 0.0011 mol/L octanesulfonic acid; flow rate: 1 mL/min; detection: suppressed conductivity; injection: 50 µL of a 0.05% detergent solution.

Fillers and Finishing Materials

Sodium sulfate is generally used as a filler in powdered detergents. Sulfate can be analyzed by means of anion exchange chromatography in the presence of chloride and orthophosphate (see Fig. 8-55) in a single run. Orthophosphate, via the diphosphate anion, represents the degradation product of triphosphate. To again demonstrate the superiority of ion chromatography over wet-chemical methods in this case, the different working steps for sulfate analysis are compared below in Table 8-6.

Table 8.6. Comparison of ion chromatography and wet-chemical methods for sulfate analysis.

Sulfate analysis	
By ion chromatography	By wet-chemical methods
1. Preparation of the sample solution 2. Injection of the sample through a membrane filter 3. Evaluation of the chromatogram	1. Preparation of the sample solution 2. Acidification of the sample solution and heating with active carbon for removal of surfactants 3. Filtration 4. Addition of a precipitant 5. Precipitation of sulfate 6. Crystallization of the precipitate overnight 7. Filtration and rinsing 8. Ashing of the filter 9. Glowing of the residue 10. Weighing and evaluation

The time expenditure per sample is about 24 hours for the wet-chemical procedure, with 2.5 hours being the effective working time. In contrast, using ion chromatography requires only 20 to 25 minutes, including calibration time, for each sulfate determination [52]. In addition, ion chromatography has the advantage that short-chain alkyl sulfonates and sulfates do not interfere with the sulfate determination and chloride and orthophosphate can be detected in the same run.

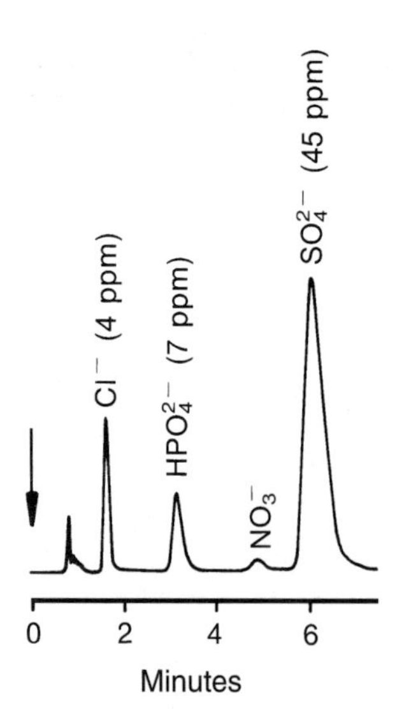

Fig. 8-55. Analysis of inorganic anions in a detergent. – Separator column: IonPac AS3; eluent: 0.0028 mol/L $NaHCO_3$ + 0.0022 mol/L Na_2CO_3; flow rate: 2.3 mL/min; detection: suppressed conductivity; injection: 50 μL of a 0.1% detergent solution (1:5 diluted).

Finishing liquid detergents and cleansers typically employ short-chain alkylbenzene sulfonates such as toluene sulfonate or cumene sulfonate, which, because of their hydrotropic properties, ensure the solubility of other detergent components in the aqueous environment. Analysis of hydrotropic compounds is performed with ion-pair chromatography. The compounds are eluted in order of increasing alkyl substitution. As revealed by Fig. 8-56, these compounds can be determined directly in the finished product without much sample preparation. The analyte samples are only diluted with de-ionized water and membrane filtered. As aromatic sulfonates may be detected both by their conductivity and their UV absorption, the choice of the suitable detection method depends mainly on the type of matrix.

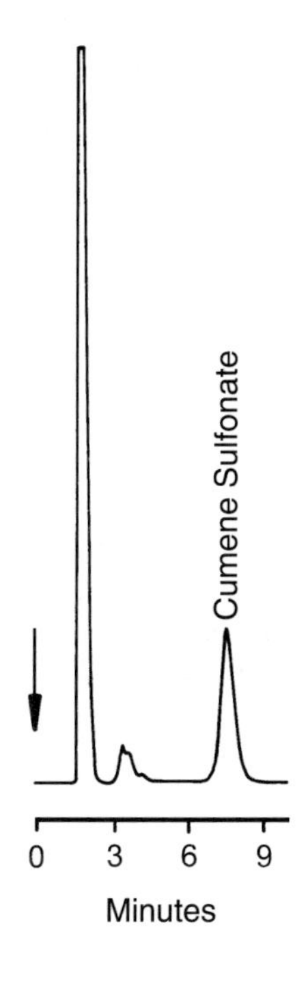

Fig. 8-56. Separation of cumene sulfonate in a cleanser. – Separator column: IonPac NS1 (10 µm); eluent: 0.01 mol/L NH_4OH / acetonitrile (91:9 v/v); flow rate: 1 mL/min; detection: suppressed conductivity; injection: 50 µL of a 0.2% sample solution.

8.5.2 Household Products

From the area of household and industrial cleansing agents, Fig. 8-57 shows the separation of gluconic acid and citric acid as key components of a weakly basic detergent. These compounds are hydroxycarboxylic acids, thus separation was performed with the aid of ion-exclusion chromatography using an IonPac ICE-AS5 as the stationary phase. As can be seen from Fig. 8-57, this stationary phase exhibits a high selectivity for the above-mentioned organic acids. They were detected conductometrically upon application of a membrane-based suppressor system.

Triethanolamine, which acts as a base in this detergent, can be separated on a neutral DVB-based polymer. An aqueous sodium hydroxide solution is employed as the eluent to prevent the protonation of the nitrogen function and as a prerequisite for the subsequent detection via pulsed amperometry. Like carbohydrates, alkanolamines may also be oxidized at a Au working electrode at pH 13. In contrast to the generally used conductometric or refractive index detection, this detection method is characterized, especially in this case, by a much higher specificity and sensitivity, allowing a strong sample dilution and ensuring independence of the matrix. Fig. 8-58 illustrates the chromatogram of a sample that has been diluted 1:1000.

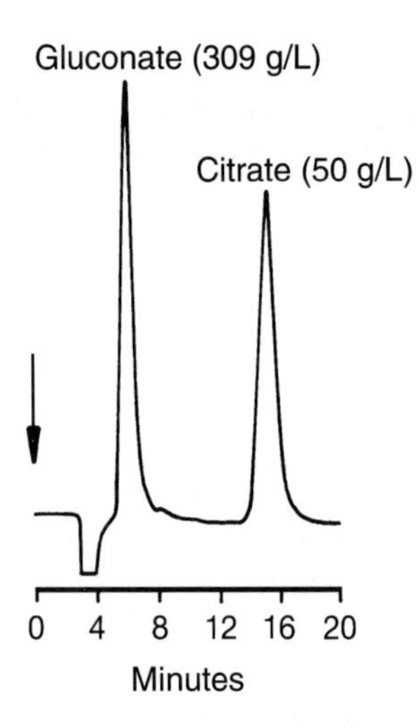

Fig. 8-57. Separation of gluconic acid and citric acid in a weakly basic cleansing agent. – Separator column: IonPac ICE-AS5; eluent: 0.0016 mol/L perfluorobutyric acid; flow rate: 0.5 mL/min; detection: suppressed conductivity; injection: 50 µL sample (1:1000 diluted).

Hydroxylamine is often substituted for triethanolamine as a base in detergents. After protonation, hydroxylamine can be separated on a low-capacity cation exchanger in an acidic medium. Because it is a weak base, detection is difficult. Amperometry may be employed for selective detection. Alternatively, the less specific and less sensitive direct

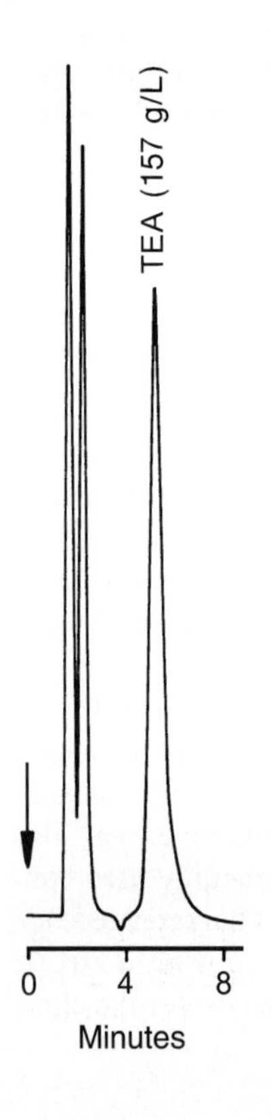

Fig. 8-58. Determination of triethanolamine in a weakly basic cleansing agent. – Separator column: IonPac NS1 (10 µm); eluent: 0.1 mol/L NaOH; flow rate: 1 mL/min; detection: pulsed amperometry at a Au working electrode; injection: 50 µL sample (1:1000 diluted).

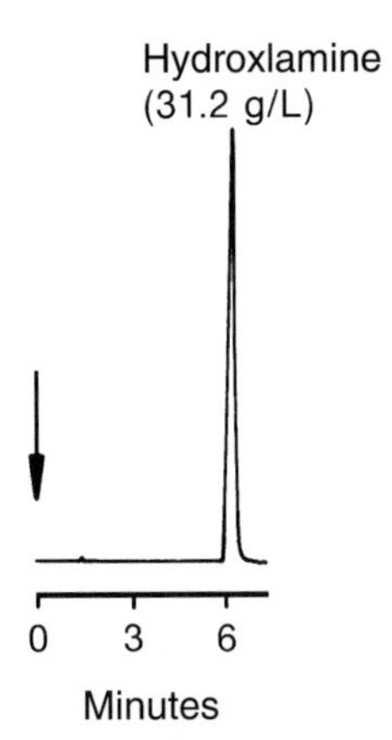

Fig. 8-59. Determination of hydroxylamine in a weakly basic cleansing agent. – Separator column: IonPac CS3; eluent: 0.03 mol/L HCl; flow rate: 1 mL/min; detection: D.C. amperometry at a Pt working electrode; oxidation potential: +0.8 V; injection: 50 μL sample (1:2000 diluted).

conductivity detection can be used. No signal is obtained upon application of the suppressor technique, as the suppressor reaction product is not sufficiently dissociated. Unlike ammonium ions, hydroxylammonium ions do not react with *o*-phthaldialdehyde, thus fluorescence detection is also not feasible. However, hydroxylamine can be easily

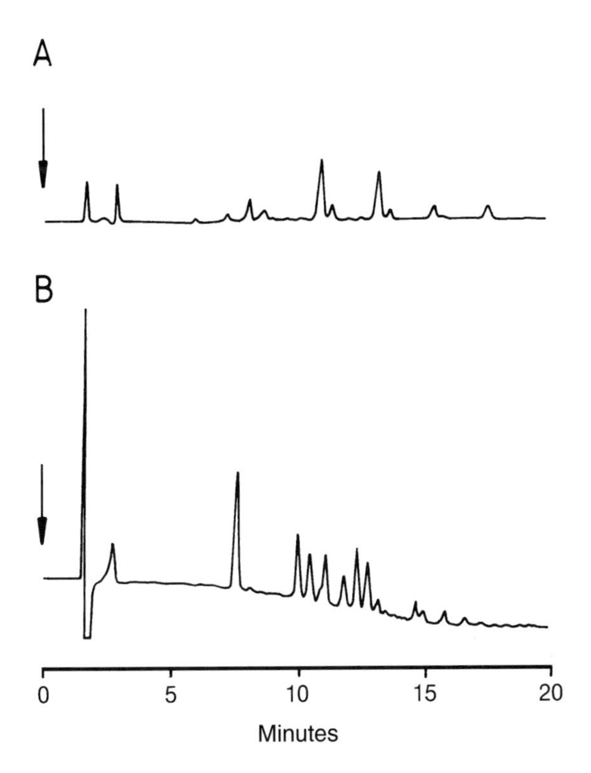

Fig. 8-60. Gradient elution of anionic surfactants in a shampoo. – Separator column: Hypersil 5 MOS; eluent: (A) 0.001 mol/L NaOAc, (B) 0.001 mol/L NaOAc – acetonitrile (10:90 v/v); gradient: linear, 34% B in 10 min to 45% B; flow rate: 1 mL/min; detection: (A) UV (254 nm), (B) direct conductivity; injection: 50 μL of a 0.2% solution.

oxidized at a Pt working electrode at potentials around +0.9 V, thus providing the required selectivity and sensitivity. The chromatogram of a sample diluted 1:2,000 (see Fig. 8-59) appears, therefore, as that of a standard.

A characteristic example in the field of cosmetics is the separation and determination of anionic surfactants in shampoos. The sole sample preparation step required is the dilution of the respective product in de-ionized water and membrane-filtration (0.45 μm) prior to injection. Fig. 8-60 illustrates the gradient elution of anionic surfactants in a commercial shampoo with simultaneous conductometric and UV detection. While the peak pattern with UV detection indicates the presence of alkylbenzene sulfonate (ABS) (see also Fig. 5-40 in Section 5.5), the characteristic peak profile for alkylether sulfate (see Fig. 5-39 in Section 5.5) is also evident in the conductometric chromatogram after visually subtracting the ABS-signal. Therefore, application of ion chromatography provides a way to unequivocally identify surfactant raw materials in finished cosmetic products without the time-consuming sample preparation.

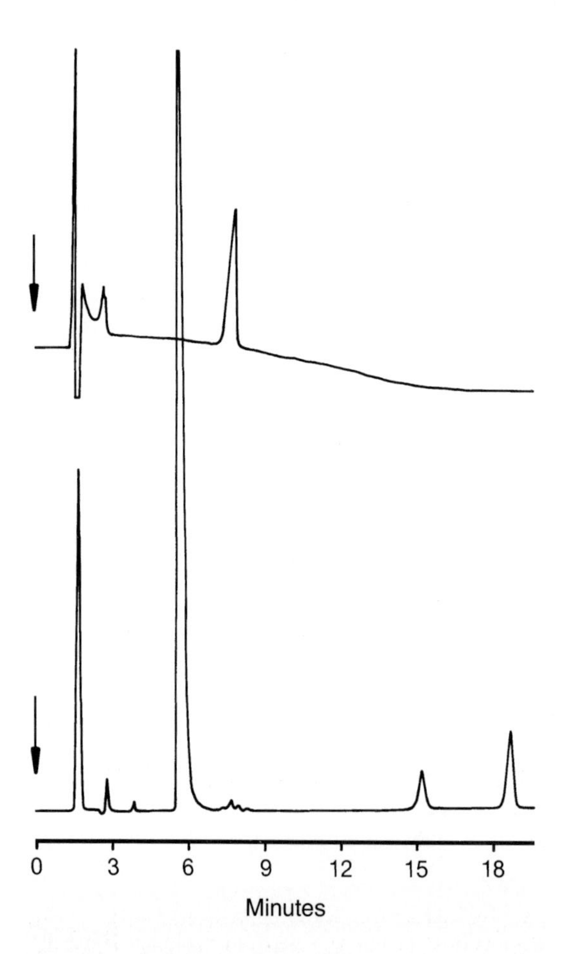

Fig. 8-61. Analysis of lauryl sulfate and other organic constituents in toothpaste. – Separator column: Hypersil 5 MOS; chromatographic conditions: see Fig. 8-60; injection: 50 μL of a 0.5% solution (Colgate); (taken from [53]).

This not only applies to shampoo but also to toothpaste, in which anionic surfactants, depending on the product, can also be identified. As is clearly demonstrated by the chromatogram shown in Fig. 8-61, the method developed for anionic surfactants analysis in shampoo is also suited for the investigation of toothpaste [53], during which the weighed amount is suspended in de-ionized water and the extract is membrane-filtered (0.45 μm) prior to injection. In the present case, lauryl sulfate was clearly identified as the surfactant component. The simultaneous application of UV detection also enables the detection of other organic constituents in the same run, the chemical structure of which cannot be attributed without knowing the basic formulation. Inorganic constituents such as the monofluorophosphate (MFP) that is included in many formulations can be directly determined by anion exchange chromatography. Aside from MFP, orthophosphate as its degradation product as well as chloride and sulfate may be analyzed simultaneously (see chromatogram in Fig. 8-62). Sample preparation is performed as described above. However, to protect the separator column the extract is passed through an OnGuard-RP-cartridge prior to injection, which serves to remove non-ionic organics.

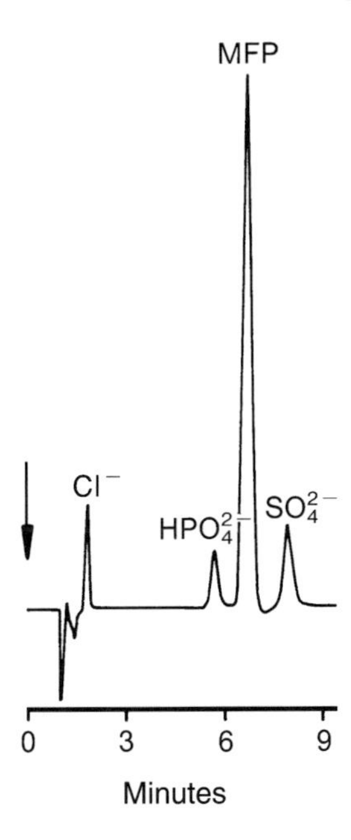

Fig. 8-62. Analysis of monofluorophosphate in toothpaste. – Separator column: IonPac AS4A; eluent: 0.0017 mol/L $NaHCO_3$ + 0.0018 mol/L Na_2CO_3; flow rate: 2 mL/min; detection: suppressed conductivity; injection: 50 μL of a 0.17% solution.

8.6 Ion Chromatography in Food and Beverage Industry

It is the task of the food and beverage industry to provide the customer with a tasty and healthy product whose nutritive value and storage quality are well defined. Manufacturing and monitoring processes have partly been automated to such an extent that only a few, simple quality tests are required. However, in some cases it is still necessary to carry out costly analyses for quality assurance purposes. In these cases, ion chromatography permits the rapid and precise determination of both inorganic and organic anions and cations even in complex matrices.

Ion chromatography is increasingly being approved by many test and research laboratories in the food and beverage industry, since only a minimal sample preparation is required for analyses utilizing this method [54]. The samples are usually extracted with de-ionized water and membrane-filtered (0.45 μm) before injecting them directly into the system. The reason ion chromatography easily deals with complex matrices lies not only in the stability and resistance to fouling of the stationary phases being used, but also in the sensitivity and specificity of the detection methods being employed. A survey of the applicability in the food and beverage industry is given in Table 8-7.

Table 8-7. Application areas and typical analytical examples in the food and beverage industry.

Application area	Analytical example
Milk products	Determination of iodide in whole milk; chloride and/or sodium in butter; lactate, pyruvate and citrate in cheese
Meat processing	Determination of the NO_2^-/NO_3^- ratio in meat products, of nitrate in the water being used
Beverages	Determination of inorganic anions and cations in the water used, in sweeteners and aromatic substances, and in the finished products; of organic acids and carbohydrates in beer, wine, and juice
Condiments	Determination of chloride, nitrate, sodium, organic acids, and heavy metals in canned fruit and canned vegetables, spices, vinegar, and fish
Baby food	Determination of choline and heavy metals
Cereal products	Determination of bromate and propionate in bakery products, Fe^{2+} and Fe^{3+}
Fats, oils, carbohydrates, and flavours	Determination of fatty acids and carbohydrates in corn syrup

Beverages

One of the key applications of ion chromatography in the food and beverage industry is the analysis of inorganic and organic anions and carbohydrates in beverages of all kinds [55,56]. This includes predominantly the investigation of wine, beer, fruit juices, and various refreshers, as well as of luxuries such as coffee and tea. Apart from inorganic anions, all these beverages contain various organic acids whose retention behavior is very similar to that of inorganic anions. However, if two anion exchangers with different selectivity are combined, inorganic anions such as chloride, nitrate, orthophosphate, and sulfate may be separated baseline-resolved under isocratic conditions together with dicarboxylic acids such as malic acid and tartaric acid. Detection is performed by measuring the electrical conductivity upon application of a membrane-based suppressor system. Figs. 8-63 and 8-64 show the application of this method to the analysis of both a blueberry juice concentrate and an old port wine. It should be noted when evaluating the orthophosphate results that only free orthophosphate but not the total phosphate is detected via ion chromatography. As some phosphate is bound in the presence of calcium, especially in fruit juices, the phosphate content determined via ion chromatography is much lower than that obtained with conventional methods after sulfuric acid digestion. Conversely, this also applies to the calcium analysis, if the samples are not treated with acid. Nitrate, eluting shortly after orthophosphate, is usually present in very low concentrations, but may be recorded via simultaneous UV detection at 215 nm in the same run without any problems.

This method is only partly applicable to the analysis of citrus juices, since the citric acid contained at high concentrations in these products is strongly retained under the given chromatographic conditions. After repeated injection, this results in a marked reduction of the exchange capacity. Therefore, a gradient technique with sodium hydroxide as the eluent must be used for analyzing such samples (s. Fig. 3-124 in Section 3.3.6).

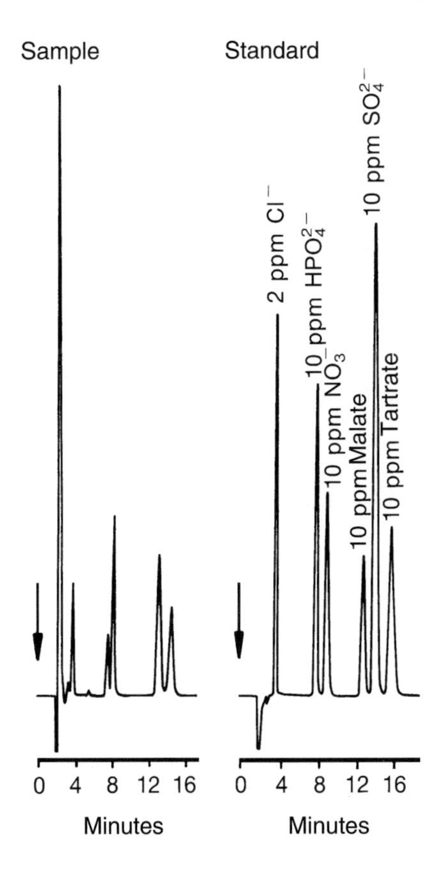

Fig. 8-63. Analysis of inorganic and organic anions in a berry juice concentrate. – Separator column: IonPac AS4 + AS4A; eluent: 0.0028 mol/L $NaHCO_3$ + 0.0022 mol/L Na_2CO_3; flow rate: 1.5 mL/min; detection: suppressed conductivity; injection: 50 µL of a 0.4% solution.

Carbohydrate analysis is much simpler in such samples. Pulsed amperometry is a very sensitive detection method and beverages to be analyzed can often be strongly diluted, so interferences caused by the matrix are usually not observed. To determine sorbitol in apple juice, for example, apple juice is simply diluted 1:1,500 with de-ionized water and injected directly (Fig. 8-65). Comparison measurements with enzymatic techniques by Baumgärtner [57] (see Table 8-8) yielded a good agreement between both methods and also for the other main components glucose and fructose, which can be determined together with sucrose in the same run.

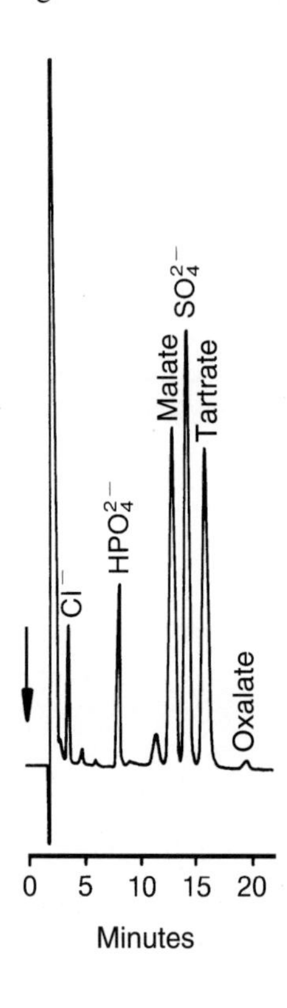

Fig. 8-64. Analysis of inorganic and organic anions in an old port wine. – Chromatographic conditions: see Fig. 8-63; injection: 50 μL sample (Messias Port, 10 years old, diluted 1:50).

Table 8-8. Comparison of the results obtained with IC and enzymatic techniques for carbohydrate analysis in apple juice; (taken from [57]).

Carbohydrate	IC [g/L]	Enzymatic [g/L]
Sorbitol	5.11	5.22
Glucose	24.7	24.5
Fructose	62.7	63.3

Higher carbohydrates may also be determined within a short time via gradient elution with NaOH/NaOAc-mixtures. Fig. 8-66 is an example with a malt beer in which maltose and maltotriose and also higher maltose oligomers could be detected. Interesting are the investigations by Ohs [58] and Baumgärtner [57] into the analysis of carbohydrates in wine, in which small quantities of sorbitol, rhamnose, arabinose and, to some extent, trehalose are present in addition to the main components glucose and fructose. The example chromatogram in Fig. 8-67 reveals that a separation of the above-mentioned sugars is possible in spite of the high concentration difference between main and minor components. Moreover, all wines that have been investigated by Baumgärtner have a common characteristic peak pattern in the oligosaccharide region.

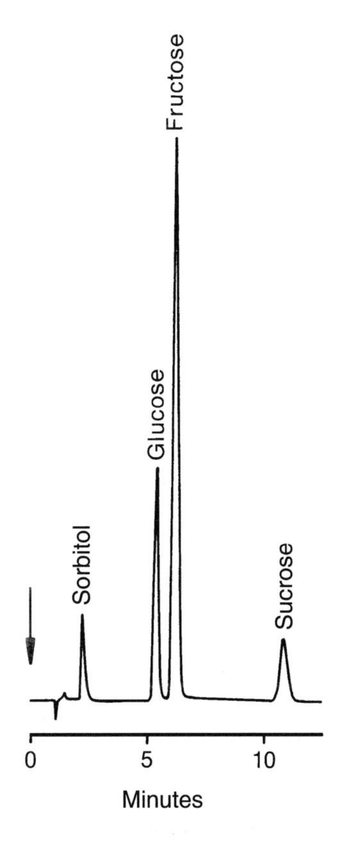

Fig. 8-65. Analysis of carbohydrates in apple juice. – Separator column: CarboPac PA1; eluent: 0.1 mol/L NaOH; flow rate: 1 mL/min; detection: pulsed amperometry at a Au working electrode; injection: 50 µL sample (diluted 1:500); (taken from [57]).

Dairy Products

A characteristic example for the analysis of dairy products is the determination of iodide in milk and whey. While the separation of iodide may be performed with both ion-exchange and ion-pair chromatography [59], amperometric detection at a Ag or GC working electrode is generally used to detect the small content in the lower ppb range with sufficient precision. Although conventional anion exchangers may not be loaded with proteins when employed for the detection of inorganic anions, whole milk can be directly injected after dilution and membrane-filtration (0.22 µm) without irreversibly fouling the separator column.

Further typical applications in the field of dairy products are the determination of sodium in butter, and the analysis of organic acids such as lactic acid, pyruvic acid, and citric acid in cheese products. The latter is performed after appropriate sample preparation via ion-exchange chromatography.

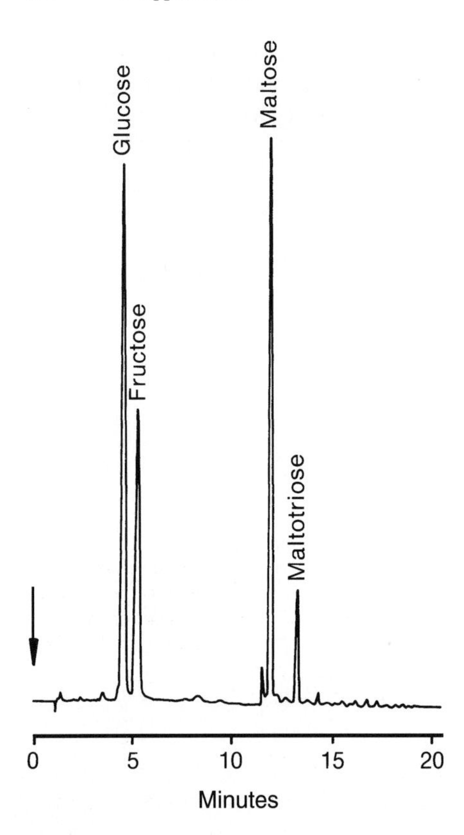

Fig. 8-66. Analysis of carbohydrates in a malt beer. – Separator column: CarboPac PA1; eluent: (A) 0.1 mol/L NaOH, (B) 0.1 mol/L NaOH + 0.5 mol/L NaOAc; gradient: 100% A isocratically for 4 min, then linearly to 100% B in 40 min; flow rate: 1 mL/min; detection: see Fig. 8-65; injection: 50 μL sample (Vitamalz, diluted 1:250); (taken from [57]).

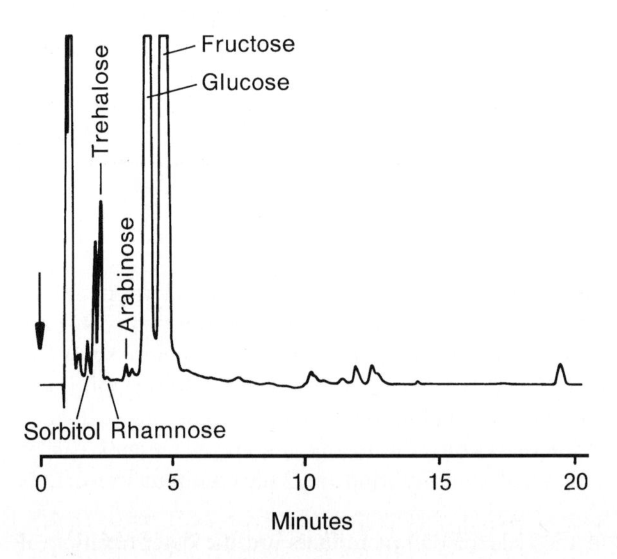

Fig. 8-67. Gradient elution of carbohydrates in wine. – Separator column: CarboPac PA1; chromatographic conditions: see Fig. 8-66; injection: 50 μL sample (Wallhäuser Pfarrgarten, Riesling 1988, diluted 1:50); (taken from [57]).

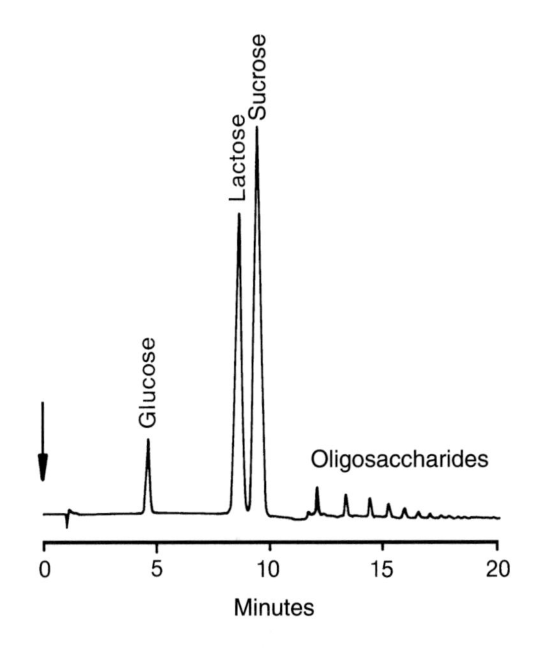

Fig. 8-68. Gradient elution of carbohydrates in soft ice. – Separator column: CarboPac PA1; chromatographic conditions: see Fig. 8-66; injection: 50 μL of a 0.16% solution of soft ice "Sundae"; (taken from [57]).

Sample preparation for the determination of carbohydrates in dairy products turns out to be very simple because of the high sensitivity of the detection methods being used. For example, soft ice was dissolved by Baumgärtner [57] with de-ionized water, membrane-filtered (0.45 μm) and directly injected. Fig. 8-68 shows the corresponding chromatogram of a 0.16% solution in which glucose, lactose, and sucrose in addition to small amounts of maltose and its oligomers could be detected by using a gradient technique. If complete resolution between the epimeric forms of glucose is desired, a dilute sodium hydroxide solution is used as the eluent. Fig. 8-69 shows the separation of galactose, glucose, and lactose in milk, that was diluted only with de-ionized water and filtered prior to injection. A disadvantage is the comparatively long analysis time, which

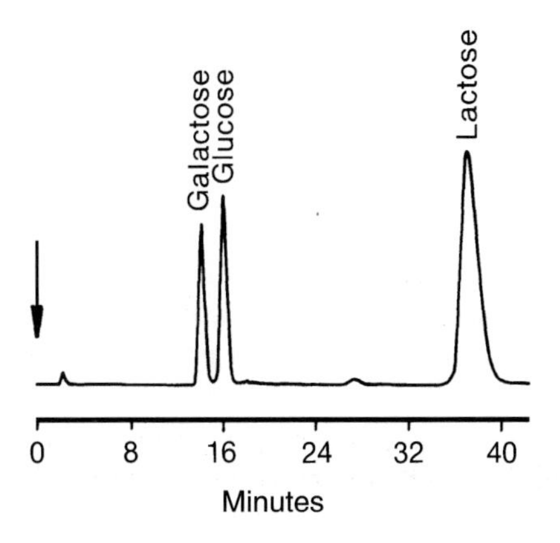

Fig. 8-69. Analysis of galactose, glucose, and lactose in milk. – Separator column: CarboPac PA1; eluent: 0.005 mol/L NaOH; flow rate: 1 mL/min; detection: see Fig. 8-68; injection: 50 μL sample (1:1000 diluted).

is attributed to the low eluent concentration. Since amperometric detection of carbohydrates can only be sensitively performed at pH 13, sodium hydroxide (c = 0.3 mol/L) was added to the separator column effluent before it enters the amperometric detector cell. Under these conditions, a pH gradient may also be applied so that the analysis time can be significantly reduced by raising the sodium hydroxide concentration during the run.

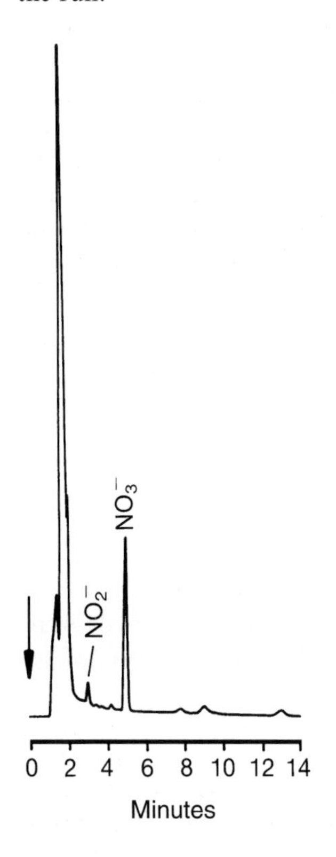

Fig. 8-70. Analysis of nitrite and nitrate in a raw sausage extract. – Separator column: IonPac AS4A; eluent: 0.0017 mol/L $NaHCO_3$ + 0.0018 mol/L Na_2CO_3; flow rate: 1.5 mL/min; detection: UV (215 nm); injection: 50 µL of a 1:2 diluted extract (10 g of sausage ad 200 mL of DI water).

Meat Processing

A key problem in the investigation of meat and sausage products is the determination of nitrite and nitrate. This could be substantially facilitated with ion chromatography [60,61]. However, this does not apply to sample preparation which has to be carried out – independent of the determination method being used – according to the procedure stipulated in §35 LMBG. This includes the sample homogenization, its extraction with a 5% borax buffer solution in a hot water bath, and the subsequent Carrez-precipitation with 15% potassium hexacyanoferrate (II) and 30% zinc sulfate solution. Depending on the concentration of the nitrogen parameters, the aqueous extracts thus obtained are further diluted with de-ionized water and membrane-filtered (0.22 µm) prior to injection. The subsequent ion chromatographic analysis is usually performed with a conventional anion exchanger applying UV detection at 215 nm. UV detection is chosen over conductometric detection because of the high sodium chloride content in these samples. This is illustrated in Fig. 8-70 with the chromatogram of a raw sausage extract with a separation of nitrite and nitrate. The identity of the signals was validated by standard addition of pure compounds to the sample that was then re-chromatographed and the added amount was quantitatively recovered. The sodium nitrite content as the sum of nitrite and nitrate was found to be 57 ppm; thus, it was clearly below the limiting value of 100 ppm.

Cutter and reddening agents used in meat processing, may also be assayed ion chromatographically for their active constituents, which include inorganic anions such as chloride and diphosphate, ascorbic acid, and carbohydrates such as glucose, sucrose, and maltose. While ascorbic acid is separated via ion-pair chromatography on a chemically bonded octadecyl-type reversed phase, anion exchange chromatography with photometric detection after derivatization with iron(III) is used to analyze diphosphate. Selectivity and sensitivity of both methods are so high that an extensive sample preparation can be eliminated. The analyte samples are only dissolved in de-ionized water and membrane-filtered (0.45 μm) prior to injection. Figs. 8-71 and 8-72 display the separation of diphosphate and ascorbic acid in a reddening agent.

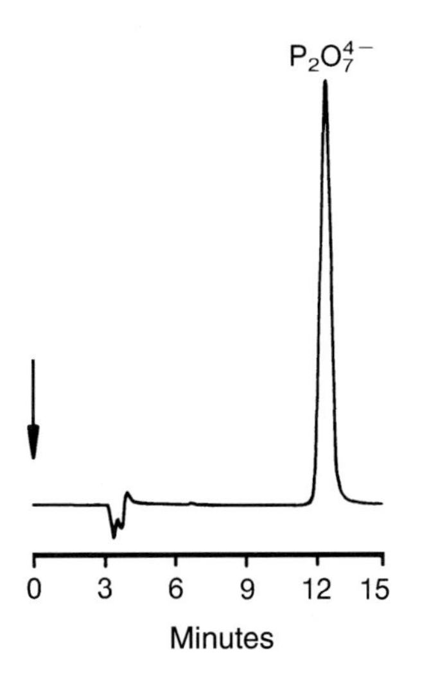

Fig. 8-71. Separation of diphosphate in a reddening agent. – Separator column: IonPac AS7; eluent: 0.03 mol/L HNO_3; flow rate: 0.5 mL/min; detection: photometry at 330 nm after reaction with ferric nitrate; injection: 50 μL of a 0.1% solution that was diluted 1:5.

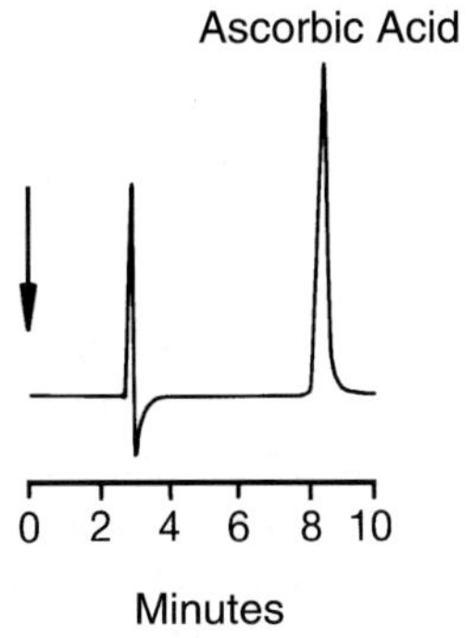

Fig. 8-72. Separation of ascorbic acid in a reddening agent. – Separator column: Hypersil 5 MOS; eluent: 0.002 mol/L TBAOH + 0.002 mol/L H_3PO_4 / acetonitrile (90:10 v/v); flow rate: 1 mL/min; detection: UV (254 nm); injection: 50 μL of a 0.1% solution that was diluted 1:25.

Sample preparation is different for investigating meat seasonings. Because of their high organic compound content, the aqueous solutions cannot be injected directly but must be treated with OnGuard-RP extraction cartridges. The cartridges contain an non-polar polymer at which matrix components are retained. Ionic species are not affected by this sample preparation. The problem-free applicability of ion chromatography to such samples with the alkaline-earth metal analysis in a meat seasoning is illustrated in Fig. 8-73.

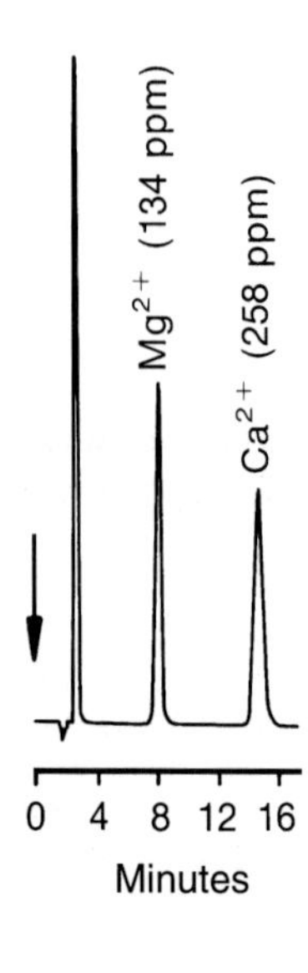

Fig. 8-73. Separation of alkaline-earth metals in a meat seasoning. – Separator column: IonPac CS3; eluent: 0.03 mol/L HCl + 0.005 mol/L 2,3-diaminopropionic acid; flow rate: 1 mL/min; detection: suppressed conductivity; injection: 50 μL of a 1% solution that was diluted 1:25.

Baby Food

A typical application in this area is the determination of choline, a biogenic amine in soy beans which are added to baby food. While the conventional choline determination procedure is very time-consuming and imprecise, this substance can be analyzed with ion-pair chromatography in less than 20 minutes (Fig. 8-74).

Recently, the determination of phytic acid (myo-inositol hexaphosphate), which is believed to have an effect on the bioavailability of mineral substances, has become important. While phytic acid is not a food constituent in the United States, in other countries it is frequently added to food as an antioxidant [62]. To elucidate the nutritive scientific relevance of phytic acid, as early as 1983 Lee and Abendroth [63] developed an ion chromatographic method for analyzing this compound. However, it required a pre-purification of the sample to be analyzed. Based on the post-column derivatization technique with iron(III), introduced by Fitchett and Woodruff [64] for determining polyphosphates in detergents, Philippy and Johnston [65] developed an analytical method for phytic acid that allows the direct injection of food extracts. They employed an IonPac AS3 latex-based anion exchanger with nitric acid as the eluent. The suitable measuring wavelength of 290 nm for the iron(III)-phytic acid complex is slightly lower than that of 330 nm used by Fitchett et al. Fig. 8-75 illustrates the application of this technique to the analysis of baby food. It was extracted for 30 minutes with 12% hydrochloric acid, the extract being centrifuged, diluted with de-ionized water to the suitable working range, and membrane-filtered (0.45 μm) prior to injection. As seen in the chromatogram in Fig. 8-75, phytic acid could be separated without any problems from the matrix constituents. Phytic acid hydrolysis products such as inositol pentaphosphate and inositol tetraphosphate do not interfere in the analysis since they precipitate with iron(III) [66].

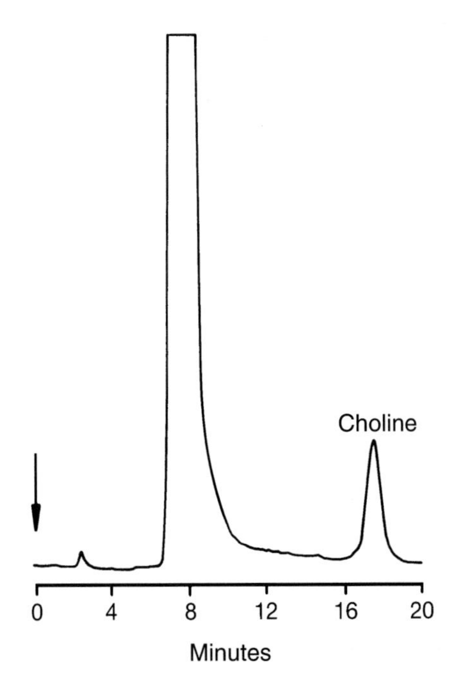

Fig. 8-74. Determination of choline in baby food. – Separator column: IonPac NS1 (10 μm); eluent: 0.002 mol/L hexanesulfonic acid / acetonitrile (95:5 v/v); flow rate: 1 mL/min; detection: suppressed conductivity; injection: 50 μL extract.

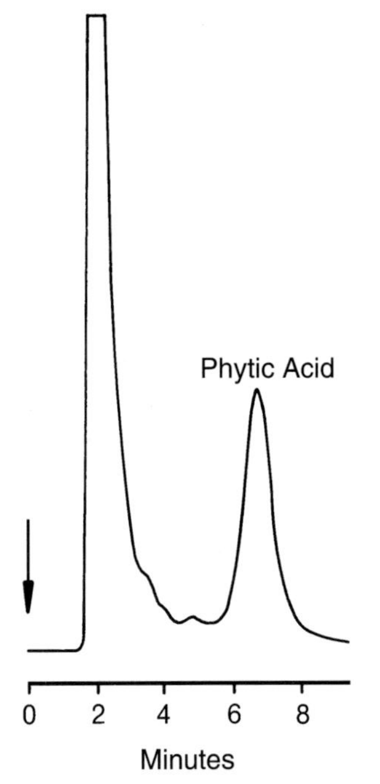

Fig. 8-75. Determination of phytic acid in baby food. – Separator column: IonPac AS3; eluent: 0.11 mol/L HNO_3; flow rate: 1 mL/min; detection: photometry at 290 nm after post-column reaction with ferric nitrate; injection: 100 μL of a diluted 1.2% HCl-extract.

Food and Tobacco

The chromatogram in Fig. 8-76 is a typical example of the efficiency of ion chromatography. A weighed amount of a canned spinach was extracted with de-ionized water,

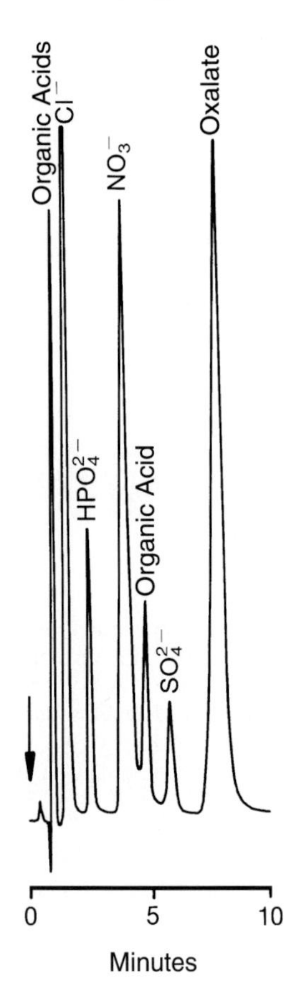

Fig. 8-76. Anion analysis of a canned spinach. – Separator column: IonPac AS4; eluent: 0.0028 mol/L $NaHCO_3$ + 0.0022 mol/L Na_2CO_3; flow rate: 2 mL/min; detection: suppressed conductivity; injection: 50 μL extract.

filtered, and injected directly. Inorganic anions and organic acids such as oxalic acid, which are contained in spinach, may be analyzed in less than 10 minutes. As in many

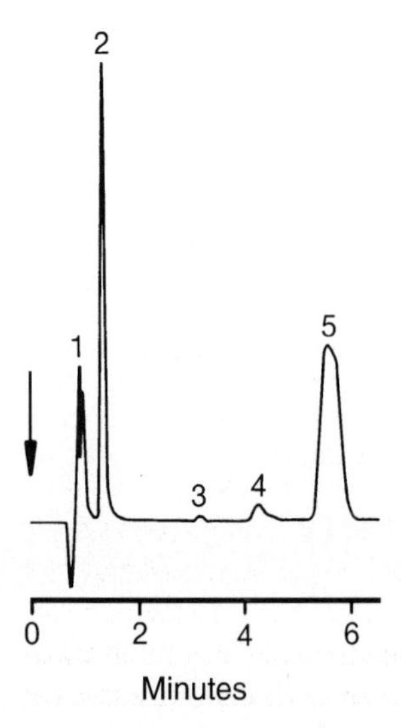

Fig. 8-77. Analysis of inorganic and organic anions in a tobacco extract. – Separator column: IonPac AS4A; eluent: 0.0017 mol/L $NaHCO_3$ + 0.0018 mol/L Na_2CO_3; flow rate: 2 mL/min; detection: suppressed conductivity; injection: 50 μL of a 1:25 diluted 1% extract; sample components: organic acids (1), chloride (2), nitrate (3), orthophosphate (4), and sulfate (5).

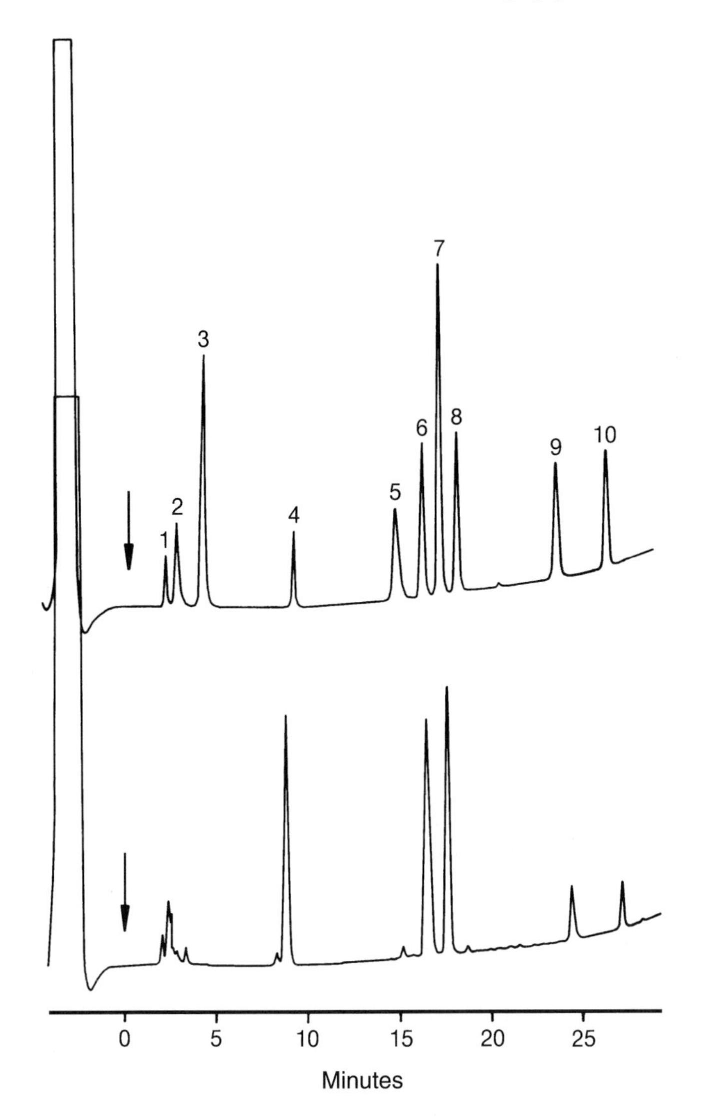

Fig. 8-78. Gradient elution of anions in a tobacco extract. – Separator column: IonPac AS5A; eluent: (A) 0.001 mol/L NaOH, (B) 0.2 mol/L NaOH; gradient: 1% B isocratically for 9 min, then linearly to 20% B in 11 min, then linearly to 45% B in 10 min; flow rate: 1 mL/min; detection: suppressed conductivity; injection: (A) 50 µL standard with 0.5 ppm fluoride (1), 10 ppm acetate (2), 10 ppm formate (3), 1 ppm chloride (4), 5 ppm nitrate (5), 8 ppm malate (6), 10 ppm sulfate (7), 10 ppm oxalate (8), 10 ppm orthophosphate (9), and 8 ppm citrate (10), (B) 50 µL of an extract diluted 1:25.

such samples, a number of organic acids are eluted within or close to the void volume of the separator column. If these components are to be separated and analyzed together with the inorganic anions in a single run, the application of a gradient technique is indispensible. An example is the analysis of an aqueous tobacco extract. If the analysis is performed under isocratic conditions with a carbonate/bicarbonate-mixture, the chromatogram represented in Fig. 8-77 is obtained. The peak group occurring in the void volume of the separator column indicates the presence of short-chain organic acids. In addition to chloride, nitrate and orthophosphate can also be identified. The peak appearing at t_{ms} = 5.4 min, usually to be attributed to sulfate, has an unusual broad peak shape, suggesting that at least one further component elutes at this position (possibly a short-chain dicarboxylic acid). To avoid the co-elution at the beginning of the chromatogram and in the retention range of sulfate, the sample was again chromatographed applying the gradient technique. The corresponding chromatogram in Fig. 8-78 is compared with that of a standard mixture of inorganic and organic anions. This representation reveals that malic acid elutes immediately before sulfate, thus overlapping the sulfate signal under isocratic conditions. Furthermore, it is advantageous for more strongly retained anions such as citric acid to be detected in the same run.

Sweeteners

More frequently, food is sweetened with sugar substitutes instead of sugar or glucose syrups. Because of the diversity of products containing artificial sweeteners, the matrices to be investigated are very complex. In addition, sweeteners may be contained in these products individually or in combination with others.

The best-known sugar substitutes include saccharin, sodium cyclamate, and acesulfam-K.

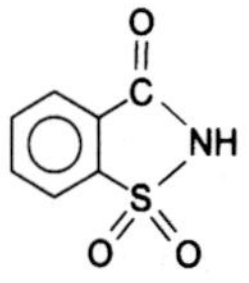

Saccharin 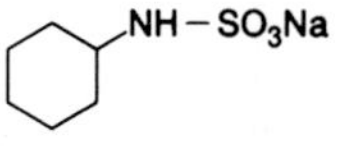   Sodium cyclamate 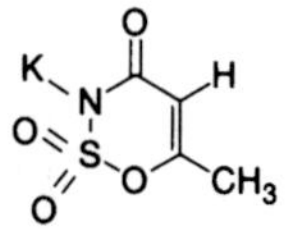   Acesulfam-K

While RPLC and RPIPC techniques applied with UV detection have already been elaborated for analyzing saccharin and acesulfam-K [67,68], the liquid chromatographic analysis of sodium cyclamate is barely substantiated in the literature because of the non-chromophoric structure of this compound. The only exception is the HPLC method introduced by Hermann et al. with direct photometric detection [69].

As the above-mentioned compounds exist as anions in alkaline environment, anion exchange chromatography with subsequent conductometric detection provides a welcome alternative to RPLC with UV detection, especially as the flavours and dyes contained in food do not represent any interference upon application of conductometric detection. Based on these discoveries, Biemer [70] recently developed an ion chromatographic technique for determining these three sugar substitutes and applied it to the analysis of chewing gum. The analyte samples were extracted with acetic acid/chloroform and the aqueous phase was injected directly after appropriate dilution. Fig. 8-79 shows the chromatogram of a chewing gum sample to which cyclamate was added. Pure sodium bicarbonate solution served as the eluent. Bromide, eluting after cyclamate, was added as an internal standard. Acesulfam-K and saccharin are much stronger retained and, thus, are eluted with pure sodium carbonate solution. The corresponding chromatogram is displayed in Fig. 8-80. In this case, fumaric acid served as the internal standard. Inorganic anions such as chloride, nitrate, orthophosphate, and sulfate do not interfere with the analysis.

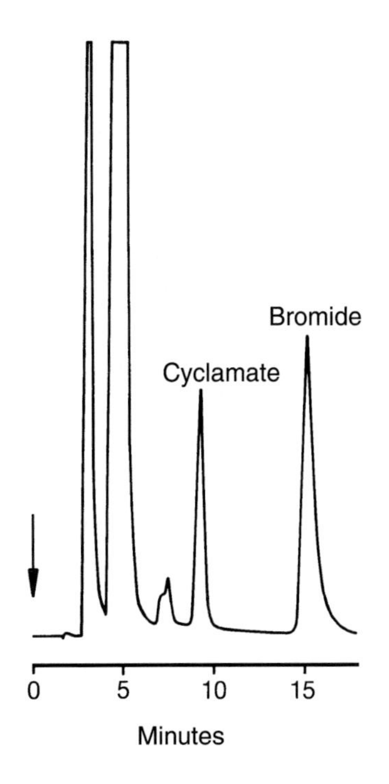

Fig. 8-79. Analysis of cyclamate in chewing gum. – Separator column: IonPac AS4A; eluent: 0.0017 mol/L $NaHCO_3$; flow rate: 2 mL/min; detection: suppressed conductivity; injection: 50 μL of a spiked chewing gum sample; (taken from [70]).

Another sugar substitute is palatinitol. The preparation of palatinitol uses sucrose as the starting material. After enzymatic rearrangement into palatinose (isomaltulose) and hydration, palatinitol is formed representing an equimolar mixture of the isomers α-D-glucopyranosido-1,6-mannitol and α-D-glucopyranosido-1,6-sorbitol:

Sucrose (α_1-β_2) → Isomaltulose (α_1-6)

Isomaltulose — α-D-Glucopyranosido-1,6-mannitol, α-D-Glucopyranosido-1,6-sorbitol

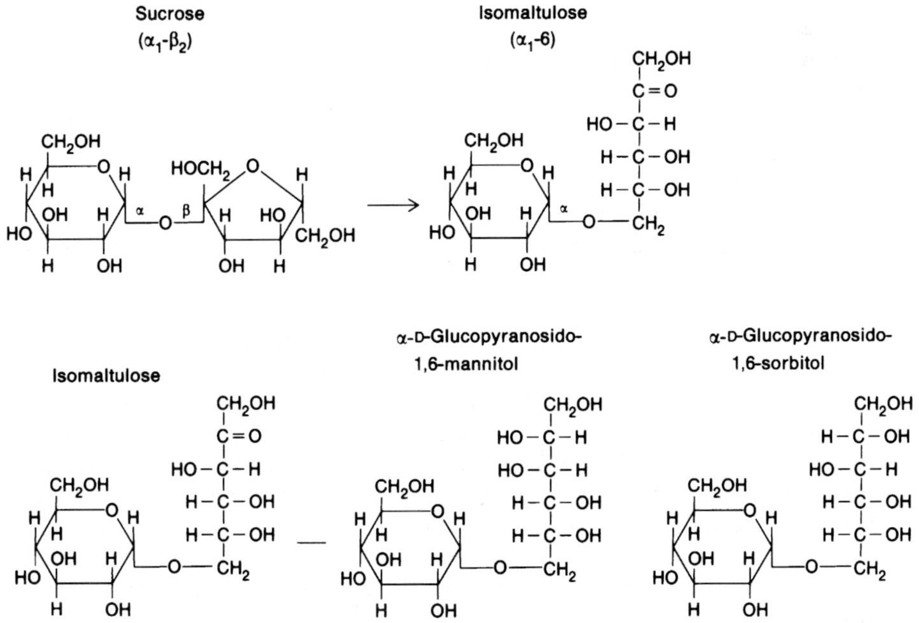

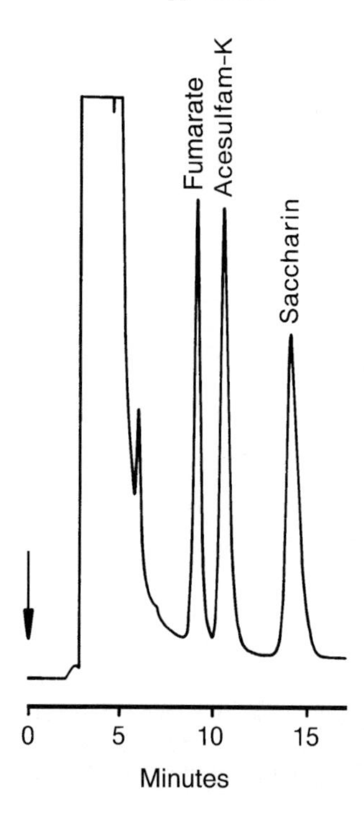

Fig. 8-80. Analysis of acesulfam-K and saccharin in chewing gum. – Separator column: IonPac AS4A; eluent: 0.0028 mol/L Na_2CO_3; flow rate: 2 mL/min; detection and injection: see Fig. 8-79; (taken from [70]).

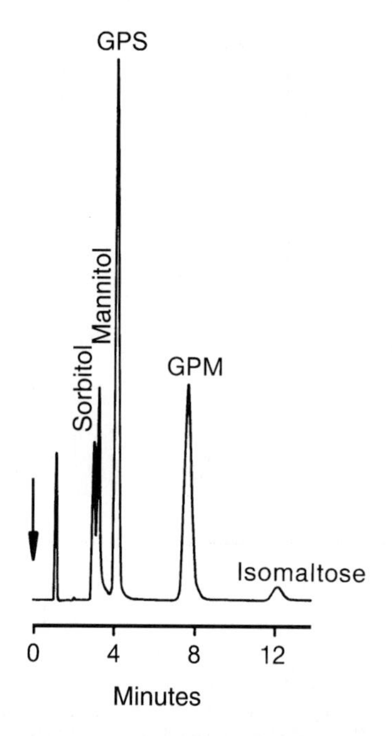

Fig. 8-81. Separation of the palatinitol isomers. – Separator column: CarboPac PA1; eluent: 0.1 mol/L NaOH; flow rate: 1 mL/min; detection: pulsed amperometry at a Au working electrode; injection volume: 50 μL; solute concentrations: 10 ppm each of α-D-glucopyranosido-1,6-sorbitol (GPS) and α-D-glucopyranosido-1,6-mannitol (GPM).

Like all carbohydrates, palatinitol may be separated in alkaline environment on a strongly basic anion exchanger in the hydroxide form and may be detected via pulsed amperometry. Fig. 8-81 shows a corresponding chromatogram with the separation of both isomers. Sorbitol, mannitol, and isomaltose can be detected as impurities in the same run.

In principle, honey can also be used to sweeten food and beverages; it is an exceptionally complex product. Nectar collected by bees contains, in addition to sucrose, varying amounts of invert sugar. In the digestive tract it is hydrolyzed to cane sugar, so that the sugar delivered by the bees contains only about 10% unchanged sucrose. In addition to these main components, various di- and trisaccharides are found in honey in different concentrations. A special feature of forest honey and fir honey is the presence of melizitose, a trisaccharide that is contained in honey dew. These excretes, found on conifers, are picked up by the bees and, consequently, are only detectable in such types of honey. Thus, the detection of melizitose in fir honey represents a quality characteristic. As seen in the chromatogram of such a forest fir honey shown in Fig. 8-82, the product investigated by Baumgärtner [57] meets this quality requirement. In addition to the main components glucose and fructose, trehalose and maltose and small concentrations of maltose oligomers can also be detected. The structures of the other carbohydrates present in traces could not be elucidated.

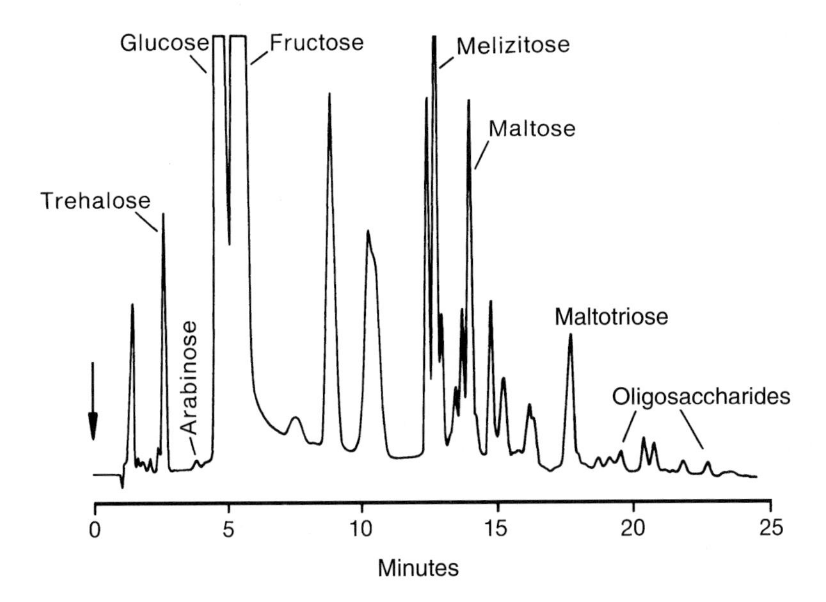

Fig. 8-82. Analysis of carbohydrates in a forest fir honey. – Separator column: CarboPac PA1; eluent: (A) 0.1 mol/L NaOH, (B) 0.1 mol/L NaOH + 0.5 mol/L NaOAc; gradient: 100% A isocratically for 2 min, then linearly to 25% B in 30 min; flow rate: 1 mL/min; detection: see Fig. 8-81; injection: 50 µL of a 0.3% solution; (taken from [57]).

8.7 Ion Chromatography in the Pharmaceutical Industry

A large number of applications for ion chromatography also exist in the pharmaceutical industry. Here, only some selected examples can be described.

Ion chromatography plays a prominent role in the characterization of pharmaceutically relevant compounds predominantly in the early stages of research. This includes the trace analysis of impurities and metabolites, the elemental analysis and structural elucidation of counter ions. These counter ions are often inorganic anions such as chloride and bromide or organic acids such as acetate, methyl sulfate, and trifluoroacetate.

Methyl sulfate is not only of analytical interest as a counter ion of pharmaceutically relevant compounds. Rychtman [71] describes the trace analysis of this substance in pharmaceutical products that have been produced via alkylation with dimethyl sulfate. Tetraalkylammonium compounds, for example, may be synthesized by quaternation of tertiary amines with dimethyl sulfate:

$$R_3N + (CH_3O)_2SO_2 \longrightarrow R_3\overset{+}{N}CH_3 + CH_3OSO_3^- \qquad (228)$$

Because methyl sulfate must not be contained in pharmaceutical products, it is later replaced by chloride via anion exchange. The exchange efficiency can be monitored by determining the remaining methyl sulfate content. The analytical procedure developed by Rychtman uses a conventional latexed anion exchanger with tetrabutylammonium hydroxide as the mobile phase. Methyl sulfate detection is performed via electrical conductivity with application of a membrane-based suppressor system. Fig. 8-83 displays the chromatogram of a research sample. Although chloride is present in 50-fold excess, the separation of methyl sulfate is baseline-resolved. The methyl sulfate peak corresponds to a sulfur content of 2.8 mg/g.

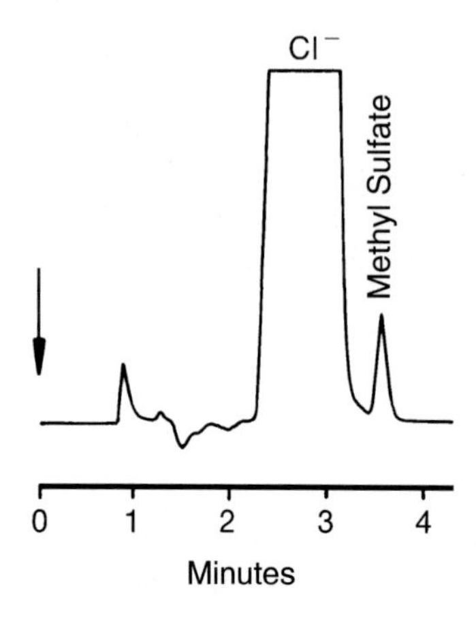

Fig. 8-83. Analysis of methyl sulfate. – Separator column: IonPac AS4A; eluent: 0.001 mol/L TBAOH; flow rate: 2 mL/min; detection: suppressed conductivity; injection: 50 μL sample; (taken from [71]).

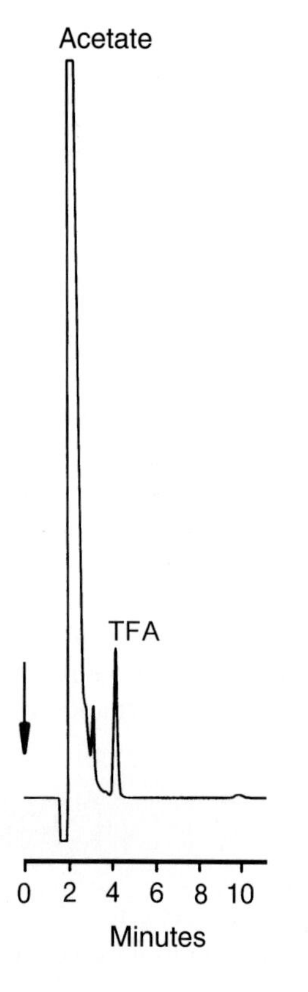

Fig. 8-84. Analysis of trifluoroacetate in a peptide sample. – Separator column: IonPac AS4A; eluent: 0.0017 mol/L $NaHCO_3$ + 0.0018 mol/L Na_2CO_3; flow rate: 1 mL/min; detection: suppressed conductivity; injection: 50 μL of a 0.1% sample.

The same stationary phase is also suited for the separation of trifluoroacetate. Trifluoroacetate may be detected, apart from chloride and sulfate, in very low concentrations in peptides and stems from the mobile phase used for the HPLC separation of the peptide. To determine all these anions in one run, it is necessary to revert to the mixture of carbonate and bicarbonate as the eluent. The chromatogram represented in Fig. 8-84 was obtained by injecting a 0.1% peptide sample in which 1.8 mg/g TFA could be detected.

Sodium sulfite, added to some pharmaceuticals as an antioxidant, can be determined ion chromatographically with high precision. However, to avoid autoxidation of sulfite to sulfate, the samples must be stabilized with formaldehyde solution. For that, 0.5 mL of a 25% formaldehyde solution is added per liter sample. Alternatively, dilution is carried out with appropriately pretreated water. The resulting formaldehyde-sulfite complex exhibits the same retention time as the sulfite ion. Fig. 8-85 shows the chromatogram of a local anaesthetic diluted 1:10. Its therapeutic agent is present as hydrochloride thus explaining the high chloride peak at the beginning of the chromatogram.

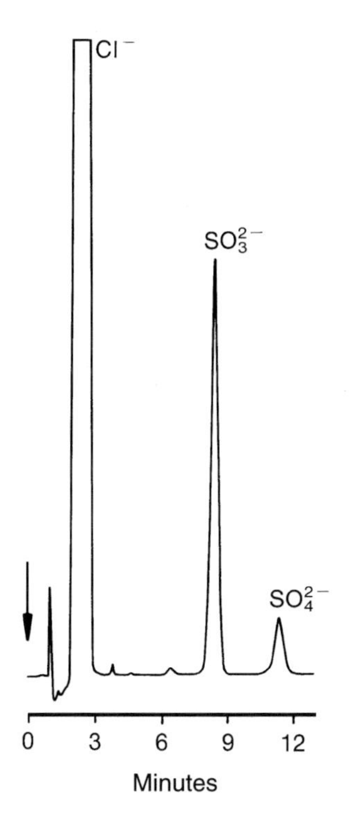

Fig. 8-85. Analysis of sulfite in a local anaesthetics. – Separator column: IonPac AS9; eluent: 0.00075 mol/L $NaHCO_3$ + 0.002 mol/L Na_2CO_3; flow rate: 2 mL/min; detection: suppressed conductivity; injection: 50 µL sample (diluted 1:10).

The last such example in Fig. 8-86 illustrates the chromatographic separation of benzalkonium chloride in nose drops.

$$\left[C_6H_5-CH_2-\overset{+}{N}(CH_3)_2-R \right] Cl^- \quad \text{with R } C_8 \text{ till } C_{18}$$

Benzalkonium chloride

Benzalkonium chloride is a quaternary ammonium base that is added to some pharmaceuticals as a disinfectant. Often, it is a mixture with varying long alkyl chains R having even-numbered members between C_{12} and C_{16}. Separation is performed with ion-pair chromatography using hydrochloric acid as the ion-pair reagent. Since benzalkonium chloride is an aromatic compound, it can be detected very sensitively with UV detection at 215 nm. In the chromatogram shown in Fig. 8-86 three signals are obtained for this compound which may be attributed to the various chain lengths of the alkyl rest. Only the signal marked by a * was used for the calibration, as the distribution of the signals is identical both in the raw material and in the sample.

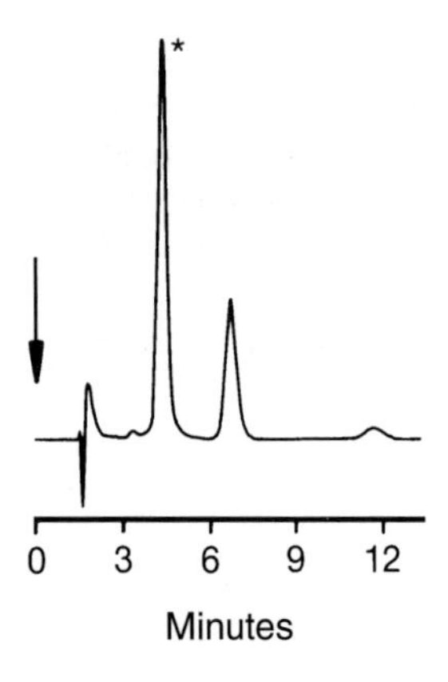

Fig. 8-86. Analysis of benzalkonium chloride in nose drops. – Separator column: IonPac NS1 (10 µm); eluent: 0.02 mol/L HCl / acetonitrile (37:63 v/v); flow rate: 1 mL/min; detection: UV (215 nm); injection: 50 µL sample (diluted 1:10).

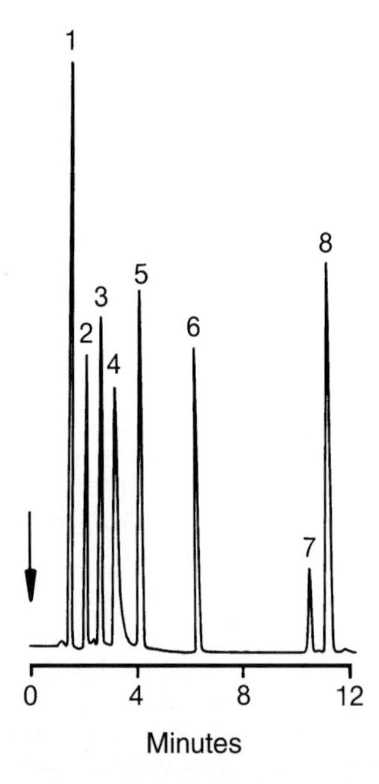

Fig. 8-87. Analysis of water-soluble vitamins. – Separator column: Spherisorb ODS 2 (5 µm); eluent: (A) 0.1 mol/L KOAc (pH 4.2 with HOAc), (B) water/methanol/acetonitrile (50:10:40 v/v/v); gradient: linear, 6% B in 30 min to 100% B; flow rate: 2 mL/min; detection: UV (254 nm); injection volume: 50 µL; solute concentrations: 5 nmol each of ascorbic acid (1), nicotinic acid (2), thiamine (3), pyridoxine (4), nicotinic acid amide (5), *p*-aminobenzoic acid (6), cyanocobalamine (7), and riboflavine (8).

In the area of vitamins, Fig. 8-87 shows the gradient elution of water-soluble vitamins at Spherisorb ODS 2, at which the most important compounds of this kind can be analyzed in less than 15 minutes. One of the lesser known vitamins is lipoic acid, also denoted as thioctic acid, which consists of eight carbon atoms and may exist in two forms: as lipoic acid – a cyclic disulfide – and in an open-chain form as dihydrolipoic acid, which carries a sulfhydryl group at the positions 6 and 8.

$(CH_2)_4$ – COOH (cyclic disulfide ring with S–S)

Lipoic acid

CH_2-SH
CH_2
CH_2-SH
$(CH_2)_4$
COOH

Dihydrolipoic acid

Lipoic acid acts as one of the coenzymes in the oxidative decarboxylation of pyruvate and other α-keto acids. It can be separated in an alkaline environment on a strongly basic anion exchanger in the hydroxide form, and can be detected like carbohydrates via pulsed amperometry at a Au working electrode. The corresponding chromatogram of a lipoic acid standard is shown in Fig. 8-88. This method allows to accurately detect 0.1 nmol lipoic acid.

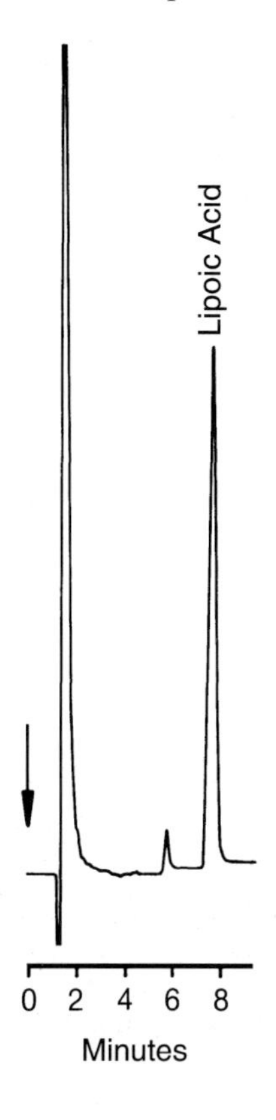

Fig. 8-88. Analysis of lipoic acid. – Separator column: CarboPac PA1; eluent: 0.1 mol/L NaOH + 0.5 mol/L NaOAc; flow rate: 1 mL/min; detection: pulsed amperometry at a Au working electrode; injection volume: 50 µL: solute concentrations: 40 mg/L.

An impressive example of the high efficiency of anion exchange chromatography resulted from the task to determine sulfate in penicillamine (α-phenoxyethyl penicillin). As revealed in the corresponding chromatogram in Fig. 8-89, in addition to the sulfate signal, three further peaks are obtained, from which the first two could be attributed to acetic acid and phenoxyacetic acid. Before the introduction of ion-exchange chromatography they were determined enzymatically to monitor the manufacturing process.

Penicillin V

Looking at the penicillin V structure, it is not surprising that this substance can also be eluted on an anion exchanger, since the carboxylate group located at the thiazolidine ring is fully dissociated in alkaline environment. The fact that it may be determined together with acetic acid, phenoxyacetic acid, and sulfate in the same run is unexpected. Thus, all three methods for characterizing penicillin V may be replaced by an ion chromatographic procedure which drastically reduces the time expenditure.

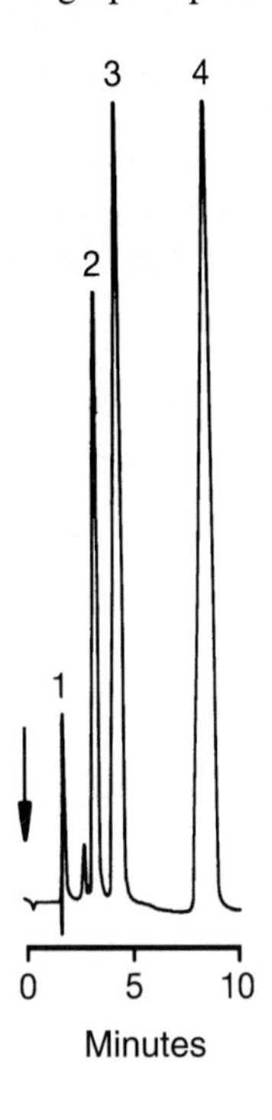

Fig. 8-89. Analysis of penicillin V. – Separator column: IonPac AS4A; eluent: 0.003 mol/L $NaHCO_3$ + 0.0024 mol/L Na_2CO_3; flow rate: 1 mL/min; detection: suppressed conductivity; injection volume: 50 µL; solute concentrations: 20 ppm acetate (1), 30 ppm phenoxyacetate (2), 500 ppm penicillin V (3), and 30 ppm sulfate (4).

Fermentation

Biotechnological techniques in the production of food and feed stuff have been known for ages, so dairy products, sour vegetables, and rising agents such as yeast and leaven for bakery products represent biotechnological products that are known to everybody. In addition to traditional fermentation techniques, gene-technological methods for producing a number of active agents and substances have become more prevalent. The best known example is the microbial production of antibiotics such as the above-mentioned penicillamine. Other pharmaceuticals, enzymes, amino acids, and organic acids such as citric acid are also produced. To optimize the yield of such products, a number of investigations are necessary during the fermentation. This concerns the nutrient solution, the analysis of components that control or induce the metabolic activity, the reprocessing of products, and the analysis of minor products and precursors.

In many cases, process control comprises only the on-line monitoring of temperature and pH value as well as an accompanying enzymatic analysis of glucose and lactic acid. Since this represents an indirect process control, product yields may only be marginally improved according to these analytical results. In contrast, a number of instrumental methods are now available which allow the analysis of essential components of the fermentation process within a short time, so that a timely intervention in the process is possible and the product yield may be optimized.

One of the fermentation processes technically used for some time is the isolation of gluconic acid, which is produced by several fungi – especially *Aspergilles niger*. Glucose is used as the substrate, from which gluconolacton is produced by glucose dehydrogenase and finally gluconic acid. Organic acids that are directly derived from carbohydrates may be separated chromatographically together with them in a strongly alkaline medium on an anion exchanger. Since organic acids, because of their carboxylate group, exhibit a much higher affinity toward the stationary phase than carbohydrates, a mixture of sodium hydroxide and sodium acetate is used as the eluent. Detection is performed via pulsed amperometry at a Au working electrode. The corresponding chromatogram is depicted in Fig. 8-90. The slightly stronger tailing of the gluconic acid peak is typical for this compound.

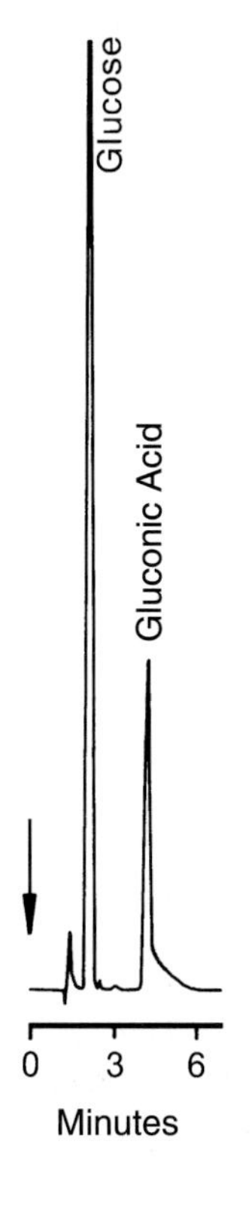

Fig. 8-90. Separation of glucose and its oxidation product gluconic acid. – Separator column: CarboPac PA1; eluent: 0.1 mol/L NaOH + 0.1 mol/L NaOAc; flow rate: 1 mL/min; detection: see Fig. 8-88; injection volume: 50 µL; solute concentrations: 100 ppm each of glucose and gluconic acid.

Carbohydrates are often used as a carbon source. Sugars and starch-containing raw materials are the most important fermentation substrates for micro-organisms in the production of antibiotics, bakery yeast, and enzymes. To optimize the fermentation, it is necessary to perform permanent time-dependent substrate checks. For this, the technique of anion exchange chromatography with pulsed amperometric detection is particularly suited because of its high sensitivity, since the matrix may be eliminated to a large extent by diluting the sample. Fig. 8-91 shows the separation of glucose, lactose, and maltose in a fermentation culture filtrate that was diluted in the ratio 1:500 with de-ionized water and filtered prior to injection. Due to the short analysis time and the high specificity of this technique, sample analysis in quasi on-line operation is conceivable.

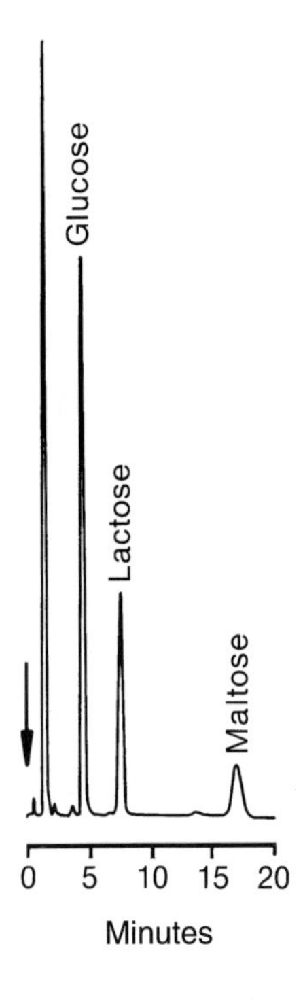

Fig. 8-91. Separation of glucose, lactose, and maltose in a fermentation culture filtrate. – Separator column: CarboPac PA1; eluent: 0.1 mol/L NaOH; flow rate: 1 mL/min; detection: see Fig. 8-88; injection: 50 μL sample (diluted 1:500).

Regarding the control of fermentation, organic acids and inorganic components play a significant role [72]. In a batch reactor for producing monoclonal antibodies, for example, an increase of the lactate value is observed three days after inoculation with the cells, while other organic acids such as acetic acid and propionic acid are reduced in their content in comparison to a control medium. The two chromatograms in Fig. 8-92 clearly show that the time required for these analyses is very short. From the area of inorganic components, Fig. 8-93 shows the analysis of transition metals in the same sample. Transition metals only play a significant role in some fermentation processes. Zinc, for example, is an important constituent in the production of the amino acid L-glutamine while iron is a key parameter in the production of citric acid. In the present case, small amounts of copper, cobalt, and zinc could be detected.

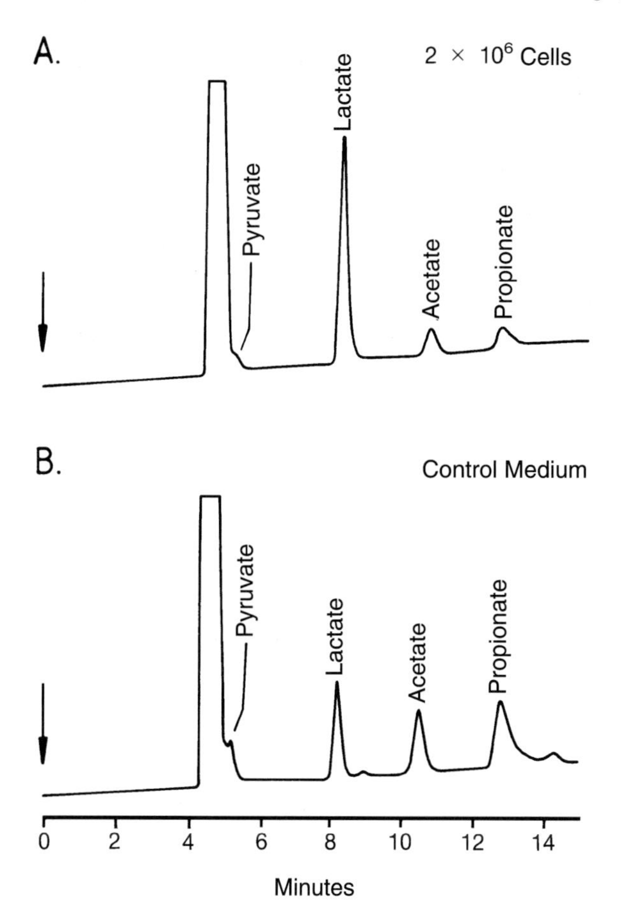

Fig. 8-92. Separation of organic acids in a fermentation culture filtrate. – Separator column: IonPac ICE-AS1; eluent: 0.001 mol/L octanesulfonic acid; flow rate: 1 mL/min; detection: suppressed conductivity; injection: 50 µL sample (diluted 1:50); (A) three days after inoculation with $2 \cdot 10^6$ cells, (B) control medium; (taken from [72]).

8.8 Ion Chromatography in Clinical Chemistry

One of the most important applications of ion chromatography in clinical chemistry is the investigation of body fluids such as saliva [73], urine, and serum both for inorganic and organic anions and cations.

In this field saliva represents one of the simplest matrices and may be directly injected without any sample preparation after appropriate dilution with de-ionized water. As seen in the respective anion chromatogram of a human saliva sample (390 mg ad 25 mL H_2O) in Fig. 8-94, the constituents may be separated without any interferences. The detection of small quantities of bromide is remarkable. The occurrence was unknown up to now and the source cannot be explained at present.

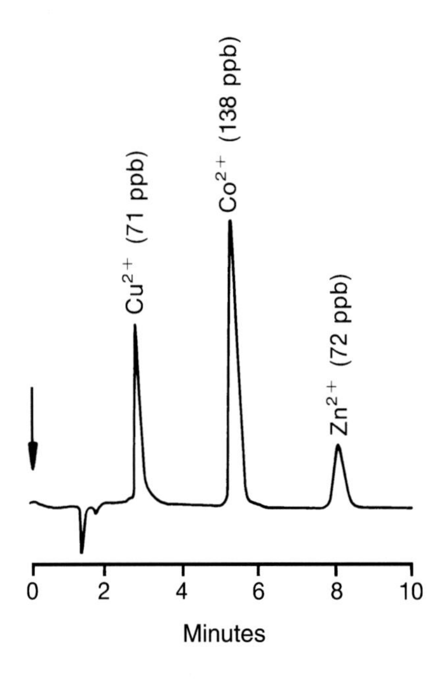

Fig. 8-93. Separation of heavy and transition metals in a fermentation culture filtrate. – Separator column: IonPac CS5; eluent: 0.035 mol/L oxalic acid (pH 4.4 with LiOH); flow rate: 1 mL/min; detection: photometry at 520 nm after reaction with PAR; injection: 50 μL sample (diluted 1:12); (taken from [72]).

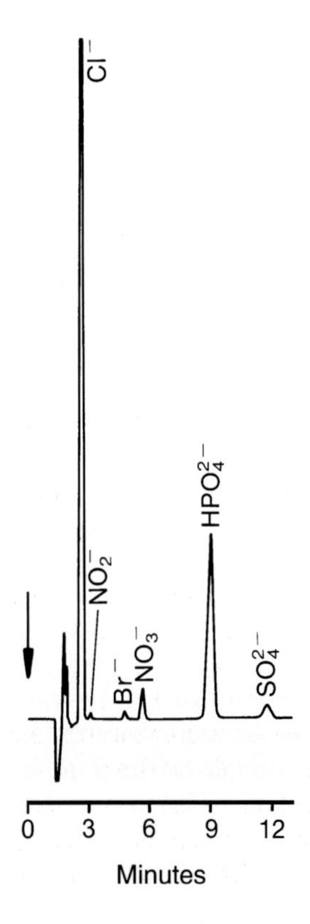

Fig. 8-94. Separation of inorganic anions in human saliva. – Separator column: IonPac AS4A; eluent: 0.0017 mol/L $NaHCO_3$ + 0.0018 mol/L Na_2CO_3; flow rate: 1.5 mL/min; detection: suppressed conductivity; injection: 50 μL of a solution of 390 mg saliva in 25 mL water.

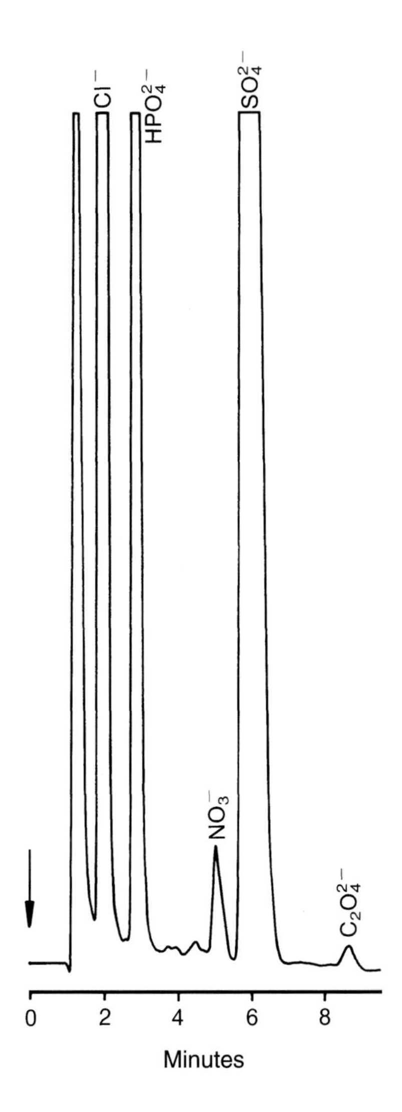

Fig. 8-95. Separation of oxalate in urine. – Separator column: IonPac AS4; eluent: 0.0028 mol/L $NaHCO_3$ + 0.0022 mol/L Na_2CO_3; flow rate: 2 mL/min; detection: suppressed conductivity; injection: 50 μL sample (diluted 1:10).

Ion chromatography is a decisive factor for treating analytical questions in the field of nephrology. Today, for example, the analysis of oxalate in urine is one of the most important determinations in the examination of patients with kidney stones [74]. Many analysis methods used such as manganometry after oxidation of oxalate to carbon dioxide, or spectrophotometry after reduction of oxalate to glyoxylic acid or glycolic acid are very labor-intensive and time-consuming, since a preceding separation of oxalate with a corresponding precipitation reaction is necessary. The direct oxalate determination by enzymatic analysis is difficult, as the enzymes are inhibited by matrix constituents such as sulfate and orthophosphate, resulting in a systematic underestimation. In the past, therefore, one had to be content with measuring the calcium content in urine to obtain evidence of irregularities in patients with kidney stones. Not until the introduction of chromatographic techniques such as GC [75,76] and HPLC [77], was it possible to determine oxalate ions directly in urine. Based on the work of Menon et al. [78], Robertson et al. [79] developed an ion chromatographic method which allows the analysis of oxalate with little sample preparation in less than 20 minutes. They employed a conventional anion exchanger as the stationary phase with a mixture of sodium carbonate and sodium bicarbonate as the eluent. To achieve the highest possible sensitivity, they passed the separator column effluent through a membrane-based suppressor systems prior to entering the detector cell. With this setup, Robertson et al. obtained a detection sensitivity of 0.5 μmol/L oxalate in acidified, diluted urine. The sample acidification with a hydrochloric acid of concentration c = 1 mol/L is necessary to redissolve calcium oxalate crystals that may have precipitated in urine. Fig. 8-95 shows a corresponding chromatogram of own work; the peak to be attributed to oxalate corresponds to a concentration of 10 μmol/L.

Under similar chromatographic conditions it is also possible to determine *p*-aminohippuric acid, which represents one of the control substances for diagnosing the kidney function. On the basis of certain analytical values, it is possible to determine the total blood circulation through the kidneys and, at higher plasma concentrations of *p*-aminohippuric acid, the maximal secretion performance of the tubular system. *p*-Aminohippuric acid is separated on a conventional anion exchanger, too, but in contrast to oxalate, it is very sensitively detected by measuring the light absorption at 254 nm. The chromatogram of a corresponding urine sample, in which a *p*-aminohippuric acid content of 730 ppm was detected, is represented in Fig. 8-96. Further urine constituents such as mandelic acid and phenylglyoxylic acid may also be determined very sensitively via UV detection. Their low affinity toward the stationary phase of an anion exchanger suggests a sodium hydroxide solution with a concentration c = 0.015 mol/L being a suitable eluent.

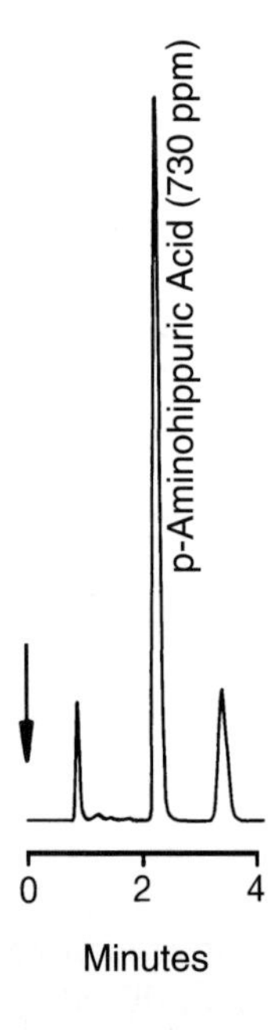

Fig. 8-96. Separation of *p*-aminohippuric acid in urine. – Separator column: IonPac AS4A; eluent: 0.0017 mol/L $NaHCO_3$ + 0.0018 mol/L Na_2CO_3; flow rate: 2 mL/min; detection: UV (254 nm); injection: 50 μL sample (diluted 1:100).

The sample preparation for ion chromatographic analysis of serums is much more elaborate. Proteins that are present in high concentration must be removed before the sample is injected, since they negatively affect the separation efficiency of the ion exchangers being used under the chromatographic conditions suited for determining inorganic ions. The precipitation method based on acetonitrile [80] has not succeeded, since proteins are not quantitatively precipitated by this method, which limits the lifetime of the separator column to a maximum of two months. In contrast, a quantitative protein precipitation is obtained with perchloric acid [81], a procedure that was successfully applied by Reiter et al. [82] to determine sulfate in human serums. High perchlorate concentrations interfere with the ion chromatographic determination of simple mineral acids, thus perchlorate must also be removed after protein precipitation. To do this, the following procedure was developed by Reiter et al.: 1 mL of the serum sample is added to 1 mL of cold perchloric acid with c = 0.665 mol/L. After shaking, the mixture is kept at 4 °C for 10 minutes to wait for complete precipitation. After centrifugation at 3,100 g for 15 minutes, 1 mL of the supernatant is added to 1 mL of cold potassium carbonate solution with c = 0.7 mol/L to precipitate the perchlorate. This mixture is centrifugated at 3,100 g for 15 minutes. The supernatant is diluted with de-ionized water according to the concentration of the analyte ions. A chromatogram (own work) of a serum sample

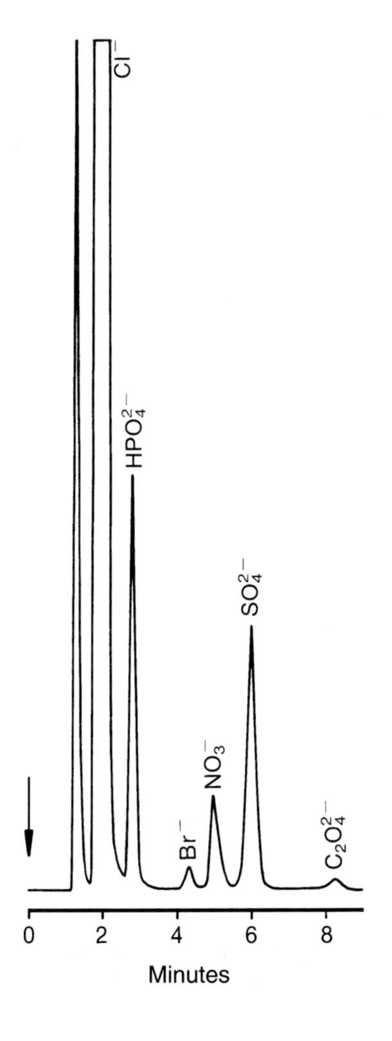

Fig. 8-97. Separation of inorganic and organic anions in serum. – Separator column: IonPac AS4A; chromatographic conditions: see Fig. 8-94; injection: 50 µL sample (diluted 1:10).

diluted 1:10 is displayed in Fig. 8-97. Apart from orthophosphate, sulfate, and small amounts of oxalate, bromide and nitrate can also be detected. This corroborates observations of de Jong [83] who, together with other authors, attributed the presence of bromide in serum to nutrition habits [84], environmental influences [85], and anaesthetics [86], and the occurence of nitrate in serum to fertilizers as indirect sources [87]. In contrast, sulfate plays a significant role in the metabolism of many endogenic compounds. So it is hoped that a better understanding of these processes is obtained by knowing the sulfate turnover as reflected by the sulfate concentrations in serum and urine.

One of the latest applications of ion chromatography is the analysis of oligosaccharides [88] in physiological samples. Particular attention is given to the biochemistry of inositol phosphates, since investigations suggest that inositol-1,4,5-triphosphate (IP_3) serves as an intracellular second messenger for a multitude of hormones, growth-stimulating factors, and neurotransmitters. Such compounds were determined by costly radiochemical methods. Smith et al. [89,90] developed an ion chromatographic method that involves the use of a conventional anion exchanger and conductometric detection with chemical suppression. Its advantage is the possibility to determine a diversity of physiologically relevant anions in one run, enabling the biochemist to follow several reactions or metabolic processes simultaneously. Radioactive labeling of the compounds can be omitted because the detection method used exhibits a high sensitivity. Thus, detection limits for the investigated samples are in the range between 20 pmol and 100 pmol. Other constituents such as phosphocreatine, amino acids, and nucleosides do not interfere with the inositol phosphate determination. In the suppressor they are converted into their cationic form and are exchanged against oxonium ions. Fig. 8-98 shows the

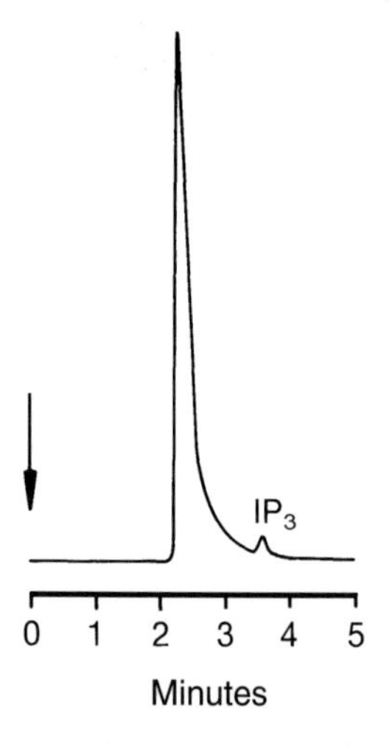

Fig. 8-98. Separation of inositol-1,4,5-triphosphate in a rat brain extract. – Separator column: IonPac AS4A; eluent: 0.0264 mol/L *p*-cyanophenol; flow rate: 2 mL/min; detection: suppressed conductivity; injection: 50 µL of a chloroform/methanol extract (2:1 v/v); (taken from [89]).

separation of IP_3 in a rat brain extract that was passed through an extraction cartridge prior to injection. This cartridge contained a silver form cation exchange resin that served to precipitate the chloride present in the sample. *p*-Cyanophenol was used as the eluent. Its elution power is high enough to elute the hexavalent IP_3 in a comparatively short time. The first experiments of Carter [91] to separate the most important isomers of the mono-, di-, and triphosphates of inositol in one run using a gradient technique are very promising. The separation represented in Fig. 8-99 was obtained with the 5-µm anion exchanger IonPac AS5A and a NaOH step gradient. The baseline shift resulting from the stepwise increase in the sodium hydroxide concentration was compensated for by baseline subtraction of a blank gradient via a data system. If radiochemical detection is used instead of conductometric detection, higher sodium hydroxide concentrations can also be used, thus enabling elution of higher inositol phosphates.

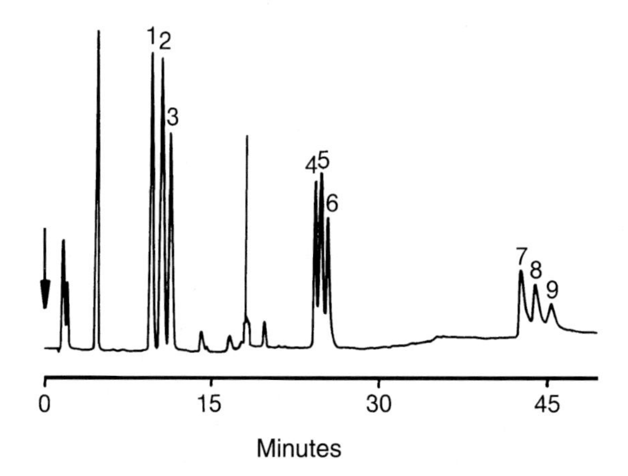

Fig. 8-99. Gradient elution of inositol phosphate isomers. – Separator column: IonPac AS5A; eluent: (A) water, (B) 0.15 mol/L NaOH; gradient: 8% B isocratically for 10 min, then 11% B isocratically for 3 min, then 52% B isocratically for 12 min, then 75% B isocratically for 35 min; flow rate: 1 mL/min; detection: suppressed conductivity; injection volume: 50 μL; solute concentrations: 30 mg/L inositol-2-phosphate (1), inositol-1-phosphate (2), inositol-4-phosphate (3), 10 mg/L inositol-1,4-diphosphate (4), inositol-2,4-diphosphate (5), inositol-4,5-diphosphate (6), inositol-1,3,4-triphosphate (7), and inositol-1,4,5-triphosphate (8).

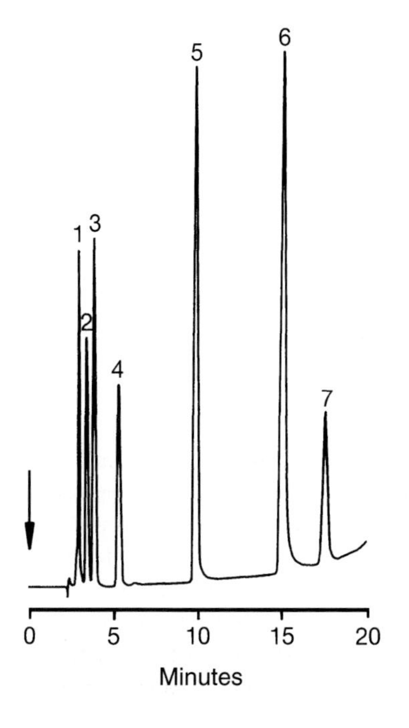

Fig. 8-100. Gradient elution of catecholamines with amperometric detection. – Separator column: Zorbax ODS; eluent: (A) 0.005 mol/L triethylamine / methanol, (B) 0.07 mol/L KH_2PO_4 + 0.005 mol/L triethylamine + 0.001 mol/L sodium butanesulfonate (pH 3); gradient: linear, 5% A in 20 min to 56% A; flow rate: 1 mL/min; detection: amperometry at a GC working electrode; oxidation potential: +0.8 V; injection volume: 50 μL; solute concentrations: 4 ppm each of norepinephrine (1), epinephrine (2), 3,4-dihydroxybenzylamine (3), dopamine (4), seratonine (5), 5-hydroxy-3-indolylacetic acid (6), and homovanillic acid (7).

Finally, mention must be made of the determination of catecholamines. Their separation is performed with ion-pair chromatography and gradient elution. Usually chemically modified ODS is used as the stationary phase. The high sensitivity required for investigating physiological samples is only accomplished by means of amperometric detection. Catecholamines can be oxidized very easily at a glassy carbon electrode at a potential of +0.8 V. A corresponding chromatogram with 4 ppm each of the most important catecholamines is depicted in Fig. 8-100.

Table 8-9. Survey of additional application areas and corresponding analytical examples.

Application area	Analytical example
Pulp and paper industry Dressing Bleaching processes Paper manufacture Waste water	Determination of sulfur and chlorine species, alkali metals, and ammonium
Agriculture Fertilizers Soil extracts	Determination of inorganic anions, alkali and alkaline-earth metals, heavy metals, and ammonium
Mining and metal processing Potash mining Coal processing Phosphatation	Determination of inorganic anions, alkali and alkaline-earth metals, heavy metals, and ammonium
Petrochemistry Scrubber solutions Crude oil Process liquors Environmental toxins	Determination of sulfur-containing compounds, cyanide, heavy metals, and inorganic anions

8.9 Other Applications

In addition to the main application areas described in detail, ion chromatography is of significant importance in other fields. From the multitude of applications summarized in Table 8-9, only those that are to be especially emphasized due to the complexity of the matrix will be dealt with here. This includes, of course, the determination of chloride and sulfate in reactive dyes, a task that seems to be trivial at first sight. However, if aqueous solutions of such dyes are injected without any sample preparation, the peak area or peak heights decrease continuously with the number of injections. Even if one takes into account that reactive dyes are anionic compounds with an aromatic backbone and that they exhibit a high affinity toward the stationary phase, this phenomenon is not completely understood. Reproducible results are only obtained when the dye compounds are removed from the analyte samples. In the simplest case, OnGuard-P solid-phase extraction cartridges are used. They contain a polyvinylpyrrolidone resin which retains interfering substances. Thus, the sample to be analyzed is injected through such a cartridge. It is important to flush these cartridges with de-ionized water prior to use to avoid chloride and sulfate contaminations. Furthermore, it should be noted that the sample should be pushed slowly through the cartridge to ensure complete extraction. If these limiting conditions are observed, the chromatogram shown in Fig. 8-101 is obtained for the dye Royal Blue. A 0.01% aqueous solution has been investigated, in which chloride and sulfate and traces of nitrate and orthophosphate could be detected.

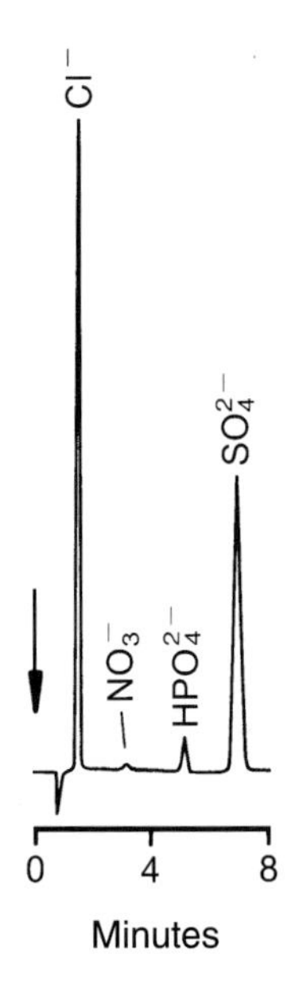

Fig. 8-101. Determination of chloride and sulfate in a reactive dye. – Separator column: IonPac AS4A; eluent: 0.0017 mol/L $NaHCO_3$ + 0.0018 mol/L Na_2CO_3; flow rate: 2 mL/min; detection: suppressed conductivity; injection: 50 μL of a 0.01% solution of Royal Blue.

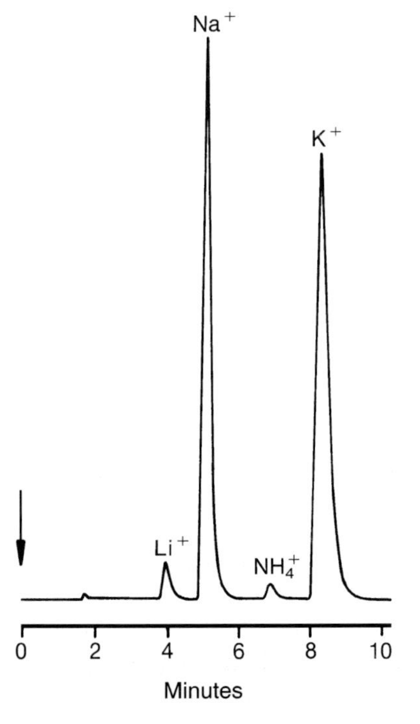

Fig. 8-102. Analysis of ammonium in cement. – Separator column: IonPac CS3; eluent: 0.03 mol/L HCl; flow rate: 1 mL/min; detection: suppressed conductivity; injection: 50 μL of a HCl-extract; (taken from [92]).

An interesting application in the field of *minerals* is the determination of ammonium in cement [92]. Ammonium ions are a natural admixture in raw products of the cement industry. In the finished product they should not exceed a certain amount, since cement is used, among other things, to line drinking water pipes. For the ion chromatographic analysis of ammonium ions, 0.1 g to 0.5 g cement are extracted with 100 mL hydrochloric acid (c = 0.1 mol/L) for about 10 minutes by shaking and stirring. After centrifugation, the extract is directly injected through a membrane filter (0.22 µm). To obtain a satisfactory separation of ammonium ions from the alkali and alkaline-earth metals, a high performance latex-based cation exchanger is used as the stationary phase. Depending on the number of samples to be analyzed every work day, it must be flushed with concentrated hydrochloric acid to remove the more strongly retained alkaline-earth metals from the column, which occupy ion-exchange functions to an increasing extent, thus, reducing the ion-exchange capacity. The chromatogram of such a cement extract is displayed in Fig. 8-102. The determined value of 193 mg/kg indicates that even small amounts of ammonium ions can be detected in cement.

For the ion chromatographic analysis of chloride in cement, that cancels out the passivation of steel surfaces in concrete and, thus, is only admitted up to a maximum content of 0.1%, Maurer et al. [93] developed a procedure in which the cement sample is extracted with nitric acid. Since a *direct* chloride determination is impossible due to the high nitrate concentration in the extract, silver nitrate is added in three-fold excess to the nitric acid extract. The precipitated silver chloride is filtered off and subsequently dissolved in 100 mL of a 0.25% ammonium hydroxide solution. This solution, according to the chloride concentration, can be further diluted with de-ionized water or injected directly.

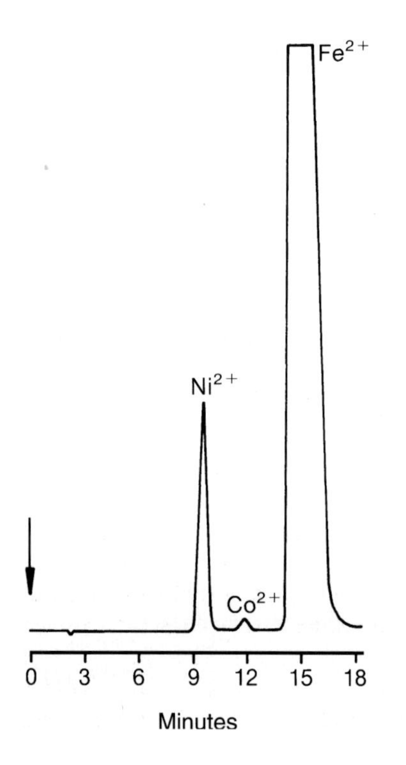

Fig. 8-103. Determination of heavy metals in crude orthophosphoric acid. – Separator column: IonPac CS5; eluent: 0.006 mol/L pyridine-2,6-dicarboxylic acid + 0.05 mol/L acetic acid + 0.05 mol/L NaOAc + 0.001 mol/L ascorbic acid; flow rate: 1 mL/min; detection: photometry at 520 nm after reaction with PAR; injection: 50 µL sample (diluted 1:10,000).

In the gypsum industry [94], ion chromatography complements the classical chemical analysis. One of the main application areas is the investigation of easily soluble accompanying compounds in the gypsum raw materials. Especially interesting are alkali and alkaline-earth metals, ammonium ions, and simple inorganic anions. Another very important application is the analysis of technical-grade gypsums, especially flue gas gypsums. These are formed in an aqueous reaction medium and, by their very nature, contain small amounts of all water-soluble accompanying compounds, which are present in both the flue gas or the absorption medium. Essentially, these are the cations sodium, magnesium, and calcium and the anions fluoride, chloride, and sulfate. Ammonium and nitrate ions may be present in flue gas denitrification. Upon application of ion chromatography these accompanying compounds can be analyzed within a short time, thus performing a fast quality control in the power plant.

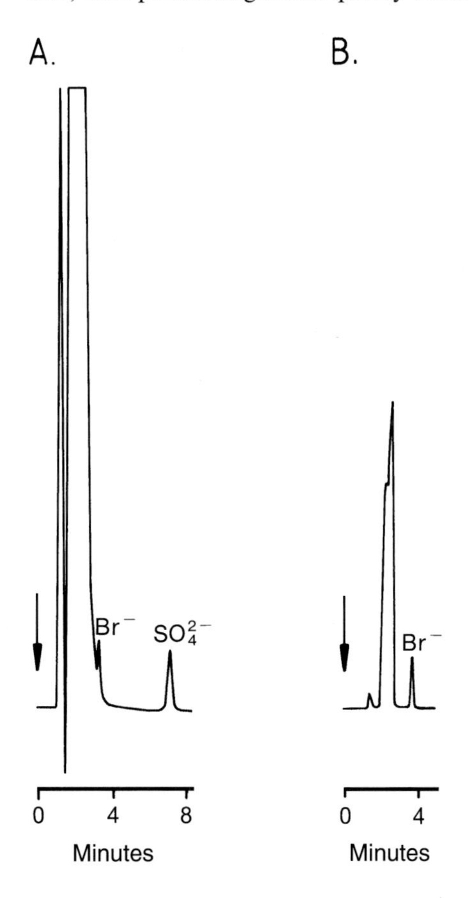

Fig. 8-104. Analysis of bromide and sulfate in brine. – Separator column: IonPac AS4A; eluent: 0.0017 mol/L $NaHCO_3$ + 0.0018 mol/L Na_2CO_3; flow rate: 1 mL/min; detection: (A) suppressed conductivity, (B) UV (200 nm); injection: 50 µL sample (diluted 1:100).

One of the most difficult application problems is the analysis of ionic impurities in all kinds of chemicals, for which absolute purity is of essential importance in some industries – above all in the semiconductor industry. The degree of difficulty to detect and to quantify ionic impurities in chemicals increases exponentially with the required degree of purity. The separation of fluoride and sulfate in crude orthophosphoric acid [95], where these ions are present in the percent range, turns out to be fairly simple. Problem-free analysis of heavy metals in these matrices is also possible. Fig. 8-103 shows the corresponding chromatogram of a 1:10,000 diluted crude orthophosphoric acid, which contains cobalt and nickel apart from higher amounts of iron. Although iron is present in orthophosphoric acid predominantly in the oxidation state +2, ascorbic acid with a concentration c = 0.001 mol/L was added to the mobile phase to convert traces of iron(III) into the reduced form, enabling a determination of total iron with ion chromatography.

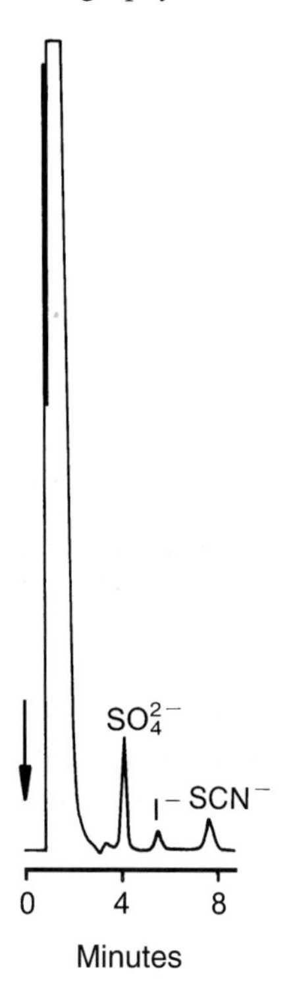

Fig. 8-105. Analysis of iodide and thiocyanate in brine. – Separator column: IonPac AS5; eluent: 0.0035 mol/L $NaHCO_3$ + 0.0029 mol/L Na_2CO_3 + 150 mg/L *p*-cyanophenol; flow rate: 2 mL/min; detection: suppressed conductivity; injection: 50 μL sample (diluted 1:100).

In contrast, the determination of ionic impurities such as bromide, iodide, thiocyanate, and sulfate in brine is much more costly because two different separation and detection methods must be applied. Non-polarizable anions such as bromide and sulfate are separated on a conventional anion exchanger. The conductometric detection usually employed is not suitable for determining bromide at very high chloride excess; therefore, detection is performed via measuring the light absorption at 200 nm. For simplicity, both detection method should be operated simultaneously as is illustrated in Fig. 8-104, which uses a brine with about 140 g/L chloride as an example. Fig 8-105 shows the corresponding chromatogram of the two polarizable anions iodide and thiocyanate, which are separated on a special anion exchanger having strongly hydrophilic functional groups to reduce the adsorption of these anions at the stationary phase. If the concentration in the injected sample solution is sufficiently high (> 0.1 mg/L), detection is performed – as in the present case – via electrical conductivity. Higher sensitivities are obtained by applying amperometric detection at an iodated Pt working electrode. Conditioning of the electrode with saturated potassium iodide solution is necessary to reach the sensitivity required for investigating brines. The silver working electrodes that have been used are hardly suited for this problem because the high chloride excess affects the baseline stability. An example chromatogram for illustrating amperometric iodide determination in brine is shown in Fig. 8-106.

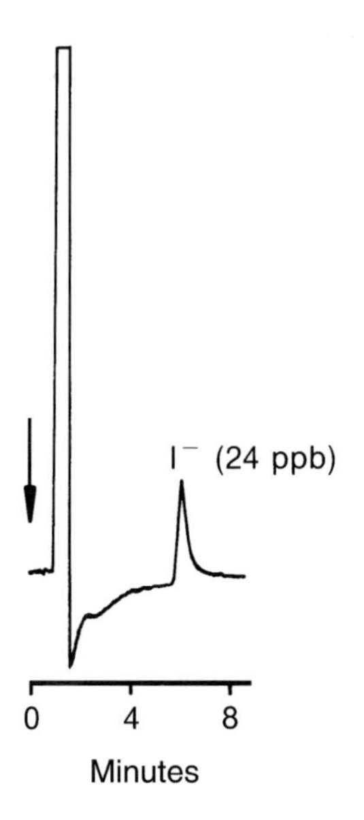

Fig. 8-106. Amperometric determination of iodide in brine. – Separator column: IonPac AS7; eluent: 0.2 mol/L HNO_3; flow rate: 1.5 mL/min; detection: amperometry at an iodated Pt working electrode; oxidation potential +0.8 V; injection: 50 µL sample (diluted 1:10).

Two examples from the chemical industry are described as representatives for many applications. For the determination of sulfate in pure alkali or alkaline-earth salts, the required detection limit can only be reached with a weighed portion of at least 10 g/L. The immense chloride excess significantly worsens the chromatographic efficiency due to overloading of the anion exchanger. Therefore, the stock solution to be analyzed is injected into the chromatograph through an OnGuard-Ag cartridge. This contains a strong-acid cation exchanger in the silver form, which precipitates most of the chloride, thereby simplifying the sulfate determination in such a matrix. Fig. 8-107 shows the corresponding chromatogram of a pure sodium chloride salt with a sulfate content of

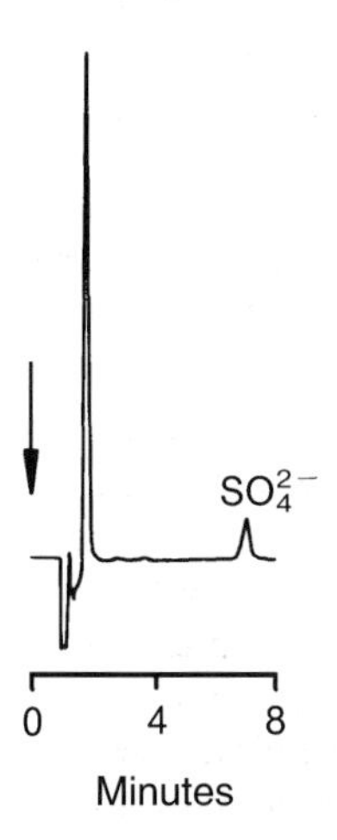

Fig. 8-107. Analysis of sulfate in a pure NaCl salt. – Separator column: IonPac AS4A; eluent: 0.0017 mol/L $NaHCO_3$ + 0.0018 mol/L Na_2CO_3; flow rate: 2 mL/min; detection: suppressed conductivity; injection: 50 µL of a 1% solution.

$2.4 \cdot 10^{-3}\%$. This representation shows that the maximum content of $10^{-3}\%$ sulfate identified for analytical-reagent quality can be easily checked by means of ion chromatography. The bromide and nitrate content is detectable directly in this matrix upon application of UV detection.

Finally, it should be pointed out that until now the application of ion chromatography to the analysis of heavy and transition metals such as iron, copper, nickel, and manganese in the ppb range in 50% sodium hydroxide solution [96, 97] was not possible. A concentration procedure developed by Kingston et al. [98] is used for the AAS analysis of heavy metals in sea water. It is based on the selective concentration of heavy metals on a macroporous iminodiacetic acid resin, where divalent cations are retained by chelation. Their affinity toward the stationary phase decreases in the order

$$Hg > Cu > Ni > Pb > Zn > Co > Cd > Fe > Mn > Ba > Ca > Sr > Mg >> Na$$

After preconcentration, the heavy and transition metals are flushed onto the analytical separator column, separated with a complexing agent as the mobile phase, and detected photometrically after derivatization with PAR. The fundamental setup of such a system is seen in Fig. 8-108. In the first step, 5 mL of the solution diluted 1:10, neutralized with nitric acid (suprapure) and being adjusted to pH 5.2 to 5.3 with ammonium acetate solution (c = 2 mol/L) is pumped through a MetPac CC-1 concentrator column, which is connected to a 3-port valve instead of an injection loop. Anions and monovalent cations are not retained at this column and are discharged. The MetPac CC-1 column is then rinsed for 1.5 minutes with ammonium acetate solution (pH 5.5) to remove the alkaline-earth metals. However, the metal ions thus concentrated cannot be directly passed to the analytical separator column, as the eluent required for this (nitric acid with a concentration c = 0.4 mol/L) is not suitable to separate the metal ions on the analytical separator column. Therefore, they are first passed through a high-capacity cation exchanger TMC-1 connected to a 4-port valve, where they are retained as a narrow band. After the TMC-1 has been converted from the oxonium form to the

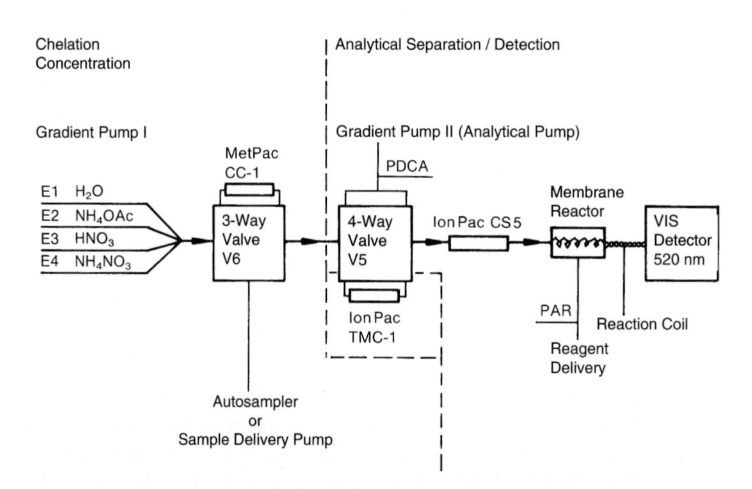

Fig. 8-108. Schematic setup of an ion chromatographic system for the analysis of heavy and transition metals in a 50% sodium hydroxide solution.

ammonium form with an ammonium nitrate solution of concentration $c = 0.1$ mol/L (pH 3.5), the metal ions are passed into the analytical separator column with the usual PDCA eluent (PDCA: pyridine-2,6-dicarboxylic acid) and are separated. In the meantime, the concentrator column can be prepared for the next sample by cleaning it with nitric acid of concentration $c = 1.5$ mol/L for a short time and then conditioning it with ammonium acetate solution.

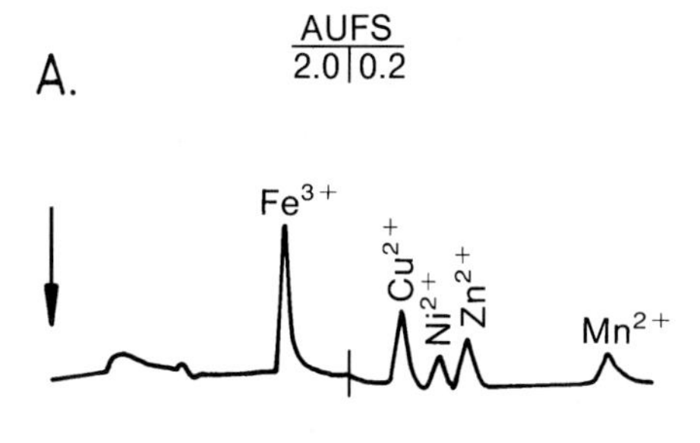

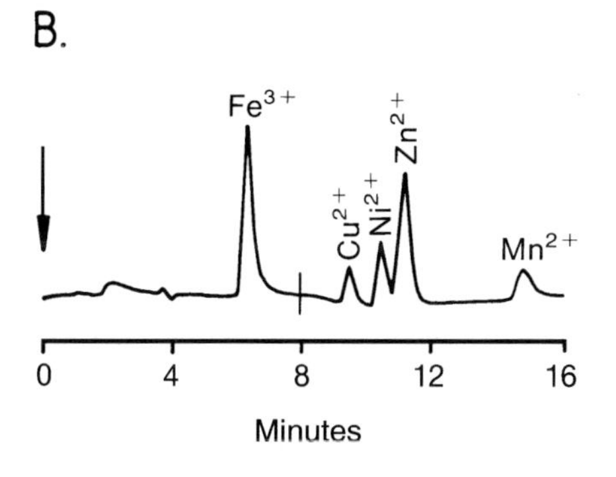

Fig 8-109. Analysis of heavy and transition metals in a 50% sodium hydroxide solution. – Separator column: IonPac CS5; eluent: 0.006 mol/L pyridine-2,6-dicarboxylic acid + 0.01 mol/L NaOH + 0.04 mol/L NaOAc + 0.05 mol/L HOAc; flow rate: 1 mL/min; detection: photometry at 520 nm after reaction with PAR; concentrated volume: 5 mL; (A) standard with 0.1 ppm Fe^{3+} and 5 ppb each of Cu^{2+}, Ni^{2+}, and Mn^{2+}, (B) sample: 50% NaOH (diluted 1:10); (taken from [97]).

This technique appears at first to be very complicated, but reproducible results are obtained by automating the rinsing processes with a gradient programmer. Fig. 8-109 shows the chromatogram of a sodium hydroxide concentrate diluted 1:10 compared to that of a blank sample. The height of the blank values for the analyte metal ions depends primarily on how carefully the chromatographic system is prepared. At the start, it should be flushed for some time with oxalic acid of concentration $c = 0.2$ mol/L to remove any metallic impurities from the system. This applies to the eluent container, all capillary tubing, as well as the two concentrator columns MetPac CC-1 and TMC-1. If all these requirements are met, only some zinc – as is seen in Fig. 109 – is detectable in the blank sample. A comparison of sample and standard reveals that the iron content of more than 100 ppb is markedly higher than that of copper, nickel, and manganese, which are in the lowest ppb range. The linearity of the method in the investigated concentration range of 0.5 ppm to 2.5 ppm for iron, and 0.025 ppm to 0.2 ppm for copper, nickel, and manganese is excellent. The recovery rate was indicated by Adams [97], with values between 98% and 106%; the relative standard deviation is surprisingly low, with values between 2% and 8%.

Of course, the procedure introduced here for analyzing heavy and transition metals is applicable not only to the investigation of strong bases such as sodium hydroxide or potassium hydroxide, but to all those matrices in which traces of heavy metals are to be determined in the presence of high alkali metal excess. These matrices include alkali salts, saline waters, and sea water [109], and also salt-rich tissue samples and urine.

Finally, Heithmar et al. [110] combined this concentration procedure with a coupled ICP/MS-technique to overcome matrix-caused interferences in the ICP/MS when environmental samples are investigated. These could be both of spectral nature [111] and of a physical-chemical nature [112]. While the former may be mathematically eliminated by introducing respective correction terms in the calibration function, physical-chemical interferences are best be removed by the above-described separation of the heavy metals to be analyzed from interfering alkali and alkaline-earth metal salts.

8.10 Sample Preparation and Matrix Problems

Sample preparation embraces all operations which help to bring the samples to be analyzed into the appropriate form. These processes include a possible crushing of the sample, its homogenization, digestion, dissolution, and filtration, all important steps that are usually interconnected. However, difficulties are caused in most cases by the sample pre-purification [99]. At this point, various techniques will be discussed in some detail, with emphasis on procedures that are typical for ion chromatographic analyses.

After the analyte sample has been dissolved, a number of working steps are frequently required before the sample can be injected into the chromatograph. In some cases, a simple filtration may suffice. However, there are numerous application problems that require removal of interfering matrix components from the sample. Also, a modification of the chemical nature of the analyte species is sometimes necessary for obtaining a better separation or detection.

In most cases, sample preparation steps take up most of the required analysis time and, thus, contribute substantially to the analysis costs. It should also be mentioned, that each manipulation of samples can falsify the analytical result; therefore, the care taken in the sample preparation directly affects the quality of the analytical result.

Sample preparation is usually performed prior to the chromatographic analysis, but it is often coupled directly to it in form of an intermediate first step. Sample preparation aims at avoiding overloading effects by appropriate dilution of the sample, removing interfering matrix constituents, and making ions which are present in very low concentrations accessible to analysis via pre-concentration.

As with all liquid chromatographic techniques, including ion chromatography, the solutions to be injected must be free of particulate matter to avoid plugging of the capillary tubings and, above all, the frits in the separator column heads. Disposable filters are generally used for the filtration of sample solutions. Owing to their Luer connector, they are also suited for manual injection, for which they are attached to the injection syringe. While membrane filters with a pore diameter of 0.45 μm are sufficient in most cases, aseptic filters with a pore diameter of 0.22 μm should be used for samples with biological activity to avoid, as much as possible, a change in sample composition by bacteriological oxidation or reduction. To preclude sample contaminations, membrane filters should be rinsed with de-ionized water prior to their use. Depending on the brand and membrane composition, nitrate ions may be released from the filters [100]. In complex matrices such as blood serums, ultrafiltration is recommended to remove proteins, a procedure in which the sample is pressed through a membrane under pressure. Such samples may not be injected directly without filtration, as the proteins present exhibit such a high affinity toward the stationary phase of an ion exchanger that cleaning the separator column causes severe difficulties.

One of the most frequent chemical sample modifications for ion chromatography is the neutralization of strongly acidic or strongly alkaline samples. This applies particularly to digestion solutions. They should also not be injected directly, as high concentrations of acid or base in the sample can result in severe baseline instabilities, which are attributed to the huge difference between the sample and eluent pH. The neutralization of samples may be accomplished in several ways. A simple addition of acid or base is only rarely possible, as the sample would be contaminated with acid anions or base cations, respectively. Even if the species is chosen which does not have to be determined, the risk of contamination by impurities contained in these chemicals is nevertheless very high. Ion-exchange resins may be used for sample neutralization. The pH value of strongly alkaline samples, for example, may be lowered by adding a high-capacity cation exchange resin in the hydronium ion form. The exchange resin used must be rinsed thoroughly with de-ionized water to prevent a sample contamination by ions that are washed out from the resin. Even if this procedure first seems to be easy to perform, two disadvantages should be noted. First, a relatively large sample volume is required, and that is not always available; second, the sample volume may change upon addition of an ion-exchange resin, as the latter is subject to swelling and shrinking processes in the presence of solvents.

Table 8-10. Commercially available packing materials for solid-phase extraction cartridges.

Packing material	
Silica	Divinylbenzene
Chemically modified silica	Chemically bonded aminopropyl phase
Aluminium oxide (acidic, basic, neutral)	Cation exchanger (H^+-form, Ag^+-form)
Anion exchanger (OH^--form)	Chelating agent
Polyvinylpyrrolidon	

Ion-exchange resins are also used as packed columns for pre-purification of samples. A characteristic example is the sample preparation of brine with the formerly used ICE suppressor (see Section 4.4), a cation exchanger in the silver form [101].

Today, the easiest way of sample preparation is the use of disposable cartridges equipped with a Luer connector that can be attached to the disposable syringe. They contain, depending on the application, different packing materials (see Table 8-10). Disposable cartridges can be employed both for the pretreatment of samples and for pre-concentration of solute ions. In the former case, it is important that only matrix components are retained, while in the latter case only solute ions should exhibit a high affinity toward the stationary phase.

Hydrophobic packing materials such as divinylbenzene [82] or chemically modified silica (C_8, C_{18}) [102] are recommended for removing aromatics, long-chain fatty acids, hydrocarbons, and surfactants from the sample. In contrast to ODS materials, divinylbenzene has the advantage of being stable over a broad pH range (pH 0 to 14). Also, divinylbenzene shows a higher selectivity for aromatic and unsaturated compounds.

Polyvinylpyrrolidone (PVP) is suited for the preparation of samples that contain humic acids, lignins, tannins, and azo dyes. In addition, PVP exhibits a high selectivity for phenolic compounds as well as for aromatic carboxylic acids and aldehydes. PVP is stable to all common HPLC solvents and is stable in the pH range between 1 and 10.

Strongly acidic cation exchangers are also available as disposable cartridges. Typically, these are materials based on polystyrene/divinylbenzene with a high degree of cross-linking (about 10%) that are stable over the entire pH range. Strongly acidic cation exchangers are offered in two forms. In the hydrogen form they serve to neutralize strongly basic samples. Typical ranges of application are soda/potash digestions or concentrated lyes, which after treatment with such a cartridge may be analyzed for inorganic anions [103]. Cation exchangers in the silver form are predominantly employed for removing high amounts of chloride from sea water, brine, and other saline samples [104]. The only disadvantage is that other halide ions such as bromide and iodide are also precipitated.

Fundamentally, when utilizing sample preparation cartridges care must be taken to ensure that they are rinsed thoroughly with de-ionized water prior to their use to avoid sample contamination by ionic constituents that possibly remained in the cartridges from the production process. In addition, hydrophobic packing materials require a rinsing with organic solvents such as methanol to activate the material's surface and to increase the binding ability for medium-polar to non-polar organics from aqueous solutions. Furthermore, it should be noted that the sample to be analyzed is passed through the cartridge as slowly as possible. Due to their low packing densities, the cartridges are actually designed for flow rates up to 50 mL/min, but it is easily recognized that the required pretreatment is much more efficient at significantly lower flow rates (< 10 mL/min). For the practical execution of the sample pretreatment it is advisable to fill the injection syringe being used with the solution to be analyzed. This solution is then pushed continuously through the cartridge that has been activated and rinsed in advance. However, the first two cartridge volumes should be discarded so that the sample does not become diluted with liquid remaining in the cartridge. The subsequent effluent can be collected or injected directly into the chromatograph.

Dialysis techniques are becoming more frequently employed in sample preparation. The diffusion of specific sample components through a membrane is common in all these techniques. If this membrane is neutral and hence permeable in a specific molecular weight range, this technique used is called *passive* dialysis. In contrast, the term *active* dialysis (also denoted as Donnan-dialysis) describes the diffusion of ions of identical charge through the membrane of an ion exchanger. If this diffusion occurs within an external electrical field, the technique is called *electrodialysis*. In general, all three techniques can be used to prepare samples for ion chromatography. While passive dialysis is a comparatively slow process that requires a larger sample volume and which leads to a substantial dilution of the sample, Donnan dialysis is used to selectively remove from the sample or add to the sample a specific species. A detailed description of these techniques is found in the paper of Hadded [99] cited earlier. Interesting in this connection is the recently published paper by Pettersen et al. [105] concerning the analysis of anionic impurities in highly concentrated sodium hydroxide via electrodialysis. In their experimental setup, the authors used an electrolysis cell with two compartments, in which the anode and cathode compartments are separated by a cation exchange membrane. The anode compartment is filled with the sodium hydroxide solution to be analyzed, the cathode compartment with dilute sodium hydroxide solution. During the electrolysis, the following reactions occur at both electrodes:

$$\text{Anode:} \quad 4\,OH^- \rightleftharpoons O_2 + H_2O + 4e^- \tag{229}$$

$$\text{Cathode:} \quad 4\,H_2O + 4e^- \rightleftharpoons 2\,H_2 + 4\,OH^- \tag{230}$$

This means that the hydroxide ion concentration decreases in the anode compartment while it increases in the cathode compartment. As diffusion of hydroxide ions through the cation exchange membrane is impossible, the sodium hydroxide concentration is thus lowered so much in the solution that it may be injected without any further pretreatment.

Finally, mention should be made about sample preparation techniques that are based on a chemical modification of the sample. This is sometimes necessary to convert the analyte component into a form suitable for separation or detection. A characteristic example is the oxidation of cyanide to cyanate with sodium hypochlorite [106], which renders an indirect cyanide determination by electrical conductivity possible. In contrast to cyanide with its low dissociation ($pK = 9.2$), the stronger dissociated cyanate ion ($pK = 3.66$) may be detected with both forms of conductometric detection. As cyanide elutes on a conventional anion exchanger in the retention range of chloride, modification of the chromatographic conditions which would be necessary for amperometric detection of cyanide is not required. However, a simultaneous analysis of mineral acids and cyanate is only possible if a separator column is used which allows cyanate to be separated from all other inorganic anions. Nitrite, which elutes on a conventional Ion Pac AS4A anion exchanger immediately before cyanate, represents a potential interference. If a satisfactory separation of both ions is impossible, nitrite may be degraded with formation of sulfate by adding amidosulfonic acid at pH 7. (Nitrite is not oxidized to nitrate by hypochlorite!) It should also be noted that the determination of bromide and nitrate is not affected in the presence of chlorate that results from hypochlorite. Chlorate usually elutes together with nitrate, but may be resolved to baseline on an IonPac AS9 stationary phase.

A similar example represents the determination of borate that is usually analyzed with ion-exclusion chromatography. However, when borate is converted into tetrafluoroborate by reacting with hydrofluoric acid [107], upon application of an anion exchanger indirect borate determination together with other inorganic anions becomes possible. Tetrafluoroborate is a polarizable anion; therefore, when using the suppressor technique some *p*-cyanophenol is added to the carbonate/bicarbonate eluent to improve the peak shape of the ion that elutes after sulfate.

A detailed description of this technique is found in a recently published book by Frei and Zech [108].

Column Maintenance

One of the most effective methods for protecting separator columns is the use of guard columns that normally contain the same stationary phase as the analytical columns. Compounds with high affinities toward the stationary phase being used are retained on the guard column, thus avoiding poisoning of the analytical separator column, which is typically indicated by a loss in the separation efficiency. Because the capacity of guard columns is limited, they must be rinsed occasionally.

The type of rinsing agent depends, first and foremost, on the stationary phase. Details are found in the guidelines of the respective column manufacturer. In general, it should be observed that guard and separator columns are stored sealed when they are removed from the chromatograph. Drying out of the column packing can lead to alterations in the column bed and result in a reduction of separation efficiency. Consult column manufacturer for appropriate storage solution. Furthermore, guard and separator columns should be protected from direct light and should be stored free from vibration. Fats, oils, surfactants, humic acids, and lignins as well as cellulose, proteins, and other high-molecular compounds are considered – particularly in ion exchangers – to be column poisons. The poisons must be removed by any of the preparation techniques described above. Organic polymer-based ion exchangers should not be rinsed with organic solvents, depending on the kind of separation material, swelling and shrinking phenomena may result. In general, only water-soluble solvents such as methanol and acetonitrile are allowed. Manufacturers' guidelines are to be followed in any case.

Concluding Remarks

A multitude of separation and detection methods developed in the last years for analyzing ionic compounds have been represented within the scope of this book. Their advantages and disadvantages have been discussed. However, it should be noted that none of the described methods is *universally* applicable and may be considered to be the method of choice in each case. Hence, for any given analytical problem in this field the factors

- type and concentration of the species to be analyzed,
- required separation and precision,
- speed and cost of analysis,
- convenience, and
- automation

should be taken into account when selecting the appropriate analysis method. It was the objective of the second edition of this book to provide a survey over the method of ion chromatography that has rapidly evolved since publication of the first edition, and to assist as much as possible in the selection of individual techniques.

Literature

Chapter 1

[1] M. Tswett, *Trav. Soc. Nat. Var.* **14** (1903) 1903.
[2] M. Tswett, *Ber. Deut. Botan. Ges.* **24** (1906) 385.
[3] N.A. Izmailov and M.S. Schraiber, *Farmatsiya* **3** (1938).
[4] E. Stahl, *Pharmazie* **11** (1956) 633.
[5] E. Stahl, *Chemiker Ztg.* **82** (1958) 323.
[6] A.J.P. Martin and R.L. Synge, *Biochem. J.* **35** (1941) 1358.
[7] A.T. James and A.J.P. Martin, *Analyst* **77** (1952) 915.
[8] C. Horvath, W. Melander, and I. Molnar, *J. Chromatogr.* **125** (1976) 129.
[9] J.H. Knox: "Theory of HPLC, Part II: Solute Interactions with the Mobile Phase and Stationary Phases in Liquid Chromatography". In: C.F. Simpson (ed.), *Practical High Performance Liquid Chromatography.* Heyden and Son, Chichester 1976.
[10] R.P.W. Scott: "Theory of HPLC, Part II: Solute Interactions with the Mobile Phase and Stationary Phases in Liquid Chromatography". In: C.F. Simpson (ed.), *Practical High Performance Liquid Chromatography.* Heyden and Son, Chichester 1976.
[11] L.R. Snyder, Chromatogr. Rev. **7** (1965) 1.
[12] G. Guiochon: "Optimization in Liquid Chromatography". In: C. Horvath (ed.). *High Performance Liquid Chromatography*, Vol. 2. Academic Press, New York 1980.
[13] H.J. Möckel, Lecture: "Instrumental Analysis I". Technical University Berlin 1974 to 1984.
[14] H. Small, T.S. Stevens, and W.C. Baumann, Anal. Chem. **47** (1975) 1801.
[15] D.T. Gjerde, J.S. Fritz, and G. Schmuckler, *J. Chromatogr.* **186** (1979) 509.
[16] R.M. Wheaton and W.C. Bauman, *Ind. Eng. Chem.* **45** (1953) 228.
[17] R.D. Rocklin and C.A. Pohl, *J. Liquid Chromatogr.* **6** (9) (1983) 1577.
[18] E.L. Johnson and B. Stevens, *Basic Liquid Chromatography.* Varian Associates Inc., Palo Alto, CA 1978, p. 92.
[19] U. Leuenberger, R. Gauch, K. Rieder, and E. Baumgartner, *J. Chromatogr.* **202** (1980) 461.
[20] DIN 38 405 Part 19 (1988) *Determination of the dissolved anions fluoride, chloride, nitrite, bromide, nitrate, orthophosphate, and sulfate in slightly contaminated water by ion chromatography.*

Chapter 2

[1] A.J.P. Martin and R.L.M. Synge, *Biochem. J.* **35** (1941) 1358.
[2] J.J. van Deemter, F.J. Zuiderweg, and A. Klinkenberg, *Chem. Eng. Sci.* **5** (1956) 271.
[3] E.L. Johnson and B. Stevens, *Basic Liquid Chromatography.* Varian Associates Inc., Palo Alto, CA 1978.
[4] A.I.M. Keulemans and A. Kwantes, in: D.K. Desty and C.L.A. Harbourn (eds.), *Vapor Phase Chromatography.* Butterworths, London 1956; p. A10.
[5] J.C. Giddings, *J. Chromatogr.* **5** (1961) 46.
[6] J.F.K. Huber and J.A.R.J. Hulsman, *Anal. Chim. Acta* **38** (1967) 305.
[7] G.J. Kennedy and J.H. Knox, *J. Chromatogr. Sci.* **10** (1972) 549.
[8] J.N. Done and J.H. Knox, *J. Chromatogr. Sci.* **10** (1972) 606.
[9] C.S. Horvath and H.-J. Lin, *J. Chromatogr.* **126** (1976) 401.

[10] C.S. Horvath and H.-J. Lin, *J. Chromatogr.* **149** (1978) 43.
[11] E. Katz, K.L. Ogan, and R.P.W. Scott, *J. Chromatogr.* **270** (1983) 51.
[12] R.E. Majors, *Anal. Chem.* **44** (1972) 1722.

Chapter 3

[1] The Bible 2. Ms. **15**, 23–25.
[2] W. Rieman and H.F. Walton, *Ion Exchange in Analytical Chemistry.* Pergamon Press 1970.
[3] H. Small, T.S. Stevens, and W.C. Baumann, *Anal. Chem.* **47** (1975) 1801.
[4] H.J. Möckel, Lecture: "Instrumental Analysis I". Technical University Berlin 1974 to 1984.
[5] H.J. Möckel, *J. Chromatogr.* **317** (1984) 589.
[6] H.J. Möckel, T. Freyholdt, J. Weiß, and I. Molnar: "The HPLC of Divalent Sulphur". In: I. Molnar (ed.), *Practical Aspects of Modern HPLC.* Walter de Gruyter & Co., Berlin – New York 1982.
[7] C. Eon and G. Guiochon, *J. Colloid. Interface Sci.* **45** (1973) 521.
[8] C. Eon, *Anal. Chem.* **47** (1975) 1871.
[9] Y. Takata und G. Muto, *Bunseki Kagaku* **28** (1979) 15.
[10] N. Yoza, K. Ito, Y. Hirai, and S. Ohashi, *J. Chromatogr.* **196** (1980) 471.
[11] H. Terada, T. Ishihara, and Y. Sakabe, *Eisei Kagaku* **26** (1980) 136.
[12] G.A. Sherwood and D.C. Johnson, *Anal. Chem. Acta* **129** (1981) 101.
[13] R.G. Gerritse, *J. Chromatogr.* **171** (1979) 527.
[14] F.A. Buytenhuys, *J. Chromatogr.* **218** (1981) 57.
[15] N. Vonk, *European Spectroscopy News* **53** (1984) 25.
[16] D.T. Gjerde, J.S. Fritz, and G. Schmuckler, *J. Chromatogr.* **186** (1979) 509.
[17] D.T. Gjerde, G. Schmuckler, and J.S. Fritz, *J. Chromatogr.* **187** (1980) 35.
[18] J.S. Fritz, D.T. Gjerde, and C. Pohlandt, *Ion Chromatography.* Dr. Alfred Hüthig Verlag, Heidelberg – Basel – New York 1982; p. 102.
[19] D.T. Gjerde and J.S. Fritz, *J. Chromatogr.* **176** (1979) 199.
[20] D.P. Lee, *J. Chromatogr. Sci. 22(8)* (1984) 327.
[21] P. Walser, *Labor Praxis*, July/August (1985) 878.
[22] K.M. Roberts, D.T. Gjerde, and J.S. Fritz, *Anal. Chem.* **53** (1981) 1691.
[23] T. Okada and T. Kuwamoto, *Anal. Chem.* **55** (1983) 1001.
[24] J.R. Benson and D.J. Woo, *J. Chromatogr. Sci.* **22** (1984) 386.
[25] P.R. Haddad, P.E. Jackson, and A.L. Heckenberg, *J. Chromatogr.* **346** (1985) 139.
[26] T.S. Stevens and M.A. Langhorst, *Anal. Chem.* **54** (1982) 950.
[27] J.S. Fritz and J.N. Story, *J. Chromatogr.* **90** (1974) 267.
[28] C. Pohl, V. Summerfelt, and J. Stillian, Presentation No. 73, *Pittsburgh Conference 1987*, Atlantic City, N.J., USA.
[29] J.E. Girard and J.A. Glatz, Am. Lab. **13** (1981) 26.
[30] R.L. Stevenson and K. Harrison, *Am. Lab.* **13** (1981) 76.
[31] S. Matsushita, Y. Tada, N. Baba, and K. Hosako, *J. Chromatogr.* **259** (1983) 459.
[32] D.R. Jenke and G.K. Pagenkopf, *Anal. Chem.* **55** (1983) 225.
[33] K. Harrison, W.C. Beckham Jr., T. Yates, and C.D. Carr, *Am. Lab.* May 1985.
[34] T. Braumann, personal communication.
[35] E. Blasius, K.P. Janzen, W. Adrian, and G. Klautke, J. Chromatogr. **96** (1974) 89.
[36] G.A. Melson, *Coordination Chemistry of Macrocyclic Compounds.* 1st Edition, Plenum Press, New York 1979.
[37] E. Blasius and K.P. Janzen, *Israel. J. Chemistry* **26** (1985) 25.
[38] M. Nakajima, K. Kimura, and T. Shono, *Anal. Chem.* **55** (1983) 463.
[39] M. Igawa, K. Saito, J. Tsukamoto, and M. Tanaka, *Anal. Chem.* **53** (1981) 1942.
[40] A. Clearfield, *Inorganic Ion Exchange Materials.* CRC Press, Boca Raton, FL, 1982.
[41] C. Laurent, H. Billiet, and L. de Galan, *Chromatographia* **17** (1983) 394.
[42] C. Laurent, H. Billiet, and L. de Galan, *J. Chromatogr.* **285** (1984) 161.
[43] G.M. Schwab and A.N.Z. Ghosh, *Z. Angew. Chem.* **53** (1940) 39.
[44] L.R. Snyder, *Principles of Adsorption Chromatography*, Marcel Dekker, New York 1968; p. 163.

[45] G.L. Schmitt and D.J. Pietrzyk, *Anal. Chem.* **57** (1985) 2247.
[46] L.G. Sillen, *Stability Constants of Metal-Ion Complexes.* The Chemical Society: London 1971, Supplement 1, No. 25.
[47] O.A. Shipgun, I.N. Voloshik, and Yu.A. Zolotov, *Anal. Sci.* **8** (1985) 335.
[48] K. Irgum, *Anal. Chem.* **59** (1987) 358.
[49] K. Irgum, *Anal. Chem.* **59** (1987) 363.
[50] D.T. Gjerde and J.S. Fritz, *Anal. Chem.* **53** (1981) 2324.
[51] J. Hertz and U. Baltensperger, *LC Magazine* **2** (1984) 600.
[52] R.P.W. Scott, *Liquid Chromatography Detectors.* Elsevier, New York 1977; p. 89.
[53] J. Stillian, R. Slingsby, and C.A. Pohl: "A Revolutionary New Suppressor For Ion Chromatography". Presentation *Pittsburgh Conference* 1985, New Orleans, LA, USA.
[54] G. Ackermann, W. Jugelt, H.-H. Möbius, H.D. Suschke, and G. Werner, *Elektrolytgleichgewichte und Elektrochemie.* VEB Verlag für Grundstoffindustrie, Leipzig 1974.
[55] DIN 38405, Part 19 (1988) *Determination of the dissolved anions fluoride, chloride, nitrite, orthophosphate, bromide, nitrate, and sulfate in slightly contaminated water by ion chromatography.*
[56] R.D. Rocklin and E.L. Johnson, *Anal. Chem.* **55** (1983) 4.
[57] K. Han and W.F. Koch, *Anal. Chem.* **59** (1987) 1016.
[58] C. Pohlandt-Watson: "A Revised Ion-Chromatographic Method for the Determination of Free Cyanide". Randburg, Council for Mineral Technology, Report M 283 1986; p. 1–3.
[59] DIN 38405, Part 13 (1981) *Determination of cyanide.*
[60] P.K. Dasgupta, K. DeCesare, and J.C. Ullrey, *Anal. Chem.* **52** (1980) 1912.
[61] M. Lindgren, A. Cedergren, and J. Lindberg, *Anal. Chem. Acta* **141** (1982) 279.
[62] H. Beyer, *Lehrbuch der organischen Chemie.* 17th Edition, S. Hirzel Verlag, Stuttgart 1973; p. 177.
[63] J.E. Tong, K. Schertenleib, and R.A. Carpio, Solid State Technology **27** (1984) 161.
[64] R.M. Merril, *LC-GC Magazine* **6** (1988) 416.
[65] Hollemann-Wiberg, *Lehrbuch der anorganischen Chemie.* 71th-80th Edition, Walter de Gruyter & Co., Berlin 1971.
[66] P.R. Haddad and A.L. Heckenberg, *J. Chromatogr.* **300** (1984) 357.
[67] D.T. Gjerde and J.S. Fritz, *Anal. Chem.* **53** (1981) 2324.
[68] J.S. Fritz, D.L. DuVal, and R.E. Barron, *Anal. Chem.* **56** (1984) 1177.
[69] D.T. Gjerde and J.S. Fritz, *Ion Chromatography.* 2nd Edition, Dr. Alfred Hüthig Verlag, Heidelberg – Basel – New York 1987; p. 148.
[70] J.G. Tarter (ed.), *Ion Chromatography,* Vol. 37. Marcel Dekker Inc., New York and Basel 1987; p. 37.
[71] J. Chang and J.S. Fritz, Iowa State University, Ames, Iowa, USA 1981.
[72] J.S. Fritz, D.T. Gjerde, and R.M. Becker, *Anal. Chem.* **52** (1980) 1519.
[73] R.M. McCormick and B.L. Karger, *J. Chromatogr.* **199** (1980) 259.
[74] B.A. Bidlingmeyer, *J. Chromatogr. Sci.* **18** (1980) 525.
[75] R.W. Melander, J.F. Erard, and Cs. Horvath, *J. Chromatogr.* **282** (1983) 229.
[76] T. Okada and T. Kuwamoto, *Anal. Chem.* **56** (1984) 2073.
[77] S. Levin and E. Grushka, *Anal. Chem.* **58** (1986) 1602.
[78] S. Levin and E. Grushka, *Anal. Chem.* **59** (1987) 1157.
[79] D.R. Jenke and G.K. Pagenkopf, *Anal. Chem.* **56** (1984) 88.
[80] D.R. Jenke and G.K. Pagenkopf, *Anal. Chem.* **56** (1984) 85.
[81] H. Small and T.E. Miller Jr., *Anal. Chem.* **54** (1982) 462.
[82] A. Ringbom, *Complexation in Analytical Chemistry.* Interscience, New York 1963; p. 198 f.
[83] J. Behnert, P. Behrend, and A. Kipplinger, *Labor Praxis* **10**(8) (1986) 872.
[84] E. Vaeth, P. Sladek, and K. Kenar, *Fresenius Z. Anal. Chem.* **329** (1987) 584.
[85] Y. Baba. N. Yoza, and S. Ohashi, *Chromatographia* **350** (1985) 119.
[86] P.D. Perrin and B. Dempsey, *Buffers for pH and Metal Ion Control.* Chapman and Hall, London 1974.
[87] A. Gaedcke, Dissertation, University of Düsseldorf 1986.
[88] U. Fischer, Dissertation, University of Düsseldorf 1986.
[89] Eur. Pat. 61106 (7.8.85) Henkel KGaA.
[90] D.P. 3111152.1 Henkel KGaA.
[91] U.S. Pat. 4,572,807 (25.2.86) Henkel KGaA.

[92] H. Waldhoff and P. Sladek, *Fresenius Z. Anal. Chem.* **320** (1986) 163.
[93] A.W. Fitchett and A. Woodruff, *LC Magazine* **1**, No.1 (1983).
[94] F. Pacholec, D.T. Rossi, L.D. Ray, and S. Vazopolos, *LC Magazine*, **3**, No.12 (1985) 1068.
[95] J. Weiß, *Fresenius Z. Anal. Chem.* **320** (1985) 679.
[96] J. Weiß and G. Hägele, *Fresenius Z. Anal. Chem.* **388** (1987) 46.
[97] P.R. Knapp, *Handbook of Analytical Derivatization Reactions.* John Wiley, New York 1979; p. 539.
[98] H.D. Scobell, K.M. Brobst, and E.M. Steele, *Cereal Chem.* **54** (1977) 1905.
[99] J. Simmer and J. Puls, *J. Chromatogr.* **156** (1978) 197.
[100] R.T. Yong, L.P. Milligan, and G.W. Mathison, *J. Chromatogr.* **209** (1981) 316.
[101] D.L. Hendrix, R.E. Lee, J.G. Baust, and H. James, *J. Chromatogr.* **210** (1981) 45.
[102] R.D. Rocklin and C.A. Pohl, *J. Liq. Chromatogr.* **6** (1983) 1577.
[103] J.A. Rendleman (1971): "Ionization of Carbohydrates in the Presence of Metal Hydroxides and Oxides". In: Carbohydrates in Solution. Advances in Chemistry, Series 117, *Am. Chem. Soc.*, Washington DC.
[104] P. Edwards and K. Haak, *Am. Lab.* April 1983.
[105] R.E. Smith and R.A. MacQuarrie, *Anal. Biochem.* **170** (1988) 308.
[106] L.A.Th. Verhaar, B.F.M. Kuster, and H.A. Clearsens, *J. Chromatogr.* **284** (1984) 1.
[107] E. Rajakylä, *J. Chromatogr.* **353** (1986) 1.
[108] B.B. Wheals and P.C. White, *J. Chromatogr.* **176** (1979) 421.
[109] E. Rajakylä and M. Palopaski, *J. Chromatogr.* **282** (1983) 595.
[110] C. Jansen, personal communication.
[111] M.R. Hardy, R.R. Townsend, and Y.C. Lee, *Anal. Biochem.* **170** (1988) 54.
[112] E. Wood, J. Lecomte, R.A. Childs, and T. Feizi, *Mol. Immunol.* **16** (1979) 813.
[113] J.C. Paulsen, J. Weinstein, and U. de Souza-E-Silva, *Eur. J. Biochem.* **140** (1984) 523.
[114] S. Honda, *Anal. Biochem.* **140** (1984) 1.
[115] V.K. Dua and C.A. Bush, *Anal. Biochem.* **137** (1984) 33.
[116] W.M. Blanken, M.L.E. Bergh, P.L. Kappen, and D.H. Van den Eijnden, *Anal. Biochem.* **145** (1985) 322.
[117] N. Tomija, N. Kurono, H. Ishihara, S. Teijima, S. Endo, Y. Arata, and N. Takahashi, *Anal. Biochem.* **163** (1987) 489.
[118] E.F. Hounsell, J.M. Rideout, N.J. Pickering, and C.K. Lim, *J. Liq. Chromatogr.* **7** (1984) 661.
[119] R.R. Townsend, M.R. Hardy, T.C. Wong, and Y.C. Lee, *Biochemistry* **25** (1986) 5716.
[120] M.R. Hardy and R.R. Townsend, *Proc. Natl. Acad. Sci. USA* **85** (1988) 3289.
[121] A. Neuberger and B.M. Wilson, *Carbohydr. Res.* **17** (1971) 89.
[122] S. Takasaki, T. Mizuochi, and A. Kobata, *Methods Enzymol.* **83** (1981) 263.
[123] W.T. Wang, N.C. LeDonne Jr., B. Ackerman, and C.C. Sweeley, *Anal. Biochem.* **141** (1984) 366.
[124] Literature citation 7 in [127].
[125] Literature citation 9 in [127].
[126] Literature citation 10 in [127].
[127] L.R. Snyder, *Chromatogr. Reviews* **7** (1965) 1.
[128] T. Sundén, H. Lindgren, A. Cedergren, and D.D. Siemer, *Anal. Chem.* **55** (1983) 2.
[129] J.G. Tarter, *Anal. Chem.* **56** (1984) 1264.
[130] R.D. Rocklin, C.A. Pohl, and J.A. Schibler, *J. Chromatogr.* **411** (1987) 107.
[131] H. Schwab, W. Rieman, and P.A. Vaughan, *Anal. Chem.* **29** (1957) 1357.
[132] R.D. Rocklin, C.A. Pohl, and R.W. Slingsby, *J. Liq. Chromatogr.* **9** (1986) 757.
[133] N.E. Good, G.D. Winget, W. Winter, T.N. Conolly, S. Izawa, and R.M.M. Singh, *Biochemistry* **5** (1966) 467.
[134] J.P. Ivey, *J. Chromatogr.* **287** (1984) 128.
[135] W.R. Jones, P. Jandik, and A.L. Heckenberg, *Anal. Chem.* **60** (1988) 1977.
[136] P. Hajos and J. Inczédy, *J. Chromatogr.* **201** (1980) 253.
[137] C.A. Pohl and E.L. Johnson, *J. Chromatogr. Sci.* **18** (1980) 442.
[138] P. Kolla, J. Köhler, and G. Schomburg, *Chromatographia* **23** (1987) 465.
[139] G. Schomburg, *LC-GC Magazine* **6** (1988) 36.
[140] K. Kimura, H. Harino, E. Hajata, and T. Shono, *Anal. Chem.* **58** (1986) 2233.
[141] K. Bächmann and K.-H. Blaskowitz, *Fresenius Z. Anal. Chem.* **333** (1989) 15.
[142] O. Samuelson, *Z. Anal. Chem.* **116** (1939) 328.

[143] J.S. Fritz and J.N. Story, *Anal. Chem.* **46** (1974) 825.
[144] S. Elchuk and R.M. Cassidy, *Anal. Chem.* **51** (1979) 1434.
[145] J.M. Riviello and W.E. Rich, Dionex Corp., Internal Report 1982.
[146] S.S. Heberling, J.M. Riviello, M.S. Taylor, S. Papanu, and M. Ebenhahn: "New and Versatile IC Columns for Metal Separations". Presentation *26th Rocky Mountain Conference* 1984, Denver, CO, USA.
[147] C.A. Pohl and J.M. Riviello, Presentation No. 108, *24th Rocky Mountain Conference* 1982, Denver, CO, USA.
[148] G.J. Sevenich and J.S. Fritz, *Anal. Chem.* **55** (1983) 12.
[149] J. Weiß, *Labor Praxis*, April/May 1987.
[150] A.E. Martell and R.M. Smith, *Critical Stability Constants*, Vol. 3. Plenum Press, New York 1977.
[151] S.S. Heberling and J.M. Riviello: "Advances in High Performance IC of Transition and Post-Transition Metals". Presentation *27th Rocky Mountain Conference* 1985, Denver, CO, USA.
[152] S.S. Heberling: "Recent Advances in Metals Determination by Ion Chromatography". Presentation *Pittsburgh Conference* 1986, Atlantic City, N.J., USA.
[153] S. Somerset, personal communication.
[154] R.M. Cassidy and S. Elchuk, *Anal. Chem.* **54** (1982) 1558.
[155] S.S. Heberling, J.M. Riviello, M. Shifen, and A.W. Ip, *Res. Dev.* September (1987)74.
[156] C.H. Knight and R.M. Cassidy, *Anal. Chem.* **56** (1984) 474.

Chapter 4

[1] R.M. Wheaton and W.C. Bauman, *Ind. Eng. Chem.* **45** (1953) 228.
[2] G.A. Harlow and D.H. Morman, *Anal. Chem.* **36** (1964) 2438.
[3] R. Wood, L. Cummings, and T. Jupille, *J. Chromatogr. Sci.* **18** (1980) 551.
[4] D.P. Lee and A.D. Lord, *LC-GC Magazine* **5** (1987) 261.
[5] V.T. Turkelson and M. Richards, *Anal. Chem.* **50** (1978) 1420.
[6] R. Wood, L. Cummings, and T. Jupille, *J. Chromatogr. Sci.* **18** (1980) 551.
[7] T. Jupille, D.W. Togami, and D.E. Burge, *Ind. Res. Dev.* **25** (1983) 151.
[8] W.E. Rich, E.L. Johnson, L. Lois, P. Kabra, B. Stafford, and L. Marton, *Clin. Chem.* **26** (1980) 1492.
[9] K. Tanaka and J.S. Fritz, *J. Chromatogr.* **361** (1986) 151.
[10] W.E. Rich, F. Smith, L. McNeill, and T. Sidebottom, "Ion Exclusion Coupled to Ion Chromatography: Instrumentation and Application". In: *Ion Chromatographic Analysis of Environmental Pollutants*, Vol. 2. Ann Arbor Science, Ann Arbor Michigan 1979; p. 17.
[11] W.E. Rich, E.L. Johnson, L. Lois, P. Kabra, and L. Marton, "Organic Acids by Ion Chromatography". In: L. Marton and P. Kabra (eds.). *Liquid Chromatography in Clinical Analysis.* The Humana Press, Inc. 1981.
[12] T. Jupille, M. Gray, B. Black, and M. Gould, *Am. Lab.* **13** (1981) 80.
[13] D.H. Spackman, W.H. Stein, and S. Moore, *Anal. Chem.* **59** (1958) 1190.
[14] M.C. Roach and M.D. Harmony, *Anal. Chem.* **59** (1987) 411.
[15] W.D. Hill, F.H. Walters, T.D. Wilson, and J.D. Stuart, *Anal. Chem.* **51** (1979) 1338.
[16] S. Einarsson, B. Josefsson, and S. Lagerkvist, *J. Chromatogr.* **282** (1983) 609.
[17] S. Einarsson, S. Folestad, B. Josefsson, and S. Lagerkvist, *Anal. Chem.* **58** (1986) 1638.
[18] J.L. Glajch and J.J. Kirkland, *J. Chromatogr. Sci.* **25** (1987) 4.
[19] B.-L. Johansson and K. Isaksson, *J. Chromatogr.* **356** (1986) 383.
[20] M. Abrahamsson and K. Gröningsson, *J. Chromatogr.* **154** (1978) 313.
[21] R.L. Heinrikson and S.C. Meredith, *Anal. Biochem.* **136** (1983) 65.
[22] H. Scholze, *J. Chromatogr.* **350** (1985) 453.
[23] J.M. Wilkinson, *J. Chromatogr. Sci.* **16** (1978) 547.
[24] N. Kaneda, M. Sato, and K. Yagi, *Anal. Biochem.* **127** (1982) 49.
[25] G. Ogden, *LC-GC Magazine* **5** (1987) 28.
[26] K. Dus, S. Lindroth, R. Pabst, and R. Smith, *Anal. Biochem.* **14** (1966) 41.
[27] P.E. Hare, *Space Life Sci.* **3** (1972) 354.
[28] J.R. Benson, *U.S. Patent Nr. 3,686,118* (1972).

[29] K. Piez and L. Morris, *Anal. Biochem.* **1** (1960) 187.
[30] M. Roth, *Anal. Chem.* **43** (1971) 880.
[31] M. Roth and A. Hampai, *J. Chromatogr.* **83** (1973) 353.
[32] J.R. Benson and P.E. Hare, *Proc. Nat. Acad. Sci. USA* **72** (1975) 619.
[33] K.S. Lee and D.G. Drescher, *Int. J. Biochem.* **9** (1978) 457.
[34] P. Bohlen and M. Mellet, *Anal. Biochem.* **94** (1974) 313.
[35] D.R. Jenke and D.S. Brown, *Anal. Chem.* **59** (1987) 1509.
[36] J.A. Polta and D.C. Johnson, *J. Liq. Chromatogr.* **6** (1983) 1727.
[37] T.L. Perry, G.H. Dixon, and S. Hansen, *Nature* (London) **206** (1965) 895.
[38] G.V. Paddock, G.B. Wilson, and A.-C. Wang, *Biochem. Biophys. Res. Commun.* **87** (1979) 946.
[39] H. Godel, Th. Graser, P. Földi, and P. Fürst, *J. Chromatogr.* **297** (1984) 49.
[40] T. Gerritsen, M.L. Rehberg, and H.A. Weisman, *Anal. Biochem.* **11** (1965) 460.
[41] H. Matsubara and R.M. Sasaki, *Biochem Biophys. Res. Commun.* **35** (1969) 175.
[42] A.P. Williams, *J. Chromatogr.* **373** (1986) 175.

Chapter 5

[1] D.P. Wittmer, N.O. Nuessle, and W.G. Haney, Jr., *Anal. Chem.* **47** (1975) 1422.
[2] S.P. Sood, L.E. Sartori, D.P. Wittmer, and W.G. Haney, *Anal. Chem.* **48** (1976) 796.
[3] "Paired Ion Chromatography, an Alternative to Ion Exchange". Waters Associates, Milford, Mass. 1975.
[4] J.H. Knox and G.R. Laird, J. Chromatogr. **122** (1976) 17.
[5] J.H. Knox and J. Jurand, *J. Chromatogr.* **125** (1976) 89.
[6] C.A. Pohl, U.S. Patent No. 4,265,634.
[7] C. Horvath, W. Melander, I. Molnar, and P. Molnar, *Anal. Chem.* **49** (1977) 2295.
[8] C. Horvath, W. Melander, and I. Molnar, *J. Chromatogr.* **125** (1976) 129.
[9] J.C. Kraak, K.M. Jonker, and J.F.K. Huber, *J. Chromatogr.* **142** (1977) 671.
[10] N.E. Hoffmann and J.C. Liao, *Anal. Chem.* **49** (1977) 2231.
[11] P.T. Kissinger, *Anal. Chem.* **49** (1977) 883.
[12] W.R. Melander and C. Horvath, *J. Chromatogr.* **201** (1980) 211.
[13] J.H. Knox and J. Jurand, *J. Chromatogr.* **103** (1975) 311.
[14] B.A. Bidlingmeyer, S.N. Deming, W.P. Price, Jr., B. Sachok, and M. Petrusek, *J. Chromatogr.* **186** (1979) 419.
[15] B.A. Bidlingmeyer, *J. Chrom. Sci.* **18** (1980) 525.
[16] C. Pohl, "Mobile Phase Ion Chromatography (MPIC). Theory and Separation." Dionex Dept. of Research & Development, IC-Exchange No. 2 (1982).
[17] N. Skelly, *Anal. Chem.* **54** (1982) 712.
[18] R.N. Reeve, *J. Chromatogr.* **177** (1979) 393.
[19] R.M. Cassidy and S. Elchuk, *Anal. Chem.* **54** (1982) 1558.
[20] R.M. Cassidy and S. Elchuk, *J. Chrom. Sci.* **21** (1983) 454.
[21] R.M. Cassidy and S. Elchuk, *J. Chromatogr.* **262** (1983) 311.
[22] B.B. Wheals, *J. Chromatogr.* **262** (1983) 61.
[23] W. Jost, R. Spatz, R. Dietz, and F. Eisenbeiss, *LaborPraxis* **11** (1984) 1184.
[24] W. Jost, R. Spatz, R. Dietz, and F. Eisenbeiss, *LaborPraxis* **10** (1984) 1016.
[25] Z. Iskandarani and D.J. Pietrzyk, *Anal. Chem.* **54** (1982) 2427.
[26] Z. Iskandarani and D.J. Pietrzyk, *Anal. Chem.* **54** (1982) 2601.
[27] I. Molnar, H. Knauer, and D. Wilk, *J. Chromatogr.* **201** (1980) 225.
[28] M. Dreux, M. Lafosse, and M. Pequinot, *Chromatographia* **15** (1982) 653.
[29] B.A. Bidlingmeyer, C.T. Santasassia and F.V. Warren, Jr., *Anal. Chem.* **59** (1987) 1843.
[30] J.P. de Kleijn, *Analyst (London)* **107** (1982) 223.
[31] U. Leuenberger, R. Gauch, K. Rieder, and E. Baumgärtner, *J. Chromatogr.* **202** (1980) 461.
[32] H.J. Cortes, *J. Chromatogr.* **234** (1982) 517.
[33] J. Weiß and M. Göbl, *Fresenius Z. Anal. Chem.* **320** (1985) 439.
[34] M. Weidenauer, P. Hoffmann, and K.H. Lieser, *Fresenius Z. Anal. Chem.* **331** (1988) 372.
[35] S.B. Rabin and D.M. Stanbury, *Anal. Chem.* **57** (1985) 1131.
[36] R. Steudel and G. Holdt, *J. Chromatogr.* **361** (1986) 379.

[37] J. Weiß, *Labor Praxis* **11** (1987) 321.
[38] J. Weiß, H.J. Möckel, A. Müller, E. Diemann, and H.-J. Walberg, *J. Chromatogr.* **439** (1988) 93.
[39] A. Müller, E. Diemann, R. Jostes, and H. Bögge, *Angew. Chem.* **93** (1981) 957.
[40] D.L. McAlese, *Anal. Chem.* **59** (1987) 541.
[41] H. Beyer, *Lehrbuch der organischen Chemie.* 17th Edition, S. Hirzel Verlag, Stuttgart 1973; p. 691.
[42] A. Marcomini and W. Giger, *Anal. Chem.* **59** (1987) 1709.
[43] J. Weiß, *J. Chromatogr.* **353** (1986) 303.
[44] B. Steinbrech, D. Neugebauer, and G. Zulauf, *Fresenius Z. Anal. Chem.* **324** (1986) 154.
[45] G. Janssen, Diploma Thesis FH Niederrhein 1987/88.
[46] G. Aced, E. Anklam, and H.J. Möckel, *J. Liq. Chromatogr.* **10** (1987) 3321.
[47] G. Aced, Dissertation Technical University Berlin 1989, D 83.
[48] H.J. Möckel. In: J.C. Giddings, E. Grushka and P.R. Brown (eds.), *Advances in Chromatography,* Vol. 26. Marcel Dekker, New York 1987.
[49] R.W. Slingsby, *J. Chromatogr.* **371** (1986) 373.
[50] J. Weiß and G. Hägele, *Fresenius Z. Anal. Chem.* **328** (1987) 46.

Chapter 6

[1] G. Ackermann, W. Jugelt, H.-H. Möbius, H.D. Suschke, and G. Werner, *Elektrolytgleichgewichte und Elektrochemie.* VEB Verlag für Grundstoffindustrie, Leipzig 1974.
[2] M. Göbl, *GIT Fachz. Lab.* **27** (1983) 261-65 and 373-75.
[3] J.S. Fritz and D.T. Gjerde, *Ion Chromatography.* Dr. Alfred Hüthig-Verlag, Heidelberg – Basel – New York 1987.
[4] D.T. Gjerde and J.S. Fritz, *Anal. Chem.* **53** (1981) 2324.
[5] D.T. Gjerde, J.S. Fritz, and G. Schmuckler, *J. Chromatogr.* **186** (1979) 509.
[6] J.E. Girard and J.E. Glatz, *Am. Lab.* **13** (1981) 26.
[7] T. Okada and T. Kuwamoto, *Anal. Chem.* **55** (1983) 1001.
[8] J.P. Ivey, *J. Chromatogr.* **287** (1984) 128.
[9] K. Irgum, *Anal. Chem.* **59** (1987) 358.
[10] R.D. Rocklin and E.L. Johnson, *Anal. Chem.* **55** (1983) 4.
[11] K. Han and W.F. Koch, *Anal. Chem.* **59** (1987) 1016.
[12] L.K. Tan and J.E. Dutrizac, *Anal. Chem.* **58** (1986) 1383.
[13] P. Edwards and K. Haak, "New Pulsed Amperometric Detector for Ion Chromatography", *Am. Lab.* April 1983.
[14] R.P. Buck, S. Singhadeja, and L.B. Rogers, *Anal. Chem.* **26** (1954) 1240.
[15] R.N. Reeve, *J. Chromatogr.* **177** (1979) 393.
[16] U. Leuenberger, R. Gauch, K. Rieder, and E. Baumgärtner, *J. Chromatogr.* **202** (1980) 461.
[17] R.J. Williams, *Anal. Chem.* **55** (1983) 851.
[18] S.H. Kola, K.A. Buckle, and M. Wooton, *J. Chromatogr.* **260** (1983) 189.
[19] P.E. Jackson, P.R. Haddad, and S. Dilli, *J. Chromatogr.* **295** (1984) 471.
[20] R.D. Rocklin, *Anal. Chem.* **56** (1984) 1959.
[21] D.-R. Yan and G. Schwedt, *Labor Praxis* January/February (1987) 48.
[22] J. Weiß, *Fresenius Z. Anal. Chem.* **328** (1987) 46.
[23] Y. Baba, N. Yoza, and S. Ohashi, *J. Chromatogr.* **295** (1984) 153.
[24] Y. Baba, N. Yoza, and S. Ohashi, *J. Chromatogr.* **348** (1985) 27.
[25] Y. Baba, N. Yoza, and S. Ohashi, *J. Chromatogr.* **350** (1985) 119.
[26] E. Vaeth, P. Sladek, and K. Kenar, *Fresenius Z. Anal. Chem.* **239** (1987) 584.
[27] J. Stillian, "Trace Analysis via Post Column Chemistry in Ion Chromatography: Silica and ppb Calcium and Magnesium in Brines", Presentation *Pittsburgh Conference* 1984, Atlantic City, N.J., USA.
[28] W. Wang, Y. Chen, and M. Wu, *Analyst (London)* **109** (1984) 281.
[29] H. Small and T.E. Miller Jr., *Anal. Chem.* **54** (1982) 462.
[30] R.A. Cochrane and D.E. Hillman, *J. Chromatogr.* **241** (1982) 392.
[31] I.M. Kolthoff and E.B. Sandell, *Textbook of Quantitative Inorganic Analysis.* Macmillan, New York 1949; p. 662–68.

[32] M. Dreux, M. Lafosse, and M. Pequinot, *Chromatographia* **15** (1982) 653.
[33] P.R. Haddad and A.L. Heckenberg, *J. Chromatogr.* **252** (1982) 177.
[34] P.R. Haddad and A.L. Heckenberg, *Chem. Aust.* **50** (1983) 275.
[35] M. Roth and H. Hampai, *J. Chromatogr.* **83** (1973) 353.
[36] P. de Montigny, J.F. Stobaugh, R.S. Givens, R.G. Carlson, K. Srinivasachar, L.A. Sternson, and T. Higuchi, *Anal. Chem.* **59** (1987) 1096.
[37] S.H. Lee and L.R. Field, *Anal. Chem.* **56** (1984) 2647.
[38] A.W. Wolkoff and R.H. Larose, *J. Chromatogr.* **99** (1974) 731.
[39] S. Katz, W.W. Pitt Jr., J.E. Mrochek, and S.J. Dinsmore, *J. Chromatogr.* **101** (1974) 193.
[40] J.H. Sherman and N.D. Danielson, *Anal. Chem.* **59** (1987) 490.
[41] J.H. Sherman and N.D. Danielson, *Anal. Chem.* **59** (1987) 1483.
[42] K. Bächmann and K.-H. Blaskowitz, *Fresenius Z. Anal. Chem.* **333** (1989) 15.
[43] S. Mho and E.S. Yeung, *Anal. Chem.* **57** (1985) 2253.
[44] T. Takeuchi and E.S. Yeung, *J. Chromatogr.* **370** (1986) 83.
[45] F.A. Buytenhuys, *J. Chromatogr.* **218** (1981) 57.
[46] P.R. Haddad und A.L. Heckenberg, *J. Chromatogr.* **300** (1984) 357.
[47] E.A. Stadlbauer, C. Trieu, H. Hingmann, H. Rohatzsch, J. Weiß, and R. Maushart, *Fresenius Z. Anal. Chem.* **330** (1988) 1.
[48] J.M. Pettersen, *Anal. Chim. Acta* **160** (1984) 263.
[49] E.A. Woolson and N. Aharonson, *J. Assoc. Off. Anal. Chem.* **63** (1980) 523.
[50] I.T. Urasa and F. Ferede, *Anal. Chem.* **59** (1987) 1563.
[51] G.J. De Menna, "The Speciation and Structure Elucidation of Transition Metal Complexes by HPLC-DCP", Presentation *Pittsburgh Conference 1986*, Atlantic City, N.J., USA.
[52] W.R. Biggs, J.T. Gano, and R.J. Brown, *Anal. Chem.* **56** (1984) 2653.

Chapter 7

[1] J. Aßhauer and H. Ullner, "Quantitative Analysis". In: H. Engelhardt and I. Halász (eds.), *Handbook of High Performance Liquid Chromatography.* Springer-Verlag, Heidelberg 1985.
[2] E. Kucera, *J. Chromatogr.* **19** (1965) 237.
[3] E. Grushka, *J. Phys. Chem.* **76** (1972) 2586.
[4] J. Jönsson, *Chromatographia* **18** (1984) 427.
[5] J.P. Foley and J.G. Dorsey, *Anal. Chem.* **55** (1983) 730.
[6] W.W. Yau, *Anal. Chem.* **49** (1977) 395.
[7] M. Doury-Berthod, P. Giampaoli, H. Pitsch, C. Sella, and C. Poitrenaud, *Anal. Chem.* **57** (1985) 2257.
[8] DIN 38405, Part 19 *Determination of the dissolved anions fluoride, chloride, nitrite, orthophosphate, bromide, nitrate, and sulfate in slightly contaminated water by ion chromatography.*
[9] R. Kaiser and G. Gottschalk, *Elementare Tests zur Beurteilung von Meßdaten.* Bibliographisches Institut Mannheim 1972.
[10] C. Vonderheid, V. Damman, W. Dürr, W. Funk, and H. Krutz, *Vom Wasser* **57** (1981) 59.
[11] W. Funk, V. Damman, C. Vonderheid, and G. Oehlmann, *Statistische Methoden in der Wasseranalytik.* VCH Verlagsgesellschaft, Weinheim 1985.
[12] K. Doerffel, *Statistik in der analytischen Chemie.* VCH Verlagsgesellschaft, Weinheim 1987.
[13] F. Ahmend, G. Bernhart, D. Rinne, M. Rogge, and B. Rüdesheim, *LaborPraxis* May (1988) 524.

Chapter 8

[1] T. Darimont, G. Schulze, and M. Sonneborn, *Fresenius Z. Anal. Chem.* **314** (1983) 383.
[2] G. Schwedt, "Ionen-Chromatographie (IC) – die High Performance-LC für Anionen und Kationen", *LaborPraxis* January/February (1984).
[3] J.A. Mosko, *Anal. Chem.* **56** (1984) 629.
[4] M.A. Tabatabai and W.A. Dick, *J. Environ. Qual.* **12** (1983) 209.

[5] O.A. Shipgun, *Trends Anal. Chem.* **4** (1985) 29.
[6] F. Schöller and F. Ollram, *Österr. Wasserwirtsch.* **35** (1983) 73.
[7] G. Resch and E. Grünschläger, *Vom Wasser* **62** (1984) 207.
[8] R. Schwabe, T. Darimont, T. Möhlmann, E. Pabel, and M. Sonneborn, *Int. J. Environ. Anal. Chem.* **14** (1983) 169.
[9] T. Okada and T. Kuwamoto, *Anal. Lett.* **17** (1984) 1743.
[10] J.P. Wilshire, *LC-Magazin* **1** (1983) 290.
[11] M. Legrand, M. De Angelis, and R.J. Delmas, *Anal. Chim. Acta* **156** (1984) 181.
[12] Ch. Siegert, "Das Chlordioxidverfahren bei der Bekämpfung von Geruch und Geschmack im Trinkwasser sowie die Bestimmung von Cl_2, ClO_2 and $NaClO_2$ nebeneinander". In: *Fortschritte der Wasserchemie,* Vol. 1. Akademie-Verlag 1964; p. 25.
[13] E. Bandi, "Untersuchungen über die Anwendungsmöglichkeit von Chlordioxid zur Entkeimung von Badewasser". In: *Mitteilungen aus dem Gebiet der Lebensmitteluntersuchung und Hygiene.* Publication by Eidg. Gesundheitsamt Bern **58** (1967) 170.
[14] H. Matusiewicz and D.F.S. Natusch, *Int. J. Environ. Anal. Chem.* **8** (1980) 227.
[15] U. Baltensperger and J. Hertz, *J. Chromatogr.* **324** (1985) 153.
[16] M.J. Willison und A.G. Clarke, *Anal. Chem.* **56** (1984) 1037.
[17] J. Forrest, D.J. Spandau, R.L. Tanner, and L. Newman, *Atmos. Environ.* **16** (1982) 1473.
[18] J.M. Lorrain, C.R. Fortune, and B. Dellinger, *Anal. Chem.* **53** (1981) 1302.
[19] T.W. Dolzine, G.G. Esposito, and D.S. Rinehart, *Anal. Chem.* **54** (1982) 470.
[20] D.V. Vinjamoori and C.S. Ling, *Anal. Chem.* **53** (1981) 1689.
[21] H. Sontheimer and M. Schnitzler, *Vom Wasser* **59** (1982) 169.
[22] M. Schnitzler, G. Lévay, W. Kühn, and H. Sontheimer, *Vom Wasser* **61** (1983) 263.
[23] M. Schnitzler, *GIT Supplement Chromatographie* **4** (1985) 32.
[24] H. Bendlin, *Chem. Labor Betrieb* **3** (1989) 108.
[25] G. Resch and E. Grünschläger, *VGB Kraftwerkstechnik* **62** (1982) 127.
[26] J. Weiß and M. Göbl, *Fresenius Z. Anal. Chem.* **320** (1985) 439.
[27] J. Weiß and G. Hägele, *Fresenius Z. Anal. Chem.* **328** (1987) 46.
[28] E. Vaeth, P. Sladek, and K. Kenar, *Fresenius Z. Anal. Chem.* **329** (1987) 584.
[29] K. Haak, *Plat. Surf. Finish* **70** (1983) 34.
[30] J. Weiß, *Galvanotechnik* **77** (1986) 2675.
[31] J.W. Dini. In: F.A. Lowenheim (ed.). *Copper Pyrophosphate Plating in Modern Electroplating.* John Wiley & Sons, New York 1984; p. 204-223.
[32] J. Böcker and Th. Bolch, *"Nickel Electroforming – Some Aspects for Process Control".* Frauenhofer Institute for Production Control and Automation Stuttgart, Presentation Int. Symposium on Electroforming/Deposition. March 1983, Los Angelos, CA, USA.
[33] G.E. Helmke, *Semiconductor Int.* **4** (1981) 119.
[34] M.M. Plechaty, *LC-Magazine* **2** (1984) 684.
[35] S. Heberling, B. Joyce, and K. Haak, "Applications of Transition Metal Ion Chromatography to High Purity and Industrial Process Waters", Presentation *5th Semiconductor Pure Water Conference,* 1986, San Francisco, CA, USA.
[36] M.E. Potts, E.J. Gavin, L.O. Angers, and E.L. Johnson, *LC-GC-Magazine* **4** (1986) 912.
[37] T.R. Dulski, *Anal. Chem.* **51** (1979) 1439.
[38] J.P. McKaveney, *Anal. Chem.* **40** (1968) 1276.
[39] L. Longwell and W.D. Maniece, *Analyst (London)* **80** (1955) 167.
[40] G.F. Longman, *The Analysis of Detergents and Detergent Products.* John Wiley & Sons, London 1975; p. 446, 455 and 480.
[41] A. Hofer, E. Brosche, and R. Heidinger, *Fresenius Z. Anal. Chem.* **253** (1971) 117.
[42] p. 403 in [40].
[43] K. Kiemstedt and W. Pfab, *Fresenius Z. Anal. Chem.* **213** (1965) 100.
[44] J. Weiß, *Tenside Detergents* **23** (1986) 237.
[45] p. 234 in [40].
[46] J.M. Rosen and H.A. Goldsmith, *Systematic Analysis of Surface Active Agents.* 2nd Edition, John Wiley & Sons, London 1972.
[47] H. Waldhoff and P. Sladek, *Fresenius Z. Anal. Chem.* **320** (1985) 163.
[48] J. Weiß, *Fresenius Z. Anal. Chem.* **320** (1985) 679.
[49] J. Weiß, *Fresenius Z. Anal. Chem.* **328** (1987) 46.
[50] E. Vaeth, P. Sladek, and K. Kenar, *Fresenius Z. Anal. Chem.* **329** (1987) 584.

[51] *Ullmanns Encyklopädie der technischen Chemie.* VCH Verlagsgesellschaft, Weinheim 1983, 4th Edition, Vol. 24; p. 97.
[52] H. Waldhoff, personal communication.
[53] G. Janssen, Diploma thesis Fachhochschule Niederrhein, Krefeld WS 1987/88.
[54] P. Edwards, *Food Technology* June (1983) 53-56.
[55] R.D. Rocklin, *LC-Magazine* **1** (1983) 504.
[56] E.J. Knudson and K.J. Siebert, *J. Am. Soc. Brew. Chem.* **42** (1984) 65.
[57] B. Baumgärtner, Diploma thesis Fachhochschule Fresenius, Wiesbaden 1989.
[58] P. Ohs, Dissertation University of Saarbrücken 1986.
[59] G. Schwedt, *GIT Supplement* **3** (1987) 76.
[60] B. Schmidt and G. Schwedt, *Dtsch. Lebensm. Rdsch.* **80** (1984) 137.
[61] B. Luckas, *Fresenius Z. Anal. Chem.* **318** (1984) 428.
[62] E. Graf, *J. Am. Oil Chem. Soc.* **60** (1983) 1861.
[63] K. Lee and J.A. Abendroth, *J. Food Sci.* **48** (1983) 1344.
[64] A.W. Fitchett and A. Woodruff, *LC-Magazine* **1** (1983) 48.
[65] B.Q. Phillippy and M.R. Johnston, *J. Food Sci.* **50** (1985) 541.
[66] A.R. De Boland, G.B. Garner, and B.L. O'Dell, *J. Agric. Food Chem.* **23** (1975) 1186.
[67] H. Terada and Y. Sakabe, *J. Chromatogr.* **346** (1985) 333.
[68] U. Zacke and H. Gründing, *Z. Lebensm. Unters. Forsch.* **184** (1987) 503.
[69] A. Herrmann, E. Damawandi, and M. Wagmann, *J. Chromatogr.* **280** (1983) 85.
[70] T.A. Biemer, *J. Chromatogr.* **463** (1989) 463.
[71] A.C. Rychtman, *LC-GC Magazine* **7** (1989) 508.
[72] W.E. Rich, R.D. Rocklin, and D.G. Gillen, "Analysis of Fermentation Broths for Carbohydrates, Metals, Organic Acids and Inorganic Ions by Ion Chromatography", Presentation *Pittsburgh Conference* 1986. Atlantic City, N.J., USA.
[73] K. Ohsawa, Y. Yoshimura, S. Watanabe, H. Tanaka, A. Yokota, K. Tamura, and K. Imaeda, *Anal. Sci.* **2** (1986) 165.
[74] W.G. Robertson, M. Peacock, P.J. Heyburn, R.W. Marshall, A. Rutherford, R.E. Williams, and P.B. Clark, "The Significance of Mild Hyperoxaluria in Calcium Stone-Formation". In: G.A. Rose, W.G. Robertson and R.W.E. Watts (eds.), *Oxalate in Human Biochemistry and Clinical Pathology.* The Wellcome Foundation Ltd., London 1979; p. 173.
[75] H.A. Moye, M.H. Malagodi, D.H. Clarke, and C.J. Miles, *Clin. Chim. Acta* **144** (1981) 173.
[76] B.G. Wolthers and M. Hayer, *Clin. Chim. Acta* **120** (1982) 87.
[77] H. Hughes, L. Hagen, and R.A.L. Sutton, *Anal. Biochem.* **119** (1982) 1.
[78] C.J. Mahle and M. Menon, *J. Urol.* **127** (1982) 159.
[79] W.G. Robertson, D.S. Scurr, A. Smith, and R.L. Orwell, *Clin. Chim. Acta* **126** (1982) 91.
[80] Applications report No. 11/82/2, Dionex GmbH Idstein.
[81] C. Neuberg, E. Strauss, and L.E. Lipkin, *Arch. Biochem.* **4** (1944) 101.
[82] Ch. Reiter, S. Müller, and Th. Müller, *J. Chromatogr.* **413** (1987) 251.
[83] P. de Jong and M. Burggraaf, *Clin. Chim. Acta* **132** (1983) 63.
[84] Z. Khalkhali and B. Parsa, "Measurement by Non-Destructive Neutran Activation Analysis of Bromine Concentrations in the Secretion of Nursing Mothers". In: *Nuclear Activation Techniques in the Life Sciences.* Vienna: Int. Atomic Energy Agency 1972, p. 461–66, cited in [83].
[85] L.-O. Plantin, Lit. [84] p. 466, cited in [83].
[86] P. Duvaldestin, *Anesthesiology* **46** (1977) 375, cited in [83].
[87] R. Rautu, V. Lupea, and Gh. Negut, *Nahrung* **18** (1974) 13, cited in [83].
[88] G.K. Grimble and A.M. Adam, *Chromatogr. Anal. February* (1985) 5.
[89] R.E. Smith, S. Howell, D. Yourtree, N. Premkumar, T. Pond, G.Y. Sun, and R.A. MacQuarrie, *J. Chromatogr.* **439** (1988) 83.
[90] R.E. Smith and R.A. MacQuarrie, *Anal. Biochem.* **170** (1988) 308.
[91] S. Carter, Dionex Corporation Sunnyvale, personal communication.
[92] K. Bussau, personal communication.
[93] J. Bruins, B. Hillebrecht, H. Monien, and W. Maurer, *Fresenius Z. Anal. Chem.* **331** (1988) 611.
[94] F. Wirsching, "Die Anwendung neuerer Analysenverfahren zur Charakterisierung von Gipsen", Presentation at *Eurogypsum*, Salzburg 1987.
[95] R.C. Sheridan, *J. Chromatogr.* **371** (1986) 383.

[96] J. Riviello, "Determination of Transition Metals by Ion Chromatography Using Chelation Concentration". Preliminary Report, Dionex Corporation, Sunnyvale 1988.
[97] L.R. Adams, "PPB Level Transition Metals (Fe, Cu, Ni, Mn) in 50% Caustic by Ion Chromatography Using Chelation Concentration", Presentation Symposium *Advances in Ion Exchange Chromatography and Electrochemical Detection.* Newport Beach, CA, USA 1989.
[98] H.M. Kingston, *Anal. Chem.* **50** (1978) 2064.
[99] P.R. Haddad, *J. Chromatogr.* **482** (1989) 267.
[100] R. Bagchi and P.R. Haddad, *J. Chromatogr.* **351** (1986) 541, cited in [99].
[101] P.F. Kehr, B.A. Leone, D.E. Harrington, and W.R. Bramstedt, *LC-GC Magazine* **4** (1986) 1118.
[102] P.E. Jackson, P.R. Haddad, and S. Dilli, *J. Chromatogr.* **295** (1984) 471.
[103] R.A. Hill, *J. High Resolut. Chromatogr. Chromatogr. Commun.* **6** (1983) 275.
[104] D.D. Siemer, *Anal. Chem.* **59** (1987) 2439.
[105] J.M. Pettersen, H.G. Johnsen, and W. Lund, *Talanta* **35** (1988) 245.
[106] M. Nonomura, *Anal. Chem.* **59** (1987) 2073.
[107] C.J. Hill and R.P. Lash, *Anal. Chem.* **52** (1980) 24.
[108] R.W. Frei and K. Zech, *Selective Sample Handling and Detection in High-Performance Liquid Chromatography*, Elsevier, Amsterdam 1988.
[109] A. Siriraks, H.M. Kingston, and J.M. Riviello, *Anal. Chem.* **62** (1990) 1185.
[110] E.M. Heithmar, T.A. Hinners, J.T. Rowan, and J.M. Riviello, *Anal. Chem.* **62** (1990) 857.
[111] J.W. McLaren, D. Beauchemin, and S.S. Berman, *Anal. Chem.* **59** (1987) 610.
[112] D. Beauchemin, J.W. McLaren, and S.S. Berman, *Spectrochim. Acta*, Part B, **42B** (1987) 467.

Index

Q

R

T

U

V